HEAT ANALYSIS AND THERMODYNAMIC EFFECTS

Edited by **Amimul Ahsan**

Heat Analysis and Thermodynamic Effects
Edited by Amimul Ahsan

Published by InTech
Janeza Trdine 9, 51000 Rijeka, Croatia

Publishing Process Manager Marija Radja

Technical Editor Teodora Smiljanic

Cover Designer InTech Design Team

Additional hard copies can be obtained from orders@intechopen.com

Heat Analysis and Thermodynamic Effects, Edited by Amimul Ahsan
p. cm.
ISBN 978-953-307-585-3

For EU product safety concerns:
IN TECH d.o.o., Prolaz Marije Krucifikse Kozulić 3, 51000 Rijeka, Croatia,
info@intechopen.com or visit our website at intechopen.com.

Meet the editor

Dr. Amimul Ahsan was born in Netrokona, Bangladesh, 1982. He is a Senior Lecturer, Department of Civil Engineering, Faculty of Engineering, University Putra Malaysia, Malaysia. He received a Ph.D in Civil Engineering from the University of Fukui, Japan. He has been involved in collaboration research with many researchers and scientists in the world since 2004 and published in diverse areas of Civil and Environmental Engineering. He has published 3 books, 6 book chapters, over 19 technical papers, and is a member of the editorial board of numerous international technical journals. He has been serving on the scientific and editorial board of InTech Open Access Publisher since 2010.

Contents

Preface

The heat transfer and analysis on heat pipe and exchanger, and thermal stress are significant issues in a design of wide range of industrial processes and devices. This book introduces advanced processes and modeling of heat transfer, gas flow, oxidation, and of heat pipe and exchanger to the international community. It includes 17 advanced and revised contributions, and it covers mainly (1) thermodynamic effects and thermal stress, (2) heat pipe and exchanger, (3) gas flow and oxidation, and (4) heat analysis.

The first section introduces spontaneous heat flow, thermodynamic effect of groundwater, stress on vertical cylindrical vessel, transient temperature fields, principles of thermoelectric conversion, and transformer performances. The second section covers thermosyphon heat pipe, shell and tube heat exchangers, heat transfer in bundles of transversly-finned tubes, fired heaters for petroleum refineries, and heat exchangers of irreversible power cycles.

The third section includes gas flow over a cylinder, gas-solid flow applications, oxidation exposure, effects of buoyancy, and application of energy and thermal performance (EETP) index on energy efficiency. The forth section presents integral transform and green function methods, micro capillary pumped loop, influence of polyisobutylene additions, synthesis of novel materials, and materials for electromagnetic launchers.

The readers of this book will appreciate the current issues of modeling on thermodynamic effects, thermal stress, heat exchanger, heat transfer, gas flow and oxidation in different aspects. The approaches would be applicable in various industrial purposes as well. The advanced idea and information described here will be fruitful for the readers to find a sustainable solution in an industrialized society.

The editor of this book would like to express sincere thanks to all authors for their high quality contributions and in particular to the reviewers for reviewing the chapters.

Acknowledgments

All praise be to Almighty Allah, the Creator and the Sustainer of the world, the Most Beneficent, Most Benevolent, Most Merciful, and Master of the Day of Judgment. He is

Omnipresent and Omnipotent. He is the King of all kings of the world. In His hand is all good. Certainly, over all things Allah has power.

The editor would like to express appreciation to all who have helped to prepare this book. The editor expresses his gratefulness to Ms. Ivana Lorkovic, Publishing Process Manager at InTech Publisher, for her continued cooperation. In addition, the editor appreciatively remembers the assistance of all authors and reviewers of this book.

Gratitude is expressed to Mrs. Ahsan, Ibrahim Bin Ahsan, Mother, Father, Mother-in-Law, Father-in-Law, and Brothers and Sisters for their endless inspiration, mental support and also necessary help whenever any difficulty occurred.

Dr. Amimul Ahsan
Department of Civil Engineering
Faculty of Engineering
University Putra Malaysia
Malaysia

Part 1

Thermodynamic and Thermal Stress

1

Enhancing Spontaneous Heat Flow

Karen V. Hovhannisyan and Armen E. Allahverdyan
A.I. Alikhanyan National Science Laboratory, Alikhanyan Brothers St. 2, 0036 Yerevan Armenia

1. Introduction

It is widely known that heat flow has a preferred direction: from hot to cold. However, sometimes one needs to reverse this flow. Devices that perform this operation need an external input of high-graded energy (work), which is lost in the process: refrigerators cool a colder body in the presence of a hotter environment, while heaters heat up a hot body in the presence of a colder one (1). The efficiency (or coefficient of performance) of these devices is naturally defined as the useful effect | for refrigerators this is the heat extracted from the colder body, while for heaters this is the heat delivered to the hotter body | divided over the work consumed per cycle from the work-source (1). The first and second laws of thermodynamics limit this efficiency from above by the Carnot value: For a refrigerator (heater) operating between two thermal baths at temperatures T_c and T_h, respectively, the Carnot efficiency reads (1)

$$\zeta_{\text{refrigerator}} = \frac{\theta}{1-\theta}, \quad \zeta_{\text{heater}} = \frac{1}{1-\theta}, \quad \theta \equiv \frac{T_c}{T_h} < 1. \tag{1}$$

There are however situations, where the spontaneous direction of the process is the desired one, but its power has to be increased. An example of such a process is perspiration (sweating) of mammals (2). A warm mammalian body placed in a colder environment will naturally cool due to spontaneous heat transfer from the body surface. Three spontaneous processes are involved in this: infrared radiation, conduction and convection (2). When the environmental temperature is not very much lower than the body temperature, the spontaneous processes are not sufficiently powerful, and the sweating mechanism is switched on: sweating glands produce water, which during evaporation absorbs latent heat from the body surface and thus cools it (2). Some amount of free energy (work) is spent in sweating glands to wet the body surface. Similar examples of heat transfer are found in the field of industrial heat-exchangers, where the external source of work is employed for mixing up the heat-exchanging fluids.

The main feature of these examples is that they combine spontaneous and driven processes. Both are macroscopic, and with both of them the work invested in enhancing the process is ultimately consumed and dissipated. Pertinent examples of enhanced transport exist in biology (4; 5). During enzyme catalysis, the spontaneous rate of a chemical reaction is increased due to interaction of the corresponding enzyme with the reaction substrate. (A chemical reaction can be regarded as particle transfer from a higher chemical potential to a lower one.) There are situations where enzyme catalysis is fueled by external sources of free energy supplied by co-enzymes (4). However, many enzymes function autonomously and cyclically: The enzyme gathers free energy from binding to the substrate, stores this free energy in slowly relaxing conformational degrees of freedom (6; 7), and then employs it for lowering the activation barrier of the reacion thereby increasing its rate (4–7). Overally, no free energy (work) is consumed for enhancing the process within this scenario. Similar situations are realized in transporting hydrophilic substances across the cell membrane (4). Since these substances are not soluble in the membrane, their motion along the (electro-chemical) potential gradient is slow, and special transport proteins are employed to enhance it (4; 5). Such a *facilitated diffusion* normally does not consume free energy (work).

These examples of enhanced processes motivate us to ask several questions. Why is that some processes of enhancement employ work consumption, while others do not? When enhancement does (not) require work consumption and dissipation? If the work-consumption does take place, how to define the efficiency of enhancement, and are there bounds for this efficiency comparable to (1)? These questions belong to thermodynamics of enhanced processes, and they are currently open. Laws of thermodynamics do not answer to them directly, because here the issue is in increasing the rate of a process. Dealing with time-scales is a weak-point of the general thermodynamic reasoning (3), a fact that motivated the development of finite-time thermodynamics (9).

Here we address these questions via analyzing a quantum model for enhanced heat transfer (8). The model describes a few-level junction immersed between two thermal baths at different temperatures; see section 2. The junction is subjected to an external field, which enhances the heat transferred by the junction along its spontaneous direction. The virtue of this model is that the optimization of the transferred heat over the junction Hamiltonian can be carried out explicitly. Based on this, we determine under which conditions the enhancement of heat-transfer does require work-consumption. We also obtain an upper bound on the efficiency of enhancement, which in several aspects is similar to the Carnot bound (1).

Heat flow in microscale and nano-scale junctions received much attention recently (10–17; 20). This is related to the general trend of technologies towards smaller scales. Needless to stress that thermodynamics of enhanced heat-transfer is relevant for this field, because it should ultimately draw the boundary between what is possible and what is not when cooling a hot body in the presence of a colder one. Brownian pumps is yet another field, where external fields are used to drive transport; see, e.g., (21; 22) and references therein. Some of the set-ups studied in this field are not far from the enhanced heat transport investigated here. However, thermodynamical quantities (such as work and enhancement efficiency) were so far not studied for these systems, though thermodynamics of Brownian motors [work-extracting devices] is a developed subject reviewed in (23).

The rest of this paper is organized as follows. The model of heat-conducting junction is introduced in section 2. Section 3 shows how the transferred heat (with and without enhancing) can be optimized over the junction structure. The efficiency of enhancing is studied in section 4. Section 5 discusses how some of the obtained results can be recovered from the formalism of linear non-equilibrium thermodynamics. We summarize in section 6. Several questions are relegated to Appendices.

2. The set-up

Our model for the heat pump (junction) consists of two quantum systems **H** and **C** with Hamiltonians $H_{\mathbf{H}}$ and $H_{\mathbf{C}}$, respectively; see Fig. 1. Each system has n energy levels and couples to its thermal bath. Similar models were employed for studying heat engines (18; 19) and refrigerators (20). It will be seen below that this model admits a classical interpretation, because all the involved initial and final density matrices will be diagonal in the energy

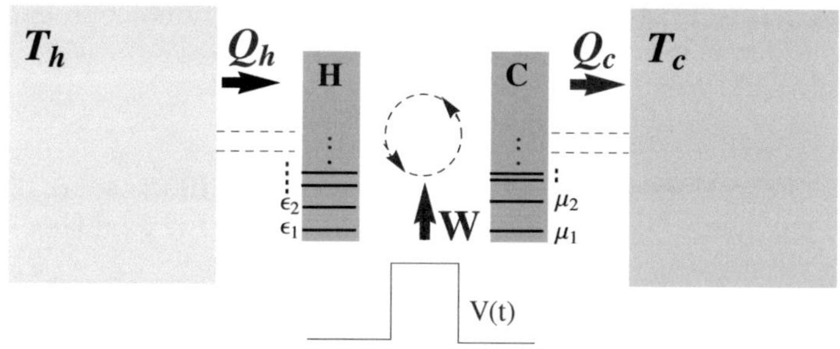

Fig. 1. The heat pump model. The few-level systems **H** and **C** operate between two baths at temperatures T_c and T_h $T_c < T_h$. During the first step of operation the two systems interact together either spontaneously or driven by a work-source at the cost of work W. During this stage couplings with the thermal baths is neglected (thermal isolation). In the second step the systems **H** and **C** do not interact with each other and freely relaxes to their equilibrium states (2) under action of the corresponding thermal bath.

representation. We shall however work within the quantum framework, since it is more intuitive.

Initially, **H** and **C** do not interact with each other. Due to coupling with their baths they are in equilibrium at temperatures $T_h = 1/\beta_h > T_c = 1/\beta_c$ [we set $k_B = 1$]:

$$\rho = e^{-\beta_h H_{\mathbf{H}}} / \mathrm{tr}\,[e^{-\beta_h H_{\mathbf{H}}}], \quad \sigma = e^{-\beta_c H_{\mathbf{C}}} / \mathrm{tr}\,[e^{-\beta_c H_{\mathbf{C}}}], \tag{2}$$

where ρ and σ are the initial Gibbsian density matrices of **H** and **C**, respectively. We write

$$\rho = \mathrm{diag}[r_n, ..., r_1], \qquad \sigma = \mathrm{diag}[s_n, ..., s_1], \tag{3}$$
$$H_{\mathbf{H}} = \mathrm{diag}[\varepsilon_n, ..., \varepsilon_1 = 0], \quad H_{\mathbf{C}} = \mathrm{diag}[\mu_n, ..., \mu_1 = 0],$$

where $\mathrm{diag}[a, .., b]$ is a diagonal matrix with entries $(a, ..., b)$, and where without loss of generality we have nullified the lowest energy level of both **H** and **C**. Thus the overall initial density matrix is

$$\Omega_{\mathrm{in}} = \rho \otimes \sigma, \tag{4}$$

and the initial Hamiltonian of the junction is

$$H_0 = H_{\mathbf{H}} \otimes 1 + 1 \otimes H_{\mathbf{C}}. \tag{5}$$

2.1 Spontaneous regime

During a spontaneous process no work is exchanged with external sources. For our situation a spontaneous heat transfer will amount to a certain interaction between **H** and **C**. Following to the approach of (25–27) we model this interaction via a Hamiltonian that conserves the (free) Hamiltonian H_0 [see (5)] for all interaction times. This then realizes the main premise of spontaneous processes: no work exchange at any time. Our model for spontaneous heat transfer consists of two steps.

1. During the first step **H** and **C** interact with each other [collision]. We assume that this interaction takes a sufficiently short time δ, and during this time the coupling with the two thermal baths can be neglected [thermal isolation]. The interaction is described by the Hamiltonian $H_{\rm int}$ added to (5):

$$H = H_{\mathbf{H}} \otimes 1 + 1 \otimes H_{\mathbf{C}} + H_{\rm int}. \tag{6}$$

The overall Hamiltonian H again lives in the n^2-dimensional Hilbert space of the junction [1]. As argued above, the interaction Hamiltonian commutes with the total Hamiltonian:

$$[H_0, H_{\rm int}] = 0, \tag{7}$$

making the energy H_0 a conserved quantity [2]. To have a non-trivial effect on the considered system, the interaction Hamiltonian $H_{\rm int}$ should not commute with the separate Hamiltonian: $[H_{\mathbf{H}} \otimes 1, H_{\rm int}] \neq 0$. For this to be the case the spectrum of H_0 should contain at least one degenerate eigenvalue. Otherwise, relations $[H_0, H_{\rm int}] = 0$ and $[H_{\mathbf{H}} \otimes 1, H_0] = 0$ will imply $[H_{\mathbf{H}} \otimes 1, H_{\rm int}] = 0$ (and thus a trivial effect of $H_{\rm int}$), because the eigen-base of H_0 will be unique (up to re-numbering of its elements and their multiplication by phase factors). The energy

$$Q_h^{[\rm sp]} = \mathrm{tr}\left(H_{\mathbf{H}} \left[\rho - \mathrm{tr}_{\mathbf{C}} \left(e^{-\frac{i\delta}{\hbar} H_{\rm int}}\, \Omega_{\rm in}\, e^{\frac{i\delta}{\hbar} H_{\rm int}} \right) \right] \right), \tag{8}$$

lost by **H** during the interaction is gained by **C**. Here $\mathrm{tr}_{\mathbf{H}}$ and $\mathrm{tr}_{\mathbf{C}}$ are the partial traces. Commutative interaction Hamiltonians (7) are applied to studying heat transfer in (25–27). Refs. (25; 26) are devoted to supporting the thermodynamic knowledge via quantum Hamiltonian models. In contrast, the approach of (27) produced new results.

2. For times larger than δ, **H** and **C** do not interact and freely relax back to their equilibrium states (2, 4) due to interaction with the corresponding thermal baths. These equilibrium states are reached after some relaxation time τ. Thus the cycle is closed | the junction returns to its initial state | and $Q_h^{[\rm sp]}$ given by (8) is the heat per cycle taken from the hot thermal bath during the relaxation (and thus during the overall cycle).

It should be obvious that once $T_h > T_c$ we get $Q_h^{[\rm sp]} > 0$: heat spontaneously flow from hot to cold. The proof of this fact is given in (19; 20; 25–27).

For times larger than τ there comes another interaction pulse between **H** and **C**, and the cycle is repeated.

2.1.1 Power

Recall that the power of heat-transfer is defined as the ratio of the transferred heat to the cycle duration τ, $Q_h^{[\rm sp]}/\tau$. For the present model τ is mainly the duration of the second stage, i.e., τ is the relaxation time, which depends on the concrete physics of the system-bath coupling. For a weak system-bath coupling τ is larger than the internal characteristic time of **H** and **C**. In contrast, for the collisional system-bath interaction, τ can be very short; see Appendix **??**.

---

[1] More precisely, we had to write the Hamiltonian (6) as $H_{\mathbf{H}} \otimes 1 + 1 \otimes H_{\mathbf{C}} + \alpha(t) H_{\rm int}$, where $\alpha(t)$ is a switching function that turns to zero both at the initial and final time. This will however not alter the subsequent discussion in any serious way.

[2] This implementation of spontaneous heat-transfer processes admits an obvious generalization: one need not require the conservation of H_0 for all interaction times, it suffices that no work is consumed or released within the overall energy budget of the process in the time-interval $[0, \delta]$. For our purposes this generalization will not be essential; see (27).

Thus the cycle time τ is finite, and the power $Q_h^{[sp]}/\tau$ does not vanish due to a large cycle time, though it can vanish due to $Q_h^{[sp]} \to 0$.

Note that some entropy is produced during the free relaxation. This entropy production can be expressed via quantities introduced in (4–8); see (20) for details. The first step does not produce entropy, because it is thermally isolated and is realized by a unitary operation that can be reversed by operating only on observable degrees of freedom ($\mathbf{H}+\mathbf{C}$). It is seen that no essential aspect of the considered model depends on details of the system-bath interaction. This is an advantage.

2.2 Driven regime

The purpose of driving the junction with an external field is to enhance [increase] the spontaneous heat $Q_h^{[sp]}$. The driven regime reduces to the above two steps, but instead of the spontaneous interaction we have the following situation: the interaction between **H** and **C** is induced by an external work-source. Thus (7) does not hold anymore. The overall interaction [between **H**, **C** and the work-source] is described via a time-dependent potential $V(t)$ in the total Hamiltonian

$$H_{\mathbf{H}} \otimes 1 + 1 \otimes H_{\mathbf{C}} + V(t) \tag{9}$$

of $\mathbf{H}+\mathbf{C}$. The interaction process is still thermally isolated: $V(t)$ is non-zero only in a short time-window $0 \le t \le \delta$ and is so large there that the influence of the couplings to the baths can be neglected.

Thus the dynamics of $\mathbf{H}+\mathbf{C}$ is unitary for $0 \le t \le \delta$:

$$\Omega_{\rm f} \equiv \Omega(\delta) = \mathcal{U}\,\Omega_{\rm i}\,\mathcal{U}^{\dagger}, \;\; \mathcal{U} = \mathcal{T}e^{-\frac{i}{\hbar}\int_0^{\delta} {\rm d}s[V(s)+H_0]}, \tag{10}$$

where $\Omega_{\rm i} = \Omega(0) = \rho \otimes \sigma$ is the initial state defined in (2), $\Omega_{\rm f}$ is the final density matrix, $\mathcal{U}$ is the unitary evolution operator, and where $\mathcal{T}$ is the time-ordering operator. The work put into $\mathbf{H}+\mathbf{C}$ reads (1; 3)

$$W = E_{\rm f} - E_{\rm i} = {\rm tr}[\,(H_{\mathbf{H}} \otimes 1 + 1 \otimes H_{\mathbf{C}})\,(\Omega_{\rm f} - \Omega_{\rm i})\,], \tag{11}$$

where $E_{\rm f}$ and $E_{\rm i}$ are initial and final energies of $\mathbf{H}+\mathbf{C}$. Due to the interaction, **H** (**C**) looses (gains) an amount of energy Q_h (Q_c)

$$Q_h = {\rm tr}(\,H_{\mathbf{H}}[\rho - \,{\rm tr}_{\mathbf{C}}\,\Omega_{\rm f}\,]\,), \tag{12}$$

$$Q_c = {\rm tr}(\,H_{\mathbf{C}}\,[{\rm tr}_{\mathbf{H}}\,\Omega_{\rm f} - \sigma\,]\,). \tag{13}$$

Eqs. (11, 12) imply an obvious relation

$$W = Q_c - Q_h. \tag{14}$$

Recall that for spontaneous processes not only the consumed work is zero, $W = 0$, but also the stronger conservation condition (7) holds.

Once $\mathbf{H}+\mathbf{C}$ arrives at the final state $\Omega_{\rm fin}$, the inter-system interaction is switched off, and **H** and **C** separately [and freely] relax to the equilibrium states (2). During this process Q_h is taken as heat from the hot bath, while Q_c is given to the cold bath. Note from section 2.1.1 that the driven operation does not change the cycle time τ, because the latter basically coincides with the relaxation time. Therefore, increasing Q_h means increasing heat transfer power.

3. Maximization of heat

3.1 Unconstrained maximization

The type of questions asked by thermodynamics concerns limiting, optimal characteristics. Sometimes the answers are uncovered directly via the basic laws of thermodynamics, an example being the Carnot bound (1). However, more frequently than not, this goal demands explicit optimization procedures (9).

We shall maximize the heat Q_h transferred from the hot bath over the full Hamiltonian of the junction. For spontaneous processes this amounts to maximizing over Hamiltonian (6) living in the n^2-dimensional Hilbert space of the junction $\mathbf{H}+\mathbf{C}$ and satisfying condition (7). For driven processes we shall maximize over Hamiltonians (9). In this case we shall impose an additional condition that the work put into $\mathbf{H}+\mathbf{C}$ in the first step does not exceed $E > 0$:

$$W \leq E. \tag{15}$$

Once the work put into the system is a resource, it is natural to operate with resources fixed from above.

Recall that the Hamiltonians (6, 9) live in the n^2-dimensional Hilbert space. The bath temperatures T_c and T_h and the dimension n^2 (the number of energy levels) will be held fixed during the maximization.

Due to (8) the maximization of the spontaneous heat $Q_h^{[\mathrm{sp}]}$ over the Hamiltonians (6, 7) amounts to maximizing over unitary operators $e^{\frac{i\delta}{\hbar}H_{\rm int}}$, and over the energies $\{\varepsilon_k\}_{k=2}^n$, $\{\mu_k\}_{k=2}^n$ of $\mathbf{H}$ and $\mathbf{C}$. Likewise, as seen from (9–11), the maximization of the driven heat Q_h amounts to maximizing under condition (15) over all unitary operators $\mathcal{U}$ living in the n^2-dimensional Hilbert space, and over the energies $\{\varepsilon_k\}_{k=2}^n$, $\{\mu_k\}_{k=2}^n$.

We should stress that the class of Hamiltonians living in the n^2-dimensional Hilbert space [with or without constraint (7)] is well-defined due to separating the heat transfer into two steps (thermally isolated interaction and isothermal relaxation). More general classes of processes can be envisaged. For instance, we may write the free Hamiltonian as $H_0 + H_{B,c} + H_{B,h}$, where H_0, $H_{B,c}$ and $H_{B,h}$ are, respectively, the Hamiltonians of the junction and the two thermal baths. Now the Hamiltonian $H_{\rm int}$ of spontaneous processes will be conditioned as $[H_{\rm int}, H_0 + H_{B,c} + H_{B,h}] = 0$. This condition is more general than (7). Then the corresponding class of driven processes can be naturally defined via the same class of Hamiltonians but without this commutation condition. We do not consider here such general processes, since we are not able to optimize them.

As seen below, the maximization of the spontaneously transferred heat (8) amounts to a particular case of maximizing Q_h. So we shall directly proceed to maximizing the driven heat Q_h; see (12).

First, take in (15) the simplest case: $E = +\infty$. This case contains the pattern of the general solution. Here we have no constraint on maximization of Q_h and it is done as follows. Since $\mathrm{tr}[H_{\mathbf{H}}\rho]$ depends only on $\{\varepsilon_k\}_{k=2}^n$, we choose $\{\mu_k\}_{k=2}^n$ and $V(t)$ so that the final energy $\mathrm{tr}[H_{\mathbf{H}}\,\mathrm{tr}_{\mathbf{C}}\Omega_{\rm f}]$ attains its minimal value zero. Then we shall maximize $\mathrm{tr}[H_{\mathbf{H}}\rho]$ over $\{\varepsilon_k\}_{k=2}^n$. Note from (3)

$$H_{\mathbf{H}} \otimes 1 = \mathrm{diag}[\,\varepsilon_1\,,\dots,\,\varepsilon_1,\dots,\,\varepsilon_n\,,\dots,\,\varepsilon_n\,],$$
$$\Omega_{\rm i} = \rho \otimes \sigma = \mathrm{diag}[\,r_1 s_1,\dots,r_1 s_n,\dots,r_n s_1,\dots,r_n s_n\,].$$

It is clear that $\mathrm{tr}\,[H_{\mathbf{H}}\,\mathrm{tr}_{\mathbf{C}}\Omega_{\rm f}] = \mathrm{tr}\,[(H_{\mathbf{H}}\otimes 1)\mathcal{U}\,\Omega_{\rm i}\,\mathcal{U}^\dagger\,]$ goes to zero when, e.g., $s_2 = \ldots = s_n \to 0$ ($\mu \equiv \mu_2 = \ldots = \mu_n \to \infty$), while $\mathcal{U}$ amounts to the SWAP operation $\mathcal{U}\rho\otimes\sigma\mathcal{U}^\dagger = \sigma\otimes\rho$. Simple symmetry considerations show that at the maximum of the initial energy $\mathrm{tr}[H_{\mathbf{H}}\sigma]$ the second level is $n-1$ fold degenerate, i.e. $\varepsilon \equiv \varepsilon_2 = \ldots = \varepsilon_n$. Denoting

$$u = e^{-\beta_h \varepsilon} \propto r_2 = .. = r_n \tag{16}$$

we obtain for $Q_h = \mathcal{Q}_h(\infty)$

$$\mathcal{Q}_h(\infty) = T_h \ln\left(\frac{1}{u}\right)\left[1 - \frac{1}{1+(n-1)u}\right] \tag{17}$$

where u is to be found from maximizing the RHS of (17) over u, i.e., u is determined via

$$1 + (n-1)u + \ln u = 0. \tag{18}$$

Note that in this case $W = +\infty$. In the $n \gg 1$ limit we have $u = \frac{\ln n}{n}\left[1 + o(1)\right]$ from (18) and $Q_h = T_h \ln n \left[1 + \mathcal{O}\left(\frac{\ln\ln n}{\ln n}\right)\right]$.

3.2 Constrained maximization

The case of a finite E in (15) is more complicated. We had to resort to numerical recipes of Mathematica 6. Denoting $\{|i_{\mathbf{H}}\rangle\}_{k=1}^{n}$ and $\{|i_{\mathbf{C}}\rangle\}_{k=1}^{n}$ for the eigenvectors of $H_{\mathbf{H}}$ and $H_{\mathbf{C}}$, respectively, we see from (11, 12) that W and Q_h feel $\mathcal{U}$ only via the matrix

$$C_{ij\,|\,kl} = |\langle i_{\mathbf{H}} j_{\mathbf{C}}|\mathcal{U}|k_{\mathbf{H}} l_{\mathbf{C}}\rangle|^2. \tag{19}$$

This matrix is double-stochastic (24):

$$\sum_{ij} C_{ij\,|\,kl} = \sum_{kl} C_{ij\,|\,kl} = 1. \tag{20}$$

Conversely, for *any* double-stochastic matrix $C_{ij\,|\,kl}$ there is *some* unitary matrix U with matrix elements $U_{ij\,|\,kl}$, so that $C_{ij\,|\,kl} = |U_{ij\,|\,kl}|^2$ (24). Thus, when maximizing various functions of W and Q_c over the unitary $\mathcal{U}$, we can directly maximize over the $(n^2-1)^2$ independent elements of $n^2 \times n^2$ double stochastic matrix $C_{ij\,|\,kl}$. This fact simplifies the problem.
Numerical maximization of Q_h over all unitaries $\mathcal{U}$ | alternatively, over all doubly stochastic C matrices (19) | and energy spectra $\{\mu_k\}_{k=2}^{n}$ and $\{\varepsilon_k\}_{k=2}^{n}$ constrained by $W \leq E$ produced the following results:

- The upper energy levels for both systems **H** and **C** are $n-1$ times degenerate [see (3)]:

$$\mu = \mu_2 = \ldots = \mu_n, \quad \varepsilon = \varepsilon_2 = \ldots = \varepsilon_n. \tag{21}$$

- The optimal unitary corresponds to SWAP:

$$\mathcal{U}\rho \otimes \sigma \mathcal{U}^{\dagger} = \sigma \otimes \rho. \tag{22}$$

- The work resource is exploited fully, i.e., the maximal Q_h is reached for $W = E$.

Though we have numerically checked these results for $n \leq 5$ only, we trust that they hold for an arbitrary n.
Denoting by $\mathcal{Q}_h$ the maximal value of Q_h, and introducing from (21)

$$v = e^{-\beta_c \mu} \quad \text{and} \quad u = e^{-\beta_h \varepsilon}, \tag{23}$$

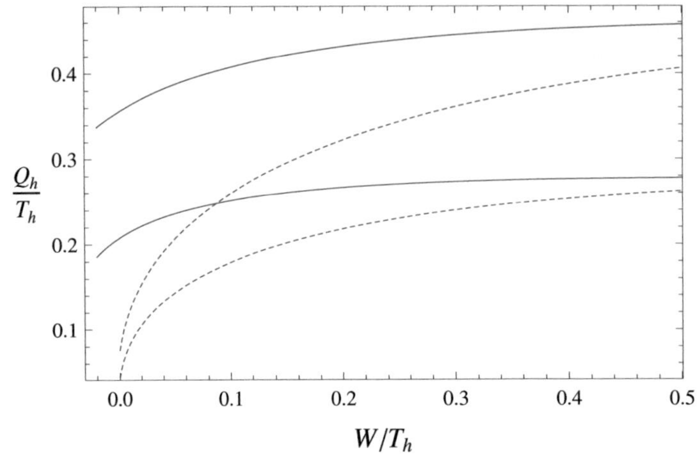

Fig. 2. The optimal transferred heat $\mathcal{Q}_h$ versus work W. Dashed curves refer to $\theta \equiv T_c/T_h = 0.9$: $n = 2$ (lower dashed curve) and $n = 3$ (upper dashed curve). Normal curves refer to $\theta = 0.5$: $n = 2$ (lower normal curve) and $n = 3$ (upper normal curve).

we have

$$\frac{\mathcal{Q}_h}{T_h} = \ln\left[\frac{1}{u}\right] \frac{(n-1)(u-v)}{[1+(n-1)v]\,[1+(n-1)u]}, \tag{24}$$

$$\frac{W}{T_h} = \frac{(\ln u - \theta \ln v)\,(n-1)(u-v)}{[1+(n-1)v]\,[1+(n-1)u]}, \tag{25}$$

where u and v in (24, 25) are determined from maximizing the RHS of (24) and satisfying constraint (25).
An important implication of (24, 25) is that $\mathcal{Q}_h(W)$ is an increasing function of W for all allowed values of W:

$$\mathcal{Q}_h(W) > \mathcal{Q}_h(W') \quad \text{if} \quad W > W'. \tag{26}$$

Fig. 2 illustrates this fact. For fixed parameters T_c, T_h and n, the allowed W's range from a certain negative value | which corresponds to work-extraction from a two-temperature system $\mathbf{H}+\mathbf{C}$ | to arbitrary $W > 0$. Eq. (26) expresses an intuitively expected, but still non-trivial fact that the best transfer of heat takes place upon consuming most of the available work. Note that this result holds only for properly optimized values of $\mathcal{Q}_h$. One can find non-optimal set-ups, where (26) is not valid [3].

3.3 Optimization of spontaneous processes

According to our discussion in section 2.1, the maximization of transferred heat $Q_h^{[sp]}$ given by (8) should proceed over all unitary operators $e^{-\frac{i\delta}{\hbar}H_{\rm int}}$ with $H_{\rm int}$ satisfying (7) and over the energies $\{\varepsilon_k\}_{k=2}^{n}$, $\{\mu_k\}_{k=2}^{n}$ of $\mathbf{H}$ and $\mathbf{C}$. This maximization has been carried out along the lines

[3] The simplest example is a junction, where the free Hamiltonian H_0 has a non-degenerate energy spectrum, and thus the condition (7) does not hold. There are no proper spontaneous processes for this case. Still there can exist a work-exracting ($W < 0$) driven processes that transfer heat from hot to cold.

described around (20). We obtained that the maximal spontaneous heat $\mathcal{Q}_h^{[\mathrm{sp}]}$ is equal to $\mathcal{Q}_h$ in (24) under condition $W = 0$:

$$\mathcal{Q}_h^{[\mathrm{sp}]} = \mathcal{Q}_h(W = 0). \tag{27}$$

Thus the optimal spontaneous processes coincide with optimal processes with zero consumed work. This result is non-trivial, because the class of unitary operators with $W = 0$ is larger than the class of unitary operators $e^{-\frac{i\delta}{\hbar}H_{\rm int}}$ with $H_{\rm int}$ satisfying (7). Nevertheless, these two classes produce the same maximal heat.

- Eqs. (26, 27) imply that if the spontaneous heat transfer process is already optimal (with respect to the junction Hamiltonian) its enhancement with help of driven processes does demand work-consumption, $W > 0$. This fact is non-trivial, because | as well known from the theory of heat-engines | also work-extraction does lead to the heat flowing from cold to hot (1; 3).

Taking $W = 0$ in (24, 25) and recalling (23) we get

$$\mu = \varepsilon, \qquad u = v^{\theta}. \tag{28}$$

The interpretation of (28) is that since there are only two independent energy gaps in the system, they have to be precisely matched for the spontaneous processes to be possible; see in this context the discussion after (7). Thus the spontaneous heat $\mathcal{Q}_h^{[\mathrm{sp}]}$ is given as

$$\frac{\mathcal{Q}_h^{[\mathrm{sp}]}}{T_c} = \ln\left[\frac{1}{v_0}\right] \frac{(n-1)(v_0^{\theta} - v_0)}{[1+(n-1)v_0^{\theta}][1+(n-1)v_0]}, \tag{29}$$

where v_0 maximizes the RHS of (29).

3.4 How much one can enhance the spontaneous process?

We would like to compare the optimal spontaneous heat (29) with the optimal heat $\mathcal{Q}_h(\infty)$ transferred under consumption of a large amount of work; see (17, 18) and recall (26). One notes that for parameters of Fig. 2 the approximate equality $\mathcal{Q}_h(\infty) \approx \mathcal{Q}_h(W)$ is reached already for $W/T_h < 1$. This figure also shows that for the temperature ratio $\theta \equiv T_c/T_h$ far from 1, the improvement of the transferred heat introduced by driving is not substantial. It is however rather sizable for $\theta \simeq 1$, because here the spontaneous heat (29) is close to zero, while the heat $\mathcal{Q}_h(\infty)$ does not depend on the temperature difference at all; see Fig. 2 and (17, 18).

4. Efficiency

We saw above that enhancing optimal spontaneous processes does require work. Once this is understood, we can ask how efficient is this work consumption. The efficiency is defined as

$$\chi(W) = \frac{\mathcal{Q}_h(W) - \mathcal{Q}_h^{[\mathrm{sp}]}}{W} > 0, \tag{30}$$

where $\mathcal{Q}_h(W)$ is the optimal heat transferred under condition that the consumed work is not larger than $W > 0$, while $\mathcal{Q}_h^{[\mathrm{sp}]}$ is the optimal spontaneous heat; see (24, 29). Note that the

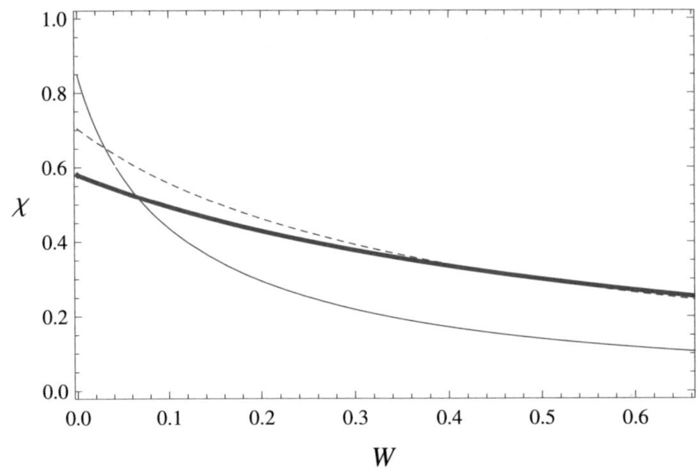

Fig. 3. The efficiency χ versus work W for $\theta \equiv T_c/T_h = 0.5$ and $n = 2$ (normal curve), $n = 10$ (dashed curve) and $n = 30$ (thick curve).

two subtracted quantities $\mathcal{Q}_h(W)$ and $\mathcal{Q}_h^{[\mathrm{sp}]}$ in (30) refer to the same junction $\mathbf{H} + \mathbf{C}$, but with different Hamiltonians; see (24, 25).
For $W \to 0$, $\chi(W)$ increases monotonically and tends to a well defined limit $\chi(0)$; see Fig. 3.

- For fixed θ and n, $\chi(0) = \chi(W \to 0)$ is the maximal possible efficiency at which the enhanced heat pump can operate. As seen from Fig. 3, this maximum is reached for

$$\mathcal{Q}_h(W) - \mathcal{Q}_h(0) \to +0 \quad \text{and} \quad W \to +0, \tag{31}$$

  where we recall that n, T_h and T_c are held fixed.
- There is thus a complementarity between the driven contribution in the heat, which according to (26) maximizes for $W \to \infty$, and the efficiency that maximizes under $W \to 0$.

Note from Fig. 4 the following aspect of the maximal efficiency $\chi(0)$: it decreases for a larger n (and a fixed θ). This is related to the fact that the optimal spontaneous heat $\mathcal{Q}_h^{[\mathrm{sp}]}$ increases for larger n.

- It is seen from Fig. 3 that

$$\chi(W) \le \chi(0) < \frac{\theta}{1-\theta}. \tag{32}$$

  We checked that this upper bound for the efficiency (30) holds for all $\theta = T_c/T_h$ and n.

It will be seen below that the upper bound $\frac{\theta}{1-\theta}$ is reached in the quasi-equilibrium limit $\theta \to 1$. Note that $\frac{\theta}{1-\theta}$ formally coicides with the Carnot limiting efficiency for ordinary refrigerators; see (2). A straightforward implication of (32) is that enhancing optimal spontaneous processes must be inefficient for $\theta \to 0$.
Let us discuss to which extent the bound (32) is similar to the Carnot bound (2) for refrigerators.
0. These two expressions are formally identical.
1. Recall that (2) is a general upper bound for the efficiency of refrigerators that transfer heat against its gradient. Such a transfer does require work-consumption. The same aspect

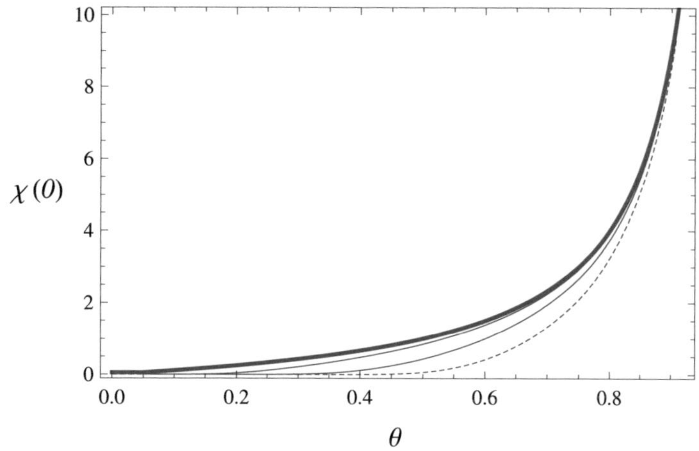

Fig. 4. The maximal efficiency $\chi(0) = \chi(W = 0)$ given by (**??**) versus $\theta = T_c/T_h$ for $n = 2$ (top normal curve), $n = 101$ (bottom normal curve), and $n = 10^5$ (dotted curve). Thick curve: the efficiency $\theta/(1-\theta)$.

is present in (30), because by its very construction the efficiency (30) refers to enhancement of the optimal spontaneous process that also demands work-consumption. To clarify this point consider a spontaneous process with the transferred heat $\mathcal{Q}_h^{[\mathrm{sp}]}$. Let this spontaneous process be non-optimal in the sense that no full optimization over the Hamiltonians (6, 7) has been carried out: $\mathcal{Q}_h^{[\mathrm{sp}]} < Q_h^{[\mathrm{sp}]}$. This non-optimal process is now enhanced via a work-consuming one. Denote by $\mathcal{Q}_h(W) > \mathcal{Q}_h^{[\mathrm{sp}]}$ the transferred heat of this process, where W is the consumed work. Following (30) one can define the efficiency of this enhancement as $\chi'(W) = [\mathcal{Q}_h(W) - \mathcal{Q}_h^{[\mathrm{sp}]}]/W$. One can now show, see Appendix 8, that $\chi'(W)$ can be arbitrary large for a sufficiently small (but non-zero) consumed work W. The reason for this unboundness is that we consider a non-optimal spontaneous process, which can be also enhanced by going to another spontaneous process.

2. We noted above that reaching bound (32) means a neglegible enhancement; see (31). The same holds for the Carnot bound (2) for refrigerators: operating sharply at the Carnot efficiency means that the heat transferred during refrigeration is zero; see (20) and references therein.

3. An obvious point where the bounds (32) and (2) differ from each other is that the latter is a straightforward implication of the first and second laws of thermodynamics, while the former is so far obtained in a concrete model only. We opine however that its applicability domain is larger than this model; some support for this opinion is discussed in section 5.

5. Enhanced heat transfer in linear non-equilibrium thermodynamics

Since the above results were obtained on a concrete model, one can naturally question their general validity. Here we indicate that these results are recovered from the formalism of linear non-equilibrium thermodynamics (28–30). This theory deals with two coupled processes: heat transfer between two thermal baths and work done by an external field. In contrast to the model studied in previous sections, the field is not time-dependent; e.g., it can be associated with the chemical potential difference (30). The difference and similarity between

set-ups, where the work is induced by a time-dependent field, and those where the work is done by a constant field is discussed in (31). For mesoscopic models which lead to linear non-equilibrium thermodynamics see, e.g., (30; 32).
In the linear regime both the driving field and the temperature difference $T_h - T_c$ are small, so that the heat Q_h and work W can be linearized (28–30); see also Appendix 8. The virtue of this approach is that it is independent from the concrete details of the studied system. The role of a free parameter | over which the setup can be optimized | is played here by the phenomenological coupling between the heat transfer and work input (28).
More specifically, we show in Appendix 8 that also within the formalism of linear non-equilibrium thermodynamics, enhancing the optimal spontaneous flow requires work-consumption. As for the efficiency, we first recall the message of (**??**): for $\theta \equiv T_c/T_h \to 1$ the maximal efficiency $\chi(0)$ does not depend on the dimension n^2 of the junction (provided that the latter is fixed) and approaches $\propto 1/(1-\theta)$. This asymptotic result is to a certain extent universal, because it is also recovered within linear non-equilibrium thermodynamics; see Appendix 8. We stress however that only the asymptotics for $\theta \to 1$ is recovered. Since this is a linear theory restricted to a small temperature difference and a small work input, it is naturally not capable of reproducing the full message of the bound (32). For this purpose one would probably need a non-linear thermodynamics theory; see, e.g., (30). Unfortunately, such theories are not so universal (system-independent) as the linear theory.

6. Summary

We started this paper by listing several representative examples of enhanced transport and posing two basic questions:

1. When enhancing a spontaneous process does require work-consumption?
2. If such a work-driven enhancement does take place, is there a general bound on its efficiency?

These questions have been answered via a quantum model for a heat-transfer junction immersed between two thermal baths at different temperatures T_c and T_h ($T_c < T_h$). The model is defined in section 2. We have chosen to work with this model of junction, because its structure is flexible enough to allow explicit maximization of the transferred heat over the junction Hamiltonian. We thus can determine the maximal heat transferred by the junction. Our basic results can be stated as follows.

1. When the spontaneous heat-transfer is already maximized over the junction Hamiltonian, its enhancement does require work-consumption.
2. The efficiency is defined with respect to the optimal spontaneous heat transfer as the heat increment due to enhancement divided over the consumed work. This efficiency is shown to be limited from above by $T_c/(T_h - T_c)$, a bound that is reached for $T_c \to T_h$. For this bound to hold it is essential that the efficiency is defined with respect to optimal spontaneous heat transfer. In its turn, the very idea of optimality refers to a class of variables to be optimized over. For the studied model this class is defined by the junctions structure and its Hamiltonian. 091338853

The main open problem with these results is whether they hold more generally than the studied model. We presented a partial evidence | partial because it is restricted to a small $T_h - T_c$ within the linear non-equilibrium thermodynamics | that they hold more generally.

7. Acknowledgement

This work has been supported by Volkswagenstiftung.
We thank B. Mehmani for several useful suggestions.

8. Appendix

Let the system $\mathbf{H}+\mathbf{C}$ function as a *non-optimal* [with respect to the Hamiltonian of $\mathbf{H}+\mathbf{C}$] spontaneous heat pump. The amount of heat transferred per cycle is $Q_h^{[\mathrm{sp}]}$. One increases the transferred heat via an external work-source acting on $\mathbf{H}+\mathbf{C}$. Now it is equal to $Q_h > Q_h^{[\mathrm{sp}]}$. The amount of work consumed per cycle is W.
The purpose of this Appendix is to show on a concrete example that there is no upper bound on the efficiency:

$$\chi' = \frac{Q_h - Q_h^{[\mathrm{sp}]}}{W}. \tag{33}$$

It can go to $+\infty$ for $W \to +0$.
Let the systems $\mathbf{H}$ and $\mathbf{C}$ be three-level systems with Hamiltonians [compare with (3)]

$$H_{\mathbf{H}} = \mathrm{diag}(0, \varepsilon, \varepsilon + \mu), \quad H_{\mathbf{C}} = \mathrm{diag}(0, \mu, \varepsilon + \mu) \tag{34}$$

where for future purposes we choose $\mu > \varepsilon > 0$. The initial Hamiltonian (5) reads

$$\mathrm{diag}[0,\, \varepsilon,\, \varepsilon+\mu,\, \mu,\, \varepsilon+\mu,\, \varepsilon+2\mu,\, \varepsilon+\mu,\, 2\varepsilon+\mu,\, 2\varepsilon+2\mu]$$

The initial state of $\mathbf{H}+\mathbf{C}$ is given by (2, 3 4):

$$\rho = \frac{\mathrm{diag}[1,\, a^\theta,\, a^\theta b^\theta]}{1 + a^\theta + a^\theta b^\theta}, \quad \sigma = \frac{\mathrm{diag}[1,\, b,\, ab]}{1 + b + ab}, \tag{35}$$

where we denoted

$$a = e^{-\beta_c \varepsilon}, \quad b = e^{-\beta_c \mu}. \tag{36}$$

The system $\mathbf{H}+\mathbf{C}$ has nine energy levels; three of them have equal energy $\varepsilon+\mu$. Thus spontaneous processes amount to unitary operations that couple these degenerate energy levels to each other, but do not induce transitions to non-degenerate energy levels [compare with (7)]. Such unitaries will produce double-stochastic matrices of the following form [compare with (19, 20)]

$$C = \begin{pmatrix} 1 & 0 & 0 & 0 & 0 & 0 & 0 & 0 & 0 \\ 0 & 1 & 0 & 0 & 0 & 0 & 0 & 0 & 0 \\ 0 & 0 & c_{11} & 0 & c_{12} & 0 & c_{13} & 0 & 0 \\ 0 & 0 & 0 & 1 & 0 & 0 & 0 & 0 & 0 \\ 0 & 0 & c_{21} & 0 & c_{22} & 0 & c_{23} & 0 & 0 \\ 0 & 0 & 0 & 0 & 0 & 1 & 0 & 0 & 0 \\ 0 & 0 & c_{31} & 0 & c_{32} & 0 & c_{33} & 0 & 0 \\ 0 & 0 & 0 & 0 & 0 & 0 & 0 & 1 & 0 \\ 0 & 0 & 0 & 0 & 0 & 0 & 0 & 0 & 1 \end{pmatrix}, \tag{37}$$

It is not difficult to see that for *fixed energy levels* (34) the spontanoues process with the largest $Q_h^{[sp]}$ amounts to $c_{13} = c_{22} = c_{31} = 1$, while other c's nullify. This means that the spontaneous process interchanges populations of the third and seventh energy levels; see (35). The heat transferred per cycle reads

$$Q_h^{[sp]} = \frac{1}{z}\left[a^\theta b^\theta - ab\right](\varepsilon + \mu), \tag{38}$$

$$z \equiv (1 + a + ab)\left(1 + a^\theta + a^\theta b^\theta\right). \tag{39}$$

Now the external field acts on the system $\mathbf{H} + \mathbf{C}$ enhancing the heat transfer, i.e., increasing $Q_h^{[sp]}$. We postulate that this action amounts to SWAP transformation. We obtain for the heat, work and efficiency (33) [compare with (12, 11)]:

$$Q_h = \frac{1}{z}\{\varepsilon\left[a^\theta - b + ba^\theta(a - b^\theta)\right]$$

$$+(\varepsilon + \mu)\left[(ab)^\theta - ab + ba^\theta(b^\theta - a)\right]\},$$

$$W = \frac{1}{z}(\mu - \varepsilon)\left[a^\theta - b - ba^\theta(b^\theta - a)\right]$$

It is seen that when $\varepsilon \to \mu$ [which means $a \to b + 0$] the difference $Q_h - Q_h^{[sp]}$ is positive and finite, while the work W is positive, but turns to zero as $W \propto \mu - \varepsilon$. Thus if one defines efficiency as $\chi' = (Q_h - Q_h^{[sp]})/W$, it will turn to infinity for $\varepsilon \to \mu$. The divergence of χ' is ultimately due to the fact that the spontaneous heat was maximized over Hamiltonian (6) only partially, i.e., it was not maximized over the energy levels of $\mathbf{H}$ and $\mathbf{C}$.

Heat-transfer enhancement from linear non-equilibrium thermodynamics

The formalism of linear non-equilibrium thermodynamics starts by introducing currents of physical quantities J_i (e.g., currents of energy or mass) and the respective conjugate forces X_i (28; 29). We introduce two such currents and forces:

$$J_1,\ J_2,\ X_1,\ X_2. \tag{40}$$

J_1 will refer to heat flowing from the high-temperature bath at temperature $T_h = 1/\beta_h$ to the lower one at temperature $T_c = 1/\beta_c$, while the second current J_2 is the work done by an external time-independent force. Thus

$$X_1 = \beta_c - \beta_h = \frac{T_h - T_c}{T^2},\ X_2 = \frac{f}{T},\ T \equiv \sqrt{T_h T_c}, \tag{41}$$

where f is the force responsible for the work (29). Once we are in the linear regime over small parameters $T_h - T_c$ and f, one can substitute T in (41) by T_c or T_h; the choice of T is conventional.

In essence, linearity means that the state which supports the currents is not far from equilibrium (29). The basic postulate of this formalism is linear relations between currents and forces (29):

$$J_1 = L_{11}X_1 + L_{12}X_2, \qquad J_2 = L_{21}X_1 + L_{22}X_2. \tag{42}$$

where the kinetic coeffcients L_{ik} do not depend on X_i due to assumed linearity of the overall process. The kinetic coeffcients L_{12} and L_{21} quantify the coupling between the two processes. The statement of the second law relevant for this composite linear process amounts to the positivity of entropy production $\mathrm{d}_iS/\mathrm{d}t$ (29)

$$\frac{\mathrm{d}_iS}{\mathrm{d}t} = X_1J_1 + X_2J_2 = \sum_{i,k=1}^{2} L_{ik}X_iX_k \geq 0. \tag{43}$$

As a consequence of the time-invariance of the underlying microscopic theories (i.e., classical or quantum mechanics), the matrix of kinetic coefficients is symmetric (29)

$$L_{12} = L_{21}. \tag{44}$$

We now assume that some work is dissipated, $J_2 > 0$, for enhancing the heat flow. The spontaneous heat transfer corresponds to no coupling between the processes: $L_{12} = L_{21} = 0$. In calculating the efficiency of enhancement we shall follow the same strategy as in sections 3 and 4: first we shall maximize the transferred heat J_1 under a fixed amount of work $W = fJ_2 = TX_2J_2$. Then the efficiency will be defined as in (30). The maximization variables are f, L_{11}, L_{22} and $L_{12} = L_{21}$. The temperatures T_c and T_h are held fixed; see also (41). Write J_1 as

$$J_1 = L_{11}X_1 + \frac{W}{TX_1} - L_{22}\frac{X_2^2}{X_1}. \tag{45}$$

During the maximization we should keep L_{11} confined by some upper limit $\mathcal{L}_{11}$; otherwise J_1 will not be finite. Eq. (44), which should hold for arbitrary X_1 and X_2, implies $L_{22} \geq 0$. Recaling that also $X_1 > 0$, we see that J_1 is maximized for $L_{11} = \mathcal{L}_{11}$ and $L_{22} = 0$:

$$\mathcal{J}_1 = \mathcal{L}_{11}X_1 + \frac{W}{TX_1}. \tag{46}$$

Thus for enhancing the optimal spontaneous heat $\mathcal{L}_{11}X_1$ we need $W > 0$ (work-consumption). Subtracting from $\mathcal{J}_1$ the spontaneous contribution $\mathcal{L}_{11}X_1$ and dividing over the consumed work W, we get for the efficiency:

$$\chi = \frac{1}{TX_1} = \frac{T}{T_h - T_c}, \tag{47}$$

which for a small temperature difference $T_h - T_c$ (recall that this is the applicability domain of the considered linear theory) coincides with (**??**).

9. References

[1] Callen,H.B. ; *Thermodynamics* (John Wiley, NY, 1985).
[2] Guyton,A. C. and Hall,J. A. , *Textbook of Medical Physiology* (Saunders Company, 1996).
[3] Lindblad, G., *Non-Equilibrium Entropy and Irreversibility* (Reidel, Dordrecht, 1983).
[4] Blumenfeld, L. A. and Tikhonov,A. N., *Biophysical Thermodynamics of Intracellular Processes: Molecular Machines of the Living Cell* (Springer, Berlin, 1994).
[5] Kovac,L., Curr. Topics in Bioenergetics, 15, 331 (1987).
[6] Lerch,H.-P. ; Mikhailov,A. S. and Hess,B. , PNAS 26, 15410 (2002).
[7] Vancelow, D. G., Biophysical Journal, 82, 2293 (2002).

[8] Hovhannisyan, K. and Allahverdyan, A. E., J. Stat. Mech. (2010) P06010
[9] Berry,R.S.; Kazakov, V.A.; Sieniutycz, S.; Szwast, Z. and Tsvilin, A.M., *Thermodynamic Optimization of Finite–Time Processes* (John Wiley & Sons, Chichester, 2000).
[10] Gemmer, J.; Michel, M. and Mahler, G., *Quantum Thermodynamics* (Springer, 2004).
[11] Lepri,S. ; Livi,R. and Politi,A. , Phys. Rep. 377, 1 (2003).
[12] Segal, D. and Nitzan, A., Phys. Rev. E 73, 026109 (2006).
[13] Marathe,R.; Jayannavar,A. M. and Dhar,A., Phys. Rev. E 75, 030103 (R) (2007).
[14] Segal,D., Phys. Rev. Lett. 101, 260601 (2008).
[15] Ren,J. and Li,B., Phys. Rev. E 81, 021111 (2010).
[16] Henrich,M. J.; Michel,M. and Mahler,G., Europhys. Lett., 76, 1057 (2006).
Henrich,M. J.; Mahler, G. and Michel,M., Phys. Rev. E 75, 051118 (2007).
[17] Feldman, T. and Kosloff, R., Phys. Rev. E, 61, 4774 (2000).
[18] Allahverdyan, A.E.; Balian, R. and Nieuwenhuizen, Th.M., J. Mod. Opt. 51, 2703 (2004).
[19] Allahverdyan, A.E.; Johal, R.S. and Mahler, G., Phys. Rev. E 77, 041118 (2008).
[20] Allahverdyan, A.E.; Hovhannisyan, K.V. and Mahler, G., *Optimal Refrigerator*, arXiv:0906.2583 (submitted to Phys. Rev. E).
[21] Parrondo, J.M.R., Phys. Rev. E 57, 7297 (1998).
Usmani, O.; Lutz, E. and Buttiker, M., Phys. Rev. E 66, 021111 (2002).
Astumian, R. D., PNAS 104, 19715 (2007).
[22] Rahav, S.; Horowitz, J. and Jarzynski, C., Phys. Rev. Lett. 101, 140602 (2008).
Chernyak, V. Y. and Sinitsyn, N. A., Phys. Rev. Lett. 101, 160601 (2008).
[23] Parrondo, J.M.R. and de Cisneros, B.J., Appl. Phys. A 75, 179 (2002).
[24] Marshall, A.W. and Olkin, I., *Inequalities: Theory of Majorization and its Applications*, (Academic Press, New York, 1979).
[25] Partovi, H. M., Phys. Lett. A 137, 440 (1989).
[26] Mityugov, V. V., Phys. Usp. 170, 681 (2000).
Brailovskii, A. B.; Vaks, V.L. and Mityugov, V.V., Phys. Usp. 166, 795 (1996).
[27] Janzing, D.; Wocjan, P.; Zeier, R.; Geiss, R. and Beth, R., Int. Jour. Theor. Phys. 39, 2217 (2000).
[28] Kedem, O. and Kaplan S. R., Trans. Faraday Soc. 61, 1897 (1965).
[29] de Groot, S. R. and Mazur, P. , *Non-equilibrium Thermodynamics* (Dover, New York, 1982).
[30] Stratonovich, R.L.,*Nonlinear Nonequilibrium Thermodynamics I, II* (Springer-Verlag, Berlin, 1992).
[31] Humphrey,T. E. and Linke,H. , Physica E 29, 390 (2005).
[32] Nakagawa,N. and Komatsu,T. S., EPL 75, 22 (2006).
Esposito,M. ; Lindenberg, K. and Van den Broeck,C., Phys. Rev. Lett. 102, 130602 (2009).

2

The Thermodynamic Effect of Shallow Groundwater on Temperature and Energy Balance at Bare Land Surface

F. Alkhaier[1], G. N. Flerchinger [2] and Z. Su[1]
[1]*Department of water resources Faculty of Geo-Information Science and Earth Observation, University of Twente*
[2]*Northwest Watershed Research Center, United States Department of Agriculture*
[1]*The Netherlands*
[2]*USA*

1. Introduction

Within the foregoing half century, several studies debated over the effect that shallow groundwater has on land surface temperature (Myers & Moore, 1972; Huntley, 1978; Quiel, 1975). As land surface temperature is a key factor when the process of energy and water exchange between land surface and atmosphere occurs, we can presume that shallow groundwater naturally affects the entire surface energy balance system.

Shallow groundwater affects thermal properties of the region below its water table. Further on, it alters soil moisture of the zone above its water table which results in affecting its thermal properties, the magnitude of evaporation, albedo and emissivity. Hence shallow groundwater affects land surface temperature and the surface energy balance in two different ways; direct and indirect (Figure 1). The direct way (henceforth referred to as thermodynamic effect) is through its distinctive thermal properties which make groundwater acts as a heat sink in summer and a heat source in winter, and affects heat propagation within soil profile. The indirect way is through its effect on soil moisture above water table and its related effects (i.e. evaporation, soil thermal properties of vadose zone, land surface emissivity and albedo).

Studies that investigated the thermodynamic effect commenced by the work of Kappelmeyer (1957), who could successfully use temperature measurements conducted at shallow depth (1.5m) to locate fissures carrying hot water from deep groundwater. Birman (1969) also found a direct relationship between shallow ground temperature and depth to groundwater. Works by Cartwright (1968, 1974), Bense & Kooi, 2004, Furuya et al. (2006) and also works by Takeuchi (1980, 1981, 1996) and Yuhara (1998) cited by Furuya et al. (2006) showed that soil temperature measurements at some depth (0.5-2 *m*) depth were useful for locating shallow aquifers in summer and winter and also for determining the depth of shallow groundwater and the velocity and direction of its flow.

On the other hand, a number of studies considered the indirect effect of shallow groundwater in terms of its effect on soil moisture of the vadose zone and at land surface (York et al., 2002; Liang & Xie, 2003; Chen & Hu, 2004; Yeh et al., 2005; Fan et al., 2007;

Gulden et al., 2007; Niu et al., 2007; Lo et al., 2008; Jiang et al., 2009). They linked shallow aquifers to land surface and atmospheric models through the effect of soil moisture in terms of its mass on the water budget and evapotranspiration at land surface.

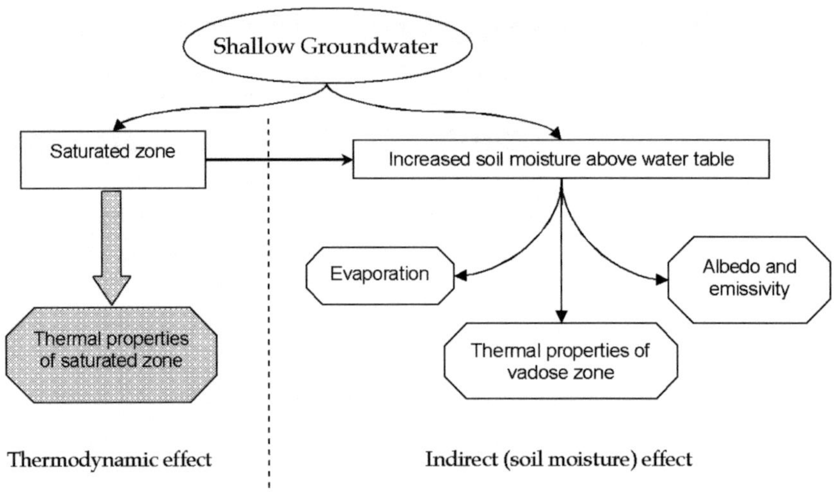

Fig. 1. Schematic description of the two different effects of groundwater

The effect of shallow groundwater on soil temperature has inspired some researchers to consider utilizing thermal remote sensing in groundwater mapping. For instance, Myers & Moore (1972) attempted to map shallow groundwater using the brightness temperature of land surface retrieved from an airborne radiometer. They found a significant correlation between land surface temperature and depths to groundwater in a predawn imagery of 26 August 1971. Huntley (1978) examined the utility of remote sensing in groundwater studies using mathematical model of heat penetration into the soil. Nevertheless, his model was not sophisticated enough to consider groundwater effect on surface energy fluxes (i.e. latent, sensible and ground heat fluxes), besides, it neglected totally the seasonal aspect of that effect. In 1982, Heilman & Moore (1982) showed that radiometric temperature measurements could be correlated to depth to shallow groundwater, but they recommended developing a technique for distinguishing water table influences from those of soil moisture to make the temperature method of value to groundwater studies.

Recently, Alkhaier et al. (2009) carried out extensive measurements of surface soil temperature in locations with variant groundwater depth, and found good correlation between soil temperature and groundwater depth. However, they also doubted about the cause of the discovered effect; was it due the indirect effect throughout soil moisture or was it because of the thermodynamic effect of the groundwater body. Furthermore, they suggested building a comprehensive numerical model that simulates the effect of shallow groundwater on land surface temperature and on the different energy fluxes at land surface.

Studies that dealt with the thermodynamic effect (Kappelmeyer, 1957; Cartwright, 1968, 1974; Birman, 1969; Furuya et al., 2006) explored that effect on soil temperature at some depth under land surface. By their deep measurements, they aimed at eliminating the indirect effect. Consequently they totally missed out considering that effect on temperature and energy fluxes at land surface. On the other hand, studies that considered the indirect effect (York et al., 2002; Liang & Xie, 2003; Chen & Hu, 2004; Yeh et al., 2005; Fan et al., 2007; Gulden et al., 2007; Niu et al., 2007; Lo et al., 2008; Jiang et al., 2009) were centered on the effect of soil moisture in terms of water mass and passed over the effect on soil thermal properties. Furthermore, studies which considered groundwater effect to be utilized in remote sensing applications (Huntley, 1978; Heilman & Moore's, 1982; Alkhaier et al., 2009) were faced with the problem of separating the effect of groundwater from that of soil moisture, there was hardly any sole study that conceptually and numerically discriminated the thermodynamic effect from the effect of soil moisture.

Quantifying the different aspects of groundwater effect can result in better understanding of this phenomenon. Further, this may advance related surface energy balance studies and remote sensing applications for shallow aquifers. This chapter centers on the thermodynamic effect which was separated out numerically from the other effects. We undertook to answer these questions: does shallow groundwater affect land surface temperature and surface energy balance at land surface regardless of its effect on soil moisture above water table? What are the magnitude and the pattern of that effect? And is that effect big enough to be detected by satellites?

With the aid of numerical modeling which progressed in complexity, we show in this chapter how the presence of groundwater, through its distinctive thermal properties within the yearly depth of heat penetration, affects directly land surface temperature and the entire surface energy balance system thereby. By applying different kinds of boundary conditions at land surface and changing the level of water table within the soil column, we observed the difference in temperature and the energy fluxes at land surface.

2. Numerical experiments

Two numerical experiments were implemented in this study. The first was simple and conducted using FlexPDE (PDE Solutions Inc.), a simulation environment which makes use of finite element technique to solve differential equations. The aim behind this experiment was to 1) prove that the thermodynamic effect of groundwater does indeed reach land surface and 2) to show that it is not appropriate to simply assign one type of boundary condition at land surface, and to explain that solving the entire surface energy balance at land surface is inevitable to realize groundwater effect. The entire surface energy balance system was simulated in the second experiment which was implemented using a well known land surface model code (Simultaneous Heat and Water model, SHAW, Flerchinger, 2000).

Initially we portray the common features among the different experiments; afterwards we describe the specific conditions for each experiment. Although the experiments were implemented within different numerical environments, they were performed using similar 1-D soil profiles. The lower boundary condition in both experiments was set at a depth of 30 m (deeper than the yearly penetration depth of heat) as a fixed temperature which is the mean annual soil temperature. Each experiment involved five simulations that were performed first for a profile with no groundwater presence, then for cases where groundwater perched at 0.5, 1, 2 and 3 meters respectively.

Groundwater presence within the soil column was introduced virtually through assigning different values of both thermal conductivity and volumetric heat capacity of saturated soil to the region below the imaginary water table. Rest of the soil in the profile was assigned the values of thermal properties for dry soil.

In the first experiment, water transfer was not considered at all; heat transfer was the only simulated process. In the second experiment water movement and soil moisture transfer were simulated normally, because SHAW simulates both heat and water transfers simultaneously and its forcing data include rainfall. Yet we adjusted the SHAW code in a way that soil thermal properties were independent from soil moisture, and were fixed and predefined as the values adopted in the first two experiments. In that way groundwater was not present actually within soil profile in SHAW simulation rather than it did exist virtually through the different thermal properties of the two imaginary zones (saturated and dry zones). By doing so, we guaranteed the harmony among the two experiments and also ensured separating the thermodynamic effect from the effect of soil moisture.

The same soil thermal properties of virtually saturated and dry zones within soil profiles were used in all experiments. Values of thermal conductivity were adopted as the values for standard Ottawa sand measured by Huntley (1978), who conducted similar modeling experiment. Volumetric heat capacity values were calculated using the expression of de Vries (1963). Accordingly, we used in all of our simulations values for thermal conductivity of 0.419 and 3.348 ($J\ m^{-1}s^{-1}{}^{\circ}C^{-1}$), and values for volumetric heat capacity of 1.10E+06 and 3.10E+06 ($J\ m^{-3}{}^{\circ}C^{-1}$) for dry and saturated sections respectively.

The first experiment involved two different simulation setups. In the first simulation setup we assigned land surface temperature as a boundary condition and observed the change in ground heat flux caused by groundwater level change within soil profile. In the second simulation setup, we applied ground heat flux as a boundary condition at land surface and observed the change in land surface temperature. The results of the two simulations suggested the indispensability of examining the effect of shallow groundwater on both temperature and ground heat flux simultaneously. To do so, it was necessary to free both of them and simulate the whole energy balance at land surface for scenarios with different groundwater levels. We accomplished that in the third experiment. All simulations were run for one year duration, after three years of pre-simulation to reach the appropriate initial boundary conditions.

2.1 Experiments 1

The experiment was conducted within FlexPDE environment. In one dimension soil column, heat transfer was simulated assuming conduction the only heat transport mechanism. Consequently, the sole considered governing equation was the diffusion equation:

$$\frac{\partial(k_S \partial T)}{\partial^2 z} = VHC \frac{\partial T}{\partial t} \tag{1}$$

where k_S is thermal conductivity ($J\ m^{-1}s^{-1}{}^{\circ}C^{-1}$), T is soil temperature ($^{\circ}C$), z is depth (m), VHC is volumetric heat capacity ($Jm^{-3}{}^{\circ}C^{-1}$) and t is time (s).

Analytically, yearly land surface temperature can be described by expanding equation (7) of Horton &Wierenga, (1983) to include both the daily and the yearly cycles and by setting the depth z to zero, hence:

$$T = \overline{T}_{avr} + A_1 \sin\left(\frac{2\pi\ t}{p_1}\right) + A_2 \sin\left(\frac{2\pi\ t}{p_2}\right) \tag{2}$$

where $\overline{T}_{avr}$ ($°C$) is the average soil temperature at all depths. A_1 and A_2 ($°C$) are the daily and yearly temperature amplitudes at land surface respectively, p_1 is one day and p_2 is one year expressed in the time unit of the equation (s).
Similarly, yearly ground heat flux at land surface can be expressed by expanding equation (10) of Horton & Wierenga (1983) to include both daily and yearly cycles and by setting the depth, z, to zero, thus:

$$G = k_s\left[A_1\sqrt{\frac{2\pi}{\alpha\, p_1}}\sin\left(\frac{2\pi\, t}{p_1}+\frac{\pi}{4}\right)+A_2\sqrt{\frac{2\pi}{\alpha\, p_2}}\sin\left(\frac{2\pi\, t}{p_2}+\frac{\pi}{4}\right)\right] \tag{3}$$

where k_s ($Jm^{-1}s^{-1}°C^{-1}$) is average soil thermal conductivity and α (m^2s^{-1}) is average thermal diffusivity.
In the first simulation, we applied land surface temperature (equation (2)) as a Dirichlet boundary condition at land surface of profiles with variant groundwater depth. As a result, FlexPDE provided the simulated ground heat flux for the different situations in terms of groundwater presence and level. Afterwards, we subtracted the resultant ground heat flux values of the profile with no-groundwater from those of profiles with groundwater and observed the differences.
On the contrary, in the second simulation we applied ground heat flux (equation (3)) as a forcing flux (Neumann boundary condition type) at land surface. Consequently, FlexPDE provided the simulated land surface temperature for the different situations in terms of groundwater presence and level. Then, we deducted the land surface temperature values of the profiles with no-groundwater from those of profiles with groundwater and observed the differences.

2.2 Experiment 2

To observe the thermodynamic effect of shallow groundwater on both land surface temperature and ground heat flux, all at once, we solved the complete balance system at land surface. This used SHAW to conduct this experiment because it presents heat and water transfer processes in detailed physics, besides, it has been successfully used to simulate land surface energy balance over a wide range of conditions and applications (Flerchinger and Cooley, 2000; Flerchinger et al., 2003, 2009; Flerchinger & Hardegree, 2004; Santanello & Friedl, 2003; Huang and Gallichand, 2006). Hereinafter, we present some of its basic features and expressions.

2.2.1 SHAW, the simultaneous heat and water model

The Simultaneous Heat and Water (SHAW) model is a one-dimensional soil and vegetation model that simulates the transfer of heat and water through canopy, residue, snow, and soil layers (Flerchinger, 2000). Surface energy balance and both water and heat transfer within the soil profile are expressed in SHAW as follows.
Surface energy balance is represented by the common equation:

$$R_n = LE + H + G \tag{4}$$

LE (Wm^{-2}) is latent heat flux, H (Wm^{-2}) is sensible heat flux and G (Wm^{-2}) is ground heat flux. R_n (Wm^{-2}) is the net radiation, which is the outcome of the incoming and outgoing radiation at the land surface as:

$$R_n = K_{in} - K_{out} + \varepsilon\ L_{in} - L_{out} \tag{5}$$

K_{in} and K_{out} are incoming and reflected short wave radiations respectively, $\varepsilon\ L_{in}$ and L_{out} are absorbed and emitted long wave radiations correspondingly, and ε is land surface emissivity.

Sensible heat flux is calculated by:

$$H = -\rho_a c_a \frac{(T_s - T_a)}{r_H} \tag{6}$$

where ρ_a ($kg\ m^{-3}$) is air density, c_a ($J\ kg^{-1}°C^{-1}$) is specific heat of air and T_a (°C) is air temperature at the measurement reference height z_{ref}; T_s is temperature (°C) of soil surface, and r_H is the resistance to surface heat transfer ($s\ m^{-1}$) corrected for atmospheric stability.

Latent heat flux is computed from:

$$LE = L\ \frac{(\rho_{vs} - \rho_{va})}{r_v} \tag{7}$$

where L is the latent heat of vaporization ($J\ kg^{-1}$), E is vapor flux ($kg\ s^{-1}m^{-2}$), ρ_{vs} ($kg\ m^{-3}$) is vapor density of soil surface and ρ_{va} ($kg\ m^{-3}$) is vapor density of air at the reference height. The resistance value for vapor transfer r_v ($s\ m^{-1}$) is taken to be equal to the resistance to surface heat transfer, r_H.

Finally, ground heat flux is expressed as:

$$G = -k_s \frac{\partial T}{\partial z} \tag{8}$$

where k_s is thermal conductivity ($J\ m^{-1}s^{-1}°C^{-1}$) and $\partial T/\partial z$ (°C m^{-1}) is soil temperature gradient. Ground heat flux is computed by solving for a surface temperature that satisfies surface energy balance, which is solved iteratively and simultaneously with the equations for heat and water fluxes within the soil profile.

The governing equation for temperature variation in the soil matrix in SHAW is:

$$VHC\frac{\partial T}{\partial t} - \rho_i L_f \frac{\partial \theta_i}{\partial t} = \frac{\partial (k_s \partial T)}{\partial^2 z} - VHC_W \frac{\partial q_l T}{\partial z} - L\left(\frac{\partial q_v}{\partial z} + \frac{\partial \rho_v}{\partial t}\right) \tag{9}$$

where ρ_i is ice density ($kg\ m^{-3}$); L_f is the latent heat of fusion ($J\ kg^{-1}$); θ_i is the volumetric ice content (m^3m^{-3}); VHC and VHC_W are the volumetric heat capacity of soil matrix and water respectively ($J\ m^{-3}C^{-1}$); q_l is the liquid water flux ($m\ s^{-1}$); q_v is the water vapor flux ($kg\ m^{-2}s^{-1}$) and ρ_v is the vapor density ($kg\ m^{-3}$).

The governing equation for water movement within soil matrix is expressed as:

$$\frac{\partial \theta_l}{\partial t} + \frac{\rho_i}{\rho_l}\frac{\partial \theta_i}{\partial t} = \frac{\partial}{\partial z}\left[k_h\left(\frac{\partial \psi}{\partial z} + 1\right)\right] + \frac{1}{\rho_l}\frac{\partial q_v}{\partial z} + U \tag{10}$$

where θ_l is the volumetric liquid water content (m^3m^{-3}), ρ_l is the liquid water density ($kg\ m^{-3}$); k_h is the unsaturated hydraulic conductivity ($m\ s^{-1}$); ψ is the soil matric potential (m) and U is a source/sink term ($m^3m^{-3}s^{-1}$).

The one-dimensional state equations describing energy and water balance are written in implicit finite difference form and solved using an iterative Newton-Raphson technique for infinitely small layers.

2.2.2 Weather and soil data

Weather conditions above the upper boundary and soil conditions at the lower boundary define heat and water fluxes into the system. Consequently, input to the SHAW model includes daily or hourly meteorological data, general site information, vegetation and soil parameters and initial soil temperature and moisture.
The forcing weather data were obtained from Ar-Raqqa, an area in northern of Syria that characterized by steppe climate (Köppen climate classification), which is semi-dry climate with an average annual rainfall of less than 200 *mm*. The simulations were run for the year 2004 after three years (2001-2003) of pre-simulation to reach appropriate initial conditions for soil profile. The daily input data includes minimum and maximum temperatures, dew point, wind speed, precipitation, and total solar radiation.
The soil for the profiles used in SHAW simulations were chosen to be standard Ottawa sand. However, since the groundwater was virtually presented within soil profile, and since the thermal properties were predefined, the type of the simulated soil is of minor importance. Basically SHAW calculates thermal conductivity and volumetric heat capacity according to the method of de Vries (de Vries, 1963). However for the sake of separating the thermodynamic effect of groundwater from the indirect one, we adjusted its FORTRAN code so the model uses the same values as used in the first experiment.
The output of the model includes surface energy fluxes, water fluxes together with temperature and moisture profiles. After solving for energy balance at the top of the different profiles, we subtracted the resultant land surface temperature, and surface heat fluxes of the no-groundwater profile from their correspondents of the profiles with the groundwater perches at 0.5, 1, 2 and 3 *m*.

3. Results

3.1 Experiment 1

By applying land surface temperature (equation(2)) as an upper boundary condition, then changing the thermal properties of the soil profile (due to the variation in the imaginary groundwater level), there was a considerable difference in the resultant simulated ground heat flux at land surface. The differences between ground heat flux of the no-groundwater profile and those of the profiles with different water table depths are shown in Figure 2a.
In winter, when the daily upshot of ground heat flux is usually directed upward (negative sign) and heat is escaping from the ground, ground heat flux of the profile with half meter groundwater depth was higher (in negative sign) than that of the no-groundwater profile. The difference in ground heat flux between the two profiles reached its peak value of almost -28 Wm^{-2} in February. The differences in ground heat fluxes between the no-groundwater profile and the profiles with groundwater at 1, 2 and 3 *m* depth behaved similarly but had smaller values of the peaks and roughly one month of delay in their occurrence between one and the next.
Quite the opposite, in summer, when the daily product of ground heat flux is usually downward (positive) and earth absorbs heat, ground heat flux of the profile with groundwater at half meter depth was also higher (but in positive sign) than that of the no-

groundwater profile, and reached similar peak value of about 28 Wm^{-2} in August. Again, the differences in ground heat flux between the no-groundwater profile and the profiles with groundwater at 1, 2 and 3 m depth behaved similarly with a delay in occurrence of the yet lower-values peaks.

Figure 2b shows the differences among the simulated land surface temperatures resulting from applying the same values of ground heat flux (equation (3)) at the surface of the profiles with different thermal properties due to variant levels of groundwater.

In winter, land surface temperature of the profile of half meter depth of groundwater was higher than that of the no-groundwater. The difference between the two, reached its peak of about 4 °C in February. Subsequently, the differences between land surface temperature of the profiles of 1, 2 and 3 m and that of the no-groundwater profile had lower peak values with a delay of almost a month between each other.

On the contrary, land surface temperature of the profile of half meter depth of groundwater was lower than that of no-groundwater in summer. The difference in temperature between the two profiles reached its peak value of about 4 °C in August. Again, the differences between land surface temperature of the profiles with groundwater at 1, 2 and 3 m depth and that of the no-groundwater profile had lower peak values with a delay in their occurrence of about month between one another (Figure 2b).

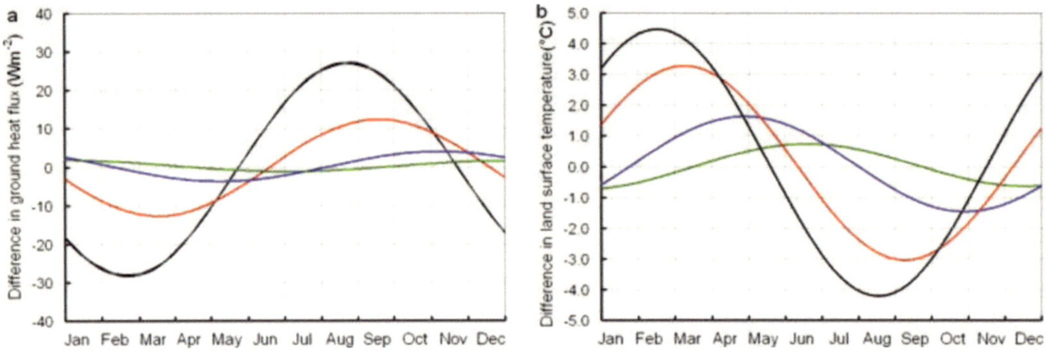

Fig. 2. a) Ground heat flux (Wm^{-2}) of the no-groundwater profile subtracted from those of profiles with water table depth of half meter (black), one meter (red) two meters (blue) and three meters (green). b) The same as (a) but for land surface temperature.

3.2 Experiment 2

With comprehensive consideration of surface energy balance and using real measured forcing data, SHAW showed more realistic results. The scattered dots in Figures 3-7 represent the differences between the no-groundwater profile and those with groundwater in terms of hourly values of the different variables which have been affected by the presence of groundwater within soil profile. The solid line drawn through the scattered dots in each figure represents the first harmonic which was computed by Fourier harmonic analysis.

Figure 3 demonstrates the surface temperature of the profile with no-groundwater subtracted from temperatures of the profiles with groundwater at 0.5, 1, 2 and 3 m depth. Land surface temperature of the profile with groundwater at half meter depth reached a value of about 1 °C higher than that of the no-groundwater profile in winter (Figure 3a). Similarly, land surface temperatures of the profiles of 1, 2 and 3 m groundwater-depth respectively reached values of roughly 0.5, 0.2 and 0.1 °C higher than that of the no-groundwater profile (Figures 3b-3d). In summer, land surface temperature of the profiles with groundwater at depths 0.5, 1, 2 and 3 *m* were lower than that of the no-groundwater profile by about 1, 0.5, 0.3 and 0.2 °C respectively.

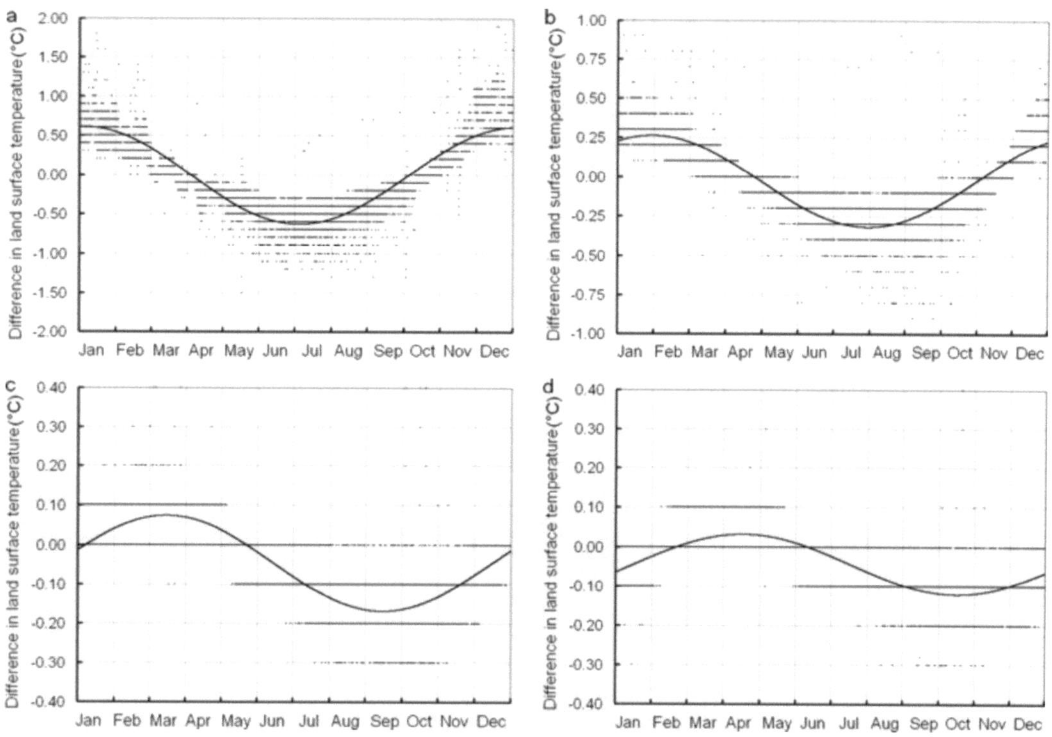

Fig. 3. Land surface temperature of the no-groundwater profile subtracted from those of profiles with groundwater at a) 0.5 *m* depth b) 1 *m* depth c) 2 *m* depth d) 3 *m* depth. Solid lines are first harmonics.

Simultaneously, ground heat flux was also influenced by the presence of groundwater as shown in Figure 4 which shows ground heat flux of the profile with no-groundwater subtracted from ground heat fluxes of the profiles with groundwater at 0.5, 1, 2 and 3 *m* depth. In wintertime, ground heat flux of the profile with half meter depth was higher (in negative sign) than that of the profile with no-groundwater by more than 11 Wm^{-2}, and also higher by about the same value (but in positive sign) in summer (Figure 4a). In the same way, ground heat fluxes of the profiles with groundwater at 1, 2 and 3 *m* depth were higher than that of the no-groundwater but with smaller peak values and with shifts in the phase (Figures 4b-4d).

Similarly, Figure 5 illustrates clear differences in sensible heat flux among the profiles of variant groundwater depths. In wintertime, sensible heat flux of the profile with groundwater at half meter depth reached a value of about 8 Wm^{-2} higher than that of the profile with no-groundwater. Quit the opposite in summertime, sensible heat flux of the profile with groundwater at half meter depth reached a value of about the same magnitude lower than that of the profile with no-groundwater (Figure 5a). Figures 5b-5d show that sensible heat fluxes of the profiles with groundwater at 1, 2 and 3 *m* depth were higher than that of the no-groundwater in wintertime but with smaller magnitudes and with shifts in the phase. In summertime, sensible heat fluxes of the profiles with groundwater at 1, 2 and 3 *m* depth were lower by similar magnitudes than that of the no-groundwater.

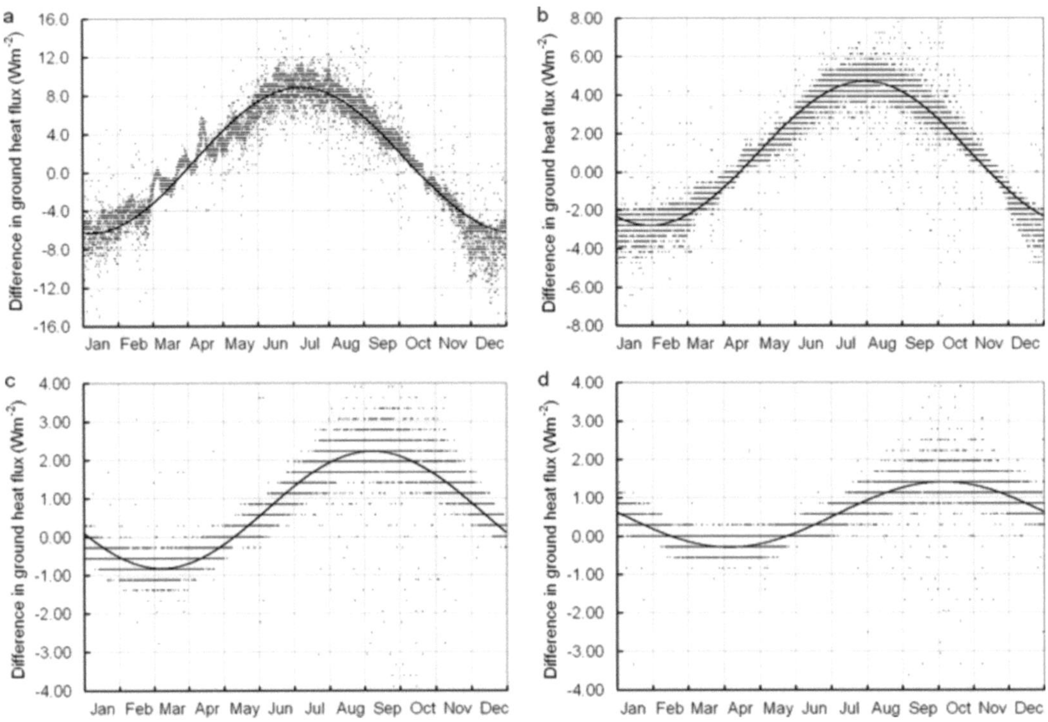

Fig. 4. Ground heat flux of the no-groundwater profile subtracted from those of profiles with groundwater at a) 0.5 *m* depth b) 1 *m* depth c) 2 *m* depth d) 3 *m* depth. Solid lines are first harmonics.

Unlike ground and sensible heat fluxes, latent heat fluxes showed very small differences among the different profiles (Figure 6). In spite of the immense amount of chaotic scattering, one can still see a small positive trend in winter and negative one in summer.

The last constituent of energy balance system which was altered by the presence of groundwater was the outgoing long-wave radiation (Figure 7). The differences looked similar to those of sensible heat flux in terms of diurnal shape and peak values but in reverse direction. Outgoing long-wave radiation of the no-groundwater profile was bigger in negative sign than that with groundwater in winter and smaller in summer.

The first harmonics sketched along of the scattered dots in Figures 3-7 demonstrated the periodic nature of the differences and were useful in pointing to the occurrence time of the differences' peaks both in winter and summer.

To have a closer look at the hourly variations (scattered dots in Figures 3-7), we zoomed in into hourly data of surface temperature and energy fluxes for two profiles: the no-groundwater profile and the profile with 50 cm groundwater depths within two different days (Figure 8). The first day was in winter (23 December, Figure 8 left side) and the second one was in summer (24 July, Figure 8 right side).

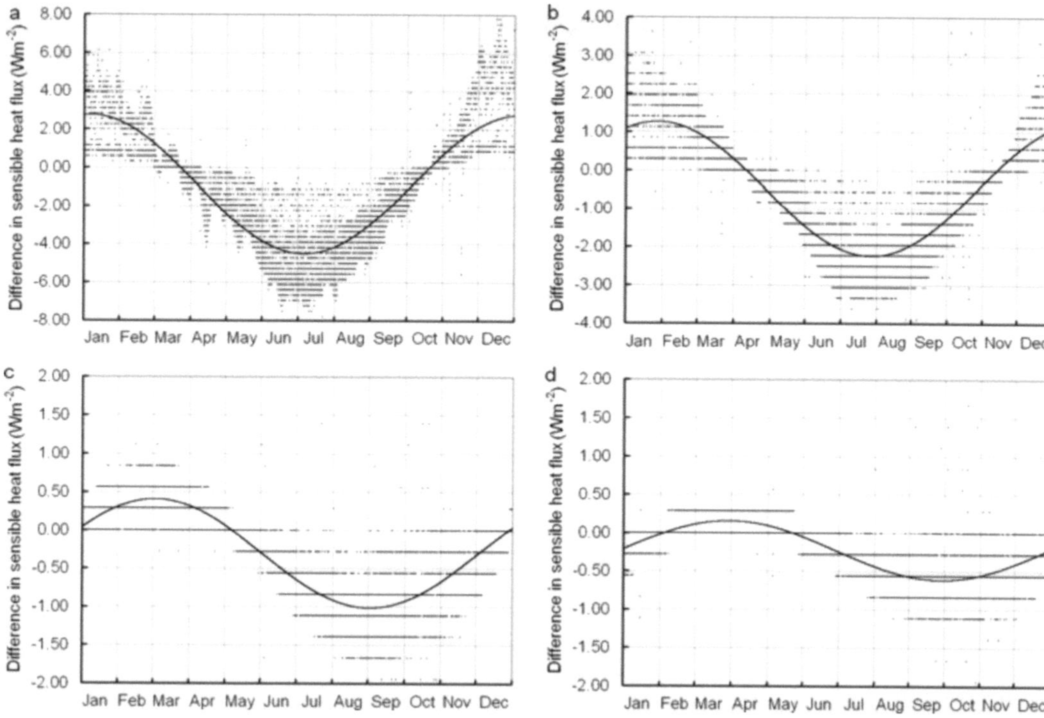

Fig. 5. Sensible heat flux of the no-groundwater profile subtracted from those of profiles with groundwater at a) 0.5 *m* depth b) 1 *m* depth c) 2 *m* depth d) 3 *m* depth. Solid lines are first harmonics.

In the winter day, land surface temperature of the no-groundwater profile was lower than that with groundwater all day long (Figure 8a). Therefore, the difference was positive. However, during nighttime the difference in land surface temperature was highest (about 1.2 °C). During daytime when the sun radiated solar energy on land surface, the difference diminished to 0.5 °C. After sunset the difference started to rise again. Oppositely, in the summer day (Figure 8b) land surface temperature of the no-groundwater profile was higher than that with the groundwater all day long; as a result, the difference was negative. Again, the difference was big at night (-1 °C) and moderated to -0.4 °C in daytime hours.

Figure 8c illustrates that in the winter day, ground heat flux of the no-groundwater profile was smaller (in negative sign) than that of the profile with groundwater during nighttime but greater than it was (in positive sign) in daytime. Hence, the difference remained negative in sign day and night. However, the difference was larger at day than it was at night. Conversely, in the summer day (Figure 8d) ground heat flux of the no-groundwater profile was bigger (in negative sign) than that of the profile with groundwater during nighttime, but smaller than it was (in positive sign) during daytime. Hence, the difference remained positive in sign during day and night, and again the difference was larger by day than it was at night.

Sensible heat flux of the no-groundwater profile was smaller than that of the profile with groundwater during day and night in the winter day. Therefore, the difference was positive all day long (Figure 8e). However, the difference was small at night (about 1 Wm^{-2}) and increased during the day up to more than 6 Wm^{-2}. In contrast, in the summer day

(Figure 8f) sensible heat flux of the no-groundwater profile was bigger than that of the profile with groundwater day and night. Therefore, the difference was negative all day long. And again the difference was small at night (about -1 Wm^{-2}) and increased during the day to more than -6 Wm^{-2}.

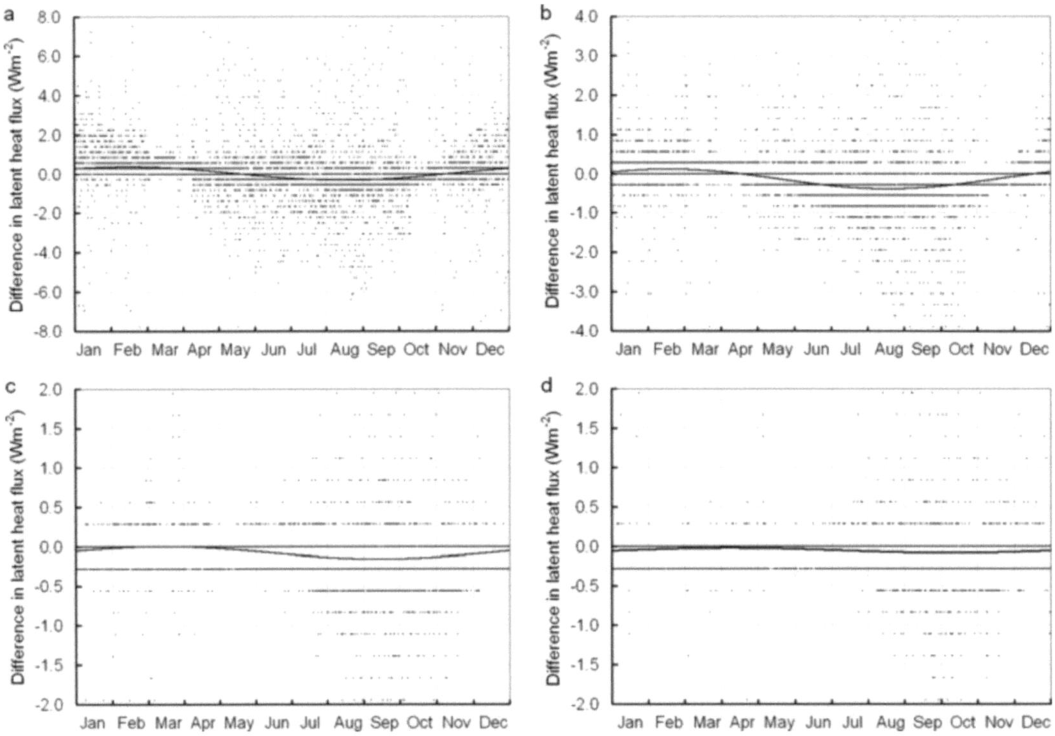

Fig. 6. Latent heat flux of the no-groundwater profile subtracted from those of profiles with groundwater at a) 0.5 m depth b) 1 m depth c) 2 m depth d) 3 m depth. Solid lines are first harmonics.

Unlike the previous two heat fluxes, latent heat flux showed very small difference between the two profiles, both in winter and summer days. In the winter day (Figure 8g) the difference in latent heat flux between the two profiles was around zero during nighttime. During daytime, latent heat flux of the profile with groundwater started to be larger than that of the no-groundwater. Oppositely, during the summer day (Figure 8h) latent heat flux of the profile with groundwater was smaller than that of the no-groundwater during daytime.

4. Discussion

In this study we show that the presence of groundwater within the yearly depth of heat penetration affects directly, and regardless of its effect on soil moisture above water table, both land surface temperature and ground heat flux, thereby affecting the entire surface energy balance system. The numerical experiments demonstrated that when we applied land surface temperature as a forcing upper boundary condition at land surface and

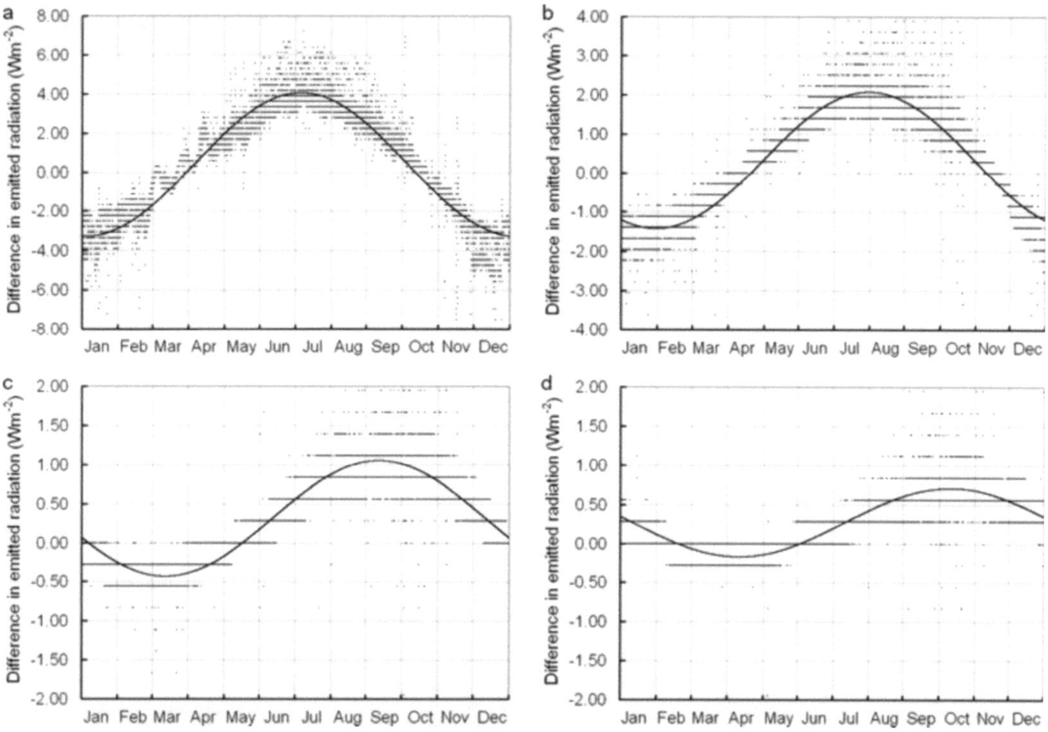

Fig. 7. Outgoing long wave radiation (Wm^{-2}) of the no-groundwater profile subtracted from those of profiles with groundwater at a) 0.5 m depth b) 1 m depth c) 2 m depth d) 3 m depth. Solid lines are first harmonics.

changed the water table depth, we obtained a significant difference in ground heat flux at land surface. On the contrary, when we applied forcing ground heat flux at land surface we obtained a considerable difference in land surface temperature by changing water table depth. Consequently, when we solved for the complete energy balance system at land surface, the thermodynamic effect of groundwater was demonstrated in simultaneous alteration of land surface temperature, ground heat flux, sensible heat flux, latent heat flux and outgoing long wave radiation at land surface.

The key reason behind this thermodynamic effect is the contrast in thermal properties within the soil profile. Resulting from the presence of groundwater, this contrast affects first and foremost heat penetration into the soil (equation (9)) which is chiefly pronounced via soil temperature and soil heat flux. Consequently, the largest difference should be marked for ground heat flux and land surface temperature.

When groundwater comes closer to land surface, it increases land surface temperature in winter and decreases it in summer (Figure 3). In this way it acts as a heat source in wintertime and a heat sink in summertime. As a result, shallow groundwater increases the intensity of ground heat flux both in winter and summer (Figure 4). In winter, it increases the upward ground heat flux which leads to further energy released from the ground. Contrarily, in summer it increases the downward ground heat flux allowing the earth to absorb more energy from the atmosphere.

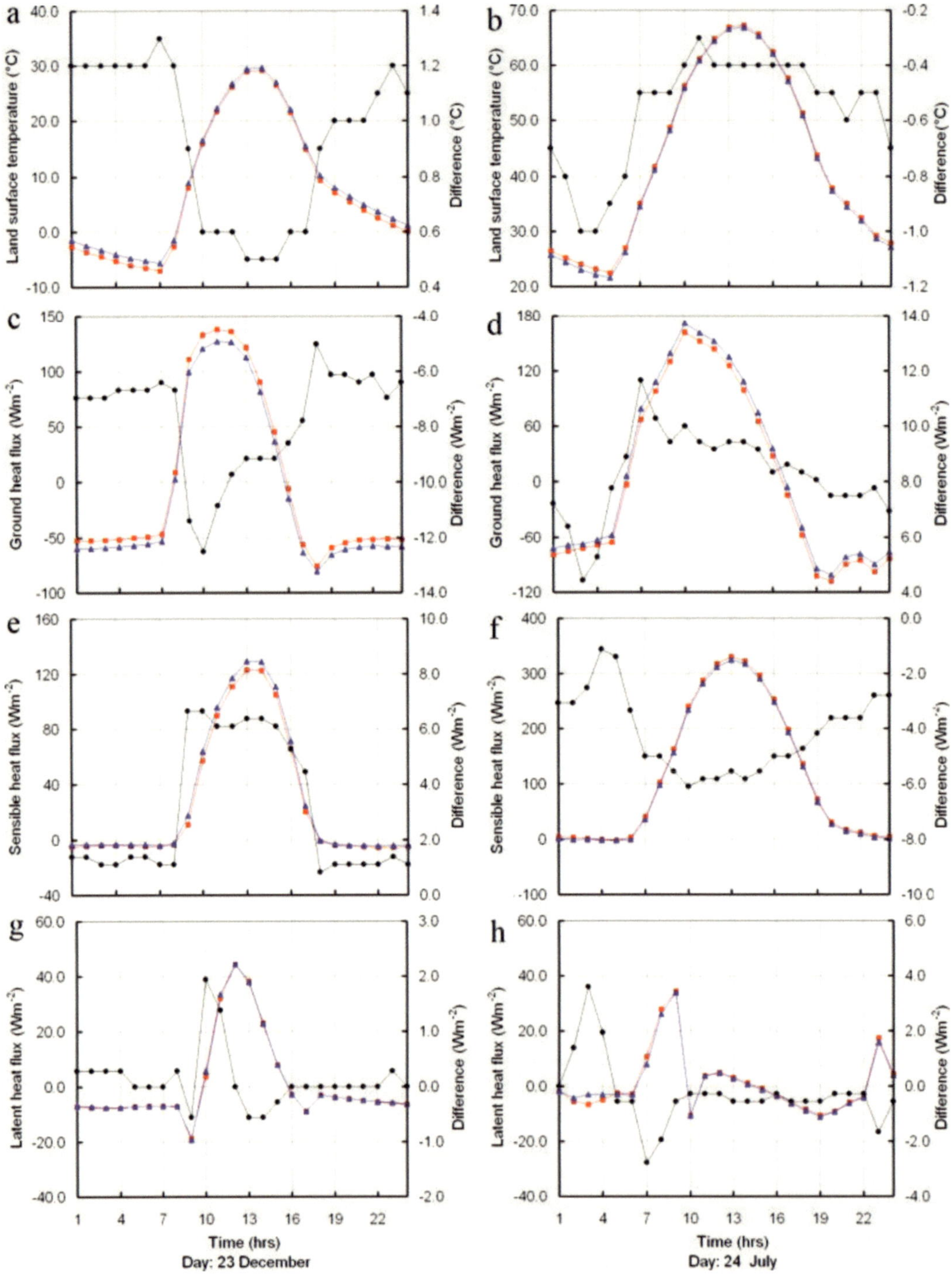

Fig. 8. Hourly values of temperature and energy fluxes of two profiles 1) with no-groundwater (red), 2) with groundwater at 50 *cm* depth (blue) and 3) the difference between them [(2)-(1)] (black), for two days: 23 Dec. (left) and 24 Jul. (right).

In the second experiment we observed a lower magnitude of temperature difference (Figure 3) than that observed in the first experiment (Figure 2b). Actually, the difference observed of land surface temperature within the first experiment (Figure 2b) was due to the fact that land surface was the single parameter which was subject to change, since the first experiment did not take into account the entire surface energy balance system. This big difference observed in the first experiment simulations were distributed among sensible and latent heat fluxes together with emitted long-wave radiation as explained by the second experiment (Figures 5-7).

Whilst sensible heat flux mitigates land surface temperature through the reciprocal swap of heat with air above land surface, latent heat flux exploits the gained heat in more evaporation, finally, outgoing long wave radiation continuously alleviates land surface temperature by emitting energy into the atmosphere. Therefore, the increase in land surface temperature in wintertime increases the amount of energy exchange between land surface and the air above it (i.e. sensible heat flux) due to the increment in temperature contrast between both of them. Contrarily, the decrease in land surface temperature in summer decreases sensible heat flux (Figure 5). Similarly the increase in land surface temperature in winter enhances evaporation, and its decrease in summer reduces evaporation (Figure 6). Yet the effect on evaporation was the smallest. Finally the increase in land surface temperature in winter increases energy emission from soil in the form of long wave radiation, and its decrease in summer causes yet smaller amount of emission (Figure 7).

Bearing in mind the convoluted interactions among energy fluxes and radiations at land surface, it is very difficult to describe how the groundwater thermodynamically affects each of them separately. Though, if we keep in mind the instantaneous nature of those interactions, we can still furnish a simplified conception of the thermodynamic effect as illustrated in Figure 9. Since the different soil thermal properties within the soil profile alter vertical heat transfer in both vertical directions (equation (9)), ground heat flux and soil temperature are the first two components to be directly affected by the thermodynamic effect. Consequently, land surface temperature affects sensible heat flux (equation (6)), latent heat fluxes (equation (7)) and the outgoing long wave radiation. The latter affects the net radiation available for the three fluxes, hence it affects again sensible and latent heat fluxes. On the other hand, ground heat flux also affects sensible and latent heat fluxes by reducing the energy left for them from the net radiation. Obviously, incoming, reflected short-wave radiation and incoming long-wave radiation stay outside the thermodynamic effect of groundwater.

The small difference in latent heat flux compared to the difference in other fluxes (Figure 6) can be justified by two reasons: Firstly, latent heat flux was originally small in this experiment due to the dry conditions in the considered area, and secondly, latent heat flux, unlike ground and sensible heat fluxes, is not a main function of land surface temperature; Whereas ground heat flux is a key function of land surface temperature and temperature of the soil beneath (equation (8)), and sensible heat flux is a primary function of land surface temperature and temperature of the air above (equation (6)), latent heat flux is a function of vapor density contrast between land surface and the atmosphere (equation (7)), and not a primary function of land surface temperature.

When groundwater depth increased, it was observed that the differences' peaks experienced a delay of about a month between one depth and the next (Figures 1-7). Similarly, it was also observed that the differences' peaks had lower values when groundwater went deeper.

The delay and the lower values can be justified by the fact that the closer the groundwater is to land surface the stronger and sooner its effect takes place on the penetrated ground heat flux.

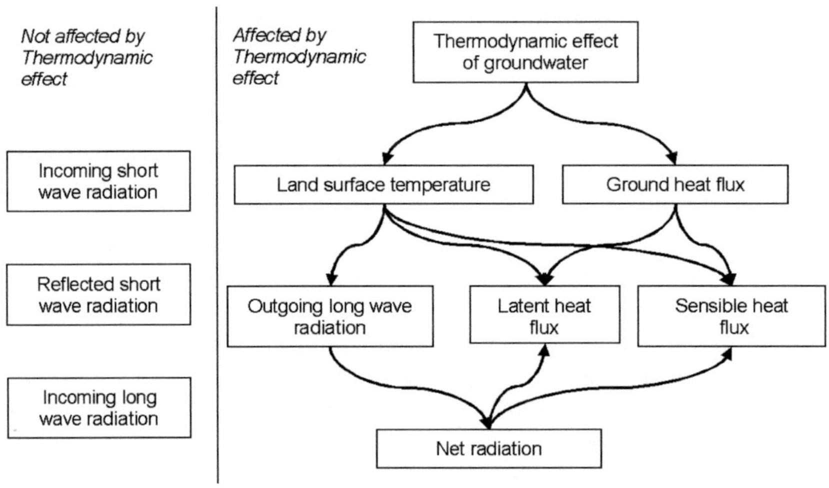

Fig. 9. Schematic description of groundwater thermodynamic effect on land surface temperature and the different components of surface energy balance.

The first experiment was simple and could not be compared to real world; therefore the observed differences in Figures 2a and 2b were sketched by neat lines without hourly fluctuations. On the contrary, the simulations in the second experiment were closer to reality and produced hourly variations presented by the scattering dots around the first harmonic lines within Figures 3-7. Samples of such hourly variations were presented in Figure 8. In both winter and summer days, the difference in land surface temperature was highest during nighttime and decreased in the daytime (Figures 8a and 8b). That was due to the fact that sensible and latent heat fluxes were stronger during daytime and had small magnitude during nighttime, in this way, they reduced the difference in land surface temperature in daytime in favor of their own differences (Figures 8e-8h).

In contrary to land surface temperature difference behavior, the difference in ground heat flux had high values in the night and had even higher values in daytime. This is explained by that the earth subsurface is the primary source of energy that drives the upward ground heat flux during nighttime, on the other hand, during daytime solar radiation provides the earth with higher amounts of energy and makes the difference in downward ground heat flux more pronounced (Figures 8c and 8d).

Alongside the normal scattering around the first harmonic lines in Figures 3-7 which presents hourly fluctuations, some outliers have been noticed. Investigating these outliers illuminated that these outliers result from the size of time-step (1 hour) used in SHAW simulations. While this can be enhanced by using smaller time step, this will require extensive simulation and numerical exertion.

In general we found that the magnitude of the thermodynamic effect on land surface temperature and surface energy balance system was small, but when considering the indirect effect, there will be two possibilities: 1) the two effects work in the same direction, then the thermodynamic effect will increase the intensity of the comprehensive effect, or 2) the two effects work in opposite directions and then the thermodynamic effect will decrease the intensity of the comprehensive effect. In this way, highlighting this aspect of groundwater effect was necessary to complete the view. As a result, it is important to study the comprehensive effect. The latter effect was studied in details by Alkhaier et al. (2011a) who took into consideration majority of the aspects through which shallow groundwater affects land surface temperature and the various components of surface energy balance system.

The recent advancement in the field of remote sensing models, e.g. Surface Energy Balance System (SEBS) (Su, 2002; van der Kwast et al., 2009; Ma et al., 2011) and Surface Energy Balance Algorithm for Land (SEBAL) (Bastiaanssen, 1995; Mohamed et al., 2004; Zwart & Bastiaanssen, 2006), has proved that satellite imagery is valuable tool in retrieving major components of surface energy balance both in day and night. With the aid of the findings in this study, together with those of the comprehensive effect of groundwater (Alkhaier et al., 2011a), one can think of utilizing those models in mapping the areal extents and depths to shallow groundwater. The best time of the year to detect that effect is most likely winter and summer time, bearing in mind the delay in the peaks' occurrence with different depths of groundwater as explained above. For the best time within a day, it is advised to investigate within the daytime hours for the difference in ground heat flux and sensible heat flux and to a minor extent for the difference in latent heat flux. Within the nighttime hours it is advised to explore the difference in land surface temperature and ground heat flux.

In a recent study, Alkhaier et al. (2011b) inspected the capacity of MODIS (Moderate-resolution Imaging Spectroradiometer), a scientific instrument on board of two currently operational satellites (Terra and Aqua), to detect the comprehensive effect of shallow groundwater on land surface temperature. Also they inspected the general features of spatial effect of shallow groundwater on surface soil moisture, surface soil temperature and surface energy balance components, at the time of image acquisition.

5. Conclusions

In summary, we conclude that shallow groundwater - regardless of its indirect effect generated via its effect on soil moisture above water table - does indeed affect directly the components of the energy balance system at land surface by its distinctive thermal properties. This thermodynamic effect is primarily obvious on land surface temperature, ground heat flux, sensible heat flux and outgoing long-wave radiation.

In terms of seasonally prospective, the thermodynamic effect on all these components is mostly pronounced in winter and summer. Whereas, in terms of hourly prospective, the difference in land surface temperature and outgoing long wave radiation is higher during nighttime, and the difference in ground and sensible heat fluxes is higher during daytime.

In spite of its small magnitudes, highlighting the different features of the thermodynamic effect is important to make the understanding of the comprehensive effect of groundwater more complete. The importance of the thermodynamic effect comes from its interaction with the indirect effect which originates from soil moisture above water table; this interaction may increase or decrease the upshot of the total effect.

Finally, it is important to give emphasis to the fact that in this study we separated numerically the thermodynamic effect from the indirect effect of groundwater on land surface and surface energy balance system. However, in real world these two effects can not be separated naturally and the image can not be complete without considering the combined effect. Nevertheless, this thermodynamic effect on land surface has not been established before and it clearly offers a more clear view of groundwater effect which is promising for enhancing the related surface energy balance studies and remote sensing applications.

6. References

Alkhaier, F., Schotting, R. J., & Su, Z. (2009). A qualitative description of shallow groundwater effect on surface temperature of bare soil. *Hydrology and Earth System Sciences*, Vol. 13, pp. 1749–1756.

Alkhaier, F., Flerchinger, G. N., & Su, Z. (2011a). The effect of shallow groundwater on land surface temperature and surface energy balance under the conditions of bare soil, I. Modeling and Description, *under review*.

Alkhaier, F., Flerchinger, G. N., & Su, Z. (2011b). The effect of shallow groundwater on land surface temperature and surface energy balance under the conditions of bare soil, II. Utilizing remote sensing in featuring that effect, *under review*.

Bastiaanssen, W. G. M. (1995). *Regionalization of surface flux densities and moisture indicators in composite terrain – A remote sensing approach under clear skies in Mediterranean climates*, Ph.D. Thesis, 273 pp., Wageningen Agricultural University, the Netherlands.

Bense, V. F., & Kooi, H. (2004). Temporal and spatial variations of shallow subsurface temperature as a record of lateral variations in groundwater flow. *Journal of Geophysical Research*, Vol. 109, B04103, doi:10.1029/2003JB002782.

Birman, H. (1969). Geothermal exploration for groundwater. *Geological Society of America Bulletin*, Vol. 80, No. 4, pp. 617–630.

Cartwright, K. (1968). Thermal prospecting for groundwater. *Water Resources Research*, Vol. 4, No. 2, pp. 395–401.

Cartwright, K. (1974). Tracing shallow groundwater systems by soil temperatures. *Water Resources Research*, Vol. 10, No. 4, pp. 847–855.

Chen, X., & Hu, Q. (2004). Groundwater influences on soil moisture and surface evaporation. Journal of Hydrology, Vol. 297, pp. 285-300.

de Vries, A. D. (1963). Thermal properties of soils, In: *Physics of plant environment*, pp. 210–235, North Holland Publication Company, Amsterdam, the Netherlands.

Fan, Y., Miguez-Macho, G., Weaver, C. P., Walko, R., & Robock, A. (2007). Incorporating water table dynamics in climate modeling: I. Water table observations and equilibrium water table simulations. *Journal of Geophysical Research*, Vol. 112, D10125, doi:10.1029/2006JD008111.

Flerchinger, G. N. (2000). *The simultaneous heat and water (SHAW) model*, Technical Report, 37 pp., Northwest Watershed Research Centre, USDA Agricultural Research Service, Boise, Idaho.

Flerchinger, G. N., & Cooley, K. R. (2000). A ten-year water balance of a mountainous semi-arid watershed. *Journal of Hydrology*, Vol. 237, pp. 86–99.

Flerchinger, G. N., & Hardegree, S. P. (2004). Modelling near-surface soil temperature and moisture for germination response predictions of post-wildfire seedbeds. *Journal of Arid Environments*, Vol. 59, pp. 369–385, doi:10.1016/j.jaridenv.2004.01.016.

Flerchinger, G. N., Sauer, T. J., & Aiken, R. A. (2003). Effects of crop residue cover and architecture on heat and water transfer at the soil surface. *Geoderma*, Vol. 116, pp. 217- 233, doi:10.1016/S0016-7061(03)00102-2.

Flerchinger, G. N., Xiao, W., Sauer, T. J., & Yu, Q. (2009). Simulation of within-canopy radiation exchange. *NJAS -Wageningen Journal of Life Sciences,* Vol. 57, pp. 5–15.

Furuya, G., Suemine, A., Sassa, K., Komatsubara, T., Watanabe, N., & Marui, H. (2006). Relationship between groundwater flow estimated by soil temperature and slope failures caused by heavy rainfall, Shikoku Island, south western Japan. *Engineering Geology*, Vol. 85, pp. 332–346, doi:10.1016/j.enggeo.2006.03.002.

Gulden, L. E., Rosero, E., Yang, Z., Rodell, M., Jackson, C. S., Niu, G., Yeh, P. J.-F., & Famiglietti, J. (2007). Improving land-surface model hydrology: Is an explicit aquifer model better than a deeper soil profile?. *Geophysical Research Letters*, Vol. 34, L09402, doi:10.1029/2007GL029804.

Heilman, J. L., & Moore, D. G. (1982). Evaluating depth to shallow groundwater using heat capacity mapping mission (HCMM) data. *Photogrammetric Engineering and Remote Sensing*, Vol. 48, No. 12, pp. 1903–1906.

Horton, R., & Wierenga, P. J. (1983). Estimating the soil heat flux from observations of soil temperature near the surface. *Soil Science Society of America,* Vol. 47, pp. 14-20.

Huang, M., & Gallichand, J. (2006). Use of the SHAW model to assess soil water recovery after apple trees in the gully region of the Loess Plateau, China. *Agricultural Water Management,* Vol. 85, pp. 67–76, doi:10.1016/j.agwat.2006.03.009.

Huntley, D. (1978). On the detection of shallow aquifers using thermal infrared imagery. *Water Resources Research*, Vol. 14, No. 6, pp. 1075– 1083.

Jiang, X., Niu, G.-Y., & Yang, Z.-L. (2009). Impacts of vegetation and groundwater dynamics on warm season precipitation over the Central United States. *Journal of Geophysical Research,* Vol. 114, D06109, doi:10.1029/2008JD010756.

Kappelmeyer, O. (1957). The use of near surface temperature measurements for discovering anomalies due to causes at depths. *Geophysical Prospective,* Vol. 5, No. 3, pp. 239–258.

Liang X., & Xie, Z. (2003). Important factors in land-atmosphere interactions: surface runoff generations and interactions between surface and groundwater. *Global and Planetary Change,* Vol. 38, pp. 101-114.

Lo, M-H., Yeh, P. J.-F., Famiglietti, J. S. (2008). Constraining water table depth simulations in a land surface model using estimated baseflow, *Advances in Water Resources*, Vol. 31, No. 12, pp. 1552-1564, ISSN 0309-1708, doi: 10.1016/j.advwatres.2008.06.007.

Ma, W., Ma, Y. & Su, Z. (2011). Feasibility of retrieving land surface heat fluxes from ASTER data using SEBS: a case study from the Namco area of the Tibetan plateau. *Arctic, Antarctic, and Alpine Research,* Vol. 43, No. 2, pp. 239-245.

Mohamed, Y. A., Bastiaanssen, W. G. M. & Savenije, H. H. G. (2004). Spatial variability of evaporation and moisture storage in the swamps of the upper Nile studied by remote sensing techniques. *Journal of Hydrology.,* Vol. 289, pp. 145-164.

Myers, V. I., & Moore D. G. (1972). Remote sensing for defining aquifers in glacial drift, *Proceedings of Eighth International Symposium on Remote Sensing of Environment*, Vol. 1, pp. 715–728, University of Michigan, October 1972.

Niu, G.-Y., Yang, Z.-L., Dickinson, R. E., Gulden, L. E., & Su, H. (2007). Development of a simple groundwater model for use in climate models and evaluation with Gravity Recovery and Climate Experiment data. *Journal of Geophysical Research.,* Vol. 112, D07103, doi:10.1029/2006JD007522.

Quiel, F. (1975). Thermal/IR in geology. *Photogrammetric Engineering and Remote Sensing*, Vol. 41, No. 3, pp. 341–346.

Santanello, J. A., & Friedl, M. A. (2003). Diurnal covariation in soil heat flux and net radiation. *Journal of Applied Meteorology*, Vol. 42, pp. 851–862.

Su, Z. (2002). The Surface Energy Balance System (SEBS) for estimation of turbulent heat fluxes. *Hydrology and Earth System Sciences*, Vol. 6, No. 1, pp. 85-99.

van der Kwast, J., et al. (2009). Evaluation of the surface energy balance system SEBS applied to ASTER imagery with flux measurements at the SPARC 2004 site, Barrax, Spain. *Hydrology and Earth System Sciences*, Vol. 13, No. 7, pp. 1337-1347.

Yeh, P. J-F., & Eltahir, E. A. B., (2005). Representation of Water Table Dynamics in a Land Surface Scheme. Part I: Model Development. *Journal of Climate*, Vol. 18, pp. 1861–1880, doi: 10.1175/JCLI3330.1

York, J. P, Person, M., Gutowski, W. J., & Winter, T. C. (2002). Putting aquifers into atmospheric simulation models: an example from the Mill Creek watershed, northeastern Kansas. *Advances in Water Resources*, Vol. 25, pp. 221–238.

Zwart, S. J., & Bastiaanssen, W. G. M. (2006). SEBAL for the description of spatial variability of water productivity in various wheat systems. *Agricultral Water Management*, Vol. 89, pp. 287-296.

3

Stress of Vertical Cylindrical Vessel for Thermal Stratification of Contained Fluid

Ichiro Furuhashi
Mito Science Analysis Intelligence Corp. Mito Ibaraki
Japan

1. Introduction

Various thermal loads are induced in elevated temperature systems, such as nuclear power plants. The load caused by the thermal stratification of contained fluid is one of those loads (Moriya et al., 1987; Bieniussa & Reck, 1996; Kimura et al., 2010). The thermal stratification is phenomenon under the condition of insufficient forced-convection mixture, where a denser fluid layer of lower temperature locates beneath a lighter fluid layer of higher temperature (Haifeng et al., 2009).

A conventional design evaluation method of vessel stress assumes an axial vessel temperature profile consisting of a straight line with the maximum fluid temperature gradient as shown in the top of Fig.1, and applies cylindrical shell theory for stress solution (Timoshenko & Woinowsky, 1959). The conventional method gives conservative solutions of thermal stresses that are proportional to the temperature gradient, and hence leads to narrower design windows.

In actual conditions, thermal stress is smaller than that from the conventional method, because of relatively moderated temperature profile due to attenuation by heat transfer on the inner surfaces and by heat conduction in vessel walls as shown in the bottom of Fig.1, as well as the cancellation of stresses at both ends of the thermal stratification section that have opposite signs generated by the reverse temperature changes. The consideration of such effects conventionally requires FEM heat conduction analyses taking the heat transfer with fluid into account and the subsequent FEM thermal stress analyses based on the above results. However, the FEM analyses are not suitable for a design work which places a high priority to get design perspective with rapid estimation.

In order to propose an accurate design method, this paper studies the steady-state vessel temperature solutions based on a model shown in Fig.2 taking the heat transfer with fluid and heat conduction into account, as well as the subsequent cylindrical shell stress ones based on the above temperature results. The obtained results are compiled into easy-to-use charts for design.

2. Theoretical analysis

The analysis model is shown in Fig.2. It is assumed that the radius of a cylindrical vessel, R, is enough larger than the vessel thickness, t ($R>>t$), so that the vessel wall can be considered as a flat plate. Here, λ is the thermal conductivity of the vessel. The positions in the plate thickness direction and the axial direction are represented by x and z, respectively.

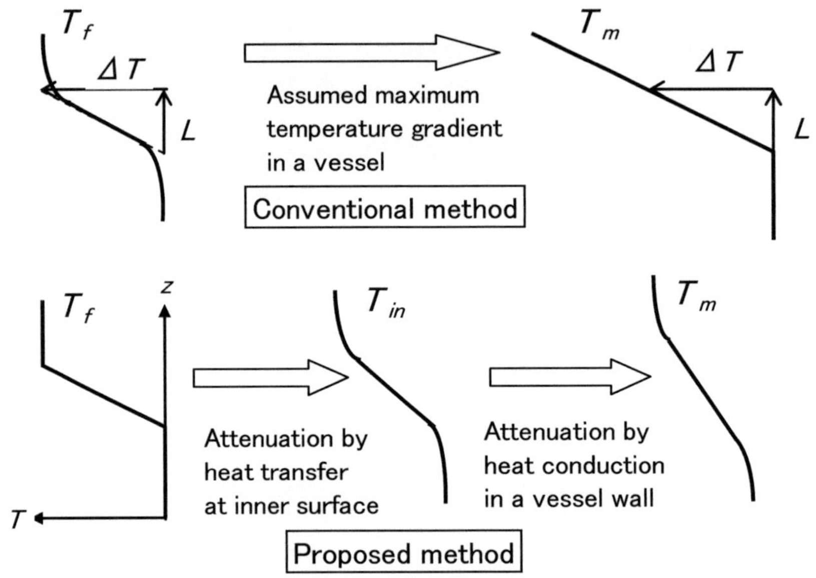

Fig. 1. Comparison of conventional and proposed methods

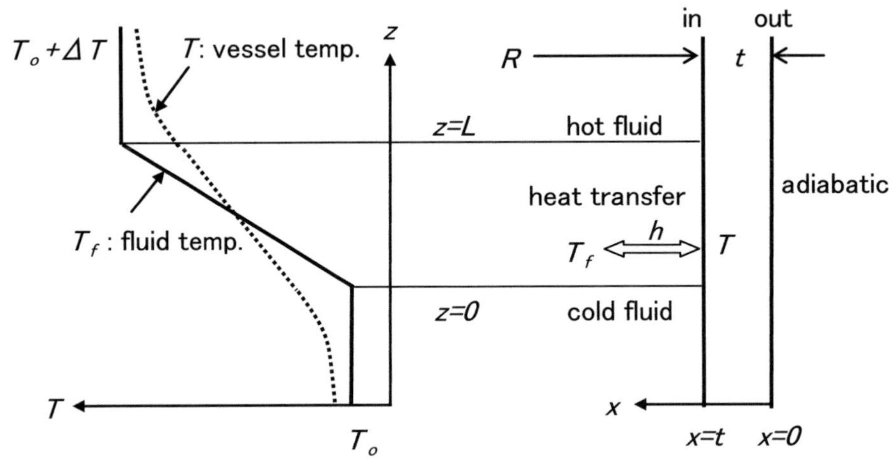

Fig. 2. Analysis model

It is assumed that the external surface of the vessel (*x=0*) is thermally-insulated and heat transfer occurs between the inner surface (*x=t*) and fluid with the heat transfer coefficient, *h*. The thermal stratification layer is represented by the range of *z=0~L*.

2.1 Step-shaped fluid temperature profile

We discuss the case in which the fluid temperature profile is given by the following step function (in the case of *L=0* in Fig.2).

$$T_f(z) = T_0 + H(z)\Delta T \tag{1}$$

Here, *H(z)* is a step function; *H(z)=0* for *z< 0,* and *H*(z)=*1* for *z>0.* Using the eigen-function expansion method (Carslaw & Jeager, 1959), the steady-state vessel wall temperature, *T(x,z),* was obtained as the following equation (Furuhashi et al., 2007, 2008).

$$T(x,z) = T_0 + H(z)\Delta T - \mathrm{sgn}(z)\sum_{n=1}^{\infty} c_n \cos(p_n x)\exp(-p_n |z|) \tag{2}$$

Here, sgn(*z*) is a sign function; sgn(*z*)=*-1* for *z<0,* and sgn(*z*)=*1* for *z>0.* Each term of the series is an eigen-function that satisfies the steady-state condition, $\partial^2 T/\partial x^2 + \partial^2 T/\partial z^2 = 0$, and the adiabatic condition at *x=0* plane, $\partial T/\partial x = 0$. The eigen-values, p_n (n=1,2,...), consist of positive roots in ascending order of eigen-value equation, Eq.(3), that is derived from the heat transfer condition at *x=t* plane, $\lambda \partial T/\partial x = h(T_f - T)$.

$$(p_n t)\tan(p_n t) = \frac{ht}{\lambda} = Bi \tag{3}$$

Here, *Bi* is the non-dimensional heat transfer coefficient (Biot number). The coefficients, c_n (n=1,2,...), are obtained from the symmetry condition, $T(x,+0) = T(x,-0) = T_0 + \Delta T/2$.

$$c_n = \frac{\Delta T \sin(p_n t)}{p_n t + \sin(p_n t)\cos(p_n t)} \tag{4}$$

The wall-averaged temperature is represented by the following equation.

$$T_m(z) = \frac{1}{t}\int_0^t T(x,z)dx = T_0 + H(z)\Delta T - \mathrm{sgn}(z)\sum_{n=1}^{\infty}\frac{c_n}{p_n t}\sin(p_n t)\exp(-p_n |z|) \tag{5}$$

The values calculated by the theoretical solution of wall-averaged temperature, Eq. (5), for *Bi*=0.1, 1, 10, 100 are plotted in Fig.3 with the note, (exact), in the legend.

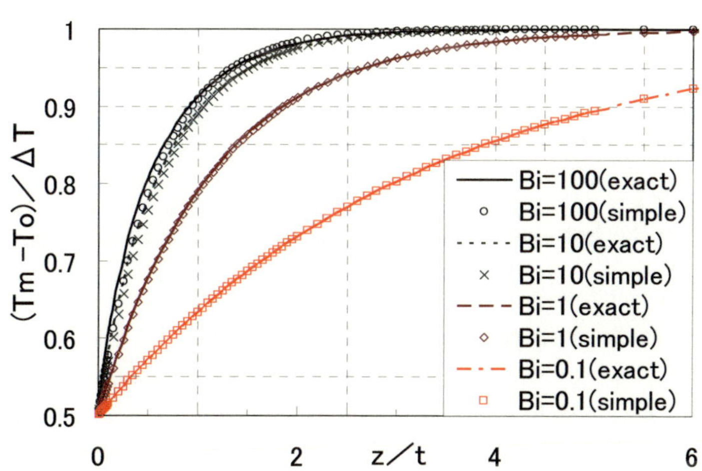

Fig. 3. Comparisons of vessel temperatures by exact solution with those by the temperature profile method

2.2 Simplified solution based on the temperature profile method

The theoretical solution, Eq.(5), is convenient for the calculation on a PC and quite useful. However, it is not a simplified equation suitable for the design evaluation because it needs a series calculation and an eigen-value calculation. Then, we tried to obtain an approximate simple solution that allows easy calculation based on the temperature profile method (Katto, 1964). The axial profile of wall-averaged temperature is approximated by the following equation (Furuhashi et al., 2007, 2008).

$$T_m(z) = T_0 + H(z)\Delta T - \mathrm{sgn}(z)\frac{\Delta T}{2}e^{-b|z|} \tag{6}$$

Here, the constant b is termed as temperature attenuation coefficient. When assuming that the temperature profile in the plate thickness direction is parabolic; $T=a_0(z)+a_1(z)x^2$, then next equations holds.

$$T_{in} = a_0(z) + a_1(z)t^2 \tag{7}$$

$$T_m = \int_0^t Tdx \Big/ t = a_0(z) + a_1(z)t^2 \Big/ 3 \tag{8}$$

Here, T_{in} represents the inner surface temperature. Temperature gradient at the inner surface in the thickness direction is given by the following equation.

$$\frac{\partial T}{\partial x}\Big|_{x=t} = 2a_1(z)t = \frac{3}{t}(T_{in} - T_m) \tag{9}$$

Using this equation, heat flux at the vessel inner surface, q, is given approximately by the following equation.

$$q = h(T_f - T_{in}) = \lambda\frac{\partial T}{\partial x}\Big|_{x=t} = \frac{3h}{3+Bi}(T_f - T_m) \tag{10}$$

Then the total heat flow from fluid to the vessel in the hot side ($z>0$), Q, is given approximately by the following equation.

$$Q = 2\pi Rh\int_0^\infty (T_f - T_{in})dz = \frac{6\pi Rh}{(3+Bi)}\int_0^\infty (T_f - T_m)dz = \frac{3\pi Rh\Delta T}{(3+Bi)b} \tag{11}$$

The heat flow from the hot side to the cold side across the $z=0$ plane, Q, is given by the following equation.

$$Q = 2\pi Rt\lambda\frac{dT_m}{dz}\Big|_{z=0} = \pi Rt\lambda\Delta Tb \tag{12}$$

Since Eqs.(11) and (12) are equivalent in the steady-state, then the coefficient b can be obtained by the following equation (Furuhashi et al., 2007, 2008).

$$b = \frac{1}{t}\sqrt{\frac{3Bi}{3+Bi}} \tag{13}$$

The values calculated by the approximate solution of wall-averaged temperature, Eq.(6), for Bi=0.1, 1, 10, 100 are plotted in Fig.3 with the note, (simple), in the legend. The maximum relative error to the theoretical solution, Eq.(5), is 0.01% for Bi=0.1, 0.3% for Bi=1, 2.0% for Bi=10 and 3.3% for Bi=100. Consequently, a high-precision easy-to-use approximate solution is obtained.

2.3 Cylindrical shell solution of steady-state thermal stress

Young's modulus, thermal expansion coefficient, and Poisson's ratio of the vessel is represented by E, a, and ν, respectively. When the vessel wall is in the context of mechanical free boundary conditions, the radial outward displacement, $u(z)$, can be obtained as the solution of the following differential equation (Timoshenko & Woinowsky, 1959; Furuhashi & Watashi, 1991).

$$\frac{d^4u}{dz^4}+4\beta^4u=\frac{p(z)}{D}+\frac{Et\alpha}{DR}T_m(z)+\frac{(1+\nu)\alpha}{t}\frac{d^2T_b(z)}{dz^2} \tag{14}$$

Here, $p(z)$ is the inner pressure, and $p(z)=0$ is assumed. $T_b(z)$ is the equivalent linear temperature difference, representing the "inner surface temperature minus outer surface temperature" in the case that the temperature profile in the wall thickness direction is linearly approximated. Here, $T_b(z)=0$ is assumed. The acceptability of this assumption will be checked in the comparisons made after. D is the flexural rigidity of the wall, and β is the stress decay coefficient.

$$D=\frac{Et^3}{12(1-\nu^2)} \tag{15}$$

$$\beta=\sqrt[4]{\frac{Et}{4DR^2}}=\frac{\sqrt[4]{3(1-\nu^2)}}{\sqrt{Rt}} \tag{16}$$

The axial bending stress σ_{zb}, circumferential membrane stress σ_{hm} and circumferential bending stress σ_{hb} is given by the following equations, respectively (Timoshenko & Woinowsky, 1959; Furuhashi & Watashi, 1991).

$$\sigma_{zb}(z)=\frac{6D}{t^2}\left\{\frac{d^2u}{dz^2}-\frac{(1+\nu)\alpha T_b}{t}\right\} \tag{17}$$

$$\sigma_{hm}(z)=E\left\{\frac{u}{R}-\alpha T_m\right\} \tag{18}$$

$$\sigma_{hb}(z)=\frac{6D}{t^2}\left\{\nu\frac{d^2u}{dz^2}-\frac{(1+\nu)\alpha T_b}{t}\right\} \tag{19}$$

The radial displacement was solved as the following equation by substituting the approximate solution of $T_m(z)$, Eq.(6), into the right side of Eq.(14) (Furuhashi et al., 2007, 2008).

$$u(z)=R\alpha T_m(z)+\mathrm{sgn}(z)\frac{mR\alpha\Delta T}{2}e^{-b|z|}-\mathrm{sgn}(z)\frac{mR\alpha\Delta T}{2}e^{-\beta|z|}\cos(\beta z)-nR\alpha\Delta Te^{-\beta|z|}\sin(\beta z) \tag{20}$$

Here, parameters, *m* and *n*, is given by the following equations, respectively.

$$m = \frac{b^4}{4\beta^4 + b^4} = \frac{(b/\beta)^4}{4 + (b/\beta)^4} \tag{21}$$

$$n = \frac{\beta^2 b^2}{4\beta^4 + b^4} = \frac{(b/\beta)^2}{4 + (b/\beta)^4} \tag{22}$$

The thermal stresses were solved as the following equations by substituting Eq.(20) into Eqs.(17), (18) and (19) (Furuhashi et al., 2007, 2008).

$$S_{zb}(z) = \frac{\sigma_{zb}(z)}{E\alpha\Delta T} = \frac{\sqrt{3}}{\sqrt{1-\nu^2}}\left\{-\mathrm{sgn}(z)n \cdot e^{-b|z|} - \frac{m}{2}e^{-\beta|z|}\sin(\beta z) + \mathrm{sgn}(z)n \cdot e^{-\beta|z|}\cos(\beta z)\right\} \tag{23}$$

$$S_{hm}(z) = \frac{\sigma_{hm}(z)}{E\alpha\Delta T} = \mathrm{sgn}(z)\frac{m}{2}e^{-b|z|} - \mathrm{sgn}(z)\frac{m}{2}e^{-\beta|z|}\cos(\beta z) - n \cdot e^{-\beta|z|}\sin(\beta z) \tag{24}$$

$$S_{hb}(z) = \frac{\sigma_{hb}(z)}{E\alpha\Delta T} = \nu S_{zb}(z) \tag{25}$$

Here, *S* represents the non-dimensional stress normalized by $E\alpha\Delta T$. These equations suggest that the non-dimensional stress profiles, the plots of *S* with regard to βz, exclusively depend on the ratio of coefficients, b/β, because both *m* and *n* are functions of b/β only.
When *t* approaches *0* (t=>0), b/β=>∞, *m*=>1, *n*=>0 and $T_m(z)$=>$T_f(z)$, and consequently, thermal stresses approach the following equations.

$$S_{zb}(z) \to -\frac{\sqrt{3}}{2\sqrt{1-\nu^2}}e^{-\beta|z|}\sin(\beta z) \tag{26}$$

$$S_{hm}(z) \to -\frac{\mathrm{sgn}(z)}{2}e^{-\beta|z|}\cos(\beta z) \tag{27}$$

Those limit solutions for b/β=>∞ yield the maximum at the *z* values shown in the following equations, lead to the upper limit of thermal stresses for free boundary conditions.

$$S_{zb,Lim}\left(z = \pm\frac{\pi}{4\beta}\right) = \mp 0.293(\nu = 0.3) \tag{28}$$

$$S_{hm,Lim}(z = \pm 0) = \mp 0.5 \tag{29}$$

2.4 Ramp-shaped fluid temperature profile

The Green functions (indicial response functions) are obtained from the temperature, displacement and stress of the vessel based on the step-shaped fluid temperature profile (Carslaw & Jeager, 1959; Morse & Feshbach, 1953). When any-shaped fluid temperature profile in z-axis direction is given, the temperature, displacement and stress of the vessel in steady-state can be obtained by convolution integral. We discuss the case in which a ramp-shaped fluid temperature profile is given as shown in Fig.2. This profile is expressed by the following equations.

$$\begin{aligned} &T_f(z) = T_0 \quad (z < 0) \\ &T_f(z) = T_0 + \Delta T \frac{z}{L} \quad (0 \le z \le L) \\ &T_f(z) = T_0 + \Delta T \quad (z > L) \end{aligned} \tag{30}$$

The $G_T(z)$ that are obtained from Eq.(6) with $T_0=0$ and $\Delta T=1$ express the approximate Green function of wall-averaged temperature of the vessel when the fluid temperature profile is given by a unit step function.

$$\begin{aligned} &G_T(z) = \frac{1}{2} e^{bz} \quad (z < 0) \\ &G_T(z) = 1 - \frac{1}{2} e^{-bz} \quad (0 \le z) \end{aligned} \tag{31}$$

When the fluid temperature profile is given by a ramp-shaped function expressed by Eq.(30), the wall-averaged temperature of the vessel was obtained by convolution integral as the following equation (Furuhashi et al., 2007, 2008).

$$\begin{aligned} T_m(z) &= T_f(-\infty) + \int_{-\infty}^{\infty} \frac{dT_f(\xi)}{d\xi} G_T(z-\xi) d\xi = T_0 + \frac{\Delta T}{L} \int_0^L G_T(z-\xi) d\xi \\ &= T_f(z) + \frac{\Delta T}{2bL} (e^{-b|z|} - e^{-b|z-L|}) \end{aligned} \tag{32}$$

Similarly, the $G_u(z)$ that are obtained from Eq.(20) with $T_0=0$ and $\Delta T=1$ express the approximate Green function of radial displacement of the vessel when the fluid temperature profile is given by a unit step function.

$$\begin{aligned} G_u(z) &= R\alpha \left\{ H(z) - \mathrm{sgn}(z) \frac{1}{2} e^{-b|z|} \right\} + \mathrm{sgn}(z) \frac{mR\alpha}{2} e^{-b|z|} \\ &- \mathrm{sgn}(z) \frac{mR\alpha}{2} e^{-\beta|z|} \cos(\beta z) - nR\alpha e^{-\beta|z|} \sin(\beta z) \end{aligned} \tag{33}$$

The radial displacement for the ramp-shaped fluid temperature profile was obtained by convolution integral as the following equation (Furuhashi et al., 2007, 2008).

$$\begin{aligned} u(z) &= R\alpha T_f(-\infty) + \int_{-\infty}^{\infty} \frac{dT_f(\xi)}{d\xi} G_u(z-\xi) d\xi = R\alpha T_0 + \frac{\Delta T}{L} \int_0^L G_u(z-\xi) d\xi \\ &= R\alpha T_m(z) - \frac{mR\alpha\Delta T}{2bL} \left(e^{-b|z|} - e^{-b|z-L|} \right) - \frac{mR\alpha\Delta T}{4\beta L} \left\{ \begin{array}{l} e^{-\beta|z|} \sin(\beta|z|) - e^{-\beta|z|} \cos(\beta|z|) \\ -e^{-\beta|z-L|} \sin(\beta|z-L|) \\ +e^{-\beta|z-L|} \cos(\beta|z-L|) \end{array} \right\} \\ &+ \frac{nR\alpha\Delta T}{2\beta L} \left\{ \begin{array}{l} e^{-\beta|z|} \sin(\beta|z|) + e^{-\beta|z|} \cos(\beta|z|) \\ -e^{-\beta|z-L|} \sin(\beta|z-L|) - e^{-\beta|z-L|} \cos(\beta|z-L|) \end{array} \right\} \end{aligned} \tag{34}$$

The thermal stresses were obtained as the following equations by substituting Eq.(34) into Eqs.(17),(18) and (19).

$$S_{zb}(z)=\frac{\sigma_{zb}(z)}{E\alpha\Delta T}=\frac{\sqrt{3}}{\sqrt{1-\nu^2}}\frac{n}{bL}\left(e^{-b|z|}-e^{-b|z-L|}\right)+\frac{\sqrt{3}}{4\sqrt{1-\nu^2}}\frac{m}{\beta L}\left\{\begin{array}{l}e^{-\beta|z|}\sin(\beta|z|)+e^{-\beta|z|}\cos(\beta|z|)\\-e^{-\beta|z-L|}\sin(\beta|z-L|)\\-e^{-\beta|z-L|}\cos(\beta|z-L|)\end{array}\right\}$$
$$+\frac{\sqrt{3}}{2\sqrt{1-\nu^2}}\frac{n}{\beta L}\left\{\begin{array}{l}e^{-\beta|z|}\sin(\beta|z|)-e^{-\beta|z|}\cos(\beta|z|)\\-e^{-\beta|z-L|}\sin(\beta|z-L|)+e^{-\beta|z-L|}\cos(\beta|z-L|)\end{array}\right\} \quad (35)$$

$$S_{hm}(z)=\frac{\sigma_{hm}(z)}{E\alpha\Delta T}=\frac{-m}{2bL}\left(e^{-b|z|}-e^{-b|z-L|}\right)-\frac{m}{4\beta L}\left\{\begin{array}{l}e^{-\beta|z|}\sin(\beta|z|)-e^{-\beta|z|}\cos(\beta|z|)\\-e^{-\beta|z-L|}\sin(\beta|z-L|)+e^{-\beta|z-L|}\cos(\beta|z-L|)\end{array}\right\}$$
$$+\frac{n}{2\beta L}\left\{\begin{array}{l}e^{-\beta|z|}\sin(\beta|z|)+e^{-\beta|z|}\cos(\beta|z|)\\-e^{-\beta|z-L|}\sin(\beta|z-L|)-e^{-\beta|z-L|}\cos(\beta|z-L|)\end{array}\right\} \quad (36)$$

$$S_{hb}(z)=\frac{\sigma_{hb}(z)}{E\alpha\Delta T}=\nu S_{zb}(z) \quad (37)$$

These solutions for the ramp-shaped fluid temperature profile, that are expressed by Eqs.(32) and (34) to (37), are symmetric with respect to the middle position, $z=L/2$, and the middle temperature, $T_0+\Delta T/2$. The z terms in the right side express the effect of the lower end ($z=0$) of stratified layer, and the ($z-L$) terms express the effect of the upper end ($z=L$) as well. The effects at both ends having opposite signs are paired and superposed, and decrease the stresses.

Since both m and n are functions of b/β only, the non-dimensional stress S exclusively depend on the ratio of coefficients, b/β, and the non-dimensional width of stratified layer, βL. When L approaches 0, the stresses approach the solutions for a step-shaped fluid temperature profile. Since one of the pair terms in the right side (either the z term or the (z-L) term) becomes negligibly small when L approaches ∞, the stresses become inversely proportional to L. Practically you may consider $S \propto 1/L$, when $bL>\pi$ and $\beta L>\pi$ ($L>2.5\sqrt{Rt}$).
When t approaches 0 ($t \Rightarrow 0$), $b/\beta \Rightarrow \infty$, $m \Rightarrow 1$, $n \Rightarrow 0$ and $T_m(z) \Rightarrow T_f(z)$. In addition, when $\beta L>\pi$ ($L>2.5\sqrt{Rt}$), thermal stresses become the maximum values at both ends of the stratified layer as expressed by the following equations.

$$S_{zb,\max}\begin{pmatrix}z=0\\z=L\end{pmatrix}=\pm\frac{\sqrt{3}}{4\beta L\sqrt{1-\nu^2}},\quad \sigma_{zb,\max}\begin{pmatrix}z=0\\z=L\end{pmatrix}=\pm\frac{3E\alpha\sqrt{Rt}}{4\{3(1-\nu^2)\}^{3/4}}\frac{\Delta T}{L} \quad (38)$$

$$S_{hm,\max}\begin{pmatrix}z=0\\z=L\end{pmatrix}=\pm\frac{1}{4\beta L},\quad \sigma_{hm,\max}\begin{pmatrix}z=0\\z=L\end{pmatrix}=\pm\frac{E\alpha\sqrt{Rt}}{4\sqrt[4]{3(1-\nu^2)}}\frac{\Delta T}{L} \quad (39)$$

Eqs.(38) and (39) are well-known solutions (Timoshenko & Woinowsky, 1959). They are often used as simple evaluation formulas in conventional structural design, because of the simplicity that the stress is proportional to the geometry parameter ($\sqrt{Rt}$) and the temperature gradient ($\Delta T/L$). However, it should be noted that these equations are applicable in the case where the vessel temperature corresponds to fluid temperature when b/β=>∞, and $\beta L>\pi$.

3. Verification by comparisons with FEM analysis

3.1 Step-shaped fluid temperature profile

FEM analyses were performed for the case where the thermal stratification of the fluid (liquid sodium) contained in a reactor vessel of a large fast reactor occurs. The analysis conditions are as follows.

1. Applied FEM code: FINAS (CRC & JAEA, 2006)
2. Reactor vessel: SUS316FR, R=5350mm, t=50mm
3. T_0=350°C, ΔT=200°C
4. Three cases for $L=0$, $4t$ and $8t$
5. Material characteristics are the values at the middle temperature of 450 °C. λ=21.512W/mK, E=164000MPa, ν=0.301 and α=1.993x10^{-5}/°C
6. Four cases for h=3000, 930, 310 and 100 W/m^{2}K (Bi=6.97, 2.16, 0.72, 0.23)

FEM results of the wall-averaged temperature for $L=0$ are shown in Fig.4 with the note, (FEM), in the legend. The results calculated by the approximate solution based on the temperature profile method, Eq.(6), are also shown in Fig.4 with the note, (simple), in the legend. Both results agree well. FEM results of the thermal stress for Bi=2.16 are shown in Fig.5 with the note, (FEM), in the legend. The results calculated by the shell solutions, Eqs.(23) to (25), are shown in Fig.5 with the note, (simple), in the legend. Both results agree well except the discrepancy of S_{hb} for $z<3t$. The discrepancy of S_{hb} results from the assumption ($T_b=0$) applied to the shell solution. Since FEM analysis is based on axisymmetric solid element, T_b is naturally taken into account. It was confirmed that the shell solution obtained once again by substituting the binomial approximation of T_b into Eqs.(14), (17) to (19) leads to good agreement with FEM analyses.

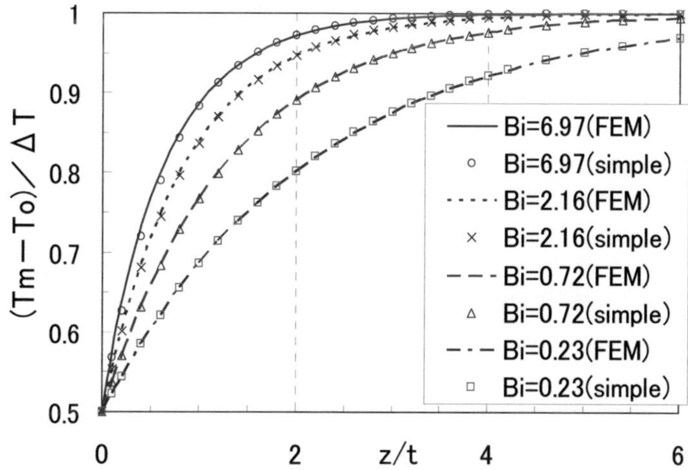

Fig. 4. Vessel temperatures for step-shaped fluid temperature

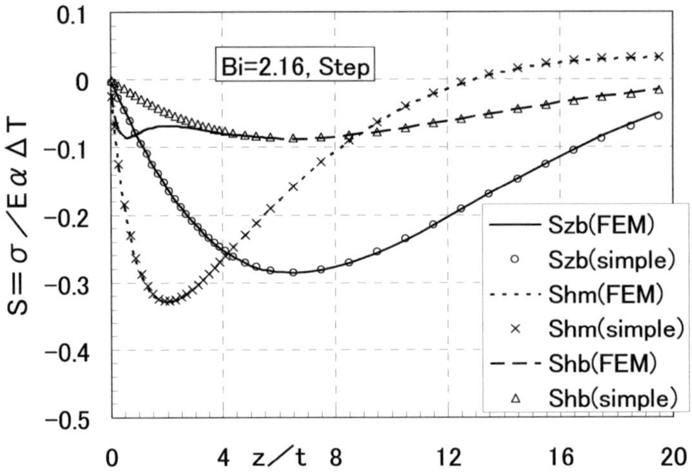

Fig. 5. Thermal stresses for step-shaped fluid temperature

However, it was decided that the shell solution with T_b is not included in this paper, because (i) it doubles the number of terms in the stress solution, requiring too much space; and (ii) the discrepancy in S_{hb} has minimal effect on the maximum stress evaluation. So we accepted the assumption (T_b= *0*) in this paper.

The temperature and stress profiles are symmetric with respect to the middle position, *z=0*, and the middle temperature. Since this figure plots only the results in the hot section (*z>0*), S_{zb} and S_{hm} shows the maximum negative (compressive) stress on the inner surface, respectively. In the cold section (*z<0*), the maximum positive (tensile) stress occurs on the inner surface.

3.2 Ramp-shaped fluid temperature profile

Analyses were performed for the width *L=4t* and *8t*, with *Bi*=6.97, 2.16, 0.72 and 0.23. Here, FEM results of the wall-averaged temperature for *L=8t* are shown in Fig.6 with the note, (FEM), in the legend. The results calculated by the approximate solution, Eq.(32) are additionally shown with the note, (simple), in the legend. Both results agree well. FEM results of the thermal stress for *Bi*=2.16 are shown in Fig.7 with the note, (FEM), in the legend. The results calculated by the shell solutions, Eqs.(35) to (37), are additionally shown with the note, (simple), in the legend. Both results agree well.

The temperature and stress profiles are symmetric with respect to the middle position, *z=L/2*, and the middle temperature. Since this figure plots only the results in the hot section (*z>L/2*), S_{zb} and S_{hm} shows the maximum negative (compressive) stress on the inner surface, respectively. In the cold section (*z<L/2*), the maximum positive (tensile) stress occurs on the inner surface.

4. Simplified thermal stress evaluation chart

The non-dimensional stress, *S*, exclusively depends on b/β and βL. Using this characteristic, we developed simple charts to estimate maximum stress, *S*, and its generating location, Δz. And we proposed a simplified thermal stress evaluation method using these charts. The *S*

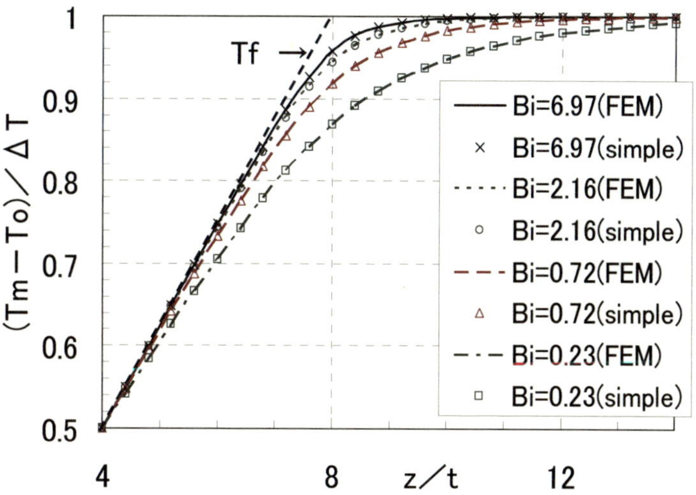

Fig. 6. Vessel temperatures for ramp-shaped fluid temperature

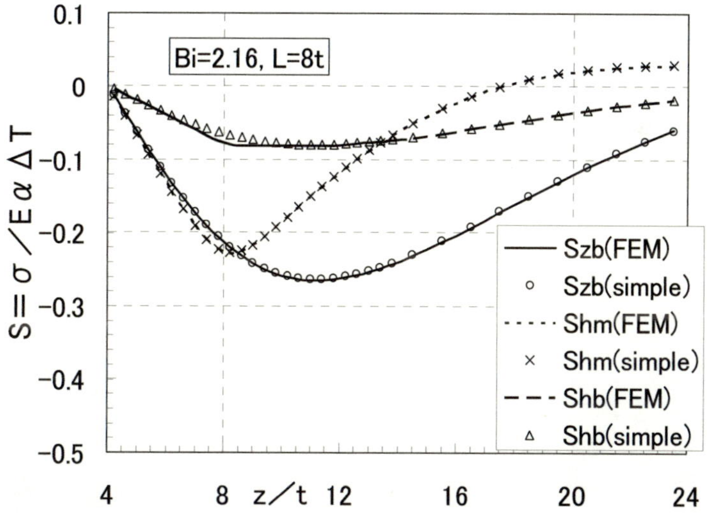

Fig. 7. Thermal stresses for ramp-shaped fluid temperature

charts were developed for $b/\beta>0.5$ and $\beta L<5$. When b/β approaches 0, S approaches 0. S is approximately inversely proportional to L for $\beta L>5$. The maximum stress location, Δz, represents the outward distance from either end of the stratified layer. In the cold section, the maximum tensile (positive) stress occurs on the inner surface at $z=-\Delta z$, while in the hot section, the maximum compressive (negative) stress occurs on the inner surface at $z=L+\Delta z$. In addition, by substituting z into Eq. (32), we can calculate the wall-averaged temperature, which is applicable to the reference temperature for material properties in structural design.

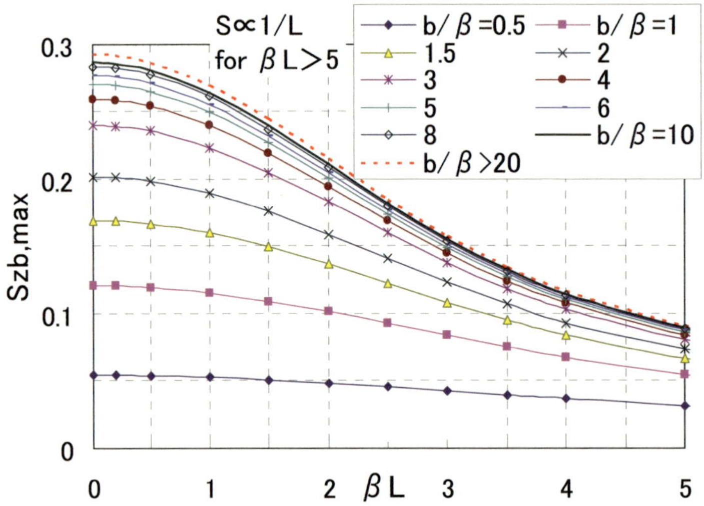

Fig. 8. The maximum bending stress

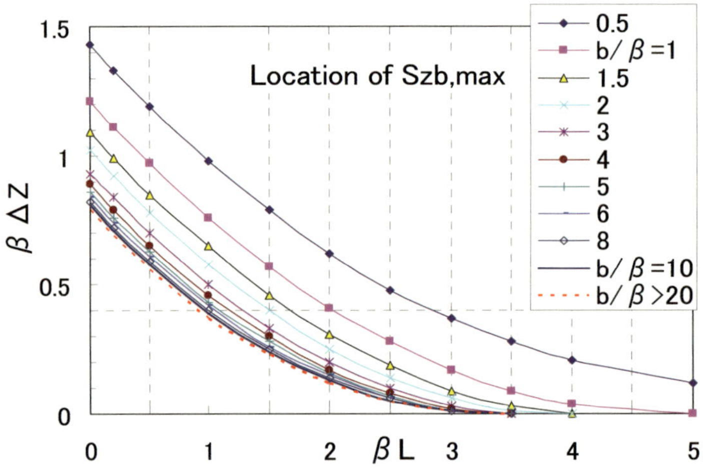

Fig. 9. Location of the maximum bending stress

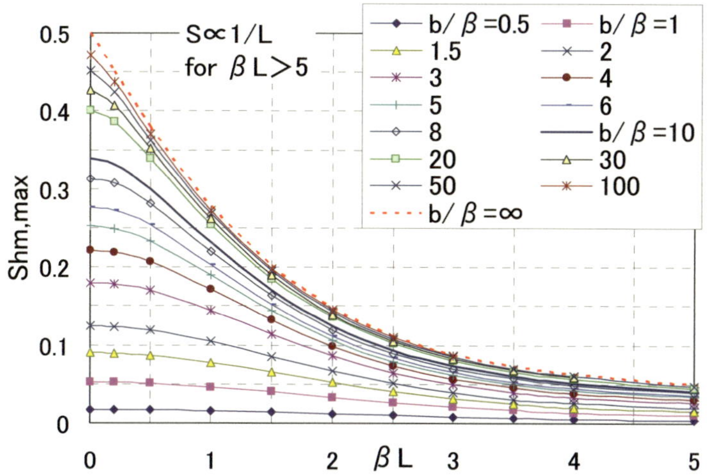

Fig. 10. The maximum membrane stress

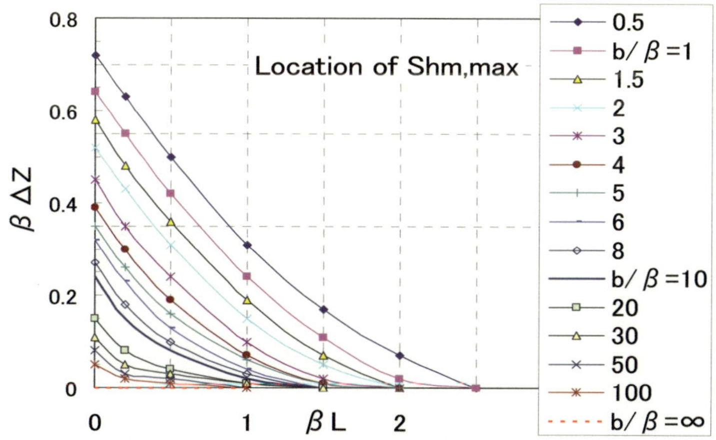

Fig. 11. Location of the maximum membrane stress

The maximum bending stress, $S_{zb,max}$ and its generating location, $\beta\Delta z$, is shown in Fig.8 and Fig.9, respectively. The maximum membrane stress, $S_{hm,max}$ and its generating location, $\beta\Delta z$, is shown in Fig.10 and Fig.11, respectively. The maximum stress intensity, $S_{n,max}$ ($=\sigma_{SI,max}/E\alpha\Delta T$) and its generating location, $\beta\Delta z$, is shown in Fig.12 and Fig.13, respectively. The stress intensity (Tresca's stress σ_{SI}) becomes the maximum value at the outer surface, where σ_z and σ_h have opposite signs.

$$\sigma_{SI} = Max\left(|\sigma_z|, |\sigma_h|, |\sigma_z - \sigma_h|\right) \tag{40}$$

A small prominence observed in Fig.13 suggests the transition from the case that $S_{n,max}$ occurs near the location of $S_{hm,max}$ to the case that $S_{n,max}$ occurs near the location of $S_{zb,max}$.
The comparisons of FEM analyses, the proposed charts and the conventional method, Eq.(38) and (39), for 2 cases, (*L=8t, Bi=6.97*) and (*L=4t, Bi=2.16*), are shown in Table 1. The parameters and S values read out from the charts for the two cases are listed below.

L	Bi	b	β	b/β	βL	S_{zb}	S_{hm}	S_n
8t	6.97	28.96	2.48	11.7	0.99	0.27	0.24	0.39
4t	2.16	22.41	2.48	9.03	0.50	0.28	0.30	0.43

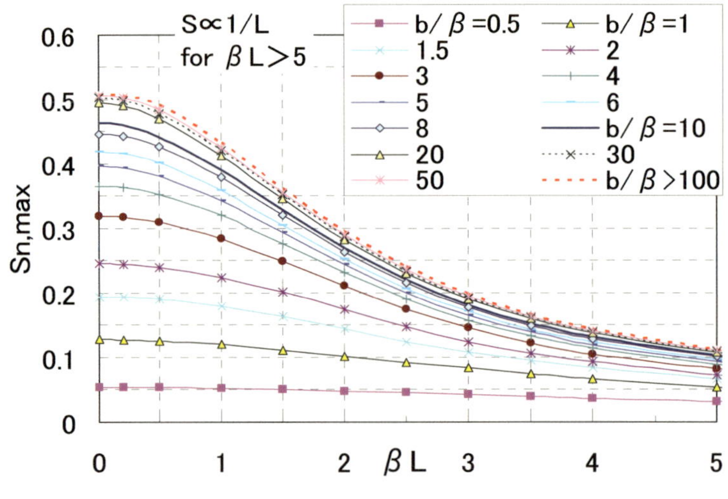

Fig. 12. The maximum stress intensity

It has been demonstrated that the proposed charts are sufficiently accurate. On the other hand, the conventional method leads to an overestimation. The main error is caused by the use of the formulas beyond the applicable range, $\beta L > \pi$ ($L > 2.5\sqrt{Rt}$). The comparison of the proposed method and the conventional method is shown in S_n-chart, Fig.14, and the above 2 cases results are plotted on the charts.

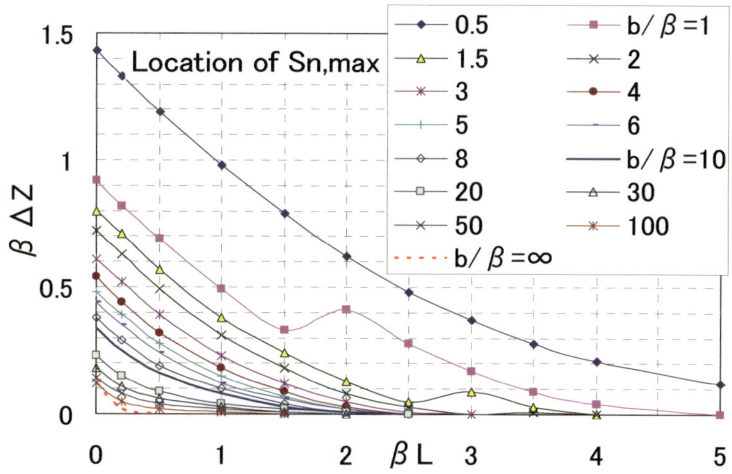

Fig. 13. Location of the maximum stress intensity

We often need to evaluate thermal stresses for observed thermal stratification phenomena in an engineering field. In most cases, axial temperature profile of interface between stratified fluid layers can be approximated by exponential curve or parabolic curve as shown in Fig.15 (Moriya et al., 1987; Haifeng et al., 2009; Kimura et al., 2010). We propose the effective width for such cases as following equation.

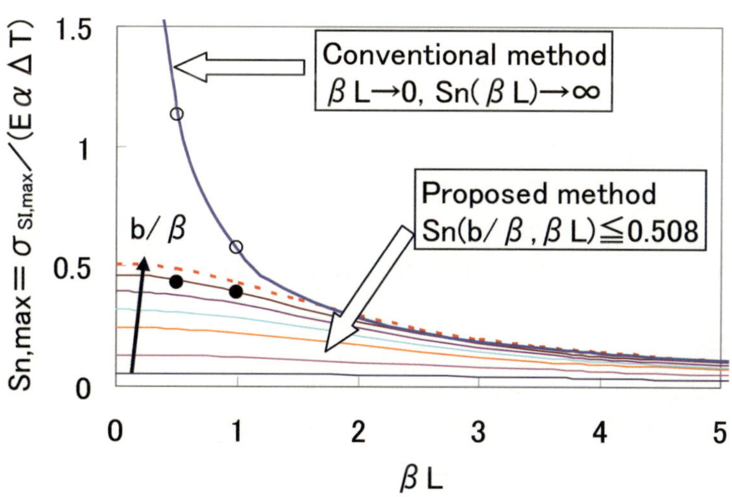

Fig. 14. Comparison of the proposed method and the conventional method

Method		Proposed method		FEM analyses		Conventional method (38)(39)	
Case	Component	σ_{max} (MPa)	Δz (mm)	σ_{max} (MPa)	Δz (mm)	σ_{max} (MPa)	Δz (mm)
$L=8t$ $Bi=6.97$	S_{zb}	177	149	175	150	300	0
	S_{hm}	157	8	157	10	165	0
	S_n	255	28	256	30	375	0
$L=4t$ $Bi=2.16$	S_{zb}	184	234	183	225	600	0
	S_{hm}	196	36	193	35	330	0
	S_n	282	73	282	75	750	0

Table 1. Comparison of stress evaluation results

$$L_{eff} = \frac{12}{\Delta T^2}\int_{T_c}^{T_h}\left(T_f - T_{med}\right)z\, dT_f \tag{41}$$

L_{eff} is nearly equal to the axial width corresponding to 90% of ΔT as shown in Fig.15. It is found that the thermal stress evaluations using the proposed charts and L_{eff} are rather conservative and good evaluation, through comparisons with the FEM analyses.

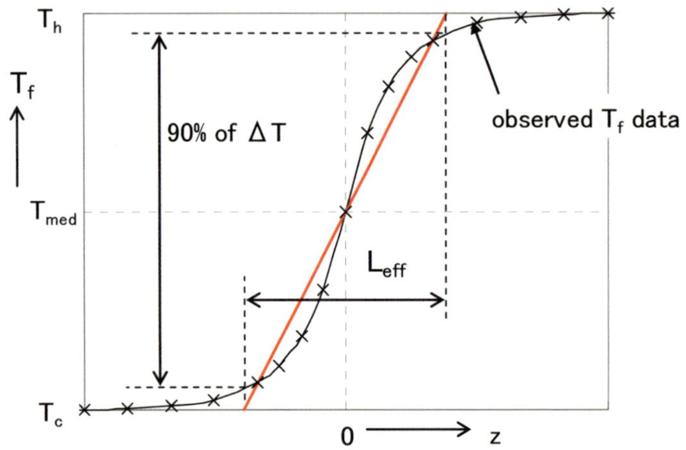

Fig. 15. Effective width of interface between stratified layers

5. Conclusion

To improve the accuracy of design evaluation methods of thermal stress induced by thermal stratification, this study have performed the theoretical analyses and FEM ones on steady-state temperature and thermal stress of cylindrical vessels, and obtained the following results.

1. The theoretical solution of steady-state temperature profiles of vessels and the approximate solution of the wall-averaged temperature based on the temperature profile method have been obtained. The wall-averaged temperature can be estimated with a high precision using the temperature attenuation coefficient, *b*.

2. The shell theory solution for thermal stress based on the approximate solution of the wall-averaged temperature has been obtained. It has been demonstrated that the non-dimensional thermal stress, $S=\sigma/E\alpha\Delta T$ exclusively depends on the ratio of coefficients, b/β, and the non-dimensional interface width between stratified layers, βL.
3. Easy-to-use charts has been developed to estimate the maximum thermal stress and its generating location using the characteristic described in (2) above. In addition, a simplified thermal stress evaluation method has been proposed.
4. Through comparison with the FEM analysis results, it has been confirmed that the proposed method is sufficiently accurate to estimate the steady-state temperature and thermal stress.
5. It has been demonstrated that the conventional simple evaluation method using the shell stress solution, which assumes axial temperature profile consisting of a straight line with the maximum fluid temperature gradient, often leads to an overestimation.
6. For the convenient application of the proposed method to engineering problems, we proposed the effective width of interface between stratified layers. The thermal stress evaluation using the proposed charts with the effective width gives slightly conservative estimations.

The proposed method enables simple evaluations of steady-state thermal stress induced by thermal stratification taking the relaxation mechanism of thermal stress into account. This method would contribute to the reduction of design cost and to the rationalization of design.

6. References

Bieniussa, K.W. and Reck, H. (1996). Piping specific analysis of stresses due to thermal stratification, *Nuclear Engineering and Design*, Vol.190, No.1, pp. 239-249, ISSN:0029-5493.

Carslaw, H.S. and Jeager, J.C. (1959). *Conduction of heat in solids*, 2nd edition, pp. 166-169, Oxford University Press.

CRC Solutions Corp. & Japan Atomic Energy Agency (2006). FINAS User's Manual version 18.0, (in Japanese).

Furuhashi, I., Kawasaki, N. and Kasahara, N. (2007). Evaluation Charts of Thermal Stresses in Cylindrical Vessels Induced by Thermal Stratification of Contained Fluid, (in Japanese), *Transactions of the Japan Society of Mechanical Engineers*, Series A, Vol.73, No.730, pp. 686-693.

Furuhashi, I., Kawasaki, N. and Kasahara, N. (2008). Evaluation Charts of Thermal Stresses in Cylindrical Vessels Induced by Thermal Stratification of Contained Fluid, *Journal of Computational Science and Technology*, Vol.2, No.4, pp. 547-558.

Furuhashi, I. and Watashi, K. (1991). A Simplified Method of Stress Calculation of a Nozzle Subjected to a Thermal Transient, *International Journal of Pressure Vessels and Piping*, Vol.45, pp. 133-162, ISSN:0308-0161.

Haifeng, G. et al. (2009). Experimental Study on the Fluid Stratification Mechanism in the Density Lock, *Journal of NUCLEAR SCIENCE and TECHNOLOGY*, Vol.46, No.9, pp. 925- 932, ISSN:0022-3131

Kimura, N. et al. (2010). Experimental Study on Thermal Stratification in a Reactor Vessel of Innovatic Sodium-Cooled Fast Reactor – Mitigation Approach of Temperature Gradient across Stratification Interface -, *Journal of NUCLEAR SCIENCE and TECHNOLOGY*, Vol.47, No.9, pp. 829- 838, ISSN:0022-3131

Katto, Y. (1964), *Conduction of Heat*, (in Japanese), (1964), p.38, Yokendo.

Moriya, S. et al. (1987). Effects of Reynolds Number and Richardson Number on Thermal Stratification in Hot Plenum, *Nuclear Engineering and Design*, Vol.99, pp. 441-451, ISSN:0029-5493.

Morse, P.M. and Feshbach, H. (1953). *Methods of Theoretical Physics*, Part.1, pp. 710-730, McGraw-Hill.

Timoshenko, S.P. and Woinowsky-Krieger, S. (1959). *Theory of plates and shells*, 2nd edition, pp. 466-501, McGraw-Hill.

4

Axi-Symmetrical Transient Temperature Fields and Quasi-Static Thermal Stresses Initiated by a Laser Pulse in a Homogeneous Massive Body

Aleksander Yevtushenko[1], Kazimierz Rozniakowski[2]
and Malgorzata Rozniakowska-Klosinska[3]
*[1]Bialystok University of Technology
Faculty of Mechanical Engineering
[2]Technical University of Lodz Faculty of Technical Physics
Information Technology and Applied Mathematics
[3]Technical University of Lodz
Poland*

1. Introduction

In the present chapter the model of a semi-infinite massive body which is heated through the outer surface by the precised heat flux, is being under study. This heat flux has the intensity directly proportional to the equivalent laser irradiation intensity. Heating of materials due to its surface irradiation by the high-power energy fluxes, which takes place during working of the laser systems, can be modelled in a specific conditions as the divided surface heat source of defined power density or heat flux of defined intensity (Rykalin et al., 1975).

Laser systems are an unusual source of electromagnetic irradiation of unique properties. These properties differ essentially from the relevant characteristics of irradiation generated by traditional sources, natural and artificial one. Laser irradiation has specific, distinguishing features: high level of spatial and time coherence, high level of monochromaticity, low divergence, high spectral intensity and continuous or impulse emission process. High level of spatial coherence gives possibility to focusing laser irradiation on the surfaces of a few to several dozens squared micrometers in a size, which correspond to very high values of power intensity even 10^8 -10^{12} W/m^2 (10^4 -10^{18} J/m^2 or 10^{23} fotons/cm^2). Effectivity of local surface heating mentioned above depends on: laser pulse duration, laser pulse structure (shape) and on irradiation intensity distribution. Three specific laser pulse structures are usually under consideration: rectangular-shape pulse, triangular-shape pulse and pulse shape approximated by some defined function. Likewise to the laser pulse structure, the spatial pulse structure (distribution of laser irradiation in a plane normal to the beam axis) is also complex and challenging for precised analytical description. In approximation the spatial distribution of laser irradiation can be described by the following relations: gaussian distribution (takes place during the working of laser beam in the single-mode regime), mixed (multi-modal) or uniform distribution. In addition, laser heat source shape can be changed by the electromagnetic or optical methods. Hence, the

optimization of the source shape problem appears on the basis of various optimisation criteria as well as the minimal losses on apparatus criterion.
In the former industrial practice, mainly the gaussian or uniform intensity distribution were applied. Various in nature thermal effects are present during industrial laser materials processing such as: laser hardening, laser surface modifications of metals and alloys. Nowadays the most significant role in the technological operations plays such formed laser beam which maximum power is achieved not in the centre but close to the edge of the heated zone (Hector & Hetnarski, 1996). That is why, in the emerging process of the new effective laser technologies, it is strongly reasonable to determine the analytical solutions and to conduct numerical analysis for the boundary value problem of transient heat conduction and quasi-static thermal stresses, which are crucial in calculations of:

- the effective absorption coefficient,
- the specific time point when surface melting occurs due to laser beam heating,
- the heating velocity and cooling,
- the controlled laser thermo cracking process (Lauriello & Chen, 1973; Yevtushenko et al., 1997),
- the other features of initiated temperature and thermal stresses fields.

2. Influence of intensity spatial distribution of laser beam on a temperature field in the irradiated massive body (semi-infinite)

2.1 Problem statement

Laser irradiation interaction of $10^4 \div 10^8\ \mathrm{W/m^2}$ power intensity on metals is equivalent to heating them by heat flux of defined intensity (Rykalin et al., 1975). If the following conditions are fulfilled:

- the power intensity generated by the laser is not sufficient to melt and evaporate the superficial layer,
- the losses because of heat emission and convection from a surface body are negligible
- the thermo-physical properties do not depend on temperature,

then the axisymmetrical boundary value problem of heat conduction for semi-infinite body in cylindrical coordinates system (r,z) with the beginning in the centre of heated surface, can be considered in the form:

$$\frac{\partial^2 T}{\partial r^2}+\frac{1}{r}\frac{\partial T}{\partial r}+\frac{\partial^2 T}{\partial z^2}=\frac{1}{k}\frac{\partial T}{\partial t},\ r\ge 0,\ z>0,\ t>0, \tag{1}$$

$$T(r,z,0)=0,\ r\ge 0,\ z\ge 0\ , \tag{2}$$

$$K\frac{\partial T}{\partial z}=-Aq(r)H(t_s-t),\ \ r\ge 0,\ z=0, t>0, \tag{3}$$

$$T(r,z,t)\to 0, \sqrt{r^2+z^2}\to\infty, t>0\,. \tag{4}$$

The uniform distribution of heat flux intensity in a circle of a radius, can be described by the formula:

$$q(r) = q_0 H(a-r),\ r \geq 0 \tag{5}$$

and for the mixed distribution – multimodal one, the heat flux intensity can be expressed by (Hector & Hetnarski, 1996):

$$q(r) = q_f [f + (1-f)K_c r^2]\, e^{-K_c r^2},\ r \geq 0\,, \tag{6}$$

where K_c – concentration coefficient, q_f – characteristic value of heat flux intensity q, $0 \leq f \leq 1$ – parameter, which characterized the irradiation intensity distribution in a plane normal to the laser beam axis. For $f = 1$ the normal (gaussian) distribution and for $f = 0$ doughnut – toroidal distribution, is obtained.

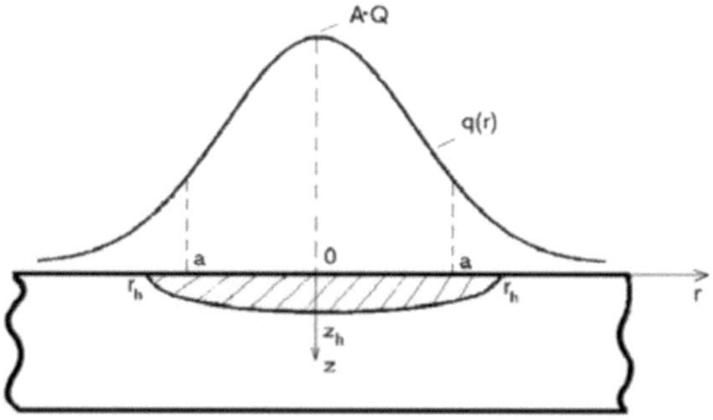

Fig. 1. Laser irradiation heating model and area shape visualization of phase transition for metals

Both distributions of laser irradiation intensity (5) and (6) are related by the following concentration coefficient (Rykalin et al., 1975)

$$K_c = B_f\, a^{-2}\,. \tag{7}$$

The numerical factor B_f in the Eq. (7) can be found from the condition below (Hector & Hetnarski, 1996):

$$\frac{Q_f}{Q} = \left(1 - \frac{1}{e}\right)100\% \approx 63.2\%\,, \tag{8}$$

where Q – total irradiation power, Q_f – irradiation power which arrives to the circle of a radius and can be easily derived after taking into consideration the distribution (6), consequently these two values can be substituted into Eq. (8). As the result the non-linear equation will be received for B_f in the form:

$$[1 + (1-f)B_f]e^{-B_f} = e^{-1} \approx 0.3678. \tag{11}$$

A numerical analysis shows that dependence of the B_f roots of Eq. (11) with respect to the f parameter is nearly linear: $B_f = B_0(1-f) + f$, where $B_0 = 2.1462$ is the value of B_f at

$f=0$. At $f=1$ from Eq. (6) is received the obvious result $B_f=1$ (Rykalin et al., 1975). By comparizing the irradiation intensity of uniform distribution (5) with the irradiation intensity of general case distribution (6) it was found

$$q_f = B_f q_0. \tag{12}$$

By introducing dimensionless variables and parameters

$$\rho=\frac{r}{a},\ \zeta=\frac{z}{a},\ \tau=\frac{kt}{a^2},\ \tau_s=\frac{kt_s}{a^2},\ T_0=\frac{q_0 a}{K},\ T^*=\frac{T}{T_0}, \tag{13}$$

and using relations (7) and (12) the boundary value heat conductivity problem (1)-(4) can be rewritten in the form:

$$\frac{\partial^2 T^*}{\partial\rho^2}+\frac{1}{\rho}\frac{\partial T^*}{\partial\rho}+\frac{\partial^2 T^*}{\partial\zeta^2}=\frac{\partial T^*}{\partial\tau},\ \rho\ge 0,\ \zeta>0,\ \tau>0, \tag{14}$$

$$T^*(\rho,\zeta,0)=0,\ \rho\ge 0,\ \zeta\ge 0, \tag{15}$$

$$\frac{\partial T^*}{\partial\zeta}=-Aq^*(\rho)H(\tau_s-\tau),\ \rho\ge 0,\ \zeta=0,\ \tau>0, \tag{16}$$

$$T^*(\rho,\zeta,\tau)\to 0,\ \sqrt{\rho^2+\zeta^2}\to\infty,\tau>0, \tag{17}$$

where

$$q^*(\rho)=B_f[f+(1-f)B_f\rho^2]e^{-B_f\rho^2},\ \text{(see Fig. 2)}. \tag{18}$$

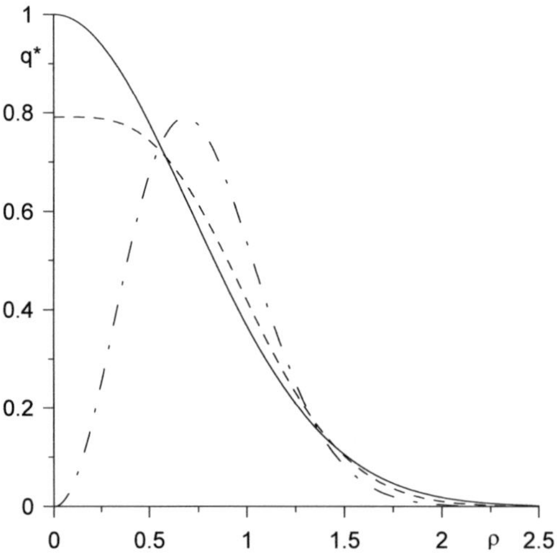

Fig. 2. Laser irradiation distribution – a function $q^*(\rho)$ for three different f parameter values (solid line corresponds to the value of $f=1$, dashed line to $f=0.5$, dot-dashed line to $f=0$).

2.2 Problem solution

The solution of the boundary value heat conduction problem (14)-(17), which is obtained by applying the integral Hankel transforms with respect to the radial variable r and Laplace transform with respect to time t, has the following form:

$$T^*(\rho,\zeta,\tau)=T^{(0)*}(\rho,\zeta,\tau)H(\tau)-T^{(0)*}(\rho,\zeta,\tau-\tau_s)H(\tau-\tau_s),\ \rho\geq 0,\zeta\geq 0,\tau\geq 0. \qquad (19)$$

$$T^{(0)*}(\rho,\zeta,\tau)=A\int_0^\infty \varphi(\xi)\Phi(\xi,\zeta,\tau)J_0(\xi\rho)d\xi\,,\ \rho\geq 0,\zeta\geq 0,\tau\geq 0\,, \qquad (20)$$

where

$$\varphi(\xi)=\int_0^\infty \rho q^*(\rho)J_0(\xi\rho)d\rho=\frac{1}{2}\left[f+(1-f)\left(1-\frac{\xi^2}{4B_f}\right)\right]e^{-\frac{\xi^2}{4B_f}},\ \xi\geq 0 \qquad (21)$$

and

$$\Phi(\xi,\zeta,\tau)=\frac{1}{2}\left[e^{-\xi\zeta}\operatorname{erfc}\left(\frac{\zeta}{2\sqrt{\tau}}-\xi\sqrt{\tau}\right)-e^{\xi\zeta}\operatorname{erfc}\left(\frac{\zeta}{2\sqrt{\tau}}+\xi\sqrt{\tau}\right)\right]. \qquad (22)$$

During transient heating of the massive body, the maximum value of temperature on the body surface is achieved at the moment $t=t_s$ ($\tau=\tau_s$) - switching laser system off, whereas inside the body at $t_h=t_s+\Delta t$ ($\tau_h=\tau_s+\Delta\tau$, where $\Delta\tau=k\Delta t/a^2$). When laser system is switched off, then laser heating source is simply cut off - as a consequence the fast cooling of the body takes place.

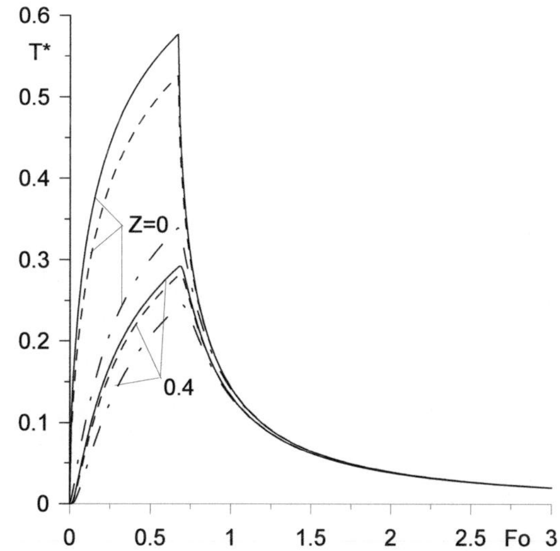

Fig. 3. Evolution of dimensionless temperature $T^*=T/(AT_0)$ on the surface in the center of heated zone ($\rho=0,\zeta=0$) and inside the body ($\rho=0,\zeta=0.4$) at $\tau_s=0.6$ (Yevtushenko et al., 2009).

Independently from the heat source intensity distribution, the retardation time Δt ($\Delta\tau$) is increasing fast with the distance from the heated surface. For a fixed value of depth from the working surface this retardation time Δt decreases with the increase of the f parameter.

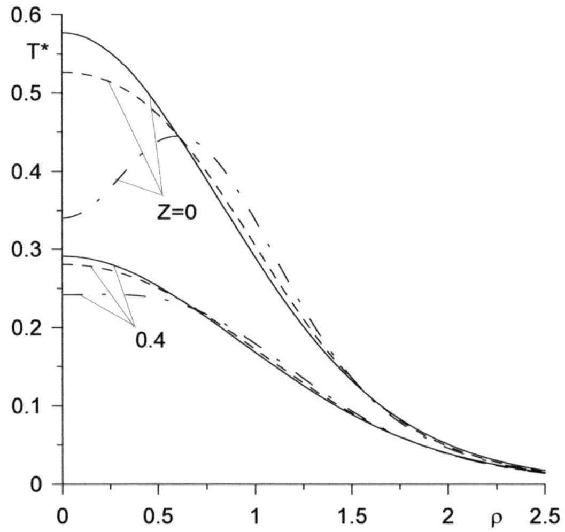

Fig. 4. Evolution of dimensionless temperature T^* along ρ variable on the body surface ($\zeta = 0$) and inside the body ($\zeta = 0.4$) at $\tau_s = 0.6$ (Yevtushenko et al., 2009).

The irradiation intensity distribution in a plane normal to the laser beam axis, see Eq.(18), affects essentially the character of surface temperature distribution. Maximum value of temperature for normal distribution of heat flux intensity is attained at the center of heated zone $r = 0$, and for a doughnut distribution is attained near a boundary of the heated zone $r = 0.6a$ (see Fig. 4).

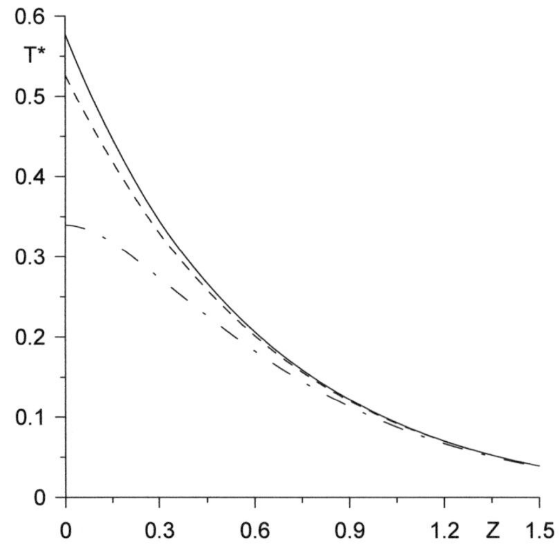

Fig. 5. Evolution of dimensionless temperature T^* along symmetry axis $\rho = 0$ at $\tau_s = 0.6$ (Yevtushenko et al., 2009).

The effective depth of heating (the distance from the boundary body surface for which temperature value equals 5% of the maximum value of temperature achieved on the surface) is independent on the form of intensity distribution of the incident heat flux and for dimensionless retardation time $\tau_s = 0.6$ is equal in approximation $1.5a$ (see Fig. 5). At work (Ashcroft & Mermin, 1986) was shown that for laser systems working in the continuous regime the above quantity does not exceed the $5a$ value (Matysiak et al., 1998). This conclusion confirms the results shown in the present chapter thanks to the numerical calculations done with use of formulas (19)-(22) at $\tau_s \to \infty$ and $\tau = 100$ (dimensionless time, when stationary temperature is achieved).

2.3 Special cases of the obtained solution

At $t_s \to \infty (\tau_s \to \infty)$ from Eq. (19) is received that $T^*(\xi,\zeta,\tau) = T^{(0)*}(\xi,\zeta,\tau)$ and formula (20) corresponds with solution presented in paper (Ready, 1997) for laser operated in the continuous regime and irradiation of semi-infinite surface. If additionally $t \to \infty (\tau \to \infty)$, then from Eq. (22) follows that $\Phi(\xi,\zeta,\infty) = e^{-\xi\zeta}$ and then stationary temperature can be derived from the relation:

$$T^{(0)*}(\rho,\zeta,\infty) = A\int_0^\infty \varphi(\xi) e^{-\xi\zeta} J_0(\xi\rho)\, d\xi, \rho \geq 0,\ \zeta \geq 0. \tag{23}$$

By substituting to the Eq. (23) $\zeta = 0$, the stationary temperature on surface for semi-infinite surface can be found

$$T^{(0)*}(\rho,0,\infty) = A\int_0^\infty \varphi(\xi) J_0(\xi\rho)\, d\xi, \rho \geq 0. \tag{24}$$

Substituting function $\phi(\xi)$ (21) under integral and with consideration of integrals:

$$\int_0^\infty e^{-\frac{\xi^2}{4B_f}} J_0(\xi\rho)\, d\xi \quad \text{and} \quad \int_0^\infty \xi^2 e^{-\frac{\xi^2}{4B_f}} J_0(\xi\rho)\, d\xi$$

with use of integrals table (Prudnikov et al., 1998) the solution for stationary temperature on semi-infinite surface was received in the following form:

$$T(r,0,\infty) = A\frac{\sqrt{\pi B_f}}{2} e^{-\frac{B_f\rho^2}{2}} \left\{ I_0\left(\frac{B_f\rho^2}{2}\right) - \frac{1-f}{2}\left[\left(1 - B_f\rho^2\right) I_0\left(\frac{B_f\rho^2}{2}\right) + B_f\rho^2 I_1\left(\frac{B_f\rho^2}{2}\right)\right]\right\}. \tag{27}$$

In similar way the distribution of the stationary temperature along axis $\rho = 0$ from the (23) solution was found:

$$T^{(0)*}(0,\zeta,\infty) = A\left\{\frac{\zeta}{2}(1-f) + \left[\frac{1}{2}(1+f) - (1-f)\zeta^2\right]\frac{\sqrt{\pi B_f}}{2} e^{B_f\zeta^2} \operatorname{erfc}(\sqrt{B_f}\zeta)\right\}. \tag{28}$$

From relations (27) and (28) follows, that stationary temperature on the body surface, in the center of heated zone, reaches the value:

$$T^{(0)*}(0,0,\infty) = A\frac{(1+f)}{4}\sqrt{\pi B_f}\,. \tag{29}$$

For normal distribution of irradiation intensity ($f=1$, $B_f=1$) from Eq. (29) the following result is obtained (Bardybahin & Czubarov, 1996) :

$$T^{(0)*}(0,0,\infty) = A\frac{\sqrt{\pi}}{2} \approx 0.8862A \tag{30}$$

and for doughnut mode structure distribution ($f=0,\ B_f=2.1462$)

$$T^{(0)*}(0,0,\infty) = A\frac{\sqrt{\pi B_f}}{2} \approx 0.6492A\,. \tag{31}$$

It should be underlined that maximum of stationary temperature is achieved not for gaussian irradiation intensity distribution in a plane normal to the laser beam axis ($f=1$), but for "almost gaussian" distribution $(f=0.93)$, when $T^{(0)*}(0,0,\infty) \approx 0.8889A$.

2.4 Time point determination when surface melting occurs due to laser beam heating

On the basis of achieved solutions (19)-(23), the time point determination when surface melting occurs due to laser beam heating, can be done. It is known that for uniform Eq. (5) and normal Eq. (6) (for $f=1$) distribution of laser heat flux intensity, the maximum temperature is achieved on the surface in the centre of the heated zone. For laser systems working in the continuous generation regime, from solutions (20)-(23) at $t_s \to \infty$ ($\tau_s \to \infty$) is derived in general

$$T^{(0)*}(0,0,\tau) = A\int_0^\infty \varphi(\xi)\,erf(\xi\sqrt{\tau})\,d\xi\,, \tag{32}$$

In case of uniform distribution (5) the function $\varphi(\xi) = J_1(\xi)/\xi$ (Matysiak et al., 1998) and then (32) formula becomes

$$T^{(0)*}(0,0,\tau) = \int_0^\infty \frac{J_1(\xi)}{\xi}\,erf(\xi\sqrt{\tau})\,d\xi\,\cdot \tag{33}$$

By differentiation of (33) solution along $\beta = \sqrt{\tau}$ variable, the following is found (Prudnikov et al., 1998):

$$\frac{dT^{(0)*}}{d\beta} = \frac{2}{\sqrt{\pi}}A\int_0^\infty J_1(\xi)e^{-\beta^2\xi^2}\,d\xi = \beta^{-1}e^{-\frac{1}{8\beta^2}}\,I_{1/2}\left(\frac{1}{8\beta^2}\right). \tag{34}$$

According to work (Abramowitz & Stegun, 1979):

$$I_{1/2}(x)=\sqrt{\frac{2}{\pi x}}shx\,, \tag{35}$$

then (34) formula can be rewritten in the form

$$\frac{dT^{(0)*}}{d\beta}=\frac{2A}{\sqrt{\pi}}\left(1-e^{-\frac{1}{4\beta^2}}\right). \tag{36}$$

By integration (36) formula along β variable with condition consideration $T^{(0)*}(0,0,0)=0$, the formula of dimensionless temperature evolution in the centre of the heated zone $(\rho=0,\,\zeta=0)$ is derived:

$$T^{(0)*}(0,0,\tau)=A\left[2\sqrt{\frac{\tau}{\pi}}\left(1-e^{-\frac{1}{4\tau}}\right)+erfc\left(\frac{1}{2\sqrt{\tau}}\right)\right]. \tag{37}$$

The respective formula for gaussian (normal distribution) laser heat flux intensity got the form (Aulyn et al., 2002):

$$T^{(0)*}(0,0,\tau)=A\left[\frac{\sqrt{\pi}}{2}-\frac{1}{\sqrt{\pi}}arctg\frac{1}{2\sqrt{\tau}}\right] \tag{38}$$

Dimensionless temperatures $T^{(0)*}(0,0,\tau)$, (37) and (38), increase in monotone mode with increase of irradiation time and reach the asymptotes respectively: 1 and $\sqrt{\pi}/2=0.8862$ (see Fig. 6).

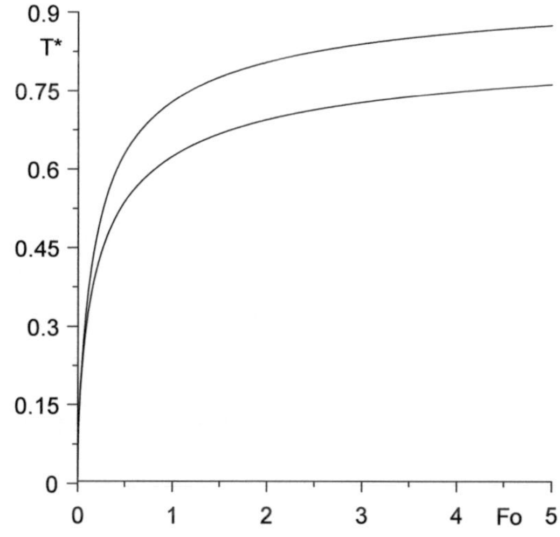

Fig. 6. Evolution of dimensionless temperature $T^{(0)*}/A$ in the centre of heated zone for laser systems working in the continuous generation regime (Yevtushenko et al., 2009).

By making assumption that at some moment τ, temperature in the centre of the heated zone reaches the melting temperature of material, then:

$$T^{(0)*}(0,0,\tau) = T_m / AT_0, \tag{39}$$

where temperature $T^{(0)*}(0,0,\tau)$ is derived from formula (37) or (38).

The right side of the Eq. (39) includes material characteristic features such as: thermal conductivity K and melting temperature T_m, the laser beam characteristic parameters: heat flux intensity q_0, laser beam radius a and effective absorption coefficient A. If all the values are known then from Eq. (39) the dimensionless boundary time point of melting beginning can be found. By choosing for a specific material the appropriate laser beam parameters, the melting start point can be calculated and damage of a surface as a result of intensive melting can be avoided.

In work (Rozniakowski, 2001), for sample of steel St45:

- $K = 33.5W/(mK)$, $k = 15\cdot 10^{-6} m^2/s$, $T_m = 1535^{\circ}C$,
- heated by the Nd:YAG laser beam in system KWANT 15 ($a = 0.64mm$ radius, laser pulse duration $t_s = 2ms$),
- $q_0 = 0.7\cdot 10^9 W/m^2$, when effective absorption coefficient is $A \approx 50\%$ (Rozniakowski, 2001),

the below calculation were conducted with use of Eqs. (13), (39) and numerical results on Fig. 6. From Eq. (13) the $T_0 = 0{,}133\cdot 10^5 K$ was determined. When the right side of Eq. (39) reaches the value $T_m/(AT_0) = 0.2308$, then on Fig. 6 the dimensionless time $\tau = 0.043$ for the uniform distribution and respectively $\tau = 0.047$ for the normal distribution can be found. These dimensionless time values have respective real time values: $t = 1.17ms$ and $t = 1.28ms$.

2.5 Determination of laser irradiation effective absorption coefficient method

It should be underlined that maximum temperature value (19) on the body surface is achieved at the moment of laser switiching off $t = t_s$ $(\tau = \tau_s)$ (Fig. 3). In order to determine the monochromatic effective absorption coefficient A, the value of the retardation time Δt $(\Delta\tau)$ is needed to be known (its value increases quickly with the distance from laser irradiated surface of the body). Retardation time Δt $(\Delta\tau)$ can be determined from the condition of the maximum temperature reached inside the semi-infinite body in the point of (r,z) (ρ,ζ) at $t = t_h$ $(\tau = \tau_h)$:

$$\left.\frac{\partial T^*(\rho,\zeta,\tau)}{\partial\tau}\right|_{\tau=\tau_h} = 0,\ \rho \ge 0,\ \zeta \ge 0. \tag{40}$$

Differentiating equations (19)-(22) with respect to dimensionless time τ we obtain

$$\frac{\partial T^*(\rho,\zeta,\tau)}{\partial\tau} = A\left[\frac{\partial T^{(0)*}(\rho,\zeta,\tau)}{\partial\tau}H(\tau) - \frac{\partial T^{(0)*}(\rho,\zeta,\tau-\tau_s)}{\partial\tau}H(\tau-\tau_s)\right], \tag{41}$$

where

$$\frac{\partial T^{(0)*}(\rho,\zeta,\tau)}{\partial\tau}=\frac{1}{\sqrt{\pi\tau}}\int_0^\infty \varphi(\xi)\xi e^{-\left(\frac{\zeta^2}{4\tau}+\xi^2\tau\right)}J_0(\xi\rho)d\xi\,. \tag{42}$$

Taking into account the form of function $\varphi(\xi)$ given by (21), the Eq. (42) can be written as:

$$\frac{\partial T^{(0)*}(\rho,\zeta,\tau)}{\partial\tau}=[f\,M_1(\rho,\tau)+(1-f)M_2(\rho,\tau)]\frac{e^{-\frac{\zeta^2}{4\tau}}}{2\sqrt{\pi\tau}}\,, \tag{43}$$

where (Prudnikov et al., 1998)

$$M_1(\rho,\tau)=\int_0^\infty \xi e^{-\xi^2\left(\frac{1}{4}+\tau\right)}J_0(\xi\rho)d\xi=\frac{2}{1+4\tau}e^{-\frac{\rho^2}{1+4\tau}}\,, \tag{44}$$

and

$$M_2(\rho,\tau)=\int_0^\infty\left(1-\frac{\xi^2}{4}\right)\xi e^{-\xi^2\left(\frac{1}{4}+\tau\right)}J_0(\xi\rho)d\xi=\left[1-\frac{1}{1+4\tau}\left(1-\frac{\rho^2}{1+4\tau}\right)\right]M_1(\rho,\tau)\,. \tag{45}$$

Substituting functions $M_i(\rho,\tau)$, $i=1,2$ (44), (45) into Eq. (43), we got:

$$\frac{\partial T^{(0)*}(\rho,\zeta,\tau)}{\partial\tau}=\frac{B_f[(1+4B_f\tau)(f+4B_f\tau)+(1-f)B_f\rho^2]}{(1+4B_f\tau)^3\sqrt{\pi\tau}}e^{\left[-\left(\frac{B_f\rho^2}{1+4B_f\tau}+\frac{\zeta^2}{4\tau}\right)\right]}\,. \tag{46}$$

By substituting the partial derivative (46) into Eq. (41), from condition (40) the nonlinear functional equation with respect to the dimensionless retardation time $\Delta\tau$ is obtained:

$$C_1C_2^3D_f(\rho)=e^{-(C_3\rho^2+C_4Z^2)}\,,\quad \rho\geq 0,\ Z\geq 0\,, \tag{47}$$

where

$$C_1=\sqrt{1-\frac{\tau_s}{\tau_h}}\,,\ C_2=\frac{4B_f\tau_s}{1+4B_f\tau_h}\,,C_3=\frac{4B_f\tau_s}{(1+4B_f\Delta\tau)(1+4B_f\tau_h)}\,,\ C_4=\frac{\tau_s}{4\tau_h\Delta\tau}\,, \tag{48}$$

and

$$D_f(\rho)=\frac{(1-f)B_f\rho^2+(1+4B_f\tau_h)(f+4B_f\tau_h)}{(1-f)B_f\rho^2+(1+4B_f\Delta\tau)(f+4B_f\Delta\tau)}\,. \tag{49}$$

By applying logarithm on (47) we have the following results:

$$C_3\rho^2+C_4\zeta^2=\ln[C_1\cdot C_2^3\cdot D_f(\rho)]^{-1}\quad \rho\geq 0,\ \zeta\geq 0\,. \tag{50}$$

Eq. (50) defines the isotherm of maximum temperature for a given values of τ_s and $\Delta\tau$ (see Fig. 7). For the case of normal distribution of irradiation intensity ($f = 1$, $B_f = 1$), the $D_f(\rho)$ function (49) is independent on dimensionless radial variable ρ and can be expressed by:

$$D_f \equiv D_1 = \frac{(1+4\tau_h)^2}{(1+4\Delta\tau)^2} = \frac{1}{C_2^2}, \tag{51}$$

where C_2 is given by (48). By using relation (51), the Eq. (50) can be rewritten in the form:

$$C_3\rho^2 + C_4\zeta^2 = \ln[C_1 C_2]^{-1}, \quad \rho \geq 0, \ \zeta \geq 0. \tag{52}$$

and consequently as:

$$\frac{\rho^2}{\alpha^2} + \frac{\zeta^2}{\beta^2} = 1, |\rho| \leq \alpha, 0 < \zeta \leq \beta, \tag{53}$$

where

$$\alpha^2 = -\frac{\ln(C_1 C_2)}{C_3}, \quad \beta^2 = -\frac{\ln(C_1 C_2)}{C_4}. \tag{54}$$

By analysing the Eqs. (53), (54) it can be found that the isotherm of maximum dimensionless temperature for the normal (gaussian) distribution of irradiation intensity has the form of half-ellipse (see Fig. 7) with the axes given by (53) – $r_h = \alpha a$, $z_h = \beta a$. In case of the toroidal (ring) distribution of the heat flux, the isotherm takes the form of curve with maximum shifted from axial axis. On symmetry axis $\rho = 0$, the Eq. (52) has form:

$$C_4\zeta^2 = \ln[C_1 C_2]^{-1}. \tag{55}$$

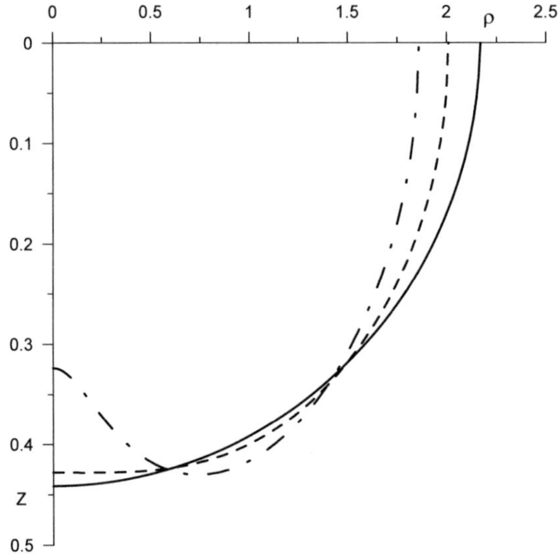

Fig. 7. Isotherms of maximum dimensionless temperature $T^*_{\max}$ at $\tau_s = 0.672$ and $\Delta\tau = 0.0054$ (Yevtushenko et al., 2009).

By substituting in Eq. (55) the coefficients C_1, C_2 and C_4 respectively (50), the following is derived:

$$\zeta_h = \left\{ \frac{4\Delta\tau(\tau_s + \Delta\tau)}{\tau_s} \ln\left[\sqrt{\frac{\Delta\tau}{\tau_s + \Delta\tau}} \left(\frac{1 + 4\Delta\tau}{1 + 4(\tau_s + \Delta\tau)} \right) \right]^{-1} \right\}^{1/2} . \quad (56)$$

From (56) equation, when retardation time $\Delta\tau$ and heating time τ_s are known, the maximum dimensionless layer depth ζ_h in irradiated material, where phase transitions occur, can be found. In the contrary case, when maximum dimensionless layer depth ζ_h in irradiated material is known (the temperature T exceeded some critical value T_h and phase transition took place), then from Eq. (56) the dimensionless retardation time $\Delta\tau$ can be found, too. From condition $T(r,z,t_h) = T_h$, where T is given from solution (19)-(22), the following formula for determination of the effective absorption coefficient, is obtained:

$$A = A^* \frac{T_h}{T_0} , \quad (57)$$

where

$$A^* = \left\{ \int_0^\infty \varphi(\xi)[\Phi(\xi,\zeta_h,\tau_s) - \Phi(\xi,\zeta_h,\Delta\tau)] J_0(\xi\rho_h) d\xi \right\}^{-1} , \quad (58)$$

and coordinates (ρ_h,ζ_h) fulfill the Eq. (53).

2.6 Experimental results

Some results of steel 45 type hardening by laser irradiation can be found in work (Rozniakowski, 2001). The application of determination of effective absorption coefficient method is also there included on the basis of effective reflection coefficient R ($A = 1 - R$). Characteristic features of mentioned above material and laser system applied in the experiment are gathered in Table 1. The steel 45 type sample had rounded shape of 20mm diameter and 6mm in thickness, the percentage of other elements had following distribution: C–$0.42 \div 0{,}50\%$, Mn–$0.5 \div 0{,}8\%$, Si–$0.17 \div 0.37\%$, P–0.040%, S–0.045%. Laser system was working in free generation regime (Nd:YAG laser type, $\lambda = 1.06\mu m$, $E_i = 1.5J$ and laser pulse time duration $t_s = 2ms$).

parameters / material	K, $[Wm^{-1}K^{-1}]$	$k \times 10^5$, $[m^2 s^{-1}]$	T_h, [K]	z_h, [μm]
steel 45	33.50	1.50	1123	40

Table 1. Characteristic features of steel 45 type.

Steel sample was heated locally by laser irradiation beam in 10 different points with changing irradiation intensity. Subsequently, the metalographic cross-section were done for the irradiated areas. By using the EPITYP-2 metalographic microscope and SEM TESLA BS300 the maximum hardened layer depth z_h were measured (see Fig. 8).

Fig. 8. Metalographic cross-section of laser irradiated steel sample.

The diameters of hardened depths were much larger than z_h, it was also observed the melting point starts when $q_0 > 8.5 \cdot 10^8 \mathrm{W/m^2}$ where:

$$q_0 = \frac{E_i}{\pi a^2 t_s}. \tag{59}$$

Moreover, it was found that for $q_0 = 5.8 \cdot 10^8 \mathrm{W/m^2}$, the hardened layer depth z_h equals $40\mu\mathrm{m}$. By using Eq. (59) the radius of irradiated area can be found as $a = 0.64\,\mathrm{mm}$ and consequently the dimensionless irradiation time τ_s. Hence from Eq. (56) at $\zeta_h = z_h / a$ the dimensionless retardation time $\Delta\tau$, was obtained (see Table 2). For such dimensionless coordinates values $(\rho_h = 0, \zeta_h)$ and dimensionless time values $(\tau_s, \Delta\tau)$, from Eq. (58) the parameter $A^* = 4.5$ was calculated, and from Eq.(57) the real effective absorption coefficient A was found, see comparison of the experimental and analytical values gathered in Table 2, (Rozniakowski, 2001).

parameters / material	$T_0 \times 10^{-5}$, $[\mathrm{K}^{-1}]$	ζ_h, [-]	τ_s, [-]	$\Delta\tau \times 10^3$ [-]	A experimental	A from Eq. (57)
steel 45	0.111	0.062	0.073	0.33	$30\% \div 50\%$	41.8%

Table 2. Comparison of the experimental and analytical values of the effective absorption coefficient A.

3. Quasi-static thermal stresses caused by laser irradiation heating

3.1 Non-stationary temperature field

By using variables and parameters (13), the axisymmetrical boundary value problem of transient heat conduction for semi-infinite body can be considered as:

$$\frac{\partial^2 T^*}{\partial \rho^2} + \frac{1}{\rho}\frac{\partial T^*}{\partial \rho} + \frac{\partial^2 T^*}{\partial \zeta^2} = \frac{\partial T^*}{\partial \tau}, \quad \rho \ge 0,\ \zeta > 0,\ \tau > 0, \tag{60}$$

$$T^*(\rho, \zeta, 0) = 0, \quad \rho \ge 0,\ \zeta \ge 0, \tag{61}$$

$$\frac{\partial T^*}{\partial \zeta} - BiT^* = -q^*(\rho)H(\tau)\,,\, \rho \ge 0,\, \zeta = 0,\ \tau > 0\,, \tag{62}$$

$$T^*(\rho,\zeta,\tau) \to 0,\, \sqrt{\rho^2 + \zeta^2} \to \infty, \tau > 0\,, \tag{63}$$

where from Eq. (18) for $f = 1$ following is given

$$q^*(\rho) = e^{-\rho^2}\,,\, \rho \ge 0\,. \tag{64}$$

By applying the Hankel integral transformation along the radial ρ (Sneddon, 1972)

$$\bar{T}^*(\xi,\zeta,\tau) = \int_0^\infty \rho T^*(\rho,\zeta,\tau) J_0(\xi\rho)\, d\rho\,, \tag{65}$$

to the boundary heat conduction problem (60)-(63) then it is denoted

$$\frac{\partial^2 \bar{T}^*}{\partial \zeta^2} - \xi^2 \bar{T}^* = \frac{\partial \bar{T}^*}{\partial \tau}\,, \tag{66}$$

$$\bar{T}^*(\xi,\zeta,0) = 0\,, \tag{67}$$

$$\bar{T}^*(\xi,\infty,\tau) = 0\,, \tag{68}$$

$$\frac{\partial \bar{T}^*}{\partial Z} = -\varphi(\xi) + Bi\bar{T}^*, \zeta = 0\,, \tag{69}$$

where after consideration of Eq. (64)

$$\varphi(\xi) = \int_0^\infty \rho e^{-\rho^2} J_0(\xi\rho)\, d\rho = \frac{1}{2} e^{-\frac{\xi^2}{4}}\,. \tag{70}$$

When applying Fourier integral transformation with generalized trigonometric kernel along ζ variable to the boundary value problem (66)-(69) (Sneddon, 1972) then

$$\bar{\bar{T}}^*(\xi,\varsigma,\tau) = \sqrt{\frac{2}{\pi}} \int_0^\infty \bar{T}^*(\xi,\zeta,\tau) N(\zeta,\varsigma)\, d\zeta\,,\ N(\zeta,\varsigma) = \varsigma\cos(\varsigma\,\zeta) + Bi\sin(\varsigma\,\zeta)\,. \tag{71}$$

As a result of transformation the Cauchy problem is obtained in the form:

$$\frac{d\bar{T}^*}{d\tau} + (\varsigma^2 + \xi^2)\,\bar{\bar{T}}^* = \sqrt{\frac{2}{\pi}}\,\varsigma\,\varphi(\xi)\,, \tau > 0\,, \tag{72}$$

$$\bar{\bar{T}}^*(\xi,\varsigma,0) = 0\,. \tag{73}$$

Solution of the ordinary differential equation (72) with initial condition (73) has form

$$\tilde{\bar{T}}^*(\xi,\varsigma,\tau)=\varphi(\xi)\,\tilde{\Phi}_0(\xi,\varsigma,\tau), \tag{74}$$

where

$$\tilde{\Phi}_0(\xi,\varsigma,\tau)=\sqrt{\frac{2}{\pi}}\frac{\varsigma}{\varsigma^2+\xi^2}\left[1-e^{-(\varsigma^2+\xi^2)\tau}\right]. \tag{75}$$

By applying to the solutions (74), (75) below listed Fourier and Hankel inverted integral transformations (Sneddon, 1972):

$$\bar{T}^*(\xi,\zeta,\tau)=\sqrt{\frac{2}{\pi}}\int_0^\infty\frac{N(\zeta,\varsigma)}{\varsigma^2+Bi^2}\tilde{\bar{T}}^*(\xi,\varsigma,\tau)\,d\varsigma, \tag{76}$$

$$T^*(\rho,\zeta,\tau)=\int_0^\infty\xi\,\bar{T}^*(\xi,\zeta,\tau)J_0(\xi\rho)\,d\xi, \tag{77}$$

the following will be obtained

$$T^*(\rho,\zeta,\tau)=\int_0^\infty\xi\,\varphi(\xi)\Phi(\xi,\zeta,\tau)J_0(\xi\rho)\,d\xi\,,\rho\ge0,\zeta>0,\tau>0, \tag{78}$$

where

$$\Phi(\xi,\zeta,\tau)=\frac{1}{2}\left[\frac{e^{-\xi\zeta}}{Bi+\xi}\text{erfc}\left(\frac{\zeta}{2\sqrt{\tau}}-\xi\sqrt{\tau}\right)+\frac{e^{\xi\zeta}}{Bi-\xi}\text{erfc}\left(\frac{\zeta}{2\sqrt{\tau}}+\xi\sqrt{\tau}\right)\right]-\frac{Bi\,e^{Bi\zeta}}{Bi^2-\xi^2}e^{(Bi^2-\xi^2)\tau}\text{erfc}\left(\frac{\zeta}{2\sqrt{\tau}}+Bi\sqrt{\tau}\right). \tag{79}$$

It should be underlined that at $\xi\to Bi$, function $\Phi(\xi,\zeta,\tau)$ (79) equals

$$\Phi(Bi,\zeta,\tau)=\frac{1}{4Bi}e^{-Bi\zeta}\text{erfc}\left(\frac{\zeta}{2\sqrt{\tau}}-Bi\sqrt{\tau}\right). \tag{80}$$

In case, when convection cooling does not occur on the semi-infinite surface of the body ($Bi=0$), function $\Phi(\xi,\zeta,\tau)$ (79) gets form (81). At $\tau\to\infty$ from solution (78), (79) the stationary temperature in the centre of heated zone $\rho=0$, $\zeta=0$, is obtained:

$$T^*_{\max}(0,0,\infty)=\frac{\sqrt{\pi}}{2}-\frac{Bi}{2}\int_0^\infty\frac{e^{-\xi^{\frac{2}{4}}}}{Bi+\xi}d\xi\,. \tag{81}$$

From solution (81) at $Bi\to0$ the maximum temperature value without convection heat exchange, can be found (89).

3.2 Quasi-static thermal stresses

Non-uniform temperature distribution (78), (79) initiates in the semi-infinite surface of the body a field of thermal stresses. Thermal displacement u_i which occurs in the elastic body as the result of temperature field interaction, without mass forces, can be found from differential equations system in partial differentials (Nowacki, 1986; Timoshenko & Goodier, 1970):

$$u_{i,jj} + \frac{1}{1-2\nu} u_{j,ji} = \frac{2(1+\nu)}{1-2\nu} \alpha_t T_{,i} \, . \tag{82}$$

By introducing thermoelastic potential $\psi(\rho,\zeta,\tau)$ with use of formulae (Nowacki, 1986):

$$u_r = \frac{1}{a}\frac{\partial \psi}{\partial \rho}, \qquad u_z = \frac{1}{a}\frac{\partial \psi}{\partial \zeta}, \tag{83}$$

then equations system (82) can be rewritten in the form of equivalent Poisson equation:

$$\nabla^2 \psi = \beta_t a^2 T_0 T^* , \tag{84}$$

where $\beta_t = \alpha_t \frac{1+\nu}{1-\nu}$, and dimensionless temperature $T^*(\rho,\zeta,\tau)$ is given by (78), (79).

In case, when thermoelastic potential $\psi(\rho,\zeta,\tau)$ is known, then respective thermal stresses can be derived from formulae (Nowacki, 1986):

$$\sigma_{rr}^{\psi} = \frac{2\mu}{a^2}\left(\frac{\partial^2 \psi}{\partial \rho^2} - \nabla^2 \psi\right), \quad \sigma_{\theta\theta}^{\psi} = \frac{2\mu}{a^2}\left(\frac{1}{\rho}\frac{\partial \psi}{\partial \rho} - \nabla^2 \psi\right), \quad \sigma_{zz}^{\psi} = \frac{2\mu}{a^2}\left(\frac{\partial^2 \psi}{\partial \zeta^2} - \nabla^2 \psi\right), \quad \sigma_{rz}^{\psi} = \frac{2\mu}{a^2}\frac{\partial^2 \psi}{\partial \rho \partial \zeta}. \tag{86}$$

Solution of the Poisson equation (84) obtained with the use of Hankel integral transformations of zero order and Laplace integral transformations, which fulfilled the boundary condtions:

$$\sigma_{zz}^{\psi} = 0 \quad \left(\frac{\partial^2 \psi}{\partial \zeta^2} - \nabla^2 \psi = 0\right), \ \zeta = 0, \ \rho \ge 0, \ \tau \ge 0 \ ,$$

$$\sigma_{rz}^{\psi} = 0 \quad \left(\frac{\partial^2 \psi}{\partial \rho \partial \zeta} = 0\right), \ \zeta = 0, \ \rho \ge 0, \ \tau \ge 0, \tag{87), (88}$$

$$u_r = 0 \quad \left(\frac{\partial \psi}{\partial \rho} = 0\right), \ \zeta \ge 0, \ \rho = 0, \ \tau \ge 0 \ ,$$

$$u_r \to 0 \ \left(\frac{\partial \psi}{\partial \rho} \to 0\right), \ u_z \to 0 \ \left(\frac{\partial \psi}{\partial \zeta} \to 0\right), \ \sqrt{\rho^2 + \zeta^2} \to \infty . \tag{89), (90}$$

has form

$$\psi(\rho,\zeta,\tau) = T_0 \beta_t \int_0^{\infty} \xi \varphi(\xi) \Phi^{\psi}(\xi,\zeta,\tau) J_0(\xi\rho) d\xi , \tag{91}$$

where

$$\Phi^{\psi}(\xi,\zeta,\tau)=\frac{1}{Bi^2-\xi^2}\left\{\left(Bi\,\tau+\frac{\zeta}{2}+\frac{Bi}{Bi^2-\xi^2}\right)\Phi^{+}(\xi,\zeta,\tau)-\right.$$
$$-\left[\xi\tau+\frac{Bi\zeta}{2\xi}-\frac{1}{2\xi}+\frac{Bi^2}{\xi(Bi^2-\xi^2)}\right]\Phi^{-}(\xi,\zeta,\tau)-\left.\sqrt{\frac{\tau}{\pi}}\,e^{-\left(\xi^2\tau+\frac{\zeta^2}{4\tau}\right)}\right\}- \tag{92}$$
$$-\frac{Bie^{Bi\,\zeta}}{Bi^2-\xi^2}e^{\left(Bi^2-\xi^2\right)\tau}\operatorname{erfc}\left(\frac{\zeta}{2\sqrt{\tau}}+Bi\sqrt{\tau}\right),$$

$$\Phi^{\pm}(\xi,\zeta,\tau)=\frac{1}{2}\left[e^{-\xi\zeta}\operatorname{erfc}\left(\frac{\zeta}{2\sqrt{\tau}}-\xi\sqrt{\tau}\right)\pm\operatorname{erfc}\left(\frac{\zeta}{2\sqrt{\tau}}+\xi\sqrt{\tau}\right)\right]. \tag{93}$$

By substituting the potential ψ (91)-(93) to formulae (86) the following is obtained:

$$\sigma_{ij}^{\psi*}(\rho,\zeta,\tau)=\int_0^\infty\varphi(\xi)S_{ij}^{\psi}(\xi,\rho,\zeta,\tau)d\xi-\delta_{ij}T^*(\rho,\zeta,\tau),\ \rho\ge0,\zeta\ge0,\tau\ge0, \tag{94}$$

where

$$S_{rr}^{\psi}(\xi,\rho,\zeta,\tau)=\Phi^{\psi}(\xi,\zeta,\tau)\left[\frac{\xi^2}{\rho}J_1(\xi\rho)-\xi^3J_0(\xi\rho)\right], \tag{95}$$

$$S_{\theta\theta}^{\psi}(\xi,\rho,\zeta,\tau)=-\Phi^{\psi}(\xi,\zeta,\tau)\frac{\xi^2}{\rho}J_1(\xi\rho), \tag{96}$$

$$S_{zz}^{\psi}(\xi,\rho,\zeta,\tau)=\Phi^{\psi}(\xi,\zeta,\tau)\xi^3J_0(\xi\rho), \tag{97}$$

$$S_{rz}^{\psi}(\xi,\rho,\zeta,\tau)=-\Phi_{,\zeta}^{\psi}(\xi,\zeta,\tau)\xi^2J_1(\xi\rho), \tag{98}$$

$$\Phi_{,\zeta}^{\psi}(\xi,\zeta,\tau)=\frac{1}{Bi^2-\xi^2}\left\{\left(\xi^2\tau+\frac{Bi\,\zeta}{2}+\frac{Bi^2}{Bi^2-\xi^2}\right)\Phi^{+}(\xi,\zeta,\tau)\right.$$
$$-\left(\xi\,Bi\,\tau+\frac{Bi}{2\xi}-\frac{\xi\zeta}{2}+\frac{\xi\,Bi}{Bi^2-\xi^2}\right)\Phi^{-}(\xi,\zeta,\tau)-\left.Bi\sqrt{\frac{\tau}{\pi}}\,e^{-\left(\xi^2\tau+\frac{\zeta^2}{4\tau}\right)}\right\} \tag{99}$$
$$-\frac{Bi^2e^{Bi\zeta}}{Bi^2-\xi^2}e^{\left(Bi^2-\xi^2\right)\tau}\operatorname{erfc}\left(\frac{\zeta}{2\sqrt{\tau}}+Bi\sqrt{\tau}\right),$$

$$\sigma_{ij}^{\psi*}=\frac{\sigma_{ij}^{\psi}}{\sigma_0},\ \sigma_0=2\mu\beta_tT_0, \tag{100}$$

and where $T^*(\rho,\zeta,\tau)$ – dimensionless temperature (78), (79), $\delta_{rr}=\delta_{\theta\theta}=1,\ \ \delta_{zz}=\delta_{rz}=0$.

It should be noticed that on semi-infinite surface of the body $\zeta = 0$ the boundary condition (88) is not fulfilled, which means that $\sigma_{rz}^{\psi *} \neq 0$. Thus, the additional problem of forces interacting on semi-infinite surface of the body, which have such distribution that stresses $\sigma_{ij}^{L*} = \sigma_{ij}^{L} / \sigma_0$ caused by them together with stresses field $\sigma_{ij}^{\psi *}$ (94)-(100), will fulfill the boundary condition, should be investigated:

$$\sigma_{zz}^{\psi *} + \sigma_{zz}^{L*} = 0 \,, \quad \sigma_{rz}^{\psi *} + \sigma_{rz}^{L*} = 0, \quad \zeta = 0, \quad \rho \geq 0, \quad \tau \geq 0 \,. \tag{101}$$

Stresses field caused by the forces axisymmetrically distributed on the elastic semi-infinite surface of the body are derived with use of the Love function $L(\rho,\zeta,\tau)$ from biharmonic equation (Abramowitz & Stegun, 1979):

$$\nabla^4 L = 0 \,. \tag{102}$$

Stresses tensor components are related with the Love function $L(\rho,\zeta,\tau)$ by relation (Nowacki, 1986)

$$\begin{aligned} \sigma_{rr}^{L} &= \frac{1}{a^3}\frac{\partial}{\partial \zeta}\left(\nu \nabla^2 L - \frac{\partial^2 L}{\partial \rho^2} \right), & \sigma_{\theta\theta}^{L} &= \frac{1}{a^3}\frac{\partial}{\partial \zeta}\left(\nu \nabla^2 L - \frac{1}{\rho}\frac{\partial L}{\partial \rho} \right), \\ \sigma_{zz}^{L} &= \frac{1}{a^3}\frac{\partial}{\partial \zeta}\left[(2-\nu)\nabla^2 L - \frac{\partial^2 L}{\partial \zeta^2} \right], & \sigma_{rz}^{L} &= \frac{1}{a^3}\frac{\partial}{\partial \rho}\left[(1-\nu)\nabla^2 L - \frac{\partial^2 L}{\partial \zeta^2} \right]. \end{aligned} \tag{103}$$

Solution of equation (102), which fulfills the boundary conditions (101), is declining in infinity and has form:

$$L(\rho,\zeta,\tau) = \frac{2\nu - 1}{2\mu}\sigma_0 a^3 \int_0^\infty \varphi(\xi) e^{-\xi\zeta} \cdot \left\{ (2\nu + 2 + \xi\zeta)\Phi^{\psi}(\xi,0,\tau) + \left(\frac{2\nu+1}{\xi} + \zeta \right)\Phi^{\psi}_{,\zeta}(\xi,0,\tau) \right\} J_0(\xi\rho) d\xi \tag{104}$$

Dimensionless stresses σ_{ij}^{L*} pertaining to $L(\rho,\zeta,\tau)$ (104) can be found by using formulae (103):

$$\sigma_{ij}^{L*}(\rho,\zeta,\tau) = \int_0^\infty \varphi(\xi)\xi^2 S_{ij}^{L}(\xi,\rho,\zeta,\tau) d\xi \,, \; \rho \geq 0, \zeta \geq 0, \tau \geq 0 \,, \tag{105}$$

where

$$\begin{aligned} S_{rr}^{L}(\xi,\rho,\zeta,\tau) = {} & [(\xi\zeta - 1)\xi\Phi^{\psi}(\xi,0,\tau) + (\xi\zeta - 2)\Phi^{\psi}_{,\zeta}(\xi,0,\tau)] J_0(\xi\rho) + \\ & + [(1 - 2\nu - \xi\zeta)\xi\Phi^{\psi}(\xi,0,\tau) + (2 - 2\nu - \xi\zeta)\Phi^{\psi}_{,\zeta}(\xi,0,\tau)] \frac{J_1(\xi\rho)}{\xi\rho} \,, \end{aligned} \tag{106}$$

$$\begin{aligned} S_{\theta\theta}^{L}(\xi,\rho,\zeta,\tau) = {} & -[2\nu\xi\Phi^{\psi}(\xi,0,\tau) + 2\nu\Phi^{\psi}_{,\zeta}(\xi,0,\tau)] J_0(\xi\rho) + \\ & + [(2\nu - 1 + \xi\zeta)\xi\Phi^{\psi}(\xi,0,\tau) + (2\nu - 2 + \xi\zeta)\Phi^{\psi}_{,\zeta}(\xi,0,\tau)] \frac{J_1(\xi\rho)}{\xi\rho} \,, \end{aligned} \tag{107}$$

$$S_{zz}^{L}(\xi,\rho,\zeta,\tau) = -[(1 + \xi\zeta)\xi\Phi^{\psi}(\xi,0,\tau) + \xi\zeta\Phi^{\psi}_{,\zeta}(\xi,0,\tau)] J_0(\xi\rho) \,, \tag{108}$$

$$S_{rz}^{L}(\xi,\rho,\zeta,\tau) = -[\xi\zeta\Phi^{\psi}(\xi,0,\tau) - (1-\xi\zeta)\Phi_{,\zeta}^{\psi}(\xi,0,\tau)]J_1(\xi\rho) \cdot \tag{109}$$

Total field of dimensionless thermal stresses $\sigma_{ij}^{*} = \sigma_{ij} / \sigma_0$ is obtained with use of superposition of stresses field $\sigma_{ij}^{\psi *}$ (94)-(100), related with thermoelastic potential ψ and with stresses σ_{ij}^{L*} (105)-(109) pertaining to the Love function L (104), in the form:

$$\sigma_{ij}^{*}(\rho,\zeta,\tau) = \int_0^{\infty} \varphi(\xi) S_{ij}(\xi,\rho,\zeta,\tau) d\xi - \delta_{ij} T^{*}(\rho,\zeta,\tau), \; \rho \geq 0, \zeta \geq 0, \tau \geq 0, \tag{110}$$

where

$$S_{ij}(\xi,\rho,\zeta,\tau) = S_{ij}^{\psi}(\xi,\rho,\zeta,\tau) + \xi^2 S_{ij}^{L}(\xi,\rho,\zeta,\tau), \tag{111}$$

and function $\varphi(\xi)$ is given by formula (70), and function S_{ij}^{ψ}, S_{ij}^{L} – by formulae (95)-(99) and (106)-(109), respectively.

3.3 Modelling of the laser thermocracking process

The non-stationary temperature field (78), (79) and the thermal stresses (110), (111) can produce an initiation and propagation of cracks in the laser irradiated body, which can lead finally to the fracture. The description of this phenomenon is based on the brittle cracking Griffith and McClintock-Walsh theory (Griffith, 1924; McClintock & Walsh, 1962). The normal stresses σ_n acting in the point of the boundary of predicted crack, can be derived from (Yevtushenko et al., 1997):

$$\sigma_n = \frac{1}{2}\left[(\sigma_1 + \sigma_3) + (\sigma_1 - \sigma_3)\cos(2\theta^{*})\right], \tag{112}$$

where σ_1 and σ_3 denote d the maximal and minimal principal stresses, $\theta^{*} = 0.5\mathrm{arctg}(1/f^{*})$ denotes the angle of crack orientation measured anticlockwise from the direction of σ_3 and f^{*} is the coefficient of surface friction.

If the normal stresses σ_n lead to opening of the crack ($\sigma_n < 0$), then the brittle fracture of body can be determined by using the original or modified Griffith criteria (Griffith, 1924). The fracture process will be started at a given point of the body if

$$\sigma_1 = \sigma_T \;\wedge 3\sigma_1 + \sigma_3 > 0, \text{ where } \sigma_T \text{ is the tensile strength of the material.} \tag{113}$$

The propagation of the cracks occurs in the plane normal to the direction of principal stress σ_1. If at $\sigma_n < 0$ in the given point of the body $3\sigma_1 + \sigma_3 < 0$, then to the initiation of fracture the modified Griffith condition (Griffith, 1924) is applied:

$$-\frac{(\sigma_1 - \sigma_3)^2}{(\sigma_1 + \sigma_3)} = 8\sigma_T \cdot \tag{114}$$

Crack propagation occurs in the plane inclined by angle θ to the direction of action of the maximum principal stress σ_1, where θ:

$$\theta = 0.5\arccos\frac{(\sigma_1 - \sigma_3)}{2(\sigma_1 + \sigma_3)} \cdot \tag{115}$$

The fracture caused by compressive stresses in the point of crack margin ($\sigma_n > 0$) can be initiated there under the McClintock and Walsh condition (McClintock & Walsh, 1962) when:

$$\sigma_3 - \sigma_1 \frac{\sqrt{1+f^{*2}}+f^*}{\sqrt{1+f^{*2}}-f^*} = \sigma_c \text{, where } \sigma_c \text{ is the compressive strength of material.} \quad (116)$$

The fracture occurs in the direction of action of the major maximum stress σ_1 .

3.4 Numerical analysis and conclusions

Dimensionless temperature T^* (78), (79) on the irradiated surface of the body $\zeta = 0$ for a different value of the ρ variable is a transient process. Maximum of the dimensionless temperature T^* is reached in the centre of the irradiated area ($\rho = 0$, $\zeta = 0$) and approaches value 0.8862 , if $\tau \to \infty$. Value $\tau = 10$ corresponds to the stationary state in the centre of irradiated area. If much closer to the centre of irradiation area $\rho = 0$, then transient process is much shorter. For $\tau = 1$ temperature in the point $\rho = 0$ equals 75% , in the point $\rho = 1.5$ equals 12% of stationary value 0.8862 .

In the experiments, with use of the Nd:YAG laser system working in quasi-stationary generation regime, where emitted energy pulse has $E_i = 35\,\mathrm{J}$ value, the radius of irradiation area was in range $0.1 \div 0.5\mathrm{mm}$ (see Fig. 9). Most of metals have thermal diffusivity coefficient around $k \approx 10^{-5}\mathrm{m}^2/\mathrm{s}$, hence on the basis of achieved solutions the characteristic time of the transient process in the centre of the irradiated area, could be found as $0.1 \div 0.5\,\mathrm{sec}$.

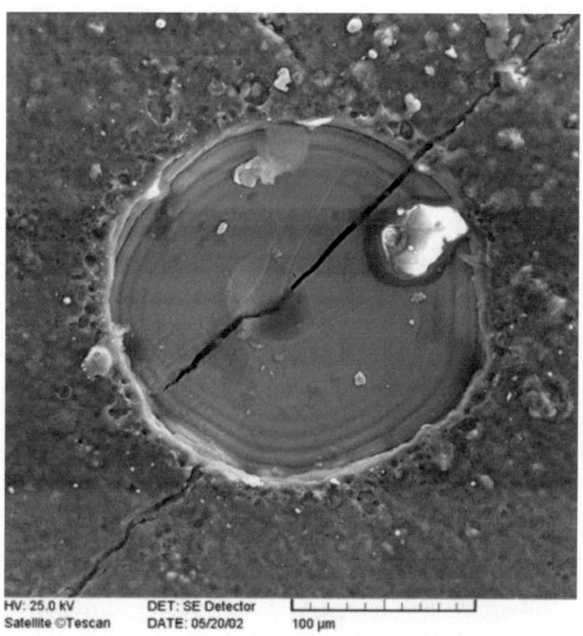

Fig. 9. SEM photograph of laser irradiated area of the Fe3%Si steel.

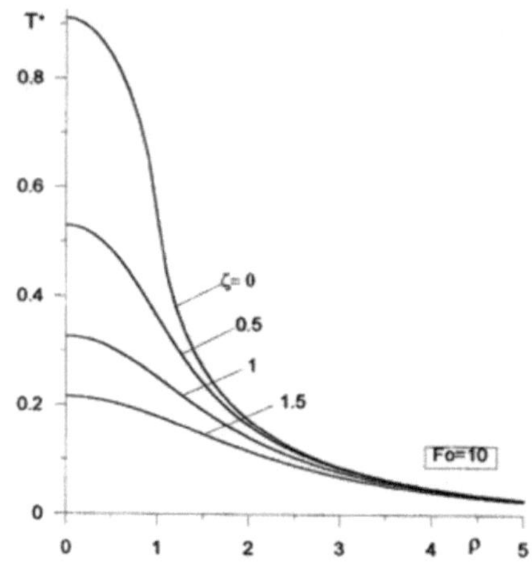

Fig. 10. Distribution of dimensionless stationary temperature T^* along the dimensionless radial variable ρ for $Bi = 0.01$, $\tau = 10$ and for different values of dimensionless variable ζ (Rozniakowski et al., 2003).

Distribution of dimensionless stationary temperature along dimensionless variable ζ for different values of dimensionless radial variable ρ is shown on Fig. 10. It can be noticed that the maximum temperatures and their higher gradients are achieved on the irradiated surface of the body in the centre of heated zone. Temperature increases quickly with the distance from the irradiated area and for $\zeta \geq 5$ $(z \geq 5a)$ reaches the starting value. Isotherms of dimensionless temperature T^* are presented on Fig. 11. It can be noticed the isotherm surfaces of temperature are the rotation surfaces along ρ axis. The highest temperature values and its highest gradients are concentrated close to the centre of laser irradiated area.

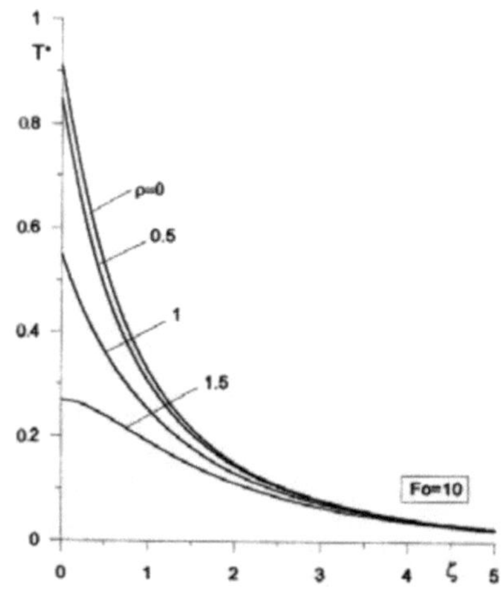

Fig. 11. Distribution of dimensionless temperature T^* along dimensionless ζ variable for $Bi = 0.01$, $\tau = 10$ and for different dimensionless values of ζ variable (Rozniakowski et al., 2003).

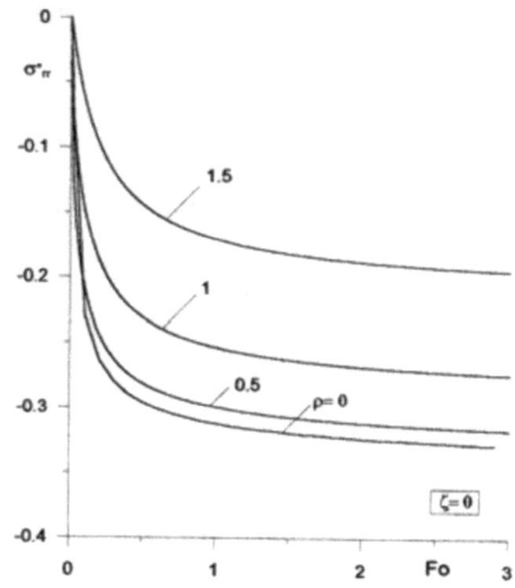

Fig. 12. Evolution of dimensionless radial stresses σ_{rr}^* on the irradiated surface of the body $\zeta = 0$ for $Bi = 0.01$ and different values of dimensionless radial variable ρ (Rozniakowski et al., 2003).

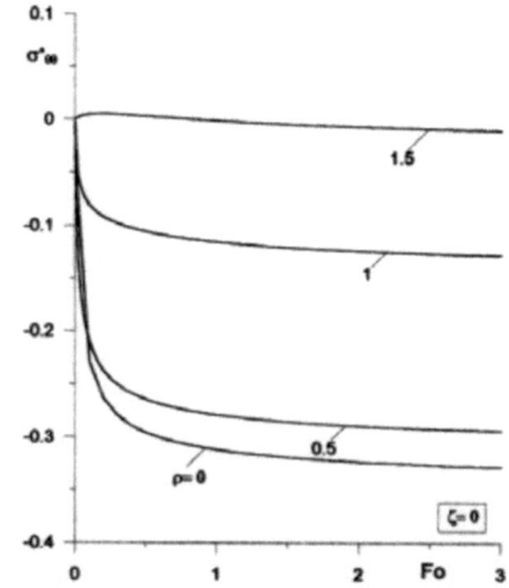

Fig. 13. Evolution of dimensionless peripheral stresses $\sigma_{\theta\theta}^*$ on the irradiated surface of the body $\zeta = 0$ for $Bi = 0.01$ and for different values of dimensionless radial variable ρ (Rozniakowski et al., 2003).

Evolution of dimensionless stresses with time σ_{rr}^* and $\sigma_{\theta\theta}^*$ on the irradiated surface of the body $\zeta = 0$ is shown on Figs. 12, 13. During irradiation process both components of stresses tensor are compressive and decrease with the distance from the centre of heated area. On the contrary, inside the body in the distance equal the radius of heated area $\zeta = 1$, the normal stresses σ_{zz}^* are stretching (see Fig. 14). With the beginning of laser irradiation process, these stresses increase quickly to the maximum value, and afterward decrease with time and reach the stationary value. The highest value of these stresses is achieved on symmetry axis $\rho = 0$.

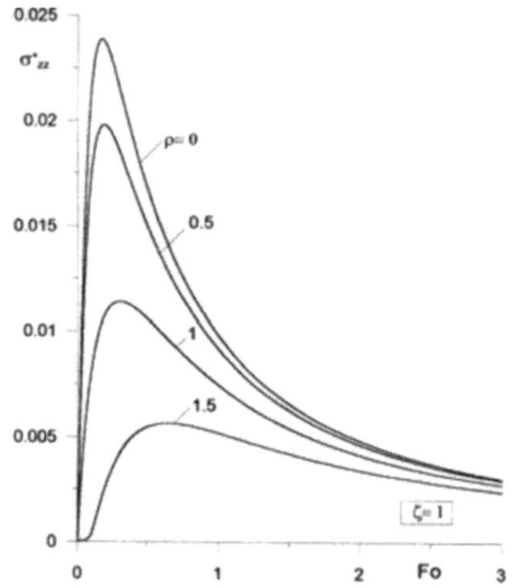

Fig. 14. Evolution of the dimensionless normal stresses σ^*_{zz} on the plane $\zeta = 1$ inside the irradiated body for $Bi = 0.01$ and for different values of dimensionless radial variable ρ (Rozniakowski et al., 2003).

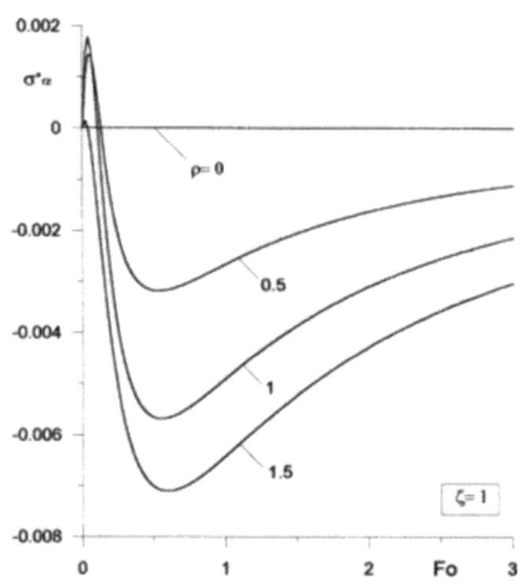

Fig. 15. Evolution of the dimensionless shear stresses σ^*_{rz} on the plane $\zeta = 1$ inside the irradiated body for $Bi = 0.01$ and for different values of dimensionless radial variable ρ (Rozniakowski et al., 2003).

On the contrary to the normal stresses σ^*_{zz}, the shear stresses σ^*_{rz} change their sign during laser irradiation process (see Fig. 15). In the very short time, after switching laser system on, the shear stresses are positive and afterward is changing to some negative value. With heating time the absolute values of stresses σ^*_{rr} and $\sigma^*_{\theta\theta}$ increase, and stresses σ^*_{zz} i σ^*_{rz}

values decrease. It should be underlined that accuracy of temperature and thermal stresses determination depends strongly on accuracy of heat exchange coefficient h determination. The relation used in present calculations $h \approx 0.02K / a$, under condition that convection heat exchange decreases the maximum temperature of the body not more than 10%, was introduced in work (Rykalin et al., 1967).

Parameters	Granite rock	Quart rock	Gabbro rock
Uniaxial tensile strength, $T^*(\rho,\zeta,0)=0$ [MPa]	9.0	13.5	16.0
Uniaxial compressive strength, $\rho \geq 0,\ \zeta \geq 0,$ [MPa]	205	190	162
Shear module, $\frac{\partial T^*}{\partial \zeta} = -q^*(\rho)I(\tau),$ [GPa]	28	36	34
Poisson coefficient ν	0.23	0.16	0.24
Thermal conduction coefficient, K [W/mK]	4.07	4.21	3.67
Thermal diffusivity coefficient $k \times 10^{-6}$ [m²/s]	0.505	2.467	0.458
Linear thermal expansion coefficient $\alpha_t \times 10^{-6}$[K⁻¹]	7.7	24.2	4.7
$T_0 \times 10^4$[K]	0.246	0.237	0.272
σ_0 [GPa]	1.69	5.70	1.42
$(\sigma_T / \sigma_0) \times 10^{-3}$	5.319	2.367	11.280

Table 3. Mechanical and thermo-physical features of granite, quart and gabbro taken from work (Yevtushenko et al., 1997).

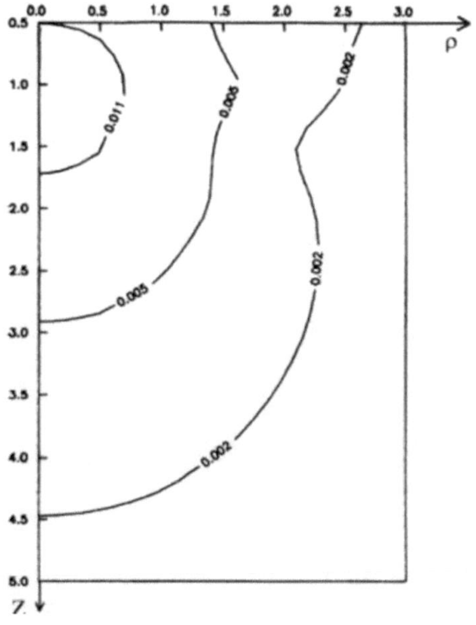

Fig. 16. Isolines of dimensionless major stresses $\sigma_1^* = \sigma_T / \sigma_0$ for features of materials from Table 3 (Yevtushenko et al., 2009).

The maximum $\sigma_1^* = \sigma_1 / \sigma_0$ and minimum $\sigma_3^* = \sigma_3 / \sigma_0$ dimensionless major stresses are changing with the distance from irradiated surface of the body for different dimensionless time values τ. The major stresses σ_1^* are stretching for $\zeta > 0$ and reach the maximum value close to the surface of semi-infinite half-space $\zeta \approx 0.8$ at the moment $\tau = 0.1$. Other major stresses σ_3^* are compressive during heating process and reach maximum value on the irradiated surface. By knowing distribution of major stresses σ_1^* and σ_3^*, with use of criterial equations (113)-(116), the initiation and cracks propagation on the surface and inside the irradiated body, can be predicted. Substituting major stresses σ_1^* and σ_3^*, calculated for $Bi = 0.01$, $\tau = 0.1$, to the criterial equations (113)-(116) it was found that space below the heated surface of the body can be divided into three specific areas, in which each one of the criterial equations is fulfilled. In area $0 \le \zeta \le 0.4$ situated directly below heated surface of the body, the McClintock-Walsh equation (116) for the cracking caused by the compressive stresses, is fulfilled. In other area, where cracking is caused by shear stresses, the modified McClintock-Walsh equations (114), (115) are applied to their prediction. The maximum thickness of this area do not exceeded $0.5a$ value. The area of stretching stresses is placed below the area in which compressive stresses are present. The Griffith criterion (113) is there applied.

On purpose of the numerical analysis three kinds of rocks were chosen: granite, quart, gabbro. The mechanical and thermo-physical features of these rocks material were taken from work (Yevtushenko et al., 1997) and gathered in Table 3. In Table 3, the constant values of T_0 (13) and σ_0 (100) were calculated for $q_0 = 10^8 \mathrm{W/m^2}$ and $a = 0.1\mathrm{mm}$. For these type materials the compressive strength σ_c is much higher than the stretching strength σ_T. Hence, cracking process of such materials can be present in area where (113) criterion is applied and maximum major stresses σ_1 are equal to the stretching strength σ_T:

$$\sigma_1^* = \sigma_T / \sigma_0 \tag{117}$$

Set of points, in area where Griffith criterion (113) is fulfilled, is given in dimensionless form by (117) and form the isolines on the $\rho\zeta$ plane. Isolines of 0.002 value (quart), 0.005 value (granite) and 0.011 (gabbro) are shown on Fig. 16.

4. Axi-symmetrical transient boundary-value problem of heat conduction and quasi-static thermoelasticity for pulsed laser heating of the semi-infinite surface of the body

4.1 Problem statement

The following axi-symmetrical boundary-value problem of heat conduction is under consideration:

$$\frac{\partial^2 T^*}{\partial \rho^2} + \frac{1}{\rho}\frac{\partial T^*}{\partial \rho} + \frac{\partial^2 T^*}{\partial \zeta^2} = \frac{\partial T^*}{\partial \tau}, \quad \rho \ge 0, \zeta > 0, \tau > 0, \tag{118}$$

$$T^*(\rho,\zeta,0) = 0, \quad \rho \ge 0, \zeta \ge 0, \tag{119}$$

$$\frac{\partial T^*}{\partial \zeta} = -q^*(\rho) I(\tau),, \quad \rho \ge 0, \zeta = 0, \tau > 0, \tag{120}$$

$$T^*(\rho,\zeta,\tau)\to 0,\ \sqrt{\rho^2+\zeta^2}\to\infty,\tau>0\,, \tag{121}$$

where dimensionless parameters were definied by formulae (13). Likewise in 3.1 sub-chapter it assumed that laser spatial irradiation intensity is normal (Hector & Hetnarski, 1996):

$$q^*(\rho)=e^{-\rho^2}\,,\ \ \rho\geq 0\,, \tag{122}$$

and function $I(\tau)$ describing the change of laser irradiation intensity with time has form

$$I(\tau)=I^*\exp\left[-\gamma(\tau^\delta-\tau_r^\delta)\right]\left(\frac{\tau}{\tau_r}\right)^\beta\,,\ \tau>0\,. \tag{123}$$

Because of the fact that accurate solution of boundary-value problem of heat conduction (118)-(121) for $I(\tau)$ (123) was not found the below method of approximation was applied.

4.2 Laser pulse of rectangular shape

Solution of the axi-symmetrical boundary-value problem of heat conduction (118)-(121) for normal spatial distribution of heat irradiation intensity (122) and constant with time

$$I(\tau)=H(\tau)\,,\ \tau\geq 0\,, \tag{124}$$

has form (Carslaw & Jaeger, 1959):

$$T^{(0)*}(\rho,\zeta,\tau)=\int_0^\infty\varphi(\xi)\Phi(\xi,\zeta,\tau)J_0(\xi\rho)d\xi\,,\ \rho\geq 0,\ \zeta\geq 0,\ \tau\geq 0\,, \tag{125}$$

where

$$\varphi(\xi)=\int_0^\infty\rho q(\rho)J_0(\xi\rho)d\rho=\frac{1}{2}e^{-\frac{\xi^2}{4}}\,,\ \xi\geq 0\,, \tag{126}$$

and function $\Phi(\xi,\zeta,\tau)$ (22).

Dimensionless quasi-static thermal stresses caused in the sem-infinite half-space by the non-stationary temperature field (125), which were achieved with use of the temperature potential methods and Love function (like in 3.2 sub-chapter) have form:

$$\sigma_{ij}^{(0)*}(\rho,\zeta,\tau)=\int_0^\infty\varphi(s)S_{ij}^{(0)}(\xi,\rho,\zeta,\tau)ds - \delta_{ij}T^{(0)*}(\rho,\zeta,\tau)\,,\ \rho\geq 0\,,\ \zeta\geq 0\,,\ \tau\geq 0\,, \tag{127}$$

where

$$\begin{aligned}S_{rr}^{(0)}(\xi,\rho,\zeta,\tau)&=\Phi^\psi(\xi,\zeta,\tau)\left[\frac{\xi}{\rho}J_1(\xi\rho)-\xi^2J_0(\xi\rho)\right]-\\&-\xi^2e^{-\xi\zeta}\Big\{-[(1-\xi\zeta)\Phi^\psi(\xi,0,\tau)+(2-\xi\zeta)\Phi^\psi_{,\zeta}(\xi,0,\tau)]J_0(\xi\rho)-\\&-[(2\nu-1+\xi\zeta)\Phi^\psi(\xi,0,\tau)+(2\nu-2+\xi\zeta)\Phi^\psi_{,\zeta}(\xi,0,\tau)]\frac{J_1(\xi\rho)}{(\xi\rho)}\Big\},\end{aligned} \tag{128}$$

$$S_{\theta\theta}^{(0)}(\xi,\rho,\zeta,\tau)=$$
$$=-\Phi^{\psi}(\xi,\zeta,\tau)\xi\frac{J_1(\xi\rho)}{\rho}-\xi^2e^{-\xi\zeta}\left\{-2\nu[\Phi^{\psi}(\xi,0,\tau)+\Phi^{\psi}_{,\zeta}(\xi,0,\tau)]J_0(\xi\rho)+\right. \tag{129}$$
$$\left.+[(2\nu-1+\xi\zeta)\Phi^{\psi}(\xi,0,\tau)+(2\nu-2+\xi\zeta)\Phi^{\psi}_{,\zeta}(\xi,0,\tau)]\frac{J_1(\xi\rho)}{(\xi\rho)}\right\},$$

$$S_{zz}^{(0)}(\xi,\rho,\zeta,\tau)=\{\Phi^{\psi}(\xi,\zeta,\tau)-e^{-\xi\zeta}[(1+\xi\zeta)\Phi^{\psi}(\xi,0,\tau)+\xi\zeta\Phi^{\psi}_{,\zeta}(\xi,0,\tau)]\}\xi^2J_1(\xi\rho), \tag{130}$$

$$S_{rz}^{(0)}(\xi,\rho,\zeta,\tau)=\{-\Phi^{\psi}_{,\zeta}(\xi,\zeta,\tau)-e^{-\xi\zeta}[\xi\zeta\Phi^{\psi}(\xi,0,\tau)-$$
$$-(1-\xi\zeta)\Phi^{\psi}_{,\zeta}(\xi,0,\tau)]\}\xi^2J_1(\xi\rho)', \tag{131}$$

$$\Phi^{\psi}(\xi,\zeta,\tau)=\left(\tau-\frac{1}{2\xi^2}\right)\Phi^{-}(\xi,\zeta,\tau)-\frac{\zeta}{2\xi}\Phi^{+}(\xi,\zeta,\tau)+\frac{1}{\xi}\sqrt{\frac{\tau}{\pi}}e^{\left[-\left(\xi^2\tau+\frac{\zeta^2}{4\tau}\right)\right]} \tag{132}$$

$$\Phi^{\psi}_{,\zeta}(\xi,\zeta,\tau)=\frac{\zeta}{2\xi}\Phi^{-}(\xi,\zeta,\tau)-\tau\Phi^{+}(\xi,\zeta,\tau), \tag{133}$$

and functions $T^{(0)*}$, $\Phi^{\pm}(\xi,\zeta,\tau)$ and factors δ_{ij} are defined in 3.2 sub-chapter. From solution (127)-(133) on the semi-infinite surface of the body $\zeta=0$ is received as follows: $\sigma_{zz}^{(0)*}(\rho,0,\tau)=\sigma_{rz}^{(0)*}(\rho,0,\tau)=0$. Solution for the rectangular-shape laser pulse:

$$I(\tau)=H(\tau_s-\tau),\ \tau\geq 0, \tag{134}$$

can be written in the form

$$T^{*}(\rho,\zeta,\tau)=T^{(0)*}(\rho,\zeta,\tau)H(\tau)-T^{(0)*}(\rho,\zeta,\tau-\tau_s)H(\tau-\tau_s),\ \rho\geq 0,\ \zeta\geq 0,\ \tau\geq 0, \tag{135}$$

$$\sigma_{ij}^{*}(\rho,\zeta,\tau)=\sigma_{ij}^{(0)*}(\rho,\zeta,\tau)H(\tau)-\sigma_{ij}^{(0)*}(\rho,\zeta,\tau-\tau_s)H(\tau-\tau_s),\ \rho\geq 0,\ \zeta\geq 0,\ \tau\geq 0, \tag{136}$$

where dimensionless temperature $T^{(0)*}$ is determined from formulae (125), (126) and dimensionless thermal stresses $\sigma_{ij}^{(0)*}$ – by using Eqs. (127)-(133).

4.3 Laser pulse of triangular shape

Solution of the axi-symmetrical boundary-value problem of heat conduction (118)-(121) for normal spatial distribution of heat irradiation intensity (122) and linearly changing with time

$$I(\tau)=\tau,\ \tau\geq 0, \tag{137}$$

has form

$$T^{(1)*}(\rho,\zeta,\tau)=\int_0^{\infty}\varphi(\xi)\Phi(\xi,\zeta,\tau)J_0(\xi\rho)d\xi,\ \rho\geq 0,\ \zeta\geq 0,\ \tau\geq 0, \tag{138}$$

where function $\varphi(\xi)$ is defined by Eq. (70), and $\Phi(\xi,\zeta,\tau)=\Phi^{\psi}(\xi,\zeta,\tau)$ (132). Dimensionless quasi-static thermal stresses generated in the semi-infinite surface of the body by the temperature field equal:

$$\sigma_{ij}^{(1)*}(\rho,\zeta,\tau)=\int_0^{\infty}\varphi(s)S_{ij}^{(1)}(\xi,\rho,\zeta,\tau)ds-\delta_{ij}T^{(1)*}(\rho,\zeta,\tau),\ \rho\geq 0,\ \zeta\geq 0,\ \tau\geq 0, \tag{139}$$

where functions $S_{ij}^{(1)}(\xi,\rho,\zeta,\tau)$ in solution (139) are derived from Eqs. (128)-(131) at:

$$\Phi^{\psi}(\xi,\zeta,\tau)=\frac{\zeta}{2\xi}\left(\tau-\frac{3}{4\xi^2}\right)\Phi^{+}(\xi,\zeta,\tau)-\left(\frac{\tau^2}{2}-\frac{\tau}{2\xi^2}+\frac{\zeta}{8\xi^2}+\frac{3}{8\xi^4}\right)\times$$
$$\times\Phi^{-}(\xi,\zeta,\tau)-\left(\frac{\tau}{2}-\frac{3}{4\xi^2}\right)\frac{1}{\xi}\sqrt{\frac{\tau}{\pi}}e^{\left[-\left(\xi^2\tau+\frac{\zeta^2}{4\tau}\right)\right]}, \tag{140}$$

$$\Phi^{\psi}_{,\zeta}(\xi,\zeta,\tau)=\left(\frac{\tau^2}{2}+\frac{\zeta^2}{8\xi^2}\right)\Phi^{+}(\xi,\zeta,\tau)+\frac{\zeta}{2\xi}\left(\frac{1}{4\xi^2}-\tau\right)\Phi^{-}(\xi,\zeta,\tau)-$$
$$-\frac{\zeta}{4\xi^2}\sqrt{\frac{\tau}{\pi}}e^{\left[-\left(\xi^2\tau+\frac{\zeta^2}{4\tau}\right)\right]}. \tag{141}$$

Dimensionless temperature and respective dimensionless thermal stresses generated in the semi-infinite surface of the body by triangle-shape laser pulse can be found as the result of solutions superposition: for the constant (125), (127) and linear (138), (139) laser pulse shape of irradiation intensity:

$$T^{*}(\rho,\zeta,\tau)=\frac{2}{\tau_r}[T^{(1)*}(\rho,\zeta,\tau)-T^{(1)*}(\rho,\zeta,\tau-\tau_r)]-$$
$$-\frac{2}{(\tau_s-\tau_r)}[T^{(1)*}(\rho,\zeta,\tau-\tau_r)-T^{(1)*}(\rho,\zeta,\tau-\tau_s)], \tag{142}$$

$$\sigma_{ij}^{*}(\rho,\zeta,\tau)=\frac{2}{\tau_r}[\sigma_{ij}^{(1)*}(\rho,\zeta,\tau)-\sigma_{ij}^{(1)*}(\rho,\zeta,\tau-\tau_r)]-$$
$$-\frac{2}{(\tau_s-\tau_r)}\sigma_{ij}^{(1)*}(\rho,\zeta,\tau-\tau_r)-\sigma_{ij}^{(1)*}(\rho,\zeta,\tau-\tau_s). \tag{143}$$

4.4 Laser pulse of any shape

In this sub-chapter the laser pulse of any shape is under consideration. Solution of the axi-symmetrical boundary-value problem of heat conduction (118)-(121) and respective thermoelasticity problem for semi-infinite surface of the body at laser pulse of any shape is found by the approximation method with the use of finite functions.

Approximation by piecewise constant functions

Closed interval $<0,\tau>$ will be divided in uniform net of points $\tau_k=k\delta\tau,\ k=0,1,\ldots,n$, gdzie $\delta\tau=\tau/n$. Set the following piecewise constant function in the form:

$$\varphi_k(\tau)=\begin{cases}1,\ \tau\in<\tau_{k-1},\tau_k>,\\0,\tau\notin<\tau_{k-1},\tau_k>,\ k=1,2,\dots,\ n.\end{cases}\tag{144}$$

Function $I(\tau)$ is approximated by the function $\varphi_k(\tau)$ (144) in the form

$$I(\tau)\approx\sum_{k=1}^{n}I(\bar{\tau}_k)\varphi_k(\tau),\ \bar{\tau}_k=\frac{\tau_{k-1}+\tau_k}{2},\ \ \tau\geq 0\cdot\tag{145}$$

The absolute accuracy of approximation given in (145) is around $O(\delta\tau)$. Hence, the solution of non-stationary boundary-value problem of heat conduction (118)-(121) with heat flux intensity of any laser pulse shape $I(\tau)$ can be written:

$$T^{*}(\rho,\zeta,\tau)=\sum_{k=1}^{n}I(\bar{\tau}_k)T_k^{(0)*}(\rho,\zeta,\tau),\ \rho\geq 0,\ \zeta\geq 0,\ \tau\geq 0,\tag{146}$$

where

$$T_k^{(0)*}(\rho,\zeta,\tau)=T^{(0)*}(\rho,\zeta,\tau-\tau_{k-1})-T^{(0)*}(\rho,\zeta,\tau-\tau_k),\tag{147}$$

and dimensionless temperature $T^{(0)*}$ is derived according to Eqs. (125), (126). Field of dimensionless thermal stresses caused in semi-infinite surface of the body by the temperature field (146), (147) is found in analogous way:

$$\sigma_{ij}^{*}(\rho,\zeta,\tau)=\sum_{k=1}^{n}I(\bar{\tau}_k)\sigma_{ij,k}^{(0)*}(\rho,\zeta,\tau),\ \rho\geq 0,\ \zeta\geq 0,\ \tau\geq 0,\tag{148}$$

$$\sigma_{ij,k}^{(0)*}(\rho,\zeta,\tau)=\sigma_{ij}^{(0)*}(\rho,\zeta,\tau-\tau_{k-1})-\sigma_{ij}^{(0)*}(\rho,\zeta,\tau-\tau_k),\tag{149}$$

and dimenionless stresses $\sigma_{ij}^{(0)*}$ are derived from Eqs. (127)-(133).

Approximation by piecewise linear functions

It is assumed that for the same time interval $<0,\tau>$ the identical uniform net of points as above is used. Set the following piecewise linear function in the form:

$$\varphi_0(\tau)=\begin{cases}\dfrac{(\tau_1-\tau)}{\delta\tau},\ \tau\in<\tau_0,\tau_1>,\\0,\ \tau\notin<\tau_0,\tau_1>,\end{cases}\qquad \varphi_k(\tau)=\begin{cases}\dfrac{(\tau-\tau_{k-1})}{\delta\tau},\ \tau\in<\tau_{k-1},\tau_k>,\\\dfrac{(\tau_{k+1}-\tau)}{\delta\tau},\ \tau\in<\tau_k,\tau_{k+1}>,\\0,\ \tau\notin<\tau_{k-1},\tau_{k+1}>,\ k=1,2,\dots,n-1\end{cases}\tag{150}$$

$$\varphi_n(\tau)=\begin{cases}\dfrac{(\tau-\tau_{n-1})}{\delta\tau},\ \tau\in<\tau_{n-1},\tau_n>,\\0,\ \tau\notin<\tau_{n-1},\tau_n>.\end{cases}$$

Thus the approximation of $I(\tau)$ is done by the following subtotal

$$I(\tau) \approx \sum_{k=0}^{n} I(\tau_k)\varphi_k(\tau), \quad \tau > 0. \tag{151}$$

Absolute approximation error (151) has order of $O(\delta\tau^2)$ (Marchuk & Agoshkov, 1981). Hence the final solution will have form:

$$T^*(\rho,\zeta,\tau) = \frac{1}{\delta\tau}\sum_{k=0}^{n} I(\tau_k)T_k^{(1)*}(\rho,\zeta,\tau), \ \rho \ge 0, \ \zeta \ge 0, \ \tau \ge 0, \tag{152}$$

where

$$T_0^{(1)*}(\rho,\zeta,\tau) = \tau_1 T^{(0)*}(\rho,\zeta,\tau) - T^{(1)*}(\rho,\zeta,\tau) + T^{(1)*}(\rho,\zeta,\tau-\tau_1), \tag{153}$$

$$T_k^{(1)*}(\rho,\zeta,\tau) = [T^{(1)*}(\rho,\zeta,\tau-\tau_{k-1}) - 2T^{(1)*}(\rho,\zeta,\tau-\tau_k) + T^{(1)*}(\rho,\zeta,\tau-\tau_{k+1}), \quad k = 1,2,\dots,n-1 \tag{154}$$

$$T_n^{(1)*}(\rho,\zeta,\tau) = [T^{(1)*}(\rho,\zeta,\tau-\tau_{n-1}) - T^{(1)*}(\rho,\zeta,\tau-\tau_n)] - (\tau_n - \tau_{n-1})T^{(0)*}(\rho,\zeta,\tau-\tau_n), \tag{155}$$

and dimensionless temperatures $T^{(0)*}$ and $T^{(1)*}$ can be derived from Eqs. (125) and (138) respectively.
Analogous quasi-static thermal stresses can be found as:

$$\sigma_{ij}^*(\rho,\zeta,\tau) = \frac{1}{\delta\tau}\sum_{k=0}^{n} I(\tau_k)\sigma_{ij,k}^{(1)*}(\rho,\zeta,\tau), \tag{156}$$

where

$$\sigma_{ij,0}^{(1)*}(\rho,\zeta,\tau) = \tau_1\sigma_{ij}^{(0)*}(\rho,\zeta,\tau) - \sigma_{ij}^{(1)*}(\rho,\zeta,\tau) + \sigma_{ij}^{(1)*}(\rho,\zeta,\tau-\tau_1) \tag{157}$$

$$\sigma_{ij,k}^{(1)*}(\rho,\zeta,\tau) = [\sigma_{ij}^{(1)*}(\rho,\zeta,\tau-\tau_{k-1}) - 2\sigma_{ij}^{(1)*}(\rho,\zeta,\tau-\tau_k) + \sigma_{ij}^{(1)*}(\rho,\zeta,\tau-\tau_{k+1}), \quad k = 1,2,\dots,n-1 \tag{158}$$

$$\sigma_{ij,n}^{(1)*}(\rho,\zeta,\tau) = [\sigma_{ij}^{(1)*}(\rho,\zeta,\tau-\tau_{n-1}) - \sigma_{ij}^{(1)*}(\rho,\zeta,\tau-\tau_n)] - (\tau_n - \tau_{n-1})\sigma_{ij}^{(0)*}(\rho,\zeta,\tau-\tau_n), \tag{159}$$

and dimensionless thermal stresses $\sigma_{ij}^{(0)*}$ and $\sigma_{ij}^{(1)*}$ can be derived from Eqs. (127) and (139) respectively.

4.5 Numerical analysis and conclusions

Determination of non-stationary temperature fields and quasi-static thermal stresses fields were done for laser irradiation of semi-infinite surface of the body with the use of laser pulse shape described by the function $I(\tau)$. It was assumed the Poisson coefficient had value of $\nu = 0{,}3$, and number of components in subtotals (145) and (151) was chosen from accuracy defined condition. Evolution of dimensionless temperature $T^* = T/T_0$ in defined points on the semi-infinite surface of the body $\zeta = 0$ is shown on Fig. 17 and along symmetry axis $\rho = 0$ on Fig. 18.

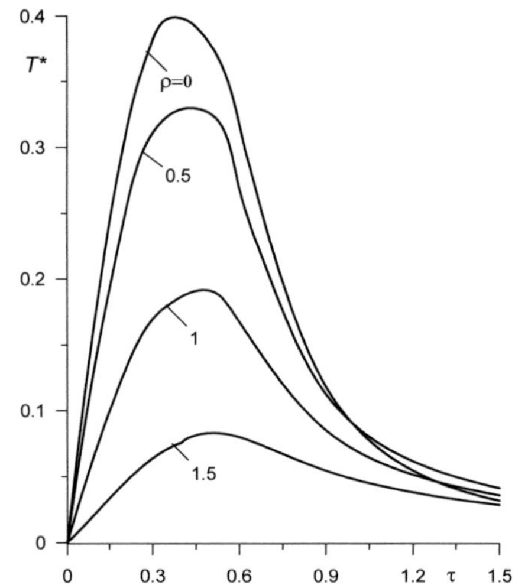

Fig. 17. Evolution of dimensionless temperature T^* on the laser irradiated semi-infinite surface of the body $\zeta = 0$ for different values of radial variable ρ (Yevtushenko & Matysiak, 2005).

Temperature in the centre of heated area ($\rho = 0, \zeta = 0$) reaches maximum value at the moment $\tau_r = 0.27$, when the laser irradiation intensity is the highest. After that, the cooling process begins as a result of decrease of laser irradiation intensity with time. With the distance from the heated centre area dimensionless time $\tau_{\max}$ of maximum temperature increases: for the values $\rho = 0.5; 1; 1.5$ equals $\tau_{\max} = 0.4; 0.48; 0.51$, respectively (see Fig. 17). Simultaneously with the dimensionless distance ζ from laser irradiated surface of the body, time of reaching the maximum temperature increases, too: for the values $\zeta = 0.1; 0.25; 0.5$ equals $\tau_{\max} = 0.1; 0.25; 0.5$, respectively (see Fig. 18). After switching laser system off ($\tau \geq 1$), temperature along symmetry axis decreases to its starting value.

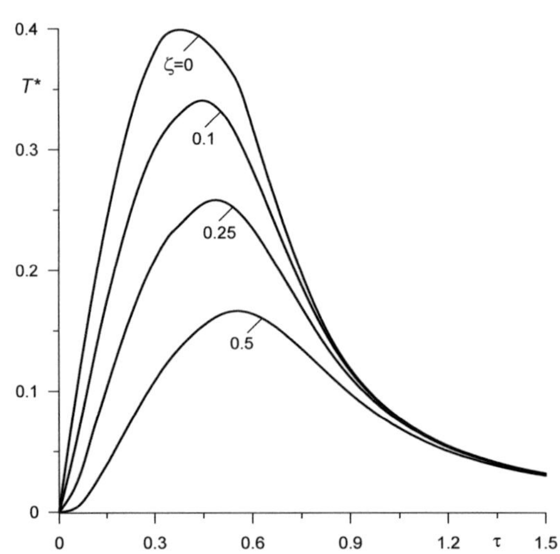

Fig. 18. Evolution of dimensionless temperature T^* along symmetry axis $\rho = 0$ for different values of dimensionless variable ζ (Yevtushenko & Matysiak, 2005).

Evolution with time of dimensionless thermal stresses $\sigma_{ij}^* = \sigma_{ij} / \sigma_0$ is shown on Fig. 19. Evolution of thermal radial stresses σ_{rr}^* and peripheral $\sigma_{\theta\theta}^*$ in the chosen four points in the distance $\zeta = 0.5$ from laser irradiated surface are very similar in nature (see Figs. 19, 20). Since switching the laser system on to the moment when $\tau_r = 0.27$, stresses are stretching and afterward change their sign (become compressive one), then their absolute value significantly increases. Maximum value of these stresses is achieved on symmetry axis $\rho = 0$ in time $\approx 2\tau_r$.
In the starting moment of laser irradiation action , the dimensionless normal stresses σ_{zz}^* is stretching but close to the moment of laser system switched off become compressive innature (see Fig. 21).

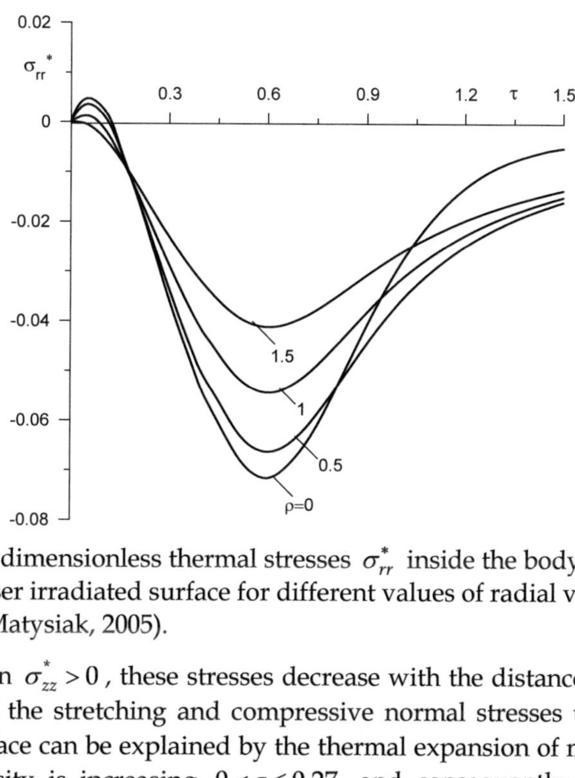

Fig. 19. Evolution of dimensionless thermal stresses σ_{rr}^* inside the body $\zeta = 0.5$ with the distance from the laser irradiated surface for different values of radial variable ρ (Yevtushenko & Matysiak, 2005).

At the moment when $\sigma_{zz}^* > 0$, these stresses decrease with the distance from the symmetry axis. Appearance of the stretching and compressive normal stresses underneath the laser irradiated body surface can be explained by the thermal expansion of material in the period of irradiation intensity is increasing $0 < \tau \leq 0.27$ and consequently by the compressing during the cooling process when $\tau > 0.27$.
Dimensionless shear stresses σ_{rz}^* are negative during almost all the heating interval and become positive after the laser system is switched off. It should be underlined that absolute value of shear stresses increases with the distance from symmetry axis $\rho = 0$.
All the tensor components of stresses have insignificant values when $\tau > 5$. Distribution of dimensionless radial stresses σ_{rr}^* and normal σ_{zz}^* along symmetry axis $\rho = 0$ for different dimensionless time values is shown on Fig. 22, 23.
Maximum value of compressive stresses σ_{rr}^* is reached when laser irradiation intensity is the highest ($\tau = 0.27$) (Fig. 22). With the distance from irradiated surface, larger then two radius of laser beam, the radial stresses negligibly small. Normal stresses σ_{zz}^* equal zero on

the laser irradiated surface $\zeta = 0$ and increase with the distance from the semi-infinite surface of the body when finally reach some maximum value (see Fig. 23). These stresses are stretching when laser system is operating and become compressive when laser system is off.

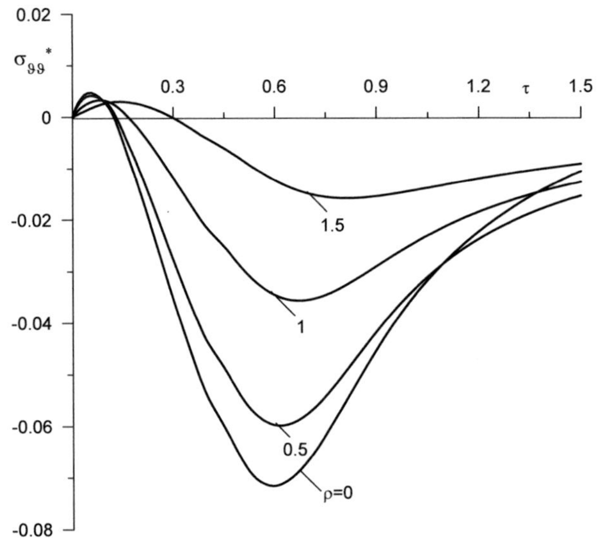

Fig. 20. Evolution of dimensionless thermal stresses $\sigma_{\theta\theta}^*$ inside the body $\zeta = 0.5$ with the distance from the laser irradiated surface for different values of radial variable ρ (Yevtushenko & Matysiak, 2005).

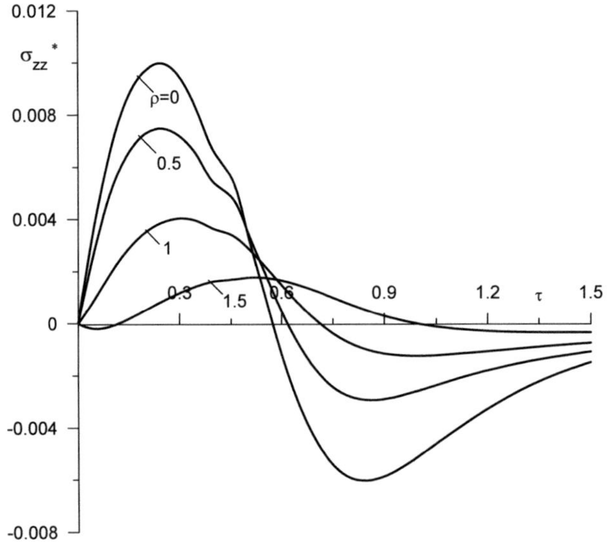

Fig. 21. Evolution of dimensionless thermal stresses σ_{zz}^* inside the body $\zeta = 0.5$ with the distance from the laser irradiated semi-infinite surface for different values of radial variable ρ (Yevtushenko & Matysiak, 2005).

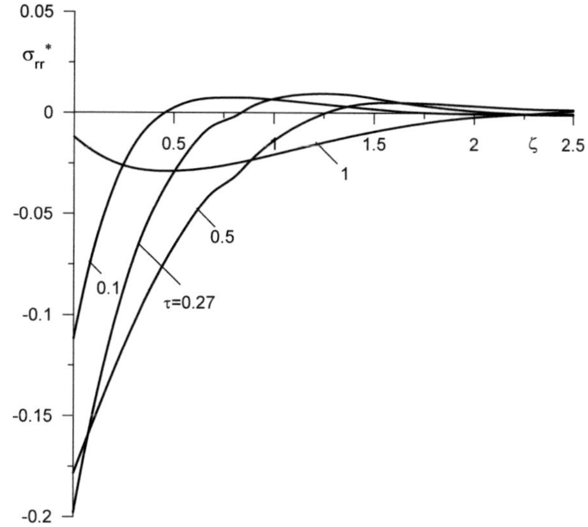

Fig. 22. Evolution of dimensionless thermal stresses σ_{rr}^* along symmetry axis $\rho = 0$ for different dimensionless time values (Yevtushenko & Matysiak, 2005).

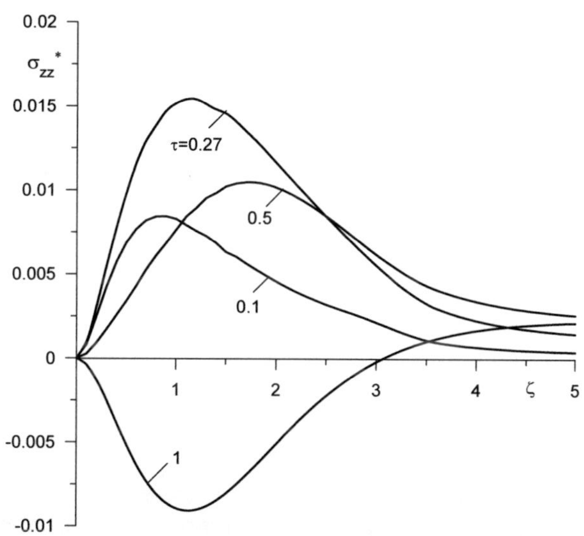

Fig. 23. Evolution of dimensionless thermal stresses σ_{zz}^* along symmetry axis $\rho = 0$ for different dimensionless time values (Yevtushenko & Matysiak, 2005).

5. References

Abramowitz, M. & Stegun, I.A. Handbook of Mathematical Functions with Formulas, Graphs and Mathematical Tables, Wiley, New York, 1972, pp. 830.

Ashcroft, N. W. & Mermin, N. D. Solid state physics, Warsaw: PWN, 1986.

Aulyn, V. et al. Розвиток і використання макро- та мікроневрівноважних процесів у матеріалах при зміцненні й відновленні деталей лазерними технологіями, Mashinoznavstvo, 3 (2002), 31-37.

Bardybahin, A.I. & Czubarov, Y.P. Influence of local irradiation intensity distribution in a plane normal to the laser beam axis on maximal temperature for the thin plate, Fizika i Chimia Obrabotki Materialov 4 (1996) 27-35.

Carslaw, H.S. & Jaeger, J.C. Conduction of heat in solids, Oxford: 2nd ed. Clarendon Press, 1959.

Griffith A.A. The theory of rupture, Proc. 1-st Int. Congress of Appl. Mech., Delft, 1924, (Delft. Waltmar) 1926, p. 55.

Hector, L.G. & Hetnarski, R.B. Thermal stresses in materials due to laser heating, in: R.B. Hetnarski, Thermal Stresses IV, Elsevier Science Publishers B.V., 1996, pp. 453-531.

Lauriello, P.J. & Chen, Y. Thermal fracturing of hard rock, Trans. ASME. J. Appl. Mech., 1973, vol. 40, no. 4, p. 909.

Marchuk, G.I. & Agoshkov, V.I. Introduction to Project-Mesh Methods (in Russian), Moskwa: Nauka, 1981.

Matysiak, S.J. et al. Temperature field in a microperiodic two-layered composite caused by a circular laser heat source, Heat Mass Tr., 1998, vol. 34, no. 1, p. 127.

McClintock F.A. & Walsh J.B. Friction on Griffith cracks under pressure, Proc. 4-th U.S. Congress of Appl. Mech., Berkeley, 1962, p. 1015.

Nowacki, W. Thermoelasticity, Oxford: Pergamon Press, 1986.

Prudnikov, A.P. et al. Integrals and series. Vol. 2. Special Functions, New York-London: Taylor & Francis, 1998, pp. 800.

Ready, J.F. Effects of high-power laser radiation, Academic Press, New York-London, 1971.

Rozniakowski, K. Application of laser radiation for examination and modification of building materials properties, BIGRAF, Warsaw, 2001, p. 198.

Rożniakowski, K. et al. Laser-induced temperature field and thermal stresses in the elstic homogeneous material, Materials Science, 2003, vol. 39, no. 3, p. 385-393.

Rykalin, N.N. et al. Laser processing of materials, (in Russian), Mashinostroenie, Moscow, 1975, pp. 296.

Sneddon, I.N. The use of integral transforms, New York: McGraw-Hill, 1972.

Timoshenko, S.P. & Goodier, J.N. Theory of Elasticity, New York: McGraw-Hill, 1970.

Yevtushenko A.A. et al. Evaluation of effective absorption coefficient during laser irradiation using of metals martensite transformation, Heat Mass Tr., 2005, vol. 41, p. 338.

Yevtushenko A.A. et al. Temperature and thermal stresses due to laser irradiation on construction materials (in Polish). Monograph, Bialystok: Technical University of Bialystok, 2009.

Yevtushenko, A.A. et al. On the modelling of laser thermal fracturing of hard rock, Engng. Trans., 1997, vol. 45, no. 3/4, p. 447.

5

Principles of Direct Thermoelectric Conversion

José Rui Camargo and Maria Claudia Costa de Oliveira
University of Taubaté
Brazil

1. Introduction

The aim of this chapter is to present some fundamental aspects of the direct thermoelectric conversion. Thermoelectric systems are solid-state heat devices that either convert heat directly into electricity or transform electric power into thermal power for heating or cooling. Such devices are based on thermoelectric effects involving interactions between the flow of heat and electricity through solid bodies. These phenomena, called Seebek effect and Peltier effect, can be used to generate electric power and heating or cooling.

The Seebeck effect was first observed by the physician Thomas Johann Seebeck, in 1821, when he was studying thermoelectric phenomenon. It consists in the production of an electric power between two semiconductors when submitted to a temperature difference. Heat is pumped into one side of the couples and rejected from the opposite side. An electrical current is produced, proportional to the temperature gradient between the hot and cold sides. The temperature differential across the converter produces direct current to a load producing a terminal voltage and a terminal current. There is no intermediate energy conversion process. For this reason, thermoelectric power generation is classified as direct power conversion.

On the other hand, a thermoelectric cooling system is based on an effect discovered by Jean Charles Peltier Athanasius in 1834. When an electric current passes through a junction of two semiconductor materials with different properties, the heat is dissipated and absorbed.

This chapter consists in eight topics. The first part presents some general considerations about thermoelectric devices. The second part shows the characteristics of the physical phenomena, which is the Seebeck and Peltier effects. The thirth part presents the physical configurations of the systems and the next part presents the mathematical modelling of the equations for evaluating the performance of the cooling system and for the power generation system. The parameters that are interesting to evaluate the performance of a cooling thermoelectric system are the coefficient of performance (COP), the heat pumping rate and the maximum temperature difference that the device will produce. It shows these parameters and also the current that maximizes the coefficient of performance, the resultant value of the applied voltage which maximizes the coefficient of performance and the current that maximizes the heat pumping rate. To evaluate the power generator performance it is presented the equations to calculate the efficiency and the power output, as well as the operating design that maximizes the efficiency, the optimum load and the load resistance that maximizes the power output. The last part of the chapter presents the selection of the proper module for a specific application. It requires an evaluation of the total system in which the thermoelectric module will be used. The overall system is dynamic and its performance is a function of several interrelated parameters, such as: the operation temperatures, the ambient temperature, the available space, the available power, among others. Finally it presentes conclusions, aknowledgment and references.

Thermoelectric modules consists of an array of *p*-type and *n*-type semiconductors elements that are heavily doped with electrical carriers. The elements are arranged into an array that is electrically connected in series but thermally connected in parallel. This array is then affixed to two ceramic plates, one on each side of the elements, that is, one covers the hot joins and the other covers the cold one.

Thermoelectric devices offer several advantages over other technologies: the absence of moving components results in an increase of reliability, a reduction of maintenance, and an increase of system life; the modularity allows for the application in a wide-scale range without significant losses in performance; the absence of a working fluid avoids environmentally dangerous leakages; and the noise reduction appears also to be an important feature.

Applications for thermoelectric modules cover a wide spectrum of product areas. These include equipment used by military, medical, industrial, consumer, scientific/laboratory, and telecommunication´s organizations. It includes a range from simple food and beverage coolers for an afternoon picnic to extremely sophisticated temperature control systems in missiles and space vehicles. Typical applications for thermoelectric modules include: avionics, calorimeters, cold chambers, cold plates, compact heat exchangers, constant temperature baths, dehumidifiers, dew point hygrometers, electronics package cooling, environmental analyzers, heat density measurement, immersion coolers, integrated circuit cooling, infrared detectors, infrared seeking missiles, microprocessor cooling, power generators, refrigerators and on-board refrigeration systems (aircraft, automobile, boat, hotel, among others).

Fig. 1. Thermoelectric modules and heat sinks

Figure 1 shows thermoelectric modules and heat sinks commercially available.
A unique aspect of thermoelectric energy conversion is that the direction of energy flow is reversible. So, for instance, if the load resistor is removed and a DC power supply is substituted, the thermoelectric device can be used to draw heat from the "heat source" element and decrease its temperature. In this configuration, the reversed energy-conversion process of thermoelectric devices is invoked, using electrical power to pump heat and produce refrigeration. This reversibility distinguishes thermoelectric energy converters from many other conversion systems. Electrical input power can be directly converted to pumped thermal power for heating or refrigerating, or thermal input power can be converted directly to electrical power for lighting, operating electrical equipment, and other work. Any thermoelectric device can be applied in either mode of operation, though the design of a particular device is usually optimized for its specific purpose.

2. The Peltier and Seebeck effects

The name "thermoelectricity" indicates a relationship between thermal and electrical phenomena. The concepts of heat, temperature and thermal balance are among the most fundamental and important to the science. Two objects are considered to be in thermal equilibrium if the exchange of heat does not exist when they both are placed in contact. This is an experimental fact. Objects in the same temperature are said to be in thermal equilibrium. This is called zeroth law of thermodynamics.
Two objects at different temperatures placed in contact exchange energy in an attempt to establish thermal equilibrium. Any work done during this process is the difference of heat lost by an object and won by another object. This is the first law of thermodynamics, in other words, energy is always conserved.
The concepts of electric charge and electric potential are also essentials. Objects are composed of positive and negative charges. Opposite electric charges attract each other and equal charges repel. These are experimental facts. Objects are said to be in electric equilibrium if there is no heat exchange when they are placed in contact. Such objects are said to be at the same electrical potential. Objects with different electrical potential exchange charges in an attempt to achieve the same electrical potential.
The electric current is the amount of electric charges which pass through a boundary of a conductor per unit time and it is related to the variation of the electrical potential, in other words, the electrical gradient. Similarly, the heat flow is the amount of heat that passes through the boundary per unit of time. Likewise, the thermal flow is related to temperature variation, in other words, to the electrical gradient.
To understand the thermoelectric effect it is needed to visualize the phenomenon in a micro scale. In the nature, the materials are made of molecules composed by atoms. Depending on the kind of interlace between the atoms, the outer electrons are more or less likely of moving around the nucleus and other electrons.
In the purer metallic conductors outer electrons, less connected to others, can move freely around all the material, as if they do not belong to any atom. These electrons transmit energy one to another through temperature variation, and this energy intensity varies depending on the nature of the material.
For this reason, if two distinct materials are placed in contact, free electrons will be transferred from the more "loaded" material to the other, so they equate themselves, such transference creates a potential difference, called contact potential, since the result will be a

pole negatively charged by the recieved electrons and another positively charged by the loss of electrons.
The following sequence of metals shows, from left to right, which is more likely to lose electrons:
(+) Rb K Na Al Zn Pb Sn Sb Bi Fe Cu Ag Au Pt (-)
In the case of semi-conductors, the transference occurs because some of the atoms that compose it are already lacking some electrons. When voltage is applied, there is a tendency to drive electrons and complete the atomic orbit. When it occurs, the atomic conduction leaves "holes" that are essentially atoms with crystalline grids that now have positive local charge. The electrons are, then, continuously drown out of the holes moving towards the next hole available. In fact, the embezzlement of these atoms is what drives the current.
Electrons move more easily in copper conductors than in semiconductors. When electrons leave the p element and entering the cold side of the copper, holes are created in the p type as the electrons go to a higher level of energy to reach an energy level of electrons that are already moving in the copper. The extra energy to create these holes come from the absorption of heat. Meanwhile, the newly-created holes move throughout the copper in the hot side. The hot side electrons of the copper move to the p element and complete the holes, releasing energy generated as heat.
The n-type conductor is doped with atoms which provide more electrons than the ones necessary to complete the atomic orbits within the crystalline grids. When the voltage is applied, these extra electrons move easily to the conduction band. However, additional energy is necessary so that the n-type electrons reach the next energy level of electrons arriving from the cold side of the copper. This extra energy comes from the heat absorbed. Finally, when the electrons leave the hot side of the n-type element, they can move freely again throughout the copper. They fall to a lower energy level, releasing heat in the process.
The information above do not cover all the details, but they can explain complex physical interactions. The main point is that the heat is always absolved in the cold side of the elements p and n, and the heat is always released in the hot side of the thermoelectric element. The pumping capacity of the module heat is proportional to electric current and depends on the geometry of the element, the number of pairs and the properties of the material.
It is also possible to form a more conductive crystal by adding impurities with less valence electron. For instance, Indium impurities (which have 3 valence electrons) are used in combination with silicon and create a crystalline structure with holes. These holes make it easier to transport electrons throughout the material when the voltage applying a voltage. In this case, the holes are considered load conductors in this conductor "positively doped" which is referred as **p**-type.

3. Thermoelectric system configurations

3.1 Thermoelectric cooling device

The pairs of thermoelectric cooling are made of two semiconductors elements, frequently made of bismuth telluride highly doped to create an excess (**n**-type) or a deficiency of electrons (**p**-type). The heat absorbed at the cold junction is transferred to the hot junction at a rate proportional to the current passing through the circuit and the number of semiconductors pairs. In practice, pairs are combined into a module which they are electrically connected in series and thermally in parallel.

A thermoelectric device consists of several **n** and **p** pellets connected electrically in series and thermally in parallel sandwiched between two ceramic plates. When the thermoelectric module is operating as a refrigerator, the bottom plate is bonded to a heat sink and, with the application of DC current of proper polarity; heat is pumped from the top plate to the bottom plate and into the heat sink, where it is dissipated to ambient. The resultant is that the top surface becomes cold. The top surface can also supply heat by simply reversing DC polarity. The same unit can be converted into a thermoelectric power generator by simply replacing the DC source with the load, or item to receive power, and apply heat to the top surface of the thermoelectric modules. Electrical power is derived from the movement of electrical carriers brought on by heat flow through the thermoelectric pellets. Holes, or positive carriers, move to the heat sink side of the **p**-type pellet making that junction electrically positive. Similarly, electron flow in the **n**-type pellets results in a net negative charge at the heat sink side of the **n**-type pellet.

A heat sink is a device that is attached to the hot side of thermoelectric module. It is used to facilitate the transfer of heat from the hot side of the module to the ambient. A cold sink is attached to the cold side of the module. It is used to facilitate heat transfer from whatever is being cooled (liquid, gas, solid object) to the cold side of the module. The most common heat sink (or cold sink) is an aluminum plate that has fins attached to it. A fan is used to move ambient air through the heat sink to pick up heat from the module.

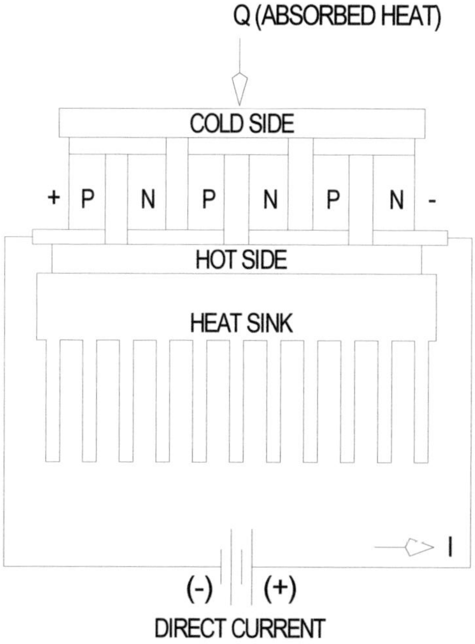

Fig. 2. Schematic of a Peltier effect (thermoelectric cooling device)

Figure 2 shows the configuration of a typical thermoelectric system that operates by the Peltier effect. The goal in this design is to collect heat from the volume of air and transfer it to an external heat exchanger and on to the external environment. It is usually done using two combinations of fan and heat sink together with one or more thermoelectric modules. The smallest sink is used together with the volume to be cooled, and cooled to a temperature lower than the volume, so using a fan the heat that passes between the fins can be collected. In its typical configuration, the insert is installed between the hot and the cold side of the sink.

When a DC current passes through the module, it transfers heat from the cold side to the hot side. At the same time, the fan in the hot side will be circulating in the ambient air the heat transferred to the heat sink fins of the hot side. It is noteworthy that the heat dissipated in the hot side does not include only the heat transferred by the application, but also the heat generated inside the module (V x I).

The heat sink transfers the heat like a steam cycle compressor system. For both, heating or cooling, it is necessary to use a sink to collect heat (heating mode) or dissipate heat (cooling mode) to the outside. Without it, the module is subject to overheating, with the hot side overheated the cold side also heats, consequently heat will not be transferred anymore. When the module reaches the temperature of reflow of the solder used, the unit will be destroyed. So a fan is always used as a heat sink to exchange heat with the external environment.

3.2 Thermoelectric power generator device

Figure 3 shows the configuration of a typical thermoelectric system that operates by the Seebeck effect.

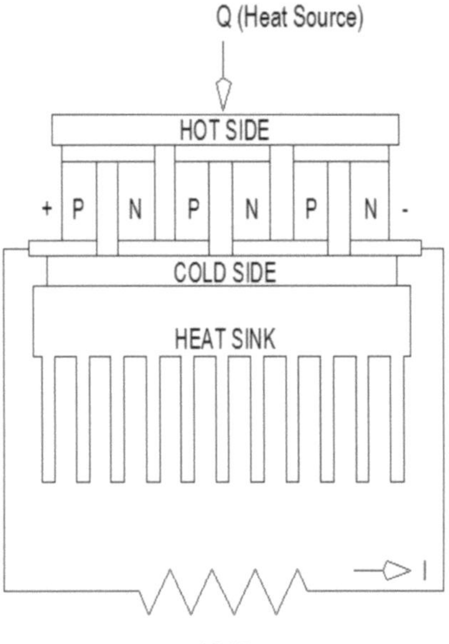

Fig. 3. Schematic of a thermoelectric power generator

There are some important practical considerations that should be made before attempting to use thermoelectric coolers in the power generation mode. Perhaps the most important consideration is the question of survivability of the module at the anticipated maximum temperature. Many standard thermoelectric cooling modules are fabricated with eutectic Bi/Sn solder which melts at approximately 138°C. However, there are some coolers being offered employing higher temperature solders designed to operate at temperatures of 200°C, even approaching 300°C. In any case, consideration should be given to operational lifetime of a thermoelectric module exposed to high temperatures. Contaminants or even constituents of the solder can rapidly diffuse into the thermoelectric material at high temperatures and degrade performance and, in extreme cases, can cause catastrophic failure. This process can be controlled by the application of a diffusion barrier onto the TE material. However, some manufactures of thermoelectric coolers employ no barrier material at all between the solder and the TE material. Although application of a barrier material is generally standard on the high temperature thermoelectric cooling modules manufactured, they are mostly intended for only short-term survivability and may or may not provide adequate MTBF´s (Mean Time Between Failures) at elevated temperatures. In summary, if one expects to operate a thermoelectric cooling module in the power generation mode, qualification testing should be done to assure long-term operation at the maximum expected operating temperature.

4. Mathematical modelling

4.1 Peltier effect

The parameters that are interesting to evaluate the performance of a cooling device are the coefficient of performance (φ), the heat pumping rate (Q_c) and the maximum temperature difference (ΔT_{max}) that the device will produce.
The coefficient of performance (COP) φ is defined as

$$\varphi = \frac{Q_c}{P} \tag{1}$$

where Q_c is the heat pumping rate from the cold side and P is the electrical power input. The "cooling effect" or "thermal load" is the heat pumping rate from the cold side and it is the sum of three terms: a) the Joule heat of each side per time unit, b) the heat transfer rate when current is equal to zero between the two sides and c) the Peltier heat rate of each side, that is, the heat removal rate is

$$Q_c = \alpha T_c I - \frac{1}{2} I^2 R - K\Delta T \tag{2}$$

where: $\alpha = (\alpha_p + \alpha_n)$ and α_p and α_n are properties of the semiconductors materials, I is the current, R is the electric resistance (Ω) and K is the total thermal conductance of the thermoelectric cooling module. The power input is

$$P = V.I = \alpha.I.\Delta T + I^2.R = \frac{V(V - \alpha.\Delta T)}{R} \tag{3}$$

where $\Delta T = (T_h - T_c)$ and T_h and T_c are the hot and cold sides temperatures. V is the applied voltage and is the sum of the electric and the Joule voltage.

$$V = \alpha \Delta T + IR \tag{4}$$

The coefficient of performance of the couple with this optimum geometry is:

$$\varphi = \frac{\left[mT_c - \frac{1}{2} m^2 - \left(\frac{\Delta T}{Z} \right) \right]}{\left(m\Delta T + m^2 \right)} \tag{5}$$

where Z is called the figure of merit of the thermoelectric association, defined by

$$Z = \frac{\alpha^2}{[(\rho_n K_n)^{1/2} + (\rho_p K_p)^{1/2}]^2} \tag{6}$$

where $m = \frac{IR}{\alpha}$

The coefficient of performance is strongly influenced by the figure of merit of the semiconductor material.
The current that maximizes the coefficient of performance is obtained by taking the derivative of the coefficient of performance with respect to m equal to zero and is

$$I_{ot} = \frac{\alpha \Delta T}{R(w-1)} \tag{7}$$

where

$$w = \left(1 + Z\overline{T}\right)^{\frac{1}{2}} \tag{8}$$

The maximum coefficient of performance is given by

$$\varphi_{\max} = \frac{T_c}{\Delta T}\left[\frac{w - \left(\frac{Th}{T_c}\right)}{(w+1)}\right] \tag{9}$$

The resultant value of the applied voltage which maximizes the coefficient of performance is

$$V_{ot} = \frac{\alpha \Delta T w}{(w-1)} \tag{10}$$

The power input is given by

$$P_{ot} = \left(\frac{w}{R}\right)\left[\frac{\alpha \Delta T}{(w-1)}\right]^2 \tag{11}$$

The current that maximizes the heat pumping rate is given by

$$I_{ot} = \frac{\alpha T_c}{R} \tag{12}$$

Thus the maximum heat pumping rate at this current is calculated by

$$Q_{c\max} = \left(\frac{\alpha^2 T_c^2}{2R}\right) - K\Delta T \tag{13}$$

4.2 Seebeck effect

The important design parameters for a power generator device are the efficiency and the power output. The efficiency is defined as the ratio of the electrical power output. The efficiency is defined as the ratio of the electrical power output P_o to the thermal power input q_h to the hot junction

$$\eta = \frac{P_o}{q_h} \tag{14}$$

The power output is the power dissipated in the load. The thermal power input to the hot junction is given by

$$q_h = \alpha T_h I + \frac{1}{2} I^2 R + K\Delta T \tag{15}$$

where: α is the Seebeck coefficient, T_h is the hot side temperature of the thermoelectric module, I is the current, R is the electric resistance (Ω) , K is the total thermal conductance of the thermoelectric cooling module and ΔT is temperature difference between hot and cold sides (T_h - T_c.). In the discussion of power generators, the positive direction for the current is from the p parameter to the n arm at the cold junction. The electrical power output is

$$P_0 = I^2 R_L = V I \tag{16}$$

where R_L is the load resistance. The current is given by

$$I = \frac{\alpha \Delta T}{(R + R_L)} \tag{17}$$

Since the open-circuit voltage is $\alpha \Delta T$. Thus the efficiency is

$$\eta = \frac{I^2 R_L}{\left(\alpha T_h I + \frac{1}{2} I^2 R + K\Delta T\right)} \tag{18}$$

The operating design, which maximizes the efficiency, will now be calculated. Let's take $S = R_L / R$. The efficiency is

$$\eta = \frac{\left(\frac{\Delta T}{T_h}\right) S}{\left\{ (1+S) - \left(\frac{\Delta T}{2T_h}\right) + \left[\frac{(1+S)^2 RK}{\alpha^2 T_h}\right]\right\}} \tag{19}$$

Again it is seen that the efficiency will be a maximum for RK minimized. Hence, the shape ratio, which maximizes the efficiency, is given by $\frac{\gamma_n}{\gamma_p} = \left(\frac{\rho_n k_p}{\rho_p k_n}\right)^{1/2}$. With this shape ratio the efficiency is

$$\eta = \frac{\left(\frac{\Delta T}{T_h}\right)s}{\left\{\left(1+s\right)-\left(\frac{\Delta T}{2T_h}\right)+\left[\frac{\left(1+s\right)^2}{zT_h}\right]\right\}} \tag{20}$$

The optimum load is calculated by setting the derivative of the efficiency with respect to s equal to zero. The efficiency with both the geometric and load resistance optimized is

$$\eta = \frac{\left(\Delta T/T_h\right)\left(\omega+1\right)}{\left[\omega+\left(T_c/T_h\right)\right]} \tag{21}$$

Under optimum load, the output current is

$$I = \frac{\alpha\Delta T}{R\left(\omega+1\right)} \tag{22}$$

and the output voltage is

$$V = \frac{\alpha\Delta T_\omega}{\left(\omega+1\right)} = \alpha\left(\Delta T\right) - IR \tag{23}$$

and the output power is

$$P_o = \left(\frac{\omega}{R}\right)\left[\frac{\alpha\Delta T}{\left(\omega+1\right)}\right] \tag{24}$$

The internal resistance R is the same as for a refrigerator and given by

$$R = \left(\frac{\alpha}{z^{1/2}}\right)\left(\frac{1}{\gamma_p}\right)\left(\frac{\rho_p}{k_p}\right)^{1/2} = \left(\frac{\alpha}{z^{1/2}}\right)\left(\frac{1}{\gamma_n}\right)\left(\frac{\rho_n}{k_n}\right)^{1/2} \tag{25}$$

or approximately by

$$R = \left(2L/A_T\right)\left(\rho_n + \rho_p\right) \tag{26}$$

In the previous equations, the load resistance and the shape ratio were adjusted to maximize the efficiency. In this section, these parameters will be selected to maximize the power output. The load resistance, which maximizes the power output, is obtained by setting equal to zero the derivative with respect to the load resistance of the power output given by Eqs. (18) and (19). The well-known result $R_L = R$ is obtained. With this load resistance, the output voltage is

$$V = \frac{1}{2}\alpha\Delta T \tag{27}$$

The current is

$$I = \frac{\alpha \Delta T}{2R} \tag{28}$$

and the power output is

$$P_o = \frac{(\alpha \Delta T)^2}{4R} \tag{29}$$

5. Selection and design

Selection of the proper thermoelectric module for a specific application requires an evaluation of the total system in which the cooler will be used. For most applications it should be possible to use one of the standard module configurations while in certain cases a special design may be needed to meet stringent electrical, mechanical, or other requirements. Although we encourage the use of a standard device whenever possible, Ferrotec America specializes in the development and manufacture of custom thermoelectric modules and we will be pleased to quote on unique devices that will exactly meet your requirements.

The overall cooling system is dynamic in nature and system performance is a function of several interrelated parameters. As a result, it usually is necessary to make a series of iterative calculations to "zero-in" on the correct operating parameters. If there is any uncertainty about which thermoelectric device would be most suitable for a particular application, we highly recommend that you contact our engineering staff for assistance.

Before starting the module selection process, the designer should be prepared to answer the following questions:

1. At what temperature must the cooled object be maintained?
2. How much heat must be removed from the cooled object?
3. Is thermal response time important? If yes, how quickly must the cooled object change temperature after DC power has been applied?
4. What is the expected ambient temperature? Will the ambient temperature change significantly during system operation?
5. What is the extraneous heat input (heat leak) to the object as a result of conduction, convection, and/or radiation?
6. How much space is available for the module and heat sink?
7. What power is available?
8. Does the temperature of the cooled object have to be controlled? If yes, to what precision?
9. What is the expected approximate temperature of the heat sink during operation? Is it possible that the heat sink temperature will change significantly due to ambient fluctuations, etc.?

Each application obviously will have its own set of requirements that likely will vary in level of importance. Based upon any critical requirements that cannot be altered, the designer's job will be to select compatible components and operating parameters that ultimately will form an efficient and reliable cooling system.

To the design of a thermoelectric system it is necessary to define the following parameters: temperature of cold surface (TC); temperature of hot surface (TH) and the amount of heat absorbed of removed by the cold surface of the thermoelectric module (QC).

If the object to be cooled is in deep contact with the cold surface of the thermoelectric module, the expected temperature of the object (TC) can be considered the temperature of the module's cold surface.
There are cases which the object to be cooled is not in deep contact with the cold side of the module, such as an amount of cooling which the heat exchanger in the cold surface of the thermoelectric module. When this kind of system is used, the cold surface required can be several degrees lower than the desired temperature of the object.
The hot surface temperature (TH) is defined by two important parameters:

1. The temperature of the environment which the heat is rejected.
2. The heat exchanger efficiency which is between the hot surface of TE and the environment.

These two temperatures TC and TH and the difference between then ΔT are very important parameters and, thus can be determined accurately if the project operated correctly.
The most difficult parameter to qualify is the amount of heat (QC) to be removed or absorbed from the cold surface of the thermoelectric module. All thermoelectric system thermal loads should be considered. These thermal loads include, but are not limited to a thermal load of the electronic device, I^2R and the conduction through any object in contact either with the cold surface or the hot surface (i.e. electrical conductors, air or gas around objects, mechanical fasteners, etc). In some cases the effects of the thermal radiation should also be considered.
Thermoelectric devices are capable of producing no-load temperature differential of 70°C. Higher temperature differentials can be achieved by cascading modules.
Once the three parameters are qualified, the selection process for a particular module starts. Basic equations of heat transfer as listed to quantify QC and T.
There are many sets of modules or group of modules that can be used in a specific application. An additional criterion frequently used to select the best module and the coefficient of performance (COP) that is defined as absorbed heat in the cold junction, divided by the total heat to be rejected by the heat exchanger.
The maximum COP has advantages of minimal power and thus minimal rejected heat by the exchanger. These advantages come at a cost which in this case is a thermoelectric device bigger to operate at a maximum COP. The major advantage of minimal COP is the lowest initial cost.
The power conversion efficiency is dependent on a variety of factors although typically it might end up at around 3%. The especific type of module that the designer needs to use is based on the heat sink, cold sink, heat flow, operating temperatures, and desired output. This requires an engineering analysis to tell exactly which module is best for each application.
Another thing to remember is that the designer will likely need power conditioning since the output power is directly related to the temperature difference. If the operating temperatures fluctuate at all, the power output will too. Generally, the system is designed such that maximum efficiency is achieved for the most common operating temperatures at the ideal voltage and current the designer require for each application.

6. Conclusion

Thermoelectric systems are solid-state heat devices that either convert heat directly into electricity or transform electric power into thermal power for heating or cooling. The

operation principles are the Seebek and Peltier effects and the devices offer several advantages over other technologies.
Applications for thermoelectric modules cover a wide spectrum of product areas. These include equipment used by military, medical, industrial, consumer, scientific/laboratory, and telecommunications organizations. Each application will have its own set of requirements that likely will vary in level of importance. Nowadays the applications are restricted to small thermal systems but the trend in recent years has been for larger thermoelectric systems. So, the termoelectricity is one important field for the development of environmentally friendly thermal systems and the researches of new thermoelectric materials with large Seebeck coefficient and appropriate technology could make a breakthrough in the applications of thermoelectric devices in many applications.

7. Aknowledgment

The authors acknowledge the financial support of The National Council for Scientific and Technological Development (CNPq), Brazil.

8. References

Abdul-Wahad, S. A., Elkamel, A., Al-Damkhi, A. M., Al-Habsi, I. A., Al-Rubai´ey´, H. S., Al-Battashi, A. K., Al-Tamimi, A. R., Al-Mamari, K. H., & Chutani, M. U. (2009). Design and experimental investigation of portable solar thermoelectric refrigerator. *Renewable Energy,* Vol.34, No.1, (January 2009), pp.(30-34), 0960-1481

Astrain, D., Viá, J. G., & Dominguez, M. (2003). Increase of COP in the thermoelectric refrigeration by the optimization of heat dissipation. *Applied Thermal Engineering,* Vol. 23, No. 18, (December 2003), pp.(2183-2200), 1359-4311

Astrain, D., Albizua, J., & Vián, J. G. (2005). Computacional model for refrigerators based on Peltier effect applications. *Applied Thermal Engineering,* Vol.25, No.17-18, (December 2005), pp.(3149-3162), 1359-4311

Bojic, M., Savanovic, G., Trifunovic, N., Radovic, L., & Daljic, D. (1997). Thermoelectric cooling of a train carriage by using a coldness-recovery device. *Energy,* Vol. 22, No. 5, (May 1997), pp.(493-500), 0360-5442

Chang, Y., Chang, C., Ke., M., & Chen, S. (2009). Thermoelectric air-cooling module for electronic devices. *Applied Thermal Engineering,* Vol.29, No.13, (September 2009), pp.(2731-2737), 1359-4311

Chen, J., Zhou, Y., Wang, H., & Wang, J. T. (2002). Comparison of the optimal performance of single-stage and two-stage thermoelectric refrigeration system. *Applied Energy,* Vol.73, No.3-4, (November/December 2002), pp.(300-312), 0306-2619.

Chen, L., Sun, F., & Wu, C. (2005). Thermoelectric-generator with Linear Phenomenological Heat-transfer Law. *Applied Energy,* Vol.81, No.4, (August 2005), pp.(358-364), 0306-2619.

Chen, L., Sun, F., & Wu, C. (2005). Performance Optimization of a Two-stage Semiconductor Thermoelectric - Generator. *Applied Energy,* Vol. 82, No.4, (December 2005), pp.(300-312), 0306-2619.

Chen, L., Gong, J., Sun, F., & Wu, C. (2002). Effect of Heat Transfer on the Performance of Thermoelectric Generators. *International Journal of Thermal Science,* Vol.41, No.1, (January 2002), pp.(95-99), 1290-0729.

Chen, L., Sun, C., & Wu, C. (2005). Thermoelectric-generator with linear phenomenological heat-transfer law. *Applied Energy, Vol.81, No.4, (august 2005), pp.(358-364), 0306-2619*

Dai, Y. J., Wang, R. Z., & Ni, L. (2003). Experimental investigation on a thermoelectric refrigerator driven by solar cells. *Renewable Energy,* Vol.28, No.1, (November 2003), pp.(949-959), 0960-1481.

Göktun, S. (1995). Design consideration for a thermoelectric refrigerator. *Energy Conversion and Management,* Vol.36, No.12, (December 1995), pp.(1197-1200), 0196-8904

Gou, X., Xiao, H., & Yang, S. (2010). Modeling, experimental study and optimization on low-temperature waste heat thermoelectric generator system. *Applied Energy,* Vol.87, No.10, (October 2010), pp.(3131-3136), 0306-2619

Hsiao, Y. Y., Chang, W. C., & Chen, S. L. (2010). A mathematical model of thermoelectric module with applications on waste heat recovery from automobile engine. *Energy,* Vol.35, No.3, (March 2010), pp.(1447-1454), 0360-5442

Huang, B. J, & Duang, C. L. (2003). System dynamic model and temperature control of a thermoelectric cooler. *International Journal of Refrigeration,* Vol.23, No.3, (May 2003), pp.(197-207), 0140-7007

Huang, B. J., Chin, C. J., & Duang, C. L. (2000). A design method of thermoelectric cooler. *International Journal of Refrigeration,* Vol. 23, No.3, (May 2000), pp.(208-218), 0140-7007

Khattab, N. M., & El Shenawy, E. T. (2006). Optimal Operation of Thermoelectric Cooler Driven by Solar Thermoelectric Generator. *Energy Conversion and Management,* Vol.47, No.4, (March 2006), pp.(407-426), 0196-8904

Kurosaki, K., Uneda, H., Muta, H., & Yamanaka, S. (2004). Thermoelectric properties of thallium antimony telluride. *Journal of Alloys and Compounds,* Vol.376, No.1-2, (August 2004), pp.(43-48), 0925-8388

Lau, P. G., & Buist, R. J. (1997). Calculation of Thermoelectric Power Generation Performance Using Finite Element Analysis, *Proceedings of the XVI International Conference on Thermoelectrics,* 9780780340572, Dresden-Germany, august 1997

Li, T., Tang, G, Gong, G., Zhang, G., Li, N., & Zhang, L. (2009). Investigation of prototype thermoelectric domestic-ventilator. *Applied Thermal Engineering,* Vol.29, No.10, (July 2009), pp.(2016-2021), 1359-4311

Lindler, K. W. (1998). Use of multi-stage cascades to improve performance of thermoelectric heat pumps. *Energy Conversion and Management,* Vol.39, No.10, (July 1998), pp.(1009-1014), 0196-8904

Luo, J., Chen, L., Sun, S., & Wu, C. (2003). Optimum allocation of heat transfer surface area for cooling load and COP optimization of a thermoelectric refrigerator. *Energy Conversion and Management,* Vol.44, No.20, (December 2003), pp.(3197-3206), 0196-8904

Riffat, S. S. B., & Ma, X. (2003). Thermoelectrics: a review of present and potential applications. *Applied Thermal Engineering,* Vol.23, No.18, (December 2003), pp.(913-935), 1359-4311

Santos, J. H., & Camargo, J. R. (2008). Aplicação de módulos termelétricos para geração de potência, *Proceedings of XIII Encontro de Iniciação Científica e IX Mostra de Pós-graduação,* 19818688, Taubaté-SP, october 2008 (in portuguese)

Sofrata, H. (1996). Heat rejection alternatives for thermoelectric refrigerators. *Energy Conversion and Management,* Vol.37, No.3, (March 1996), pp.(269-280), 0196-8904

Wu, C. (1996). Analysis of Waste-heat Thermoelectric Power Generators. *Applied Thermal Engineering,* Vol.16, No.1, (January 1996), pp.(63-69), 1359-4311

Zhang, H. Y., Mui, Y. C., & Tarin, C. (2010). Analysis of thermoelectric cooler performance for high power electronic packages. *Applied Thermal Engineering,* Vol.30, No.6-7, (May 2010), pp.(561-568), 1359-4311

6

On the Thermal Transformer Performances

Ali Fellah and Ammar Ben Brahim
Gabes university, Engineers National School,
Applied Thermodynamic Research Unit, Gabes
Tunisia

1. Introduction

Different approaches are considered to select optimum criteria for technical process analysis. The maximization of the efficiency and the minimization of the total cost enclosing capital and running costs are the main purposes e.g. Munoz and Von Spakovsky (2003). Physical and thermodynamic criteria and technical and economic considerations have to be joined while analyzing energetic conversion processes (Berlitz et al. 1999; Chen 1995). Thus, deducing economic findings is the common objective of all intentions. Furthermore, the use of the interdisciplinary modeling methods has recently constituted the most important orientation of the technical system studies and process analyses. The simplification of both mathematical description and hypothesis definition of the interaction effects due to internal irreversibilities lead to the development of interesting simple but universal models. Internal irreversibilities due to heat transfer, throttling, mixing and internal dissipation of the working fluid, which are responsible for entropy generation are always present in a real heat driven refrigerator (Chen et Schouten, 1998). However, many works do not satisfy all the futures because the distribution of heat transfer properties between the components is taken as inputs and no as a result to be deduced from the optimization procedure.
In the other hand, the interaction effects in the internal processes do not favor separated studies. Therefore, it will be necessary to consider discreet parts of the whole system as they were independent. Then, the characteristics will be treated according to mathematical and physical couplings. To perform these approaches, the theory of finite time thermodynamic is mainly used. Seeing that the heat transfer processes are defined according to temperature finite difference method and the inner and outer reversibilities should be taken into consideration (Fellah et al., 2010). According to the study's finality, the decomposition of an overall system into subsystems may constitute a helpful tool, for which the physical and mathematical couplings would permit the efficient application of the investigation methods than a whole problem with a unique task. This could reduce the size of the mathematical problem. In fact, many attempts have been made to reduce the size of the problem, using the decomposition method on stage or/and on block models (Berlitz et al. 1999; Feidt and Lang 2002; Chen 1995; Chen and Wu 1996; Fellah, 2008 and Fellah et al., 2010).
The results obtained for various thermodynamic cycle analyses using FTT are closer to real device performance than those obtained using classical thermodynamics. During the last two decades, many optimization studies for refrigerators based on endoreversible and irreversible models have been performed by considering various objective functions. Wijeysundera, 1997;

Sahin et Kodal, 2003; Fellah et al. 2010, analyzed the performance of the three heat reservoir endoreversible cycle. Goktun, 1997 and Chen et Wu, 1996 analyzed the performance of three heat reservoir irreversible cycle. In both analyses Newton's heat transfer law is used.
From this context, Sokolov and Hersagal, (1993) optimized the system performance of a solar driven year-round ejector refrigeration system. Vargas et al. 1996 investigated the optimal condition for a refrigerator driven by a solar collector considering the three heat transfer irreversibility. Wijeysundera, 1997 and Bejan, (1995) analyzed a solar powered absorption cooling system using the three heat reservoir cycle including external heat transfer irreversibility. Later, Chen and Schouten, (1998) discussed the optimum performance of an irreversible absorption refrigeration cycle in which three external heat transfer irreversibilities have been considered.
Nevertheless, all those studies focus on the systems steady-state behavior. Vargas et al., (1998) and (2000) studied a transient endoreversible model of a heat driven refrigeration plant. The optimization is done from the point of view of the heat that drives the cycle. In their analyses, they concluded that the optimal fuel flow rate and minimum time to reach a prescribed cold space temperature are influenced by the thermal load in the refrigerated space and the thermal conductance of the walls. Recently, Vargas et al., (2000) optimized a solar collector driven water heating and absorption cooling plant using the three heat reservoir cycle.
Usually, real difficulties solving the optimization problem may appear owing to the significant number of unknowns. As a resolution to the above queries, an optimization of a thermodynamic cycle is carried out in the purpose to define the process optimal design parameters. As a case in point, the energetic process considered is a solar absorption refrigeration cycle. Two facets are considered for the investigation. Firstly, the methodology is based on the hierarchical decomposition, the endoreversibility principles and the Lagrange multipliers optimization method in a permanent regime. Secondly, the transient regime is considered.

2. Hypothesis

The analysis method, based on the endoreversible model, could be performed according to the following hypotheses which correspond to the nominal working conditions considered in the most process conceptual studies:

- The heat source temperatures are constant,
- The heat sources are reservoirs with fixed heat capacities,
- The overall heat transfer coefficients U_i of the heat exchangers and between the cycle and the surrounding reservoirs are constant,
- The system operates in a steady state,
- The heat transfer process between the work fluid and the source depends only on their temperatures.
- The heat transfers with the sources and the barriers are permanent and linear,
- The heat transfers between the subsystems and through the boundaries are the only sources of irreversibilities,
- All quantity entering in the system is considered as positive and all quantity coming out as negative,
- The heat exchangers operate in counter flow.

3. Hierarchical decomposition

There are three technical system decomposition types. The first is a physical decomposition (in equipment) used for macroscopic conceptual investigations. The second method is a disciplinary decomposition, in tasks and subtasks, used for microscopic analysis of mass and heat transfer processes occurring in different components. The third method is a mathematical decomposition associated to the resolution procedure of the mathematical model governing the system operating mode (Aoltola, 2003).
The solar absorption refrigeration cycle, presented on Fig. 1 (Fellah et al., 2010), is one of many interesting cycles for which great efforts have been consecrated. The cycle is composed by a solar concentrator, a thermal solar converter, an intermediate source, a cold source and four main elements: a generator, an absorber, a condenser and an evaporator. The thermal solar converter constitutes a first thermal motor TM_1 while the generator and the absorber constitute a second thermal motor TM_2 and the condenser and the evaporator form a thermal receptor TR. The exchanged fluxes and powers that reign in the different compartments of the machine are also mentioned. The parameterization of the cycle comprises fluxes and powers as well as temperatures reigning in the different compartments of the machine.
The refrigerant vapor, stemmed from the generator, is condensed and then expanded. The cooling load is extracted from the evaporator. The refrigerant vapor, stemmed from the evaporator, is absorbed by the week solution in the absorber. The rich solution is then decanted from the absorber into the generator through a pump.
The number of the decomposition levels must be in conformity with the physical bases of the installation operating mode. The mathematical identification of the subsystem depends on the establishment of a mathematical system with nil degree of freedom (DoF). Here, the decomposition consists in a four levels subdivision. The first level presents the compact global system which is a combination of the thermal motors TM_1 and TM_2 with the thermal receptor TR. After that, this level is decomposed in two sublevels the thermal converter TM_1 and the command and refrigeration system TM_2+TR. This last is subdivided itself to give the two sublevels composed by the thermal engine TM_2 and the thermal receptor TR. The fourth level is composed essentially by the separated four elements the generator, the absorber, the condenser and the evaporator. For more details see Fellah et al., 2010.

4. Optimization problem formulation

For heat engines, power-based analysis is usually used at maximum efficiency and working power, whereas the analysis of refrigerators is rather carried out for maximal cooling load. Therefore, there is no correspondence with the maximal value of the coefficient of performance COP. According to the objectives of the study, various concepts defined throughout the paper of Fellah et al. 2006 could be derived from the cooling load parameter e.g. the net Q_e, the inverse $1/Q_e$, the inverse specific A/Q_e cooling load.
For an endoreversible heat transformer (Tsirlin et Kasakov 2006), the optimization procedure under constraints can be expressed by:

$$\max_{u_i>0}\left[P=\sum_{i=1}^{n}Q_i(T_i,u_i)\right] \tag{1}$$

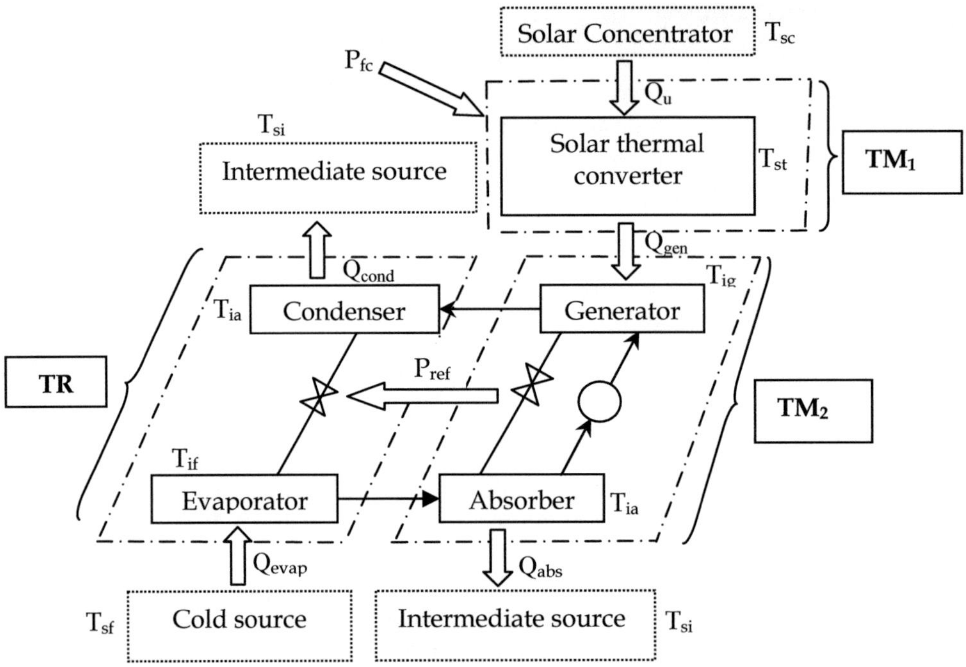

Fig. 1. Working principle and decomposition of a solar absorption refrigerator cycle

Under the constraints:

$$\sum_{i=1}^{n}\frac{Q_i(T_i,u_i)}{u_i}=0 \tag{2}$$

And

$$\sum_{j=1}^{n}Q_{ij}(T_j,T_i)=Q_i(T_i,u_i)\quad \mathrm{i}=1,\ldots,\mathrm{m} \tag{3}$$

where T_i : temperature of the ith subsystem

Q_{ij} : the heat flux between the ith and the jth subsystem

$Q(T_i, u_i)$: the heat flux between the ith subsystem and the transformer

P: the transformer power.

The optimization is carried out using the method of Lagrange multipliers where the thermodynamic laws constitute the optimization constraints. The endoreversible model takes into account just the external irreversibility of the cycle, consequently there is a minimization of the entropy production comparing to the entropy production when we consider internal and external irreversibilities.

For a no singular problem described by equations (1 to 3), the Lagrange function can be expressed as follows:

$$L=\sum_{i=1}^{m}Q_i+\sum_{i=m+1}^{n}Q_i-\Lambda\sum_{i=1}^{m}Q_i/u_i-\Lambda\sum_{i=m+1}^{n}Q_i/u_i+\sum_{i=1}^{m}\lambda_i\left(\sum_{j=1}^{n}Q_{ij}-Q_i\right) \tag{4}$$

Where λ_i and Λ are the Lagrange multipliers, m is the number of subsystems and n is the number of contacts.
According to the selected constraint conditions, the Lagrange multipliers λi are of two types. Some are equivalent to temperatures and other to dimensionless constants. The refrigerant temperatures in the condenser and the absorber are both equal to T_{ia}. Thus and with good approximation, the refrigeration endoreversible cycle is a three thermal sources cycle. The stability conditions of the function L for i> m are defined by the Euler-Lagrange equation as follows:

$$\frac{\partial L}{\partial u_i} = \frac{\partial}{\partial u_i}\left[Q_i(T_i, u_i)(1 - \Lambda/u_i)\right] = 0 \text{ Where } (i = m+1, \dots, n) \quad (5)$$

5. Endoreversible behavior in permanent regime

5.1 Optimal characteristics

Analytical resolution delivers the following temperature distributions:

$$T_{ig}/T_{ia} = (T_{st}/T_{int})^{1/2} \quad (6)$$

$$T_{ie}/T_{ia} = (T_{cs}/T_{int})^{1/2} \quad (7)$$

$$T_{st}/T_{ia} = (T_{sc}/T_{int})^{1/2} \quad (8)$$

Expressions (6 to 8) relay internal and external temperatures. Generalized approaches (e.g. Tsirlin et Kasakov, 2006) and specific approaches (e.g. Tozer and Agnew, 1999) have derived the same distributions.
The thermal conductances UA_i, constitute the most important parameters for the heat transformer analysis. They permit to define appropriate couplings between functional and the conceptual characteristics. Considering the endoreversibility and the hierarchical decomposition principles, the thermal conductance ratios in the interfaces between the different subsystems and the solar converter, are expressed as follows:

$$UA_e \ / \ UA_{st} = I_{st}T_{ie}^{1/2}(T_{int}^{1/2}-T_{st}^{1/2}) \ / \ I_eT_{sc}^{1/2}(T_{ie}^{1/2}-T_{int}^{1/2}) \quad (9)$$

$$UA_g \ / \ UA_{st} = I_{st}T_{st}^{1/2} \ / \ I_gT_{sc}^{1/2} \quad (10)$$

$$UA_c \ / \ UA_{st} = I_{st}T_{int}^{1/2}(T_{int}^{1/2}-T_{st}^{1/2}) \ / \ I_aT_{sc}^{1/2}(T_{ie}^{1/2}-T_{int}^{1/2}) \quad (11)$$

$$UA_a/UA_{st}=I_{st}T_{int}^{1/2}/I_aT_{sc}^{1/2} \quad (12)$$

Where I_i represents the ith interface temperature pinch.
The point of merit is the fact that there is no need to define many input parameters while the results could set aside many functional and conceptual characteristics. The input parameters for the investigation of the solar refrigeration endoreversible cycle behaviors could be as presented by Fellah, 2008:

- The hot source temperature T_{sc} for which the transitional aspect is defined by Eufrat correlation (Bourges, 1992; Perrin de Brichambaut, 1963) as follows:

$$T_{sc} = -1.11t^2 + 31.34t + 1.90 \quad (13)$$

where t represents the day hour.

- The cold source temperature T_{sf}, $0°C \leq T_{sf} \leq 15°C$
- The intermediate source temperature T_{si}, $25°C \leq T_{si} \leq 45°C$.

For a solar driven refrigerator, the hot source temperature T_{sc} achieves a maximum at midday. Otherwise, the behavior of T_{sc} could be defined in different operating, climatic or seasonal conditions as presented in Boukhchana et al.,2011.

The optimal parameters derived from the simulation are particularly the heating and refrigerant fluid temperatures in different points of the cycle:

- The heating fluid temperature at the generator inlet T_{if},
- The ammonia vapor temperature at the generator outlet T_{ig},
- The rich solution and ammonia liquid temperatures at both the absorber and the condenser outlets T_{ia},
- The ammonia vapor temperature at the evaporator outlet T_{ie},

Relative stability is obtained for the variations of the indicated temperatures in terms of the coefficient of performance COP. However, a light increase of T_{ig} and T_{if} and a light decrease of Tia are observed. These variations affect slightly the increase of the COP. Other parameters behaviors could be easily derived and investigated. The cooling load Q_e increases with the thermal conductance increase reaching a maximum value and then it decreases with the increase of the COP. The decrease of Q_e is more promptly for great T_{sc} values. Furthermore, the increase of COP leads to a sensible decrease of the cooling load. It has been demonstrated that a COP value close to 1 could be achieved with a close to zero cooling load. Furthermore, there is no advantage to increase evermore the command hot source temperature

Since the absorption is slowly occurred, a long heat transfer time is required in the absorber. The fluid vaporization in the generator requires the minimal time of transfer. Approximately, the same time of transfer is required in the condenser and in the evaporator. The subsystem TM_2 requires a lower heat transfer time than the subsystem TR.

5.2 Power normalization

A normalization of the maximal power was presented by Fellah, 2008. Sahin and Kodal (1995) demonstrated that for a subsystem with three thermal reservoirs, the maximal power depends only on the interface thermal conductances. The maximal normalized power of the combined cycle is expressed as:

$$\tilde{P} = UA_2(UA_1 + UA_3)/\left[UA_2(UA_1 + UA_3) + UA_1\ UA_3\right] \tag{14}$$

Thus, different cases can be treated.

a. If $UA_1 \neq UA_2 \neq UA_3$ then $\tilde{P} < 1$. The power deduced from the optimization of a combined cycle is lower than the power obtained from the optimization of an associated endoreversible compact cycle.
b. If, for example $UA_1 = UA_3$; Then $\tilde{P}$ can be expressed as:

$$\tilde{P} = 1/\left[1 + 1/2\kappa^2\right] \tag{15}$$

where: $\kappa = \sqrt{UA_2/UA_1}$.

For important values of κ, equation (7) gives $\tilde{P} \approx 1$. The optimal power of the combined cycle is almost equal to the optimal power of the simple compact cycle.

c. If $UA_1 = UA_2 = UA_3$ then $\tilde{P} = 2/3$. It is a particular case and it is frequently used as simplified hypothesis in theoretical analyses of systems and processes.

5.3 Academic and practical characteristics zones
5.3.1 Generalities
Many energetic system characteristics variations present more than one branch e.g. Summerer, 1996; Fellah et al.2006; Fellah, 2008 and Berrich, 2011. Usually, academic and theoretical branches positions are different from theses with practical and operational interest ones. Both branches define specific zones. The most significant parameters for the practical zones delimiting are the high COP values or the low entropy generation rate values. Consequently, researchers and constructors attempt to establish a compromise between conceptual and economic criteria and the entropy generation allowing an increase of performances. Such a tendency could allow all-purpose investigations.
The Figure 2 represents the COP variation versus the inverse specific cooling load (A_t/Q_{evap}) the curve is a building block related to the technical and economic analysis of absorption refrigerator. For the real ranges of the cycle operating variables, the curve starts at the point M defined by the smallest amount of (A/Q_e) and the medium amount of the COP. Then, the curve leaves toward the highest values in an asymptotic tendency. Consequently, the M point coordinates constitute a technical and economic criterion for endoreversible analyses in finite time of solar absorption refrigeration cycles Berlitz et al.(1999), Fellah 2010 and Berrich, 2011 . The medium values are presented in the reference Fellah, 2010 as follow:

$$0,4 \leq A / Q_e \leq 0,5\, m^2 / kW \tag{16}$$

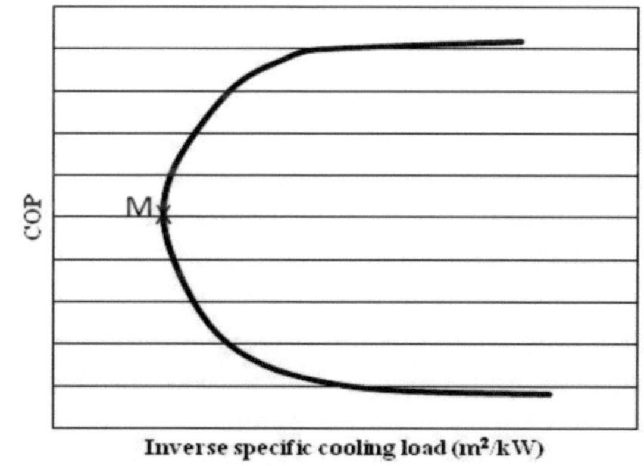

Fig. 2. Inverse specific cooling load versus the COP.

5.3.2 Optimal zones characteristics
The Figure 3 illustrates the effect of the ISCL on the entropy rate for different temperatures of the heat source. Thus, for a Neat Cooling Load Q_e and a fixed working temperature T_{sc}, the total heat exchange area A and the entropy produced could be deduced.
The minimal entropy downiest zones are theses where the optimal operational zones have to be chosen. The point M is a work state example. It is characterized by a heat source

temperature of about 92°C and an entropy rate of 0.267kW/K and an A/Q_e equal to 24.9%. Here, the domain is decomposed into seven angular sectors. The point M is the origin of all the sectors.

The sector R is characterized by a decrease of the entropy while the heat source temperature increases. The result is logic and is expected since when the heat source temperature increases, the COP increases itself and eventually the performances of the machine become more interesting. In fact, this occurs when the irreversibility decreases. Many works have presented the result e.g. Fellah et al. 2006. However, this section is not a suitable one for constructors because the A/Q_e is not at its minimum value.

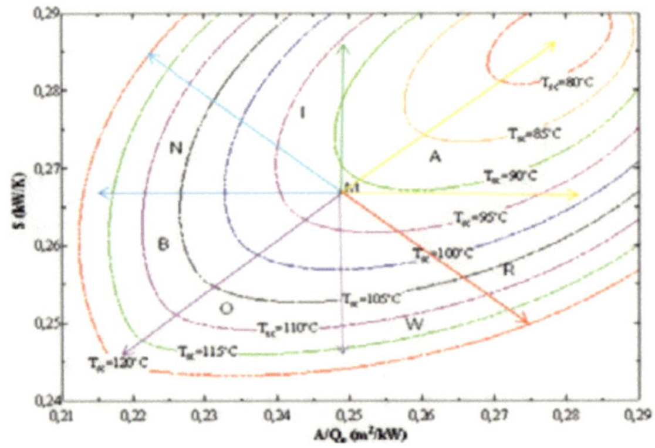

Fig. 3. Entropy rate versus the inverse specific cooling load.

The sector A is characterized by an increase of the entropy while the heat source temperature decreases from the initial state i.e. 92°C to less than 80°C. The result is in conformity with the interpretation highly developed for the sector R.

The sector I is characterized by an increase of the entropy rate while the heat source temperature increases. The reduction of the total area by more than 2.5% of the initial state is the point of merit of this sector. This could be consent for a constructor.

The sector N presents a critical case. It is characterized by a vertical temperature curves for low T_{sc} and a slightly inclined ones for high T_{sc}. Indeed, it is characterized by a fixed economic criterion for low source temperature and an entropy variation range limited to maximum of 2% and a slight increase of the A/Q_e values for high values of the heat source temperature with an entropy variation of about 6.9%.

The sector B is characterized by slightly inclined temperature curves for low T_{sc} and vertical ones for high T_{sc}, opposing to the previous zone. Indeed, the A/Q_e is maintained constant for a high temperature. The entropy variation attains a maximum value of 8.24%. For low values of the temperature, A/Q_e increases slightly. The entropy gets a variation of 1.7%. The entropy could be decreased by the increase of the heat source temperature. Thus it may be a suitable region of work.

As well, the sector O represents a suitable work zone.

The sector W is characterized by horizontal temperature curves for low T_{sc} and inclined ones for high T_{sc}. In fact, the entropy is maintained fixed for a low temperature. For high values of the temperature, the entropy decreases of about 8.16%. For a same heat source temperature, an increase of the entropy is achievable while A/Q_e increases. Thus, this is not the better work zone.

It should be noted that even if it is appropriate to work in a zone more than another, all the domains are generally good as they are in a good range:

$$0.21 < A/Q_e < 0.29\ m^2/kW \tag{19}$$

A major design is based on optimal and economic finality which is generally related to the minimization of the machine's area or to the minimization of the irreversibility.

5.3.3 Heat exchange areas distribution

For the heat transfer area allocation, two contribution types are distinguished by Fellah, 2006. The first is associated to the elements of the subsystem TM_2 (command high temperature). The second is associated to the elements of the subsystem TR (refrigeration low temperature). For COP low values, the contribution of the subsystem TM_2 is higher than the subsystem TR one. For COP high values, the contribution of the subsystem TR is more significant. The contribution of the generator heat transfer area is more important followed respectively, by the evaporator, the absorber and the condenser.

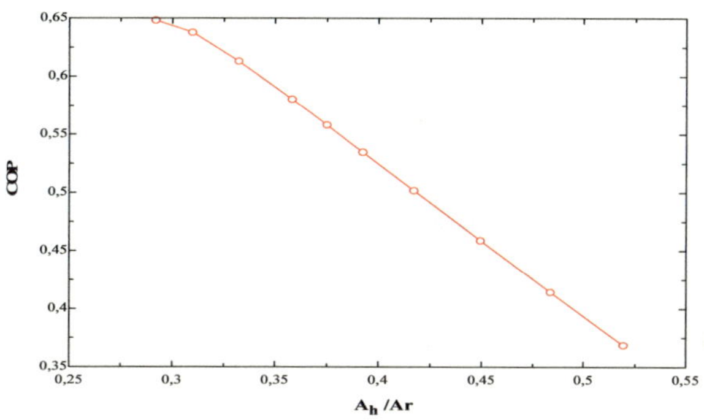

Fig. 4. Effect of the areas distribution on the COP

The increase of the ratio U_{MT2}/U_{RT} leads to opposite variations of the area contributions. The heat transfer area of MT_2 decreases while the heat transfer of TR increases. For a ratio U_{MT2}/U_{RT} of about 0.7 the two subsystems present equal area contributions.

The figure 4 illustrates the variation of the coefficient of performance versus the ratio A_h/A_r. For low values of the areas ratio the COP is relatively important. For a distribution of 50%, the COP decreases approximately to 35%.

6. Endoreversible behavior in transient regime

This section deals with the theoretical study in dynamic mode of the solar endoreversible cycle described above. The system consists of a refrigerated space, an absorption refrigerator and a solar collector. The classical thermodynamics and mass and heat transfer balances are used to develop the mathematical model. The numerical simulation is made for different operating and conceptual conditions.

6.1 Transient regime mathematical model

The primary components of an absorption refrigeration system are a generator, an absorber, a condenser and an evaporator, as shown schematically in Fig.5. The cycle is driven by the heat transfer rate Q_H received from heat source (solar collector) at temperature T_H to the generator at temperature T_{HC}. Q_{Cond} and Q_{Abs} are respectively the heat rejects rates from the condenser and absorber at temperature T_{0C}, i.e.T_{0A}, to the ambient at temperature T_0 and Q_L is the heat input rate from the cooled space at temperature T_{LC} to the evaporator at temperature T_L. In this analysis, it is assumed that there is no heat loss between the solar collector and the generator and no work exchange occurs between the refrigerator and its environment. It is also assumed that the heat transfers between the working fluid in the heat exchangers and the external heat reservoirs are carried out under a finite temperature difference and obey the linear heat-transfer law ''Newton's heat transfer law''.

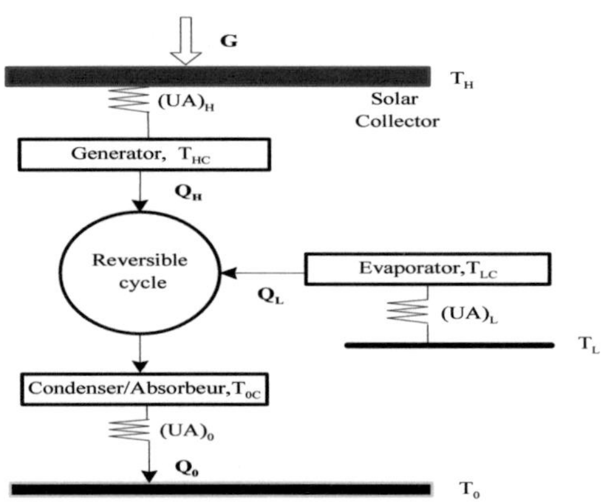

Fig. 5. The heat transfer endoreversible model of a solar driven absorption refrigeration system.

Therefore, the steady-state heat transfer equations for the three heat exchangers can be expressed as:

$$\begin{aligned} Q_L &= UA_L(T_L - T_{LC}) \\ Q_H &= UA_H(T_H - T_{HC}) \\ Q_0 &= UA_0(T_{0C} - T_0) \end{aligned} \tag{20}$$

From the first law of thermodynamics:

$$Q_H + Q_L = Q_0 \tag{21}$$

According to the second law of thermodynamics and the endoreversible property of the cycle, one may write:

$$\frac{Q_H}{T_{HC}} + \frac{Q_L}{T_{LC}} = \frac{Q_0}{T_{0C}} \tag{22}$$

The generator heat input Q_H can also be estimated by the following expression:

$$Q_H = \eta_{sc} A_{sc} G_T \tag{23}$$

Where A_{sc} represents the collector area, G_T is the irradiance at the collector surface and η_{sc} stands for the collector efficiency. The efficiency of a flat plate collector can be calculated as presented by Sokolov and Hersagal, (1993):

$$Q_H = A_{st} G_T b(T_{st} - T_H) \tag{24}$$

Where b is a constant and T_{st} is the collector stagnation temperature.
The transient regime of cooling is accounted for by writing the first law of thermodynamics, as follows:

$$mCv_{air} \frac{dT_L}{dt} = UA_w (T_0 - T_L) + Q_1 - Q_L \tag{25}$$

Where UA_w $(T_0\text{-}T_L)$ is the rate of heat gain from the walls of the refrigerated space and Q_1 is the load of heat generated inside the refrigerated space.
The factors UA_H, UA_L and UA_0 represent the unknown overall thermal conductances of the heat exchangers. The overall thermal conductance of the walls of the refrigerated space is given by UA_W. The following constraint is introduced at this stage as:

$$UA = UA_H + UA_L + UA_0 \tag{26}$$

According to the cycle model mentioned above, the rate of entropy generated by the cycle is described quantitatively by the second law as:

$$\frac{dS}{dt} = \frac{Q_0}{T_{0C}} - \frac{Q_H}{T_{HC}} - \frac{Q_L}{T_{LC}} \tag{27}$$

In order to present general results for the system configuration proposed in Fig. 5, dimensionless variables are needed. Therefore, it is convenient to search for an alternative formulation that eliminates the physical dimensions of the problem. The set of results of a dimensionless model represent the expected system response to numerous combinations of system parameters and operating conditions, without having to simulate each of them individually, as a dimensional model would require. The complete set of non dimensional equations is:

$$\left\{ \begin{aligned} \overline{Q}_L &= z(\tau_L - \tau_{LC}) \\ \overline{Q}_H &= y(\tau_H - \tau_{HC}) \\ \overline{Q}_0 &= (1 - y - z)(\tau_{0C} - 1) \\ \overline{Q}_H &= B(\tau_{st} - \tau_H) \\ \overline{Q}_H + \overline{Q}_L &= \overline{Q}_0 \\ \frac{\overline{Q}_H}{\tau_{HC}} + \frac{\overline{Q}_L}{\tau_{LC}} &= \frac{\overline{Q}_0}{\tau_{0C}} \\ \frac{d\tau_L}{d\theta} &= w(\tau_0 - \tau_L) + \overline{Q}_1 - \overline{Q}_L \\ \frac{d\overline{S}}{d\theta} &= \overline{Q}_0 - \frac{\overline{Q}_H}{\tau_H} - \frac{\overline{Q}_L}{\tau_L} \end{aligned} \right. \tag{28}$$

Where the following group of non-dimensional transformations is defined as:

$$\tau_H = \frac{T_H}{T_0}, \quad \tau_L = \frac{T_L}{T_0}, \tau_{st} = \frac{T_{st}}{T_0},$$
$$\tau_{LC} = \frac{T_{LC}}{T_0}, \tau_{OC} = \frac{T_{OC}}{T_0}, \tau_{HC} = \frac{T_{HC}}{T_0},$$
$$\overline{Q}_H = \frac{Q_H}{UA.T_0}, \overline{Q}_L = \frac{Q_L}{UA.T_0}, \overline{Q}_0 = \frac{Q_0}{UA.T_0}, \overline{Q}_1 = \frac{Q_1}{UA.T_0}, \tag{29}$$
$$B = \frac{A_{sc} G_T b}{UA}, \theta = \frac{t.UA}{mCv_{air}}$$

B describes the size of the collector relative to the cumulative size of the heat exchangers, and y, z and w are the conductance allocation ratios, defined by:

$$y = \frac{UA_H}{UA}, z = \frac{UA_L}{UA}, w = \frac{UA_w}{UA} \tag{30}$$

According to the constraint property of thermal conductance UA in Eq. (26), the thermal conductance distribution ratio for the condenser can be written as:

$$x = \frac{UA_0}{UA} = 1 - y - z \tag{31}$$

The objective is to minimize the time θ_{set} to reach a specified refrigerated space temperature, $\tau_{L,set}$, in transient operation. An optimal absorption refrigerator thermal conductance allocation has been presented in previous studies e.g. Bejan, 1995 and Vargas et al., (2000) for achieving maximum refrigeration rate, i.e.,$(x,y,z)_{opt}$ =(0.5,0.25,0.25), which is also roughly insensitive to the external temperature levels (τ_H, τ_L). The total heat exchanger area is set to A=4 m^2 and an average global heat transfer coefficient to U=0.1 kW/m^{2}K in the heat exchangers and U_w=1.472 kW /m^{2}K across the walls which have a total surface area of A_w=54 m^2, T_0= 25°C and Q_1=0.8 kW. The refrigerated space temperature to be achieved was established at $T_{L,set}$=16°C.

6.2 Results

The search for system thermodynamic optimization opportunities started by monitoring the behavior of refrigeration space temperature τ_L in time, for four dimensionless collector size parameter B, while holding the other as constants, i.e., dimensionless collector temperature τ_H=1.3 and dimensionless collector stagnation temperature τ_{st}=1.6. Fig.6 shows that there is an intermediate value of the collector size parameter B, between 0.01 and 0.038, such that the temporal temperature gradient is maximum, minimizing the time to achieve prescribed set point temperature ($\tau_{L,set}$=0.97). Since there are three parameters that characterize the proposed system (τ_{st}, τ_H, B), three levels of optimization were carried out for maximum system performance.

The optimization with respect to the collector size B is pursued in Fig. 7 for time set point temperature, for three different values of the collector stagnation temperature τ_{st} and heat source temperatures τ_H=1.3. The time θ_{set} decrease gradually according to the collector size parameter B until reaching a minimum $\theta_{set,min}$ then it increases. The existence of an optimum with respect to the thermal energy input $\overline{Q}_H$ is not due to the endoreversible model aspects.

However, an optimal thermal energy input $\overline{Q}_H$ results when the endoreversible equations are constrained by the recognized total external conductance inventory, UA in Eq. (26), which is finite, and the generator operating temperature T_H.

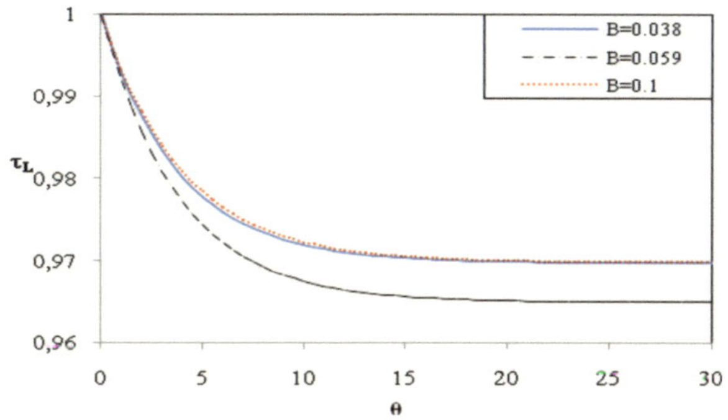

Fig. 6. Low temperature versus heat transfer time for B=0.1,0.059,0.038.

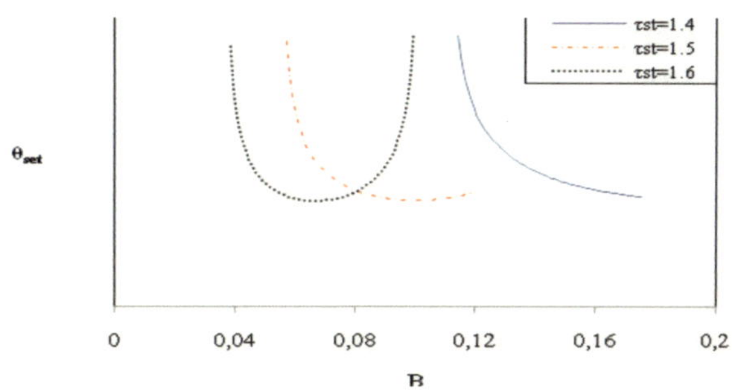

Fig. 7. The effect of dimensionless collector size B on time set point temperature.

These constraints are the physical reasons for the existence of the optimum point. The minimum time to achieve prescribed temperature is the same for different values of stagnation temperature τ_{st}. The optimal dimensionless collector size B decreases monotonically as τ_{st} increases and the results are shown in Fig. 8. The parameter τ_{st} has a negligible effect on B_{opt} if τ_{st} is greater than 1.5 and B_{opt} is less than 0.1. Thus, τ_{sc} has more effect on the optimal collector size parameter B_{opt} than that on the relative minimum time.
The results plotted in Figures 8, 9 and 10 illustrate the minimum time $\theta_{set,min}$ and the optimal parameter B_{opt} respectively against dimensionless collector temperature τ_H , thermal load inside the cold space $\overline{Q}_1$ and conductance fraction w. The minimum time $\theta_{set,min}$ decrease and the optimal parameter B_{opt} increase as τ_H increase. The results obtained accentuate the importance to identify B_{opt} especially for lower values of τ_H. $\overline{Q}_1$ has an almost negligible effect on B_{opt} . B_{opt} remains constant, whereas an increase in $\overline{Q}_1$ leads to an increase in $\theta_{set,min}$.. Obviously, a similar effect is observed concerning the behaviors of B_{opt} and $\theta_{set,min}$ according to conductance allocation ratios w.
During the transient operation and to reach the desired set point temperature, there is total entropy generated by the cycle. Figure 11 shows its behavior for three different collector size parameters, holding τ_H and τ_{st} constant, while Fig.12 displays the effect of the collector size

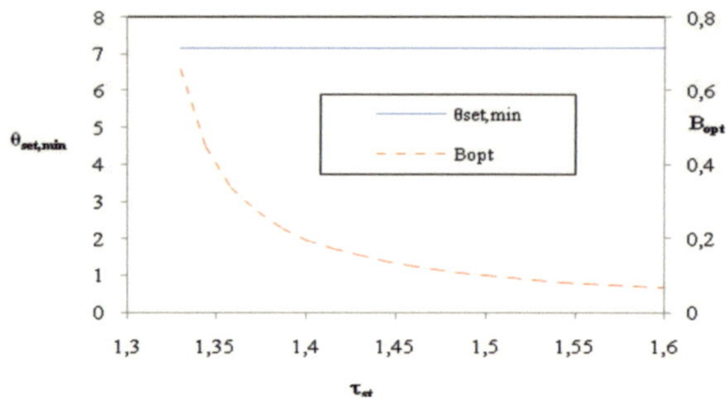

Fig. 8. The effect of the collector stagnation temperature τ_{st} on minimum time set point temperature and optimal collector size.

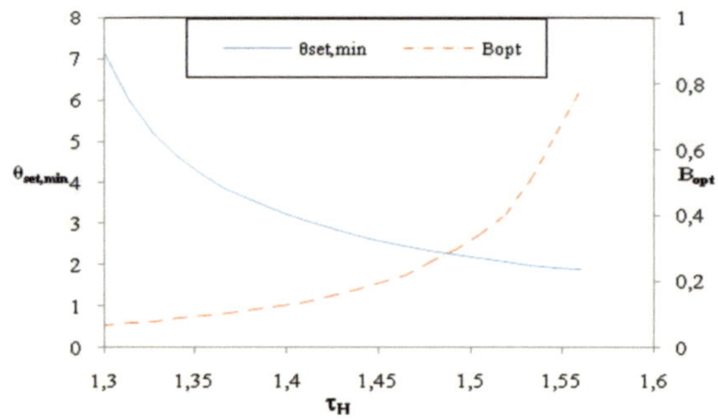

Fig. 9. The effect of dimensionless heat source temperatures τ_H on minimum time set point temperature and optimal collector size (τ_{st}=1.6).

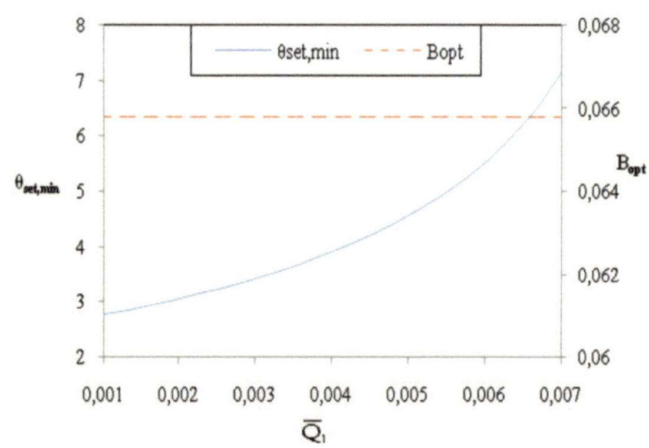

Fig. 10. The effect of thermal load in the refrigerated space on minimum time set point temperature and optimal collector size (τ_H=1.3 and τ_{st}=1.6).

on the total entropy up to θ_{set}. The total entropy increases with the increase of time and this is clear on the basis of the second law of thermodynamics, the entropy production is always positive for an externally irreversible cycle. There is minimum total entropy generated for a certain collector size. Note that B_{opt}, identified for minimum time to reach $\tau_{L,set}$, does not coincide with B_{opt} where minimum total entropy occurs.

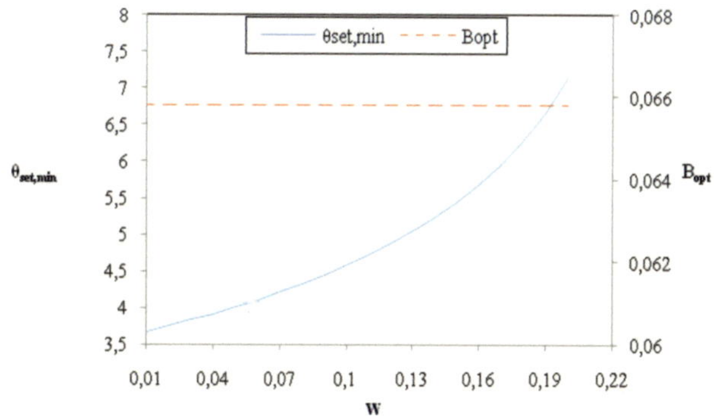

Fig. 11. The effect of conductance fraction on minimum time set point temperature and optimal collector size (τ_H=1.3 and τ_{st}=1.6).

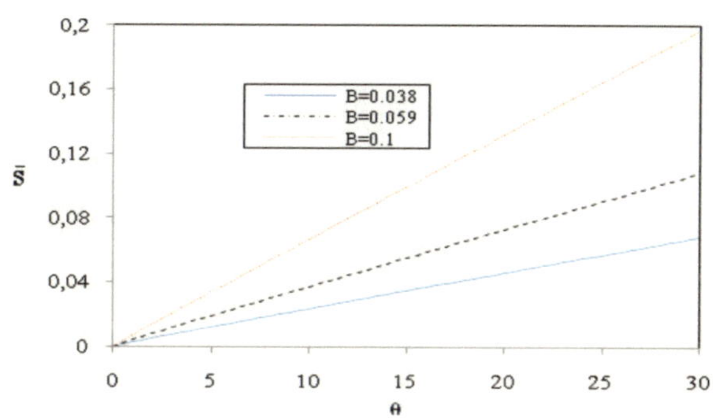

Fig. 12. Transient behavior of entropy generated during the time (τ_H=1.3 and τ_{st}=1.6).

Stagnation temperature and temperature collector effects on minimum total entropy generated up to θ_{set} and optimal dimensionless collector size are shown respectively in Figs.13 and 14. $\bar{S}_{set,min}$ is independent of τ_{st}, but, as the temperature stagnation increase B_{opt} decrease. This behavior is different from what was observed in the variation of temperature collector. An increase of stagnation temperature leads to a decrease of $\bar{S}_{set,min}$ and to an increase of B_{opt}. This result brings to light the need for delivering towards the greatest values of τ_{st} to approach the real refrigerator.
The optimization with respect to the size collector parameters for different values of τ_{st} is pursued in Figure 15 for evaporator heat transfer. There is an optimal size collector to attain maximum refrigeration.

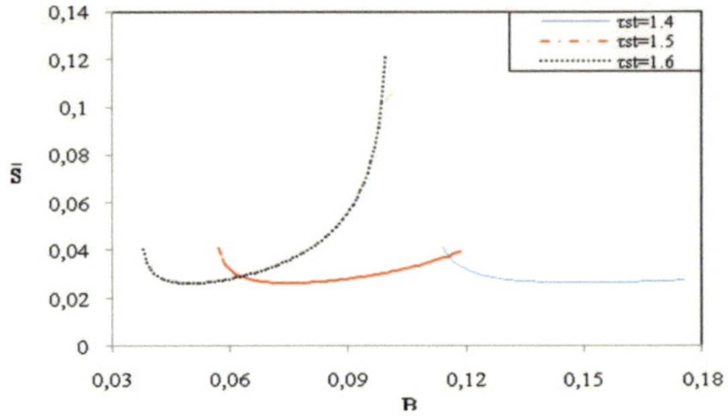

Fig. 13. Total entropy generated to reach a refrigerated space temperature set point temperature (τ_H=1.3)

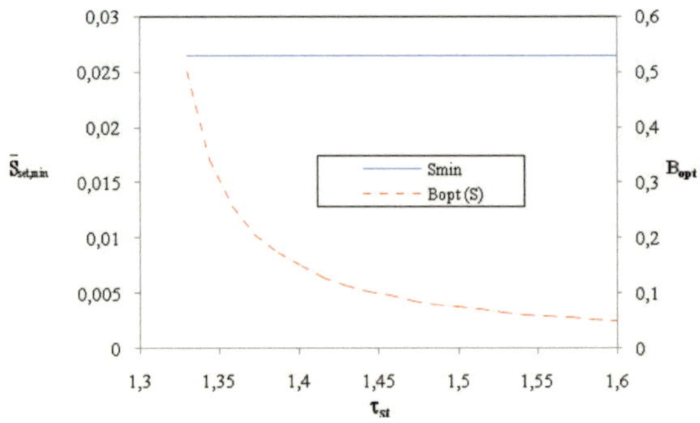

Fig. 14. The effect of dimensionless collector stagnation temperature, τst, on minimum entropy set point temperature and optimal collector size (τ_H=1.3).

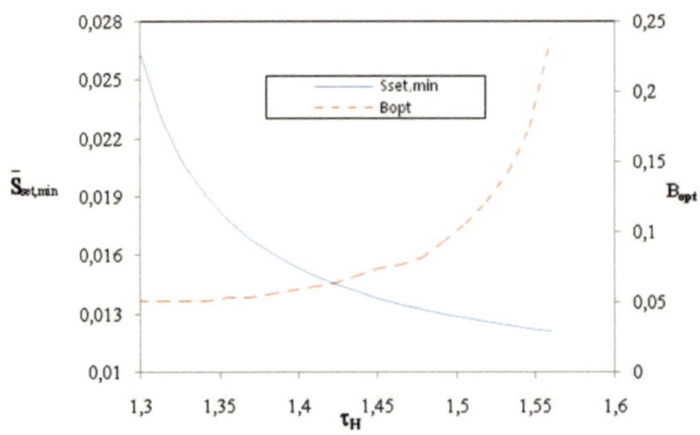

Fig. 15. The effect of dimensionless collector stagnation temperature, τ_H, on minimum entropy set point temperature and optimal collector size (τ_{st}=1.6).

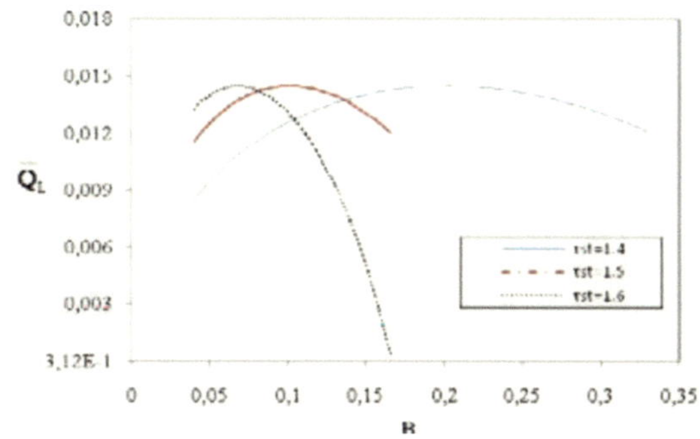

Fig. 16. The effect of dimensionless collector size, B on heat exchanger Q_L (τ_H=1.3 and τ_L=0.97).

Finally, Figures 17 and 18 depict the maximization of the heat input to evaporator and optimal size collector with stagnation temperature and temperature collector, respectively. $\overline{Q}_{L,max}$ remains constant and B_{opt} decreases. On the other hand, the curves of Fig. 15 indicate that as τ_H increases, $\overline{Q}_{L,max}$ and B_{opt} increases. For a τ_H value under 1.35, B_{opt} is lower than 0.1.

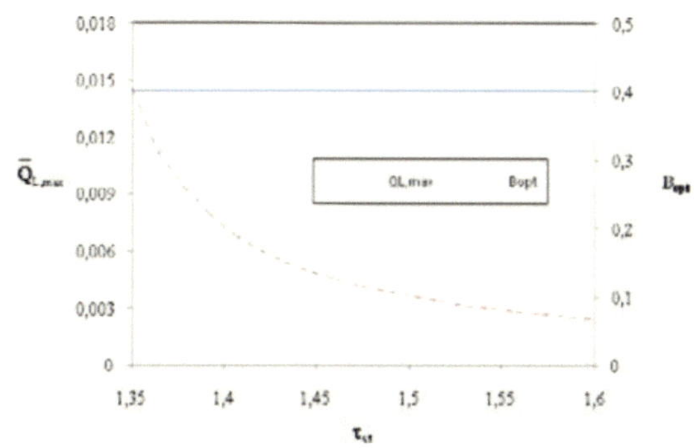

Fig. 17. Maximum heat exchanger, $Q_{L,max}$ to reached a refrigerated space temperature set point temperature (τ_H=1.3 and τ_L=0.97).

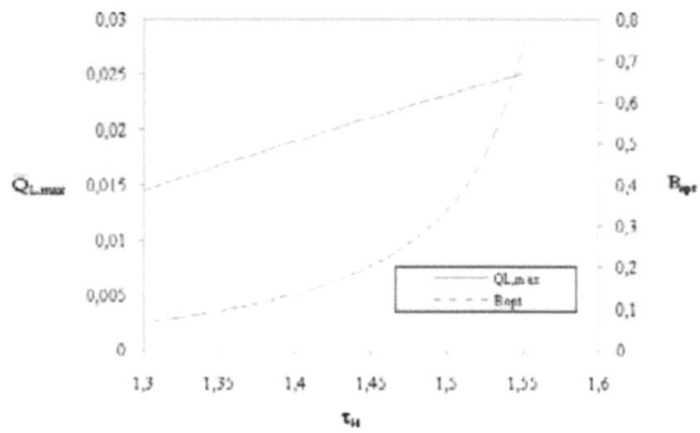

Fig. 18. Maximum heat exchanger, $Q_{L,max}$ to reached a refrigerated space temperature set point temperature (τ_{st}=1.3 and τ_L=0)

7. Conclusion

This chapter has presented an overview of the energy conversion systems optimization. Regarding the permanent regime, the functional decomposition and the optimization under constraints according to endoreversibility principles were the basis of the methodology. This procedure leads to a simple mathematical model and presents the advantage to avoid the use of equations with great number of unknowns. In so doing and as an example, the optimization of solar absorption refrigerator is investigated. The conceptual parameters are less sensible to temperature variations but more sensible to overall heat transfer coefficients variations. The couplings between the functional and conceptual parameters have permitted to define interesting technical and economical criteria related to the optimum cycle performances. The results confirm the usefulness of the hierarchical decomposition method in the process analyze and may be helpful for extended optimization investigations of other conversion energy cycles.

Also, the analysis in transient regime is presented. An endoreversible solar driven absorption refrigerator model has been analyzed numerically to find the optimal conditions. The existence of an optimal size collector for minimum time to reach a specified temperature in the refrigerated space, minimum entropy generation inside the cycle and maximum refrigeration rate is demonstrate. The model accounts for the irreversibilities of the three heat exchangers and the finiteness of the heat exchanger inventory (total thermal conductance).

8. References

Aoltola, J. (2003). Simultaneous synthesis of flexible heat exchanger networks. Thesis, Helsinky University of Technology

Bejan, A. (1995). Optimal allocation of a heat exchanger inventory in heat driven refrigerators", Heat Mass Transfer, vol.38, pp. 2997-3004,

Berrich, E.; Fellah, A.; Ben Brahim, A. & Feidt, M. (2011). Conceptual and functional study of a solar absorption refrigeration cycle. Int. J. Exergy vol.8,3, 265-280.

Boukhchana, Y.; Fellah, A.; & Ben Brahim, A. (2010). Modélisation de la phase génération d'un cycle de réfrigération par absorption solaire à fonctionnement intermittent. Int J Refrig. 34, 159-167

Bourges, B. (1992). Climatic data handbook for Europe. Kluwer,Dordrecht

Chen, J. (1995). The equivalent cycle system of an endoreversible absorption refrigerator and its general performance characteristics. Energy 20:995–1003

Chen, J. & Wu, C. (1996). General performance characteristics of an n stage endoreversible combined power cycle system at maximum specific power output. Energy Convers Manag 37:1401–1406

Chen, J. & Schouten, A. (1998). Optimum performance characteristics of an irreversible absorption refrigeration system", Energy Convers Mgmt, vol.39, pp. 999-1007,

Feidt, M. & Lang, S. (2002). Conception optimale de systèmes combinés à génération de puissance, chaleur et froid. Entropie 242:2–11

Fellah, A. ; Ben Brahim, A. ; Bourouis, M. & Coronas, A. (2006). Cooling loads analysis of an equivalent endoreversible model for a solar absorption refrigerator. Int J Energy 3:452–465

Fellah, A. (2008). Intégration de la décomposition hiérarchisée et de l'endoréversibilité dans l'étude d'un cycle de réfrigeration par absorption solaire: modélisation et optimisation. Thesis, Université de Tunis-Elmanar, Ecole nationale d'ingénieurs, Tunis, Tunisia

Fellah, A.; Khir, T.; & Ben Brahim, A. (2010). Hierarchical decomposition and optimization of thermal transformer performances. Struct Multidisc Optim 42(3):437–448

Goktun, S. (1997). Optimal Performance of an Irreversible Refrigerator with Three Berlitz, J.T.; Satzeger, V.; Summerer, V.; Ziegler, F. & Alefeld, G. (1999). A contribution to the evaluation of the economic perspectives of absorption chillers. Int J Refrig 22:67–76

Martinez, P.J. & Pinazo, J.M. (2002). A method for design analysis of absorption machines. Int J Refrig 25:634–639

Munoz, J.R. & Von Spakovsky, M.R. (2003). Decomposition in energy system synthesis/design optimization for stationary and aerospace applications. J Aircr 40:35–42 Heat Sources", Energy, vol. 22, pp. 27-31,

Perrin de Brichambaut, Ch. (1963). Rayonnement solaire: échanges radiatifs naturels. Editions Gautier-Villars, Paris

Sahin, B. & Kodal, A. (1995). Steady state thermodynamic analysis of a combined Carnot cycle with internal irreversibility. Energy 20:1285–1289

Summerer, F. (1996). Evaluation of absorption cycles with respect to COP and economics, *Int. J. Refrig.*, Vol. 19, No. 1, pp.19–24

Sokolov, M. & Hersagal, D. (1996). Optimal coupling and feasibility of a solar powered year-round ejector air conditioner", Solar Energy vol.50, pp. 507-516, 1993.

Tozer, R. & Agnew, B. (March 1999). Optimization of ideal absorption cycles with external irreversibilities. Int. Sorption Heat Pump Conference pp. 1-5, Munich,.

Tsirlin, A.M.; Kazakov, V.; Ahremenkov, A.A. & Alimova, N. A. (2006). Thermodynamic constraints on temperature distribution in a stationary system with heat engine or refrigerator. J.Phys.D: Applied Physics 39 4269-4277.

Vargas, J.V.C.; Horuz, I.; Callander, T. M. S.; Fleming, J. S. & Parise, J. A. R. (1998). Simulation of the transient response of heat driven refrigerators with continuous temperature control. Int. J. Refrig., vol.21, pp. 648–660,.

Vargas, J.V.C.; Ordonez, J. C.; Dilay, A. & Parise. J. A. R. (2000). Modeling, simulation and optimization of a solar collector driven water heating and absorption cooling plant. Heat Transfer Engineering, vol.21, pp. 35-45,

Wijeysundera, N.E. (1997). Thermodynamic performance of solar powered ideal absorption cycles. Solar energy, pp.313-319

Part 2

Heat Pipe and Exchanger

7

Optimal Shell and Tube Heat Exchangers Design

Mauro A. S. S. Ravagnani[1], Aline P. Silva[1] and Jose A. Caballero[2]
[1]*State University of Maringá*
[2]*University of Alicante*
[1]*Brazil*
[2]*Spain*

1. Introduction

Due to their resistant manufacturing features and design flexibility, shell and tube heat exchangers are the most used heat transfer equipment in industrial processes. They are also easy adaptable to operational conditions. In this way, the design of shell and tube heat exchangers is a very important subject in industrial processes. Nevertheless, some difficulties are found, especially in the shell-side design, because of the complex characteristics of heat transfer and pressure drop. Figure 1 shows an example of this kind of equipment.

In designing shell and tube heat exchangers, to calculate the heat exchange area, some methods were proposed in the literature. Bell-Delaware is the most complete shell and tube heat exchanger design method. It is based on mechanical shell side details and presents more realistic and accurate results for the shell side film heat transfer coefficient and pressure drop. Figure 2 presents the method flow model, that considers different streams: leakages between tubes and baffles, bypass of the tube bundle without cross flow, leakages between shell and baffles, leakages due to more than one tube passes and the main stream, and tube bundle cross flow. These streams do not occur in so well defined regions, but interacts ones to others, needing a complex mathematical treatment to represent the real shell side flow.

In the majority of published papers as well as in industrial applications, heat transfer coefficients are estimated, based, generally on literature tables. These values have always a large degree of uncertainty. So, more realistic values can be obtained if these coefficients are not estimated, but calculated during the design task. A few number of papers present shell and tube heat exchanger design including overall heat transfer coefficient calculations (Polley *et al.*, 1990, Polley and Panjeh Shah, 1991, Jegede and Polley, 1992, and Panjeh Shah, 1992, Ravagnani, 1994, Ravagnani et al. (2003), Mizutani *et al.*, 2003, Serna and Jimenez, 2004, Ravagnani and Caballero, 2007a, and Ravagnani *et al.*, 2009).

In this chapter, the work of Ravagnani (1994) will be used as a base to the design of the shell and tube heat exchangers. A systematic procedure was developed using the Bell-Delaware method. Overall and individual heat transfer coefficients are calculated based on a TEMA (TEMA, 1998) tube counting table, as proposed in Ravagnani *et al.* (2009), beginning with the smallest heat exchanger with the biggest number of tube passes, to use all the pressure drop and fouling limits, fixed before the design and that must be satisfied. If pressure drops or fouling factor are not satisfied, a new heat exchanger is tested, with lower tube passes number or larger shell diameter, until the pressure drops and fouling are under the fixed limits. Using a trial and error systematic, the final equipment is the one that presents the minimum heat exchanger area for fixed tube length and baffle cut, for a counting tube TEMA table including 21 types of shell and tube bundle diameter, 2 types of external tube diameter, 3 types of tube pitch, 2 types of tube arrangement and 5 types of number of tube passes.

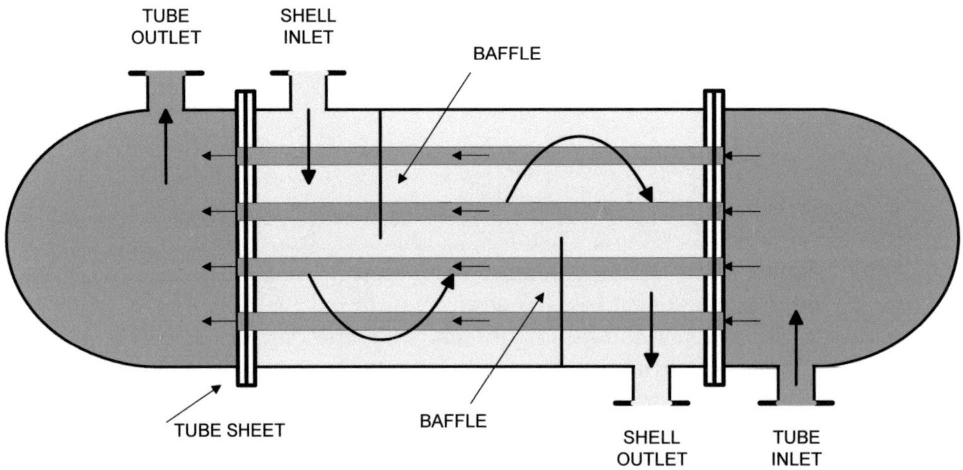

Fig. 1. Heat exchanger with one pass at the tube side

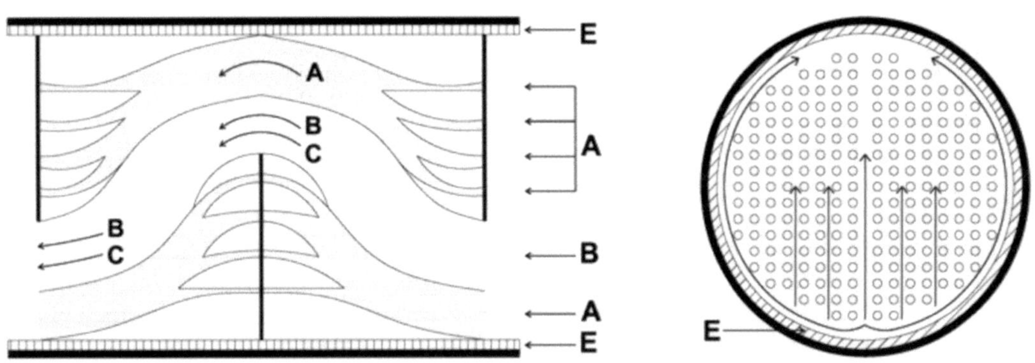

Fig. 2. Bell-Delaware streams considerations in the heat exchanger shell side

Two optimisation models will be considered to solve the problem of designing shell and tube heat exchangers. The first one is based on a General Disjunctive Programming Problem (GDP) and reformulated to a Mixed Integer Nonlinear Programming (MINLP) problem and solved using Mathematical Programming and GAMS software. The second one is based on the Meta-Heuristic optimization technique known as Particle Swarm Optimization (PSO). The differences between both models are presented and commented, as well as its applications in Literature problems.

2. Ravagnani and Caballero (2007a) model formulation

The model for the design of the optimum shell and tube equipment considers the objective function as the minimum cost including exchange area cost and pumping cost, rigorously following the Standards of TEMA and respecting the pressure drop and fouling limits. Parameters are: T_{in} (inlet temperature), T_{out} (outlet temperature), m (mass flowrate), ρ (density), Cp (heat capacity), μ (viscosity), k (thermal conductivity), ΔP (pressure drop), rd (fouling factor) and area cost data. The variables are tube inside diameter (d_{in}), tube outside diameter (d_{ex}), tube arrangement (arr), tube pitch (pt), tube length (L), number of tube passes (N_{tp}) and number of tubes (N_t), the external shell diameter (Ds), the tube bundle diameter (D_{otl}), number of baffles (N_b), baffles cut (l_c) and baffles spacing (l_s), heat exchange area (A), tube-side and shell-side film coefficients (h_t and h_s), dirty and clean global heat transfer coefficient (U_d and U_c), pressure drops (ΔP_t and ΔP_s), fouling factor (rd) and the fluids location inside the heat exchanger. The model is formulated as a General Disjunctive Programming Problem (GDP) and reformulated to a Mixed Integer Nonlinear Programming problem and is presented below.

Heat exchanger fluids location:

Using the GDP formulation of Mizutani et al. (2003), there are two possibilities, either the cold fluid is in the shell side or in the tube side. So, two binary variables must be defined, y_1^f and y_2^f. If the cold fluid is flowing in the shell side, or if the hot fluid is on the tube side, $y_1^f = 1$. It implies that the physical properties and hot fluid mass flowrate will be in the tube side, and the cold fluid physical properties and mass flowrate will be directed to the shell side. If $y_1^f = 0$, the reverse occurs. This is formulated as:

$$y_1^f + y_2^f = 1 \tag{1}$$

$$m^h = m_1^h + m_2^h \tag{2}$$

$$m^c = m_1^c + m_2^c \tag{3}$$

$$m^t = m_1^h + m_1^c \tag{4}$$

$$m^s = m_2^h + m_2^c \tag{5}$$

$$m_1^h \le m^{upper} y_1^f \tag{6}$$

$$m_1^c \le m^{upper} y_2^f \tag{7}$$

$$m_2^h \le m^{upper} y_2^f \tag{8}$$

$$m_2^c \le m^{upper} y_1^f \tag{9}$$

$$\mu^t = y_1^f \mu^h + y_2^f \mu^c \tag{10}$$

$$\mu^s = y_2^f \mu^h + y_1^f \mu^c \tag{11}$$

$$Cp^t = y_1^f Cp^h + y_2^f Cp^c \tag{12}$$

$$Cp^s = y_2^f Cp^h + y_1^f Cp^c \tag{13}$$

$$k^t = y_1^f k^h + y_2^f k^c \tag{14}$$

$$k^s = y_2^f k^h + y_1^f k^c \tag{15}$$

$$\rho^t = y_1^f \rho^h + y_2^f \rho^c \tag{16}$$

$$\rho^s = y_2^f \rho^h + y_1^f \rho^c \tag{17}$$

For the definition of the shell diameter (D_s), tube bundle diameter (D_{otl}), tube external diameter (d_{ex}), tube arrangement (*arr*), tube pitch (*pt*), number of tube passes (N_{tp}) and the number of tubes (N_t), a table containing this values according to TEMA Standards is constructed, as presented in Table 1. It contains 2 types of tube external diameter, 19.05 and 25.4 mm, 2 types of arrangement, triangular and square, 3 types of tube pitch, 23.79, 25.4 and 31.75 mm, 5 types of number of tube passes, 1, 2, 4, 6 and 8, and 21 different types of shell and tube bundle diameter, beginning on 205 mm and 173.25 mm, respectively, and finishing in 1,524 mm and 1,473 mm, respectively, with 565 rows. Obviously, other values can be aggregated to the table, if necessary.

D_s	D_{otl}	d_{ex}	*arr*	*pt*	N_{tp}	N_t
0.20500	0.17325	0.01905	1	0.02379	1	38
0.20500	0.17325	0.01905	1	0.02379	2	32
0.20500	0.17325	0.01905	1	0.02379	4	26
0.20500	0.17325	0.01905	1	0.02379	6	24
0.20500	0.17325	0.01905	1	0.02379	8	18
0.20500	0.17325	0.01905	1	0.02540	1	37
0.20500	0.17325	0.01905	1	0.02540	2	30
0.20500	0.17325	0.01905	1	0.02540	4	24
0.20500	0.17325	0.01905	1	0.02540	6	16
.	.	.	.	.	.	.
.	.	.	.	.	.	.
.	.	.	.	.	.	.
1.52400	1.47300	0.02540	1	0.03175	6	1761
1.52400	1.47300	0.02540	1	0.03175	8	1726
1.52400	1.47300	0.02540	2	0.03175	1	1639
1.52400	1.47300	0.02540	2	0.03175	2	1615
1.52400	1.47300	0.02540	2	0.03175	4	1587
1.52400	1.47300	0.02540	2	0.03175	6	1553
1.52400	1.47300	0.02540	2	0.03175	8	1522

Table 1. Tube counting table proposed

To find D_s, D_{otl}, d_{ex}, *arr*, *pt*, *ntp* and *Nt*, the following equations are proposed:

$$D_s = \sum_{i=1}^{565} d_{si} . ynt(i) \tag{18}$$

$$D_{otl} = \sum_{i=1}^{565} d_{otli} . ynt(i) \tag{19}$$

$$d_{ex} = \sum_{i=1}^{565} d_{exi} . ynt(i) \tag{20}$$

$$arr = \sum_{i=1}^{565} arri . ynt(i) \tag{21}$$

$$pt = \sum_{i=1}^{565} pti . ynt(i) \tag{22}$$

$$ntp = \sum_{i=1}^{565} ntpi . ynt(i) \tag{23}$$

$$nt = \sum_{i=1}^{565} nti . ynt(i) \tag{24}$$

$$\sum_{i=1}^{565} ynt(i) = 1 \tag{25}$$

Definition of the tube arrangement (*arr*) and the arrangement (*pn* and *pp*) variables:

$$pn = pn^1 + pn^2 \tag{26}$$

$$pp = pp^1 + pp^2 \tag{27}$$

$$pt = pt^1 + pt^2 \tag{28}$$

$$pn^1 = 0{,}5 . pt^1 \tag{29}$$

$$pn^2 = pt^2 \tag{30}$$

$$pp^1 = 0{,}866 . pt^1 \tag{31}$$

$$pp^2 = pt^2 \tag{32}$$

$$pt^1 \geq 0{,}02379 y_{tri}^{arr} \tag{33}$$

$$pt^2 \geq 0{,}02379 y_{cua}^{arr} \tag{34}$$

$$pt^1 \leq 0{,}03175 y_{tri}^{arr} \tag{35}$$

$$pt^2 \leq 0{,}03175 y_{cua}^{arr} \tag{36}$$

$$y_{tri}^{arr} + y_{cua}^{arr} = 1 \tag{37}$$

$$y_1^{dex} + y_2^{dex} = 1 \tag{38}$$

Definition of tube internal diameter (d_{in}):
This value, according to TEMA (1988) can be found for different values of d_{ex} and BWG. In this case, just two tube external diameters will be considered, which implies just two set of possibilities of d_{in}, as can be seen on Tables 2 and 3. However, other values can be aggregated. BWG determination can be formulated as:
for d_{ex}=0.01905, $BWG = \{10,11,12,...,18\}$
for d_{ex}=0.0254, $BWG = \{8,9,10,11,12,...,18\}$

$$BWG_1 = \sum_{j=1}^{9} y_{1j}^{bwg} BWG_j^1 \tag{39}$$

$$\sum_{j=1}^{9} y_{1j}^{bwg} \leq 1 \tag{40}$$

$$BWG_2 = \sum_{j=1}^{11} y_{2j}^{bwg} BWG_j^2 \tag{41}$$

$$\sum_{j=1}^{11} y_{2j}^{bwg} \leq 1 \tag{42}$$

$$BWG = BWG_1 + BWG_2 \tag{43}$$

d_{in} can be found by the following equations:

$$d_{in1} = \sum_{j=1}^{9} y_{1j}^{bwg} . d_{in^1} \tag{44}$$

$$d_{in_2} = \sum_{j=1}^{11} y_{2j}^{bwg} . d_{in^2} \tag{45}$$

$$d_{in} = d_{in^1} + d_{in^2} \tag{46}$$

BWG	d_{in}(m)
10	0.0122
11	0.0129
12	0.0135
13	0.0142
14	0.0148
15	0.0154
16	0.0157
17	0.0161
18	0.0166

Table 2. Determination of d_{in} for d_{ex} = 0.01905 m

BWG	d_{in}(m)
8	0.0170
9	0.0179
10	0.0186
11	0.0193
12	0.0199
13	0.0206
14	0.0212
15	0.0217
16	0.0221
17	0.0225
18	0.0229

Table 3. Determination of *din* for d_{ex} = 0.0254 m

$$y_1^{dex} \Rightarrow \vee_j y_{1j}^{bwg} \quad \text{or} \quad 1 - y_1^{dex} + \sum_j y_{1j}^{bwg} \geq 1 \tag{47}$$

$$y_2^{dex} \Rightarrow \vee_j y_{2j}^{bwg} \quad \text{or} \quad 1 - y_2^{dex} + \sum_j y_{2j}^{bwg} \geq 1 \tag{48}$$

$$y_{1j}^{bwg} \Rightarrow y_1^{dex} \quad \text{or} \quad 1 - y_{1j}^{bwg} + y_1^{dex} \geq 1 \tag{49}$$

$$y_{2j}^{bwg} \Rightarrow y_2^{dex} \quad \text{or} \quad 1 - y_{2j}^{bwg} + y_2^{dex} \geq 1 \tag{50}$$

Definition of tube length (L):
Five kinds of tube length are considered, according with TEMA (1988):

$$NL = \{2{,}438;\ 3{,}658;\ 4{,}877;\ 6{,}096;\ 6{,}706\}$$

$$L = 2{,}438 y_1^l + 3{,}658 y_2^l + 4{,}877 y_3^l + 6{,}096.y_4^l + 6{,}706.y_5^l \tag{51}$$

$$y_1^l + y_2^l + y_3^l + y_4^l + y_5^l = 1 \tag{52}$$

Definition of baffle spacing (*ls*):
According to TEMA (1988), baffle spacing must be between D_s and $D_s/5$. In this case, the following values will be considered:

$$ls \leq D_s \tag{53}$$

$$ls \geq D_s / 5 \tag{54}$$

Cross-flow at or near centerline for one cross-flow section (*Sm*):

$$Sm \leq ls.\left(D_s - D_{otl} + \frac{(pt - d_{ex}).(D_{otl} - d^{ex})}{pt} \right) + M(1 - y_{tri}^{arr}) \tag{55}$$

$$Sm \geq ls.\left(D_s - D_{otl} + \frac{(pt - d_{ex}).(D_{otl} - d_{ex})}{pt} \right) - M(1 - y_{tri}^{arr}) \tag{56}$$

$$Sm \leq ls.\left(D_s - D_{otl} + \frac{(pt - d_{ex}).(D_{otl} - d_{ex})}{pn}\right) + M(1 - y_{sq}^{arr}) \quad (57)$$

$$Sm \geq ls.\left(D_s - D_{otl} + \frac{(pt - d_{ex}).(D_{otl} - d_{ex})}{pn}\right) - M(1 - y_{sq}^{arr}) \quad (58)$$

Definition of the flow regimen in the shell side:
Reynolds number (Re_s) is given by:

$$\mathrm{Re}_s = \frac{m_s.d_{ex}}{\mu_s.Sm} \quad (59)$$

The shell side fluid velocity (v_s) is given by:

$$v_s = \frac{m_s / \rho_s}{(D_s / pt).(pt - d_{ex}).l_s} \quad (60)$$

According to Smith (2005), the velocity limits must be:

$$0.5 \leq v_s \leq 2,\ v_s \text{ in m/s} \quad (61)$$

The flow regimen is defined as a function of Reynolds number. Considering that in real heat exchangers the Reynolds number are generally high, laminar flow can be neglected, and Reynolds number, in this work, will be considered just for values greater then 100.

$$\mathrm{Re}_{s1} \geq 10^4.y_1^{res} \quad (62)$$

$$\mathrm{Re}_{s1} \leq 10^6.y_1^{res} \quad (63)$$

$$\mathrm{Re}_{s2} \geq 10^3.y_2^{res} \quad (64)$$

$$\mathrm{Re}_{s2} \leq 10^4.y_2^{res} \quad (65)$$

$$\mathrm{Re}_{s3} \geq 10^2.y_3^{res} \quad (66)$$

$$\sum_r y_r^{res} = 1 \quad (67)$$

$$\mathrm{Re}_s = \sum_r \mathrm{Re}_{sr} \quad (68)$$

Colburn factor (j_i) and Fanning factor (fl_s) determination:
Both, Colburn and Fanning factor are functions of Reynolds number and the tube arrangement, as shown on Table 4, extracted from Mizutani *et al.* (2003).
According to Mizutani *et al.* (2003), the DGP formulation is:

$$\sum_r \sum_s y_{r,s}^{rearr}.A_{r,s}^{a_1} = a_1 \quad (69)$$

$$\sum_r \sum_s y_{r,s}^{rearr}.A_{r,s}^{a_2} = a_2 \tag{70}$$

$$\sum_r y_r^{arr}.A_r^{a_3} = a_3 \tag{71}$$

$$\sum_r y_r^{arr}.A_r^{a_4} = a_4 \tag{72}$$

$$\sum_r \sum_s y_{r,s}^{rearr}.A_{r,s}^{b_1} = b_1 \quad 73)$$

$$\sum_r \sum_s y_{r,s}^{rearr}.A_{r,s}^{b_2} = b_2 \tag{74}$$

$$\sum_r y_r^{arr}.A_r^{b_3} = b_3 \tag{75}$$

$$\sum_r y_r^{arr}.A_r^{b_4} = b_4 \tag{76}$$

arr	Re_s	a_1	a_2	a_3	a_4	b_1	b_2	b_3	b_4
tri	10^5-10^4	0.321	-0.388	1.450	0.519	0.372	-0.123	7.00	0.500
tri	10^4-10^3	0.321	-0.388	1.450	0.519	0.486	-0.152	7.00	0.500
tri	10^3-10^2	0.593	-0.477	1.450	0.519	4.570	-0.476	7.00	0.500
tri	10^2-10	1.360	-0.657	1.450	0.519	45.100	-0.973	7.00	0.500
tri	< 10	1.400	-0.657	1.450	0.519	48.000	-1.000	7.00	0.500
sq	10^5-10^4	0.370	-0.395	1.187	0.370	0.391	-0.148	6.30	0.378
sq	10^4-10^3	0.107	-0.266	1.187	0.370	0.082	0.022	6.30	0.378
sq	10^3-10^2	0.408	-0.460	1.187	0.370	6.090	-0.602	6.30	0.378
sq	10^2-10	0.900	-0.631	1.187	0.370	32.100	-0.963	6.30	0.378
sq	< 10	0.970	-0.667	1.187	0.370	35.000	-1.000	6.30	0.378

Table 4. Empirical coefficients for equations (69) to (80) as function of Reynolds number and tube arrangement

$$a = \frac{a_3}{1+0{,}14.(\mathrm{Re}_s)^{a4}} \tag{77}$$

$$ji = a_1.1{,}064^a.(\mathrm{Re}_s)^{a2} \tag{78}$$

$$b = \frac{b_3}{1+0{,}14.(\mathrm{Re}_s)^{b4}} \tag{79}$$

$$fl^s = b_1.1{,}064^b.(\mathrm{Re}_s)^{b2} \tag{80}$$

$$1 - y_r^{res} - y_s^{arr} + y_{r,s}^{rearr} \geq 1 \tag{81}$$

$$\sum_r \sum_s y_{r,s}^{rearr} = 1 \tag{82}$$

Number of baffles (*Nb*):

$$Nb = \frac{L}{ls} - 1 \tag{83}$$

Number of tube rows crossed by the ideal cross flow (*Nc*):

$$Nc = \frac{D_s \lfloor 1 - 2.(l_c / D_s) \rfloor}{pp} \tag{84}$$

l_c is the baffle cut. The most used value is

$$l_c = 0{,}25.D_s \tag{85}$$

Fraction of total tubes in cross flow (*Fc*):

$$Fc = \frac{1}{\pi}\left[\pi + 2.\lambda.\sin(\arccos(\lambda)) - 2.\arccos(\lambda)\right] \quad 86)$$

where:

$$\lambda = \frac{D_s - 2.l_c}{D_{otl}} \tag{87}$$

Number of effective cross-flow tube rows in each windows (*Ncw*):

$$Ncw = \frac{0{,}8.l_c}{pp} \tag{88}$$

Fraction of cross-flow area available for bypass flow (*Fsbp*):

$$Fsbp = \frac{ls.\lfloor D_s - D_{otl} \rfloor}{Sm} \tag{89}$$

Shell-to-baffle leakage area for one baffle (*Ssb*):

$$Ssb = \frac{D_s.\delta_{sb}}{2}\left[\pi - \arccos\left(1 - \frac{2.l_c}{D_s}\right)\right] \tag{90}$$

where

$$\delta_{sb} = \left(\frac{3{,}1 + 0{,}004.(D_s.1000)}{1000}\right)$$

Angle values are in radians.
Tube-to-baffle leakage area for one baffle (*Stb*):

$$Stb = 0{,}0006223.dex.N_t\,(1 + Fc)\text{, m}^2 \tag{91}$$

Area for flow through window (*Sw*):
It is given by the difference between the gross window area (*Swg*) and the window area occupied by tubes (*Swt*):

$$Sw = Swg - Swt \tag{92}$$

where:

$$Swg = \frac{(D_s)^2}{4}\left[\arccos\left(1-2\frac{l_c}{D_s}\right)-\left(1-2\frac{l_c}{D_s}\right)\sqrt{1-\left(1-2\frac{l_c}{D_s}\right)^2}\right] \tag{93}$$

and:

$$Swt = (N_t/8)(1-Fc).\pi.(D_s)^2 \tag{94}$$

Shell-side heat transfer coefficient for an ideal tube bank (ho_i):

$$ho_i = \frac{j_i.Cp_s.m^s}{Sm}.\left(\frac{k_s}{Cp_s.\mu_s}\right)^{2/3} \tag{95}$$

Correction factor for baffle configuration effects (*Jc*):

$$Jc = Fc + 0{,}54.(1-Fc)^{0{,}345} \tag{96}$$

Correction factor for baffle-leakage effects (*Jl*):

$$Jl = \alpha + (1-\alpha).\exp\left(-2{,}2.\frac{Ssb+Stb}{Sm}\right) \tag{97}$$

where:

$$\alpha = 0{,}44.\left(1-\frac{Ssb}{Ssb+Stb}\right) \tag{98}$$

Correction factor for bundle-bypassing effects (*Jb*):

$$Jb = \exp(-0{,}3833.Fsbp) \tag{99}$$

Assuming that very laminar flow is neglected ($Re_s < 100$), it is not necessary to use the correction factor for adverse temperature gradient buildup at low Reynolds number.
Shell-side heat transfer coefficient (h_s):

$$h_s = ho_i.Jc.Jl.Jb \tag{100}$$

Pressure drop for an ideal cross-flow section (ΔP_{bi}):

$$\Delta P_{bi} = \frac{2.fl_s.Nc.(m_s)^2}{\rho_s.Sm^2} \tag{101}$$

Pressure drop for an ideal window section (ΔP_{wi}):

$$\Delta P_{wi} = (2 + 0{,}6.Ncw).\frac{(m_s)^2}{2.Sw.\rho_s.Sm} \tag{102}$$

Correction factor for the effect of baffle leakage on pressure drop (Rl):

$$Rl = \exp\left[-1{,}33.\left(1 + \frac{Ssb}{Ssb + Stb}\right).\left(\frac{Stb + Ssb}{Sm}\right)^k\right] \tag{103}$$

where:

$$k = -0{,}15.\left(1 + \frac{Ssb}{Ssb + Stb}\right) + 0{,}8 \tag{104}$$

Correction factor for bundle bypass (Rb):

$$Rb = \exp[-1{,}3456.Fsbp] \tag{105}$$

Pressure drop across the Shell-side (ΔP_s):

$$\Delta P_s = 2.\Delta P_{bi}.\left(1 + \frac{Ncw}{Nc}\right).Rb + (Nb+1).\Delta P_{bi}.R_b.R_l + Nb.\Delta P_{wi}.Rl \tag{106}$$

This value must respect the pressure drop limit, fixed before the design:

$$\Delta P_s \leq \Delta P_s design \tag{107}$$

Tube-side Reynolds number (Re_t):

$$\mathrm{Re}_t = \frac{4.m_t.N_{tp}}{\pi.din.\mu_t.N_t} \tag{108}$$

Friction factor for the tube-side (fl_t):

$$\frac{1}{\sqrt{fl_t}} = -4\log\left[\frac{0.27\varepsilon}{d_{ex}} + (7/\mathrm{Re}_t)^{0.9}\right] \tag{109}$$

where ε is the roughness in mm.
Prandtl number for the tube-side (Pr_t):

$$\mathrm{Pr}_t = \frac{\mu_t.Cp_t}{k_t} \tag{110}$$

Nusselt number for tube-side (Nu_t):

$$Nu_t = 0{,}027.(\mathrm{Re}_t)^{0{,}8}.(\mathrm{Pr}_t)^{1/3} \tag{111}$$

Tube-side heat transfer coefficient (h_t):

$$h_t = \frac{Nu_t.k_t}{d_{in}}.\frac{d_{in}}{d_{ex}} \tag{112}$$

Tube-side velocity (v_t):

$$v_t = \frac{\mathrm{Re}_t.\mu_t}{\rho_t.d_{in}} \tag{113}$$

The velocity limits are:

$$1 \le v_t \le 3, \quad v_t \text{ in m/s} \tag{114}$$

Tube-side pressure drop (including head pressure drop) (ΔP_t):

$$\Delta P_t = \rho_t.\left(\frac{2.fl_t.N_{tp}.L.(v_t)^2}{d_{in}} + 1{,}25.N_{tp}.(v_t)^2\right) \tag{115}$$

This value must respect the pressure drop limit, fixed before the design:

$$\Delta P_t \le \Delta P_t design \tag{116}$$

Heat exchanged:

$$Q = m_s.Cp_s(Ten_h - Tsai_h)_s \text{ or: } Q = m_s.Cp_s(Tsai_c - Ten_c)_s \tag{117.a}$$

$$Q = m_t.Cp_t(Ten_h - Tsai_h)_t \text{ or: } Q = m_t.Cp_t(Tsai_c - Ten_c)_t \tag{117.b}$$

Heat exchange area:

$$Area = N_t.\pi.d_{ex}.L \tag{118}$$

LMTD:

$$t_1 = Tout_h - Tin_c \tag{119}$$

$$t_2 = Tin_h - Tout_c \tag{120}$$

Chen (1987) LMDT approximation is used:

$$LMTD = [t_1 t_2 (t_1 + t_2)/2]^{1/3} \tag{121}$$

Correction factor for the LMTD (F_t):
For the F_t determination, the Blackwell and Haydu (1981) is used:

$$R = \frac{Tin_h - Tout_h}{Tout_c - Tin_c} \tag{122}$$

$$S = \frac{Tout_c - Tin_c}{Tin_h - Tin_c} \tag{123}$$

$$F_t = f_1(R,S) = \left(\frac{\sqrt{R^2+1}}{R-1}\right)\frac{\log[(1-P_{x1})/(1-R.P_{x1})]}{\log\left[\frac{2/P_{x1}-1-R+\sqrt{R^2+1}}{2/P_{x1}-1-R-\sqrt{R^2+1}}\right]} \tag{124}$$

where

$$P_{x1} = \frac{1-\left[\frac{R.S-1}{S-1}\right]^{1/NS}}{R-\left[\frac{R.S-1}{S-1}\right]^{1/NS}} \tag{125}$$

NS is the number of shells.
or, if $R = 1$,

$$F_t = f_2(R,S) = \frac{P_{x2}-\sqrt{R^2+1}/(1-P_x)}{\log\left[\frac{2/P_{x1}-1-R+\sqrt{R^2+1}}{2/P_{x1}-1-R-\sqrt{R^2+1}}\right]} \tag{126}$$

where

$$P_{x2} = P/(NS - NS.S + P) \tag{127}$$

$$R_1 \le R + M(1-y_{ft}^1) \tag{128}$$

$$R_1 \ge R - M(1-y_{ft}^1) \tag{129}$$

$$R \le 0.99 + M(1-y_{ft}^1) \tag{130}$$

$$F_t \le f_1(R,S) + M(1-y_{ft}^1) \tag{131}$$

$$F_t \ge f_1(R,S) - M(1-y_{ft}^1) \tag{132}$$

$$R \ge 0.99 - M(1-y_{ft}^2) \tag{133}$$

$$R \le 1.01 + M(1-y_{ft}^2) \tag{134}$$

$$F_t \le f_2(R,S) + M(1-y_{ft}^2) \tag{135}$$

$$F_t \ge f_2(R,S) - M(1-y_{ft}^2) \tag{136}$$

$$R \ge 1.01 - M(1-y_{ft}^3) \tag{137}$$

$$R_2 \le R + M(1-y_{ft}^3) \tag{138}$$

$$R_2 \geq R - M(1 - y_{fi}^3) \tag{139}$$

$$F_t \leq f_1(R,S) + M(1 - y_{fi}^3) \tag{140}$$

$$F_t \geq f_1(R,S) - M(1 - y_{fi}^3) \tag{141}$$

$$y_{fi}^1 + y_{fi}^2 + y_{fi}^3 = 1 \tag{142}$$

According to Kern (1950), practical values of F_t must be greater than 0.75. This constraint must be aggregated to the model:

$$F_t \geq 0.75 \tag{143}$$

Dirty overall heat transfer coefficient (U_d):

$$U_d = \frac{Q}{Area.LMTD} \tag{144}$$

Clean overall heat transfer coefficient (U_c):

$$U_c = \frac{1}{\left(\frac{d_{ex}}{d_{in}.h_t} + \frac{r_{in}.d_{ex}}{d_{in}} + \frac{d_{ex}.\log(d_{ex}/d_i n)}{2.k_{tube}} + r_{out} + \frac{1}{h_s}\right)} \tag{145}$$

Fouling factor calculation (r_d):

$$r_d = \frac{U_c - U_d}{U_c.U_d} \tag{146}$$

This value must respect the fouling heat exchanger limit, fixed before the design:

$$r_d \geq r_{d\,design} \tag{147}$$

For fluids with high viscosity, like the petroleum fractions, the wall viscosity corrections could be included in the model, both on the tube and the shell sides, for heat transfer coefficients as well as friction factors and pressure drops calculations, since the viscosity as temperature dependence is available. If available, the tubes temperature could be calculated and the viscosity estimated in this temperature value. For non-viscous fluids, however, this correction factors can be neglected.

Two examples were chosen to apply the Ravagnani and Caballero (2007a) model.

2.1 Example 1

The first example was extracted from Shenoy (1995). In this case, there is no available area and pumping cost data, and the objective function will consist in the heat exchange area minimization. Temperature and flow rate data as well as fluids physical properties and limits for pressure drop and fouling are in Table 5. It is assumed also that the tube thermal conductivity is 50 W/mK and the roughness factor is 0.0000457. Pressure drop limits are 42 kPa for the tube-side and 7 kPa for the shell-side. A dirt resistance factor of 0.00015 m^2K/W should be provided on each side.

Stream	T_{in} (K)	T_{out} (K)	m (kg/s)	μ (kg/ms)	ρ (kg/m³)	Cp (J/kgK)	K (W/mK)	r_d (W/mK)
Kerosene	371.15	338.15	14.9	.00023	777	2684	0.11	1.5e-4
Crude oil	288.15	298.15	31.58	.00100	998	4180	0.60	1.5e-4

Table 5. Example 1 data

With these fluids temperatures the LMTD correction factor will be greater than 0.75 and one shell is necessary to satisfy the thermal balance.
Table 6 presents the heat exchanger configuration of Shenoy (1995) and the designed equipment, by using the proposed MINLP model. In Shenoy (1995) the author uses three different methods for the heat exchanger design; the method of Kern (1950), the method of Bell Delaware (Taborek, 1983) and the rapid design algorithm developed in the papers of Polley *et al.* (1990), Polley and Panjeh Shah (1991), Jegede and Polley (1992) and Panjeh Shah (1992) that fixes the pressure drop in both, tube-side and shell-side before the design. The author fixed the cold fluid allocation on the tube-side because of its fouling tendency, greater than the hot fluid. Also some mechanical parameters as the tube outlet and inlet diameters and the tube pitch are fixed. The heat transfer area obtained is 28.4 m^2. The other heat exchanger parameters are presented in Table 6 as well as the results obtained in present paper with the proposed MINLP model, where two situations were studied, fixing and not fixing the fluids allocation. It is necessary to say that Shenoy (1995) does not take in account the standards of TEMA. According to Smith (2005), this type of approach provides just a preliminary specification for the equipment. The final heat exchanger will be constrained to standard parameters, as tube lengths, tube layouts and shell size. This preliminary design must be adjusted to meet the standard specifications. For example, the tube length used is 1.286 m and the minimum tube length recommended by TEMA is 8 ft or 2.438 m. If the TEMA recommended value were used, the heat transfer area would be at least 53 m^2.
If the fluids allocation is not previously defined, as commented before, the MINLP formulation will find an optimum for the area value in 28.31 m^2, with the hot fluid in the tube side and in a triangular arrangement. The shell diameter would be 0.438 m and the number of tubes 194. Although with a higher tube length, the heat exchanger would have a smaller diameter. Fouling and shell side pressure drops are very close to the fixed limits.
If the hot fluid is previously allocated on the shell side, because of the cold fluid fouling tendency, the MINLP formulation following the TEMA standards will find the minimum area equal to 38.52 m^2. It must be taken into account that when compared with the Shenoy (1995) value that would be obtained with the same tube length of 2.438 m (approximately 53 m^2), the area would be smaller, as well as the shell diameter and the number of tubes.

2.2 Example 2

As previously commented, the objective function in the model can be the area minimization or a cost function. Some rigorous parameters (usually constants) can be aggregated to the cost equation, considering mixed materials of construction, pressure ratings and different types of exchangers, as proposed in Hall *et al.* (1990).

The second example studied in this chapter was extracted from Mizutani *et al.* (2003). In this case, the authors proposed an objective function composed by the sum of area and pumping cost. The pumping cost is given by the equation:

$$P_{\cos t} = c_{\cos t} \cdot \left(\frac{\Delta P_t . m_t}{\rho_t} + \frac{\Delta P_s . m_s}{\rho_s} \right) \tag{148}$$

The objective function to be minimized is the total annual cost, given by the equation:

$$Min \;\; totalannual\cos t = a_{\cos t} \left(Area \right)^{b_{\cos t}} + P_{\cos t} \tag{149}$$

Table 7 presents costs, temperature and flowrate data as well as fluids physical properties. Also known is the tube thermal conductivity, 50 W/mK. As both fluids are in the liquid phase, pressure drop limits are fixed to 68.95 kPa, as suggested by Kern (1950). As in Example 1, a dirt resistance factor of 0.00015 m²K/W should be provided on each side.

Table 8 presents a comparison between the problem solved with the Mizutani et al. (2003) model and the model of Ravagnani and Caballero (2007a). Again, two situations were studied, fixing and not fixing the fluids allocation. In both cases, the annual cost is smaller than the value obtained in Mizutani et al. (2003), even with greater heat transfer area. It is because of the use of non-standard parameters, as the tube external diameter and number of tubes. If the final results were adjusted to the TEMA standards (the number of tubes would be 902, with d_{ex} = 19.05 mm and Ntp = 2 for square arrangement) the area should be approximately 264 m². However, the pressure drops would increase the annual cost. Using the MINLP proposed in the present paper, even fixing the hot fluid in the shell side, the value of the objective function is smaller.

Analysing the cost function sensibility for the objective function studied, two significant aspects must be considered, the area cost and the pumping cost. In the case studied the proposed MINLP model presents an area value greater (264.15 and 286.15 m² vs. 202.00 m²) but the global cost is lower than the value obtained by the Mizutani *et al.* (2003) model (5250.00 $/year vs. 5028.29 $/year and 5191.49 $/year, respectively). It is because of the pumping costs (2424.00 $/year vs. 1532.93 $/year and 1528.24 $/year, respectively).

Obviously, if the results obtained by Mizutani *et al.* (2003) for the heat exchanger configuration (number of tubes, tube length, outlet and inlet tube diameters, shell diameter, tube bundle diameter, number of tube passes, number of shells and baffle spacing) are fixed the model will find the same values for the annual cost (area and pumping costs), area, individual and overall heat transfer coefficients and pressure drops as the authors found. It means that it represents a local optimum because of the other better solutions, even when the fluids allocation is previously fixed.

The two examples were solved with GAMS, using the solver SBB, and Table 9 shows a summary of the solver results. As can be seen, CPU time is not high. As pointed in the Computational Aspects section, firstly it is necessary to choose the correct tool to solve the problem. For this type of problem studied in the present paper, the solver SBB under GAMS was the better tool to solve the problem. To set a good starting point it is necessary to give all the possible flexibility in the lower and upper variables limits, prior to solve the model, i.e., it is important to fix very lower low bounds and very higher upper limits to the most influenced variables, as the Reynolds number, for example.

	Shenoy (1995)	Ravagnani and Caballero (2007a) (Not fixing fluids allocation)	Ravagnani and Caballero (2007a) (fixing hot fluid on the shell side)
Area (m^2)	28.40	28.31	38.52
Q (kW)	1320	1320	1320
D_s (m)	0.549	0.438	0.533
D_{otl} (m)	0.516	0.406	0.489
Nt	368	194	264
Nb	6	6	19
ls (m)	0.192	0.105	0.122
Ntp	6	4	2
d_{ex} (mm)	19.10	19.05	19.05
d_{in} (mm)	15.40	17.00	17.00
L (m)	1.286	2.438	2.438
pt (mm)	25.40	25.40	25.40
h_t (W/m^2K)	8649.6	2759.840	4087.058
h_s (W/m^2K)	1364.5	3831.382	1308.363
U_d (W/m^2K)	776	779.068	572.510
U_c (W/m^2K)	1000.7	1017.877	712.422
ΔP_t (kPa)	42.00	26.915	7.706
ΔP_s (kPa)	3.60	7.00	7.00
r_d ($m^{2o}C/W$)	4.1e-3	3.01e-4	3.43e-4
NS	1	1	1
F_t	0.9	0.9	0.9
DTML (K)	88.60	88.56	88.56
arr	square	triangular	Square
v_t *(m/s)*	---	1.827	1.108
v_s *(m/s)*	---	0.935	1.162
hot fluid allocation	shell	tube	Shell

Table 6. Results for example 1

Stream	T_{in} (K)	T_{out} (K)	m (kg/s)	μ (kg/ms)	ρ (kg/m^3)	Cp (J/kgK)	k (W/mK)	ΔP (kPa)	r_d (W/mK)
1	368.15	313.75	27.78	3.4e-4	750	2840	0.19	68.95	1.7e-4
2	298.15	313.15	68.88	8.0e-4	995	4200	0.59	68.95	1.7e-4

$a_{cost} = 123$, $b_{cost} = 0.59$, $c_{cost} = 1.31$

Table 7. Example 2 data

3. The model of Ravagnani et al. (2009) PSO algorithm

Alternatively, in this chapter, a Particle Swarm Optimization (PSO) algorithm is proposed to solve the shell and tube heat exchangers design optimization problem. Three cases extracted from the literature were also studied and the results shown that the PSO algorithm for this type of problems, with a very large number of non linear equations. Being a global optimum heuristic method, it can avoid local minima and works very well with highly nonlinear problems and present better results than Mathematical Programming MINLP models.

	Mizutani et al. (2003)	Ravagnani and Caballero (2007a) (Not fixing fluids allocation)	Ravagnani and Caballero (2007a) (fixing hot fluid on the shell side)
Total annual cost ($/year)	5250.00	5028.29	5191.47
Area cost ($/year)	2826.00	3495.36	3663.23
Pumping cost ($/year)	2424.00	1532.93	1528.24
Area (m^2)	202.00	264.634	286.15
Q (kW)	4339	4339	4339
D_s (m)	0.687	1.067	0.838
D_{otl} (m)	0.672	1.022	0.796
N_t	832	680	713
Nb	8	7	18
ls (m)	0.542	0.610	0.353
N_{tp}	2	8	2
d_{ex} (mm)	15.90	25.04	19.05
d_{in} (mm)	12.60	23.00	16.00
L (m)	4.88	4.88	6.71
h_t (W/m^2°C)	6,480.00	1,986.49	4,186.21
h_s (W/m^2°C)	1,829.00	3,240.48	1,516.52
U_d (W/m^2°C)	----	655.298	606.019
U_c (W/m^2°C)	860	826.687	758.664
ΔP_t (kPa)	22.676	23.312	13.404
ΔP_s (kPa)	7.494	4.431	6.445
r_d (m^2°C/W)	-----	3.16e-4	3.32e-4
v_t (m/s)	----	1.058	1.003
v_s (m/s)	----	0.500	0.500
NS	----	1	1
arr	square	square	square
Hot fluid allocation	shell	tube	shell

Table 8. Results for example 2

	Example 1	Example 2
Equations	166	157
Continuous variables	713	706
Discrete variables	53	602
CPU time [a] Pentium IV 1 GHz (s)	.251	.561

Table 9. Summary of Solver Results

Kennedy and Elberhart (2001), based on some animal groups social behavior, introduced the Particle Swarm Optimization (PSO) algorithm. In the last years, PSO has been successfully applied in many research and application areas. One of the reasons that PSO is attractive is that there are few parameters to adjust. An interesting characteristic is its global search character in the beginning of the procedure. In some iteration it becomes to a local search method when the final particles convergence occur. This characteristic, besides of increase the possibility of finding the global optimum, assures a very good precision in the obtained value and a good exploration of the region near to the optimum. It also assures a good representation of the parameters by using the method evaluations of the objective function during the optimization procedure.

In the PSO each candidate to the solution of the problem corresponds to one point in the search space. These solutions are called particles. Each particle have also associated a velocity that defines the direction of its movement. At each iteration, each one of the particles change its velocity and direction taking into account its best position and the group best position, bringing the group to achieve the final objective.

In the present chapter, it was used a PSO proposed by Vieira and Biscaia Jr. (2002). The particles and the velocity that defines the direction of the movement of each particle are actualised according to Equations (153) and (154):

$$v_i^{k+1} = w \cdot v_i^k + c_1 \cdot r_1 \cdot \left(p_i^k - x_i^k\right) + c_2 \cdot r_2 \cdot \left(p_{GLOBAL}^k - x_i^k\right) \tag{150}$$

$$x_i^{k+1} = x_i^k + v_i^{k+1} \tag{151}$$

Where $\boldsymbol{x}_k^{(i)}$ and $\boldsymbol{v}_k^{(i)}$ are vectors that represent, respectively, position and velocity of the particle i, ω_k is the inertia weight, c1 and c2 are constants, r1 and r2 are two random vectors with uniform distribution in the interval [0, 1], $\boldsymbol{p}_k^{(i)}$ is the position with the best result of particle i and $\boldsymbol{p}_k^{global}$ is the position with the best result of the group. In above equations subscript k refers to the iteration number.

In this problem, the variables considered independents are randomly generated in the beginning of the optimization process and are modified in each iteration by the Equations (153) and (154). Each particle is formed by the follow variables: tube length, hot fluid allocation, position in the TEMA table (that automatically defines the shell diameter, tube bundle diameter, internal and external tube diameter, tube arrangement, tube pitch, number of tube passes and number of tubes).

After the particle generation, the heat exchanger parameters and area are calculated, considering the Equations from the Ravagnani and Caballero (2007a) as well as Equations (155) to (160). This is done to all particles even they are not a problem solution. The objective function value is obtained, if the particle is not a solution of the problem (any constraint is violated), the objective function is penalized. Being a heuristic global optimisation method, there are no problems with non linearities and local minima. Because of this, some different equations were used, like the MLTD, avoiding the Chen (1987) approximation.

The equations of the model are the following:

Tube Side :

Number of Reynolds (Re_t): Equation (108);

Number of Prandl (Pr_t): Equation (110);

Number of Nusselt (Nu_t): Equation (111);

Individual heat transfer coefficient (h_t): Equation (112);

Fanning friction factor (fl_t): Equation (109);

Velocity (v_t): Equation (113);

Pressure drop (ΔP_t): Equation (115);

Shell Side:
Cross-flow area at or near centerline for one cross-flow section (S_m):

$$\begin{bmatrix} triangular \Rightarrow \left[Sm = ls \cdot \left(D^s - Dft + \frac{\left(pt - d^t_{ex}\right)\cdot\left(Dft - d^t_{ex}\right)}{pt} \right) \right] \\ square \Rightarrow \left[Sm = ls \cdot \left(D^s - Dft + \frac{\left(pt - d^t_{ex}\right)\cdot\left(Dft - d^t_{ex}\right)}{pn} \right) \right] \end{bmatrix} \quad (152)$$

Number of Reynolds (Re_s): Equation (59);
Velocity (v_s): Equation (60);
Colburn factor (j_i): Equations (77) and (78);
Fanning friction factor (fl^s): Equations (79 and 80);
Number of tube rows crossed by the ideal cross flow (Nc): Equation (84);
Number of effective cross-flow tube rows in each window (Ncw): Equation (88);
Fraction of total tubes in cross flow (Fc): Equations (86) and (87);
Fraction of cross-flow area available for bypass flow ($Fsbp$): Equation (89);
Shell-to-baffle leakage area for one baffle (Ssb): Equation (90);
Tube-to-baffle leakage area for one baffle (Stb): Equation (91);
Area for flow through the windows (Sw): Equation (92);
Shell-side heat transfer coefficient for an ideal tube bank (ho_i): Equation (94);
Correction factor for baffle configuration effects (Jc): Equation (95);
Correction factor for baffle-leakage effects (Jl): Equations (96) and (97);
Correction factor for bundle-bypassing effects (Jb): Equation (98);
Shell-side heat transfer coefficient (h_s): Equation (99);
Pressure drop for an ideal cross-flow section (ΔP_{bi}): Equation (100);
Pressure drop for an ideal window section (ΔP_{wi}): Equation (101);
Correction factor for the effect of baffle leakage on pressure drop (Rl): Equations (102) and (103);
Correction factor for bundle bypass (Rb): Equation (104);
Pressure drop across the Shell-side (ΔP_s): Equation (105);
General aspects of the heat exchanger:
Heat exchanged (Q): Equations (117a) and (117b);
LMTD:

$$\begin{aligned} \Delta T1 &= T^h_{in} - T^c_{out} \\ \Delta T2 &= T^h_{out} - T^c_{in} \\ LMTD &= \frac{(\Delta T1 - \Delta T2)}{ln\left(\frac{\Delta T1}{\Delta T2}\right)} \end{aligned} \quad (153)$$

Correction factor for the LMTD: Equations (122) to (127);
Tube Pitch (pt):

$$pt = 1.25 \cdot d^t_{ex} \quad (154)$$

Bafles spacing (ls):

$$ls = \frac{L^t}{(Nb+1)} \tag{155}$$

Definition of the tube arrangement (*pn* and *pp*) variables:

$$\left[\begin{array}{l} triangular \Rightarrow \begin{bmatrix} pn = 0.5 \cdot pt \\ pp = 0.866 \cdot pt \end{bmatrix} \\ square \Rightarrow \begin{bmatrix} pn = pt \\ pp = pt \end{bmatrix} \end{array}\right. \tag{156}$$

Heat exchange area (*Area*):

$$\mathrm{Area} = \mathrm{n}^t \cdot \pi \cdot \mathrm{d}_{ex}^t \cdot \mathrm{L}^t \tag{157}$$

Clean overall heat transfer coefficient (*Uc*): Equation (145);
Dirty overall heat transfer coefficient (*Ud*): Equation (144);
Fouling factor (*rd*): Equation (146).
The Particle Swarm Optimization (PSO) algorithm proposed to solve the optimization problem is presented below. The algorithm is based on the following steps:

i. Input Data
- Maximum number of iterations
- Number of particles of the population (Npt)
- c1, c2 and w
- Maximum and minimum values of the variables (lines in TEMA table)
- Streams, area and cost data (if available)

ii. Random generation of the initial particles
There are no criteria to generate the particles. The generation is totally randomly done.
- Tube length (just the values recommended by TEMA)
- Hot fluid allocation (shell or tube)
- Position in the TEMA table (that automatically defines the shell diameter, the tube bundle diameter, the internal and the external tube diameter, the tube arrangement, the tube pitch, the number of tube passes and the number of tubes)

iii. Objective function evaluation in a subroutine with the design mathematical model
With the variables generated at the previous step, it is possible to calculate:
- Parameters for the tube side
- Parameters for the shell side
- Heat exchanger general aspects
- Objective Function

All the initial particles must be checked. If any constraint is not in accordance with the fixed limits, the particle is penalized.

iv. Begin the PSO
Actualize the particle variables with the PSO Equations (150) and (151), re-evaluate the objective function value for the actualized particles (step iii) and verify which is the particle with the optimum value;

v. Repeat step iv until the stop criteria (the number of iterations) is satisfied.
During this PSO algorithm implementation is important to note that all the constraints are activated and they are always tested. When a constraint is not satisfied, the objective

function is weighted and the particle is automatically discharged. This proceeding is very usual in treating constraints in the deterministic optimization methods.
When discrete variables are considered if the variable can be an integer it is automatically rounded to closest integer number at the level of objective function calculation, but maintained at its original value at the level of PSO, in that way we keep the capacity of changing from one integer value to another.
Two examples from the literature are studied, considering different situations. In both cases the computational time in a Pentium(R) 2.8 GHz computer was about 18 min for 100 iterations. For each case studied the program was executed 10 times and the optima values reported are the average optima between the 10 program executions. The same occurs with the PSO success rate (how many times the minimum value of the objective function is achieved in 100 iterations).
The examples used in this case were tested with various sets of different parameters and it was evaluated the influence of each case in the algorithm performance. The final parameters set was the set that was better adapted to this kind of problem. The parameters used in all the cases studied in the present paper are shown in Table 10.

c1	c2	w	Npt
1.3	1.3	0.75	30

Table 10. PSO Parameters

3.1 Example 3

This example was extracted from Shenoy (1995). The problem can be described as to design a shell and tube heat exchanger to cool kerosene by heating crude oil. Temperature and flow rate data as well as fluids physical properties and limits for pressure drop and fouling are in Table 11. In Shenoy (1995) there is no available area and pumping cost data, and in this case the objective function will consist in the heat exchange area minimization, assuming the cost parameters presented in Equation (04). It is assumed that the tube wall thermal conductivity is 50 WmK^{-1}. Pressure drop limits are 42 kPa for the tube-side and 7 kPa for the shell-side. A fouling factor of 0.00015 m^2KW^{-1} should be provided on each side.
In Shenoy (1995) the author uses three different methods for the heat exchanger design; the method of Kern (1950), the method of Bell Delaware (Taborek, 1983) and the rapid design algorithm developed in the papers of Polley *et al.* (1990), Polley and Panjeh Shah (1991), Jegede and Polley (1992) and Panjeh Shah (1992) that fixes the pressure drop in both, tube-side and shell-side before the design. Because of the fouling tendency the author fixed the cold fluid allocation on the tube-side. The tube outlet and inlet diameters and the tube pitch are fixed.
Table 12 presents the heat exchanger configuration of Shenoy (1995) and the designed equipment, by using the best solution obtained with the proposed MINLP model of Ravagnani and Caballero (2007a) and the PSO algorithm proposed by Ravagnani et al. (2009). In Shenoy (1995) the standards of TEMA are not taken into account. This type of approach provides just a preliminary specification for the equipment. The final heat exchanger will be constrained by standard parameters, as tube lengths, tube layouts and shell size. This preliminary design must be adjusted to meet the standard specifications. For example, the tube length used is 1.286 m and the minimum tube length recommended by TEMA is 8 ft or 2.438 m. As can be seen in Table 12, the proposed methodology with

the PSO algorithm in the present paper provides the best results. Area is 19.83 m², smaller than 28.40 m² and 28.31 m², the values obtained by Shenoy (1995) and Ravagnani and Caballero (2007a), respectively, as well as the number of tubes (102 vs. 194 and 368). The shell diameter is the same as presented in Ravagnani and Caballero (2007a), i.e., 0.438 m, as well as the tube length. Although with a higher tube length, the heat exchanger would have a smaller diameter. Fouling and shell side pressure drops are in accordance with the fixed limits.

The PSO success rate (how many times the minimum value of the objective function is achieved in 100 executions) for this example was 78%.

Stream	T_{in} (K)	T_{out} (K)	m (kg/s)	μ (kg/ms)	ρ (kg/m³)	Cp (J/kgK)	K (W/mK)	r_d (W/mK)
Kerosene	371.15	338.15	14.9	.00023	777	2684	0.11	1.5e-4
Crude oil	288.15	298.15	31.58	.00100	998	4180	0.60	1.5e-4

Table 11. Example 3 data

	Shenoy (1995)	Ravagnani and Caballero (2007a) best solution	Ravagnani et al. (2009)
Area (m²)	28.40	28.31	19.83
D_s (m)	0.549	0.438	0.438
Tube lenght (mm)	1286	2438	2438
d_{out}^t (mm)	19.10	19.10	25.40
d_{in}^t (mm)	15.40	17.00	21.2
Tubes arrangement	Square	Triangular	Square
Baffle spacing (mm)	0.192	0.105	0.263
Number of baffles	6	6	8
Number of tubes	368	194	102
tube passes	6	4	4
shell passes	1	1	1
ΔP^s (kPa)	3.60	7.00	4.24
ΔP^t (kPa)	42.00	26.92	23.11
h_s (kW/m²°C)	8649.6	3831.38	5799.43
h_t (kW/m²°C)	1364.5	2759.84	1965.13
U (W/m²°C)	1000.7	1017.88	865.06
r_d (m²°C/W)	0.00041	0.00030	0.00032
Ft factor	0.9	0.9	0.9
Hot fluid allocation	Shell	Tube	Tube
v_t (m/s)	**	1.827	2.034
v_s (m/s)	**	0.935	0.949

Table 12. Results for the Example 2

3.2 Example 4

The next example was first used for Mizutani *et al.* (2003) and is divided in three different situations.

Part A: In this case, the authors proposed an objective function composed by the sum of area and pumping cost. Table 13 presents the fluids properties, the inlet and outlet temperatures and pressure drop and fouling limits as well as area and pumping costs. The objective function to be minimized is the global cost function. As all the temperatures and flow rates are specified, the heat load is also a known parameter.

Part B: In this case it is desired to design a heat exchanger for the same two fluids as those used in Part A, but it is assumed that the cold fluid target temperature and its mass flow rate are both unknown. Also, it is considered a refrigerant to achieve the hot fluid target temperature. The refrigerant has a cost of $7.93/1000 tons, and this cost is added to the objective function.

Part C: In this case it is supposed that the cold fluid target temperature and its mass flow rate are unknowns and the same refrigerant used in Part B is used. Besides, the hot fluid target temperature is also unknown and the exchanger heat load may vary, assuming a cost of $20/kW.yr to the hot fluid energy not exchanged in the designed heat exchanged, in order to achieve the same heat duty achieved in Parts A and B.

Fluid	T_{in} (K)	T_{out} (K)	m (kg/s)	μ (kg/ms)	ρ (kg/m³)	Cp (J/kgK)	k (W/mK)	ΔP_{max} (kPa)	rd (W/mK)
A	368.15	313.75	27.78	3.4e-4	750	2,840	0.19	68.95	1.7e-4
B	298.15	313.15	68.88	8.0e-4	995	4,200	0.59	68.95	1.7e-4

$$A_{cost} = 123 \cdot A^{0.59}$$

$$Pump_{cost} = 1.31 \cdot \left(\frac{\Delta P^t \cdot m^t}{\rho^t} + \frac{\Delta P^s \cdot m^s}{\rho^s} \right)$$

$$\$/year, \ A = m^2 \ \ \Delta P = Pa \ \ m = kg/s \ \ \rho = kg/m^3$$

Table 13. Data for Example 6

All of the three situations were solved with the PSO algorithm proposed by Ravagnani et al. (2009) and the results are presented in Table 14. It is also presented in this table the results of Mizutani *et al.* (2003) and the result obtained by the MINLP proposition presented in Ravagnani and Caballero (2007a) for the Part A. It can be observed that in all cases the PSO algorithm presented better results for the global annual cost. In Part A the area cost is higher than the presented by Mizutani *et al* (2003) but inferior to the presented by Ravagnani and Caballero (2007a). Pumping costs, however, is always lower. Combining both, area and pumping costs, the global cost is lower. In Part B the area cost is higher than the presented by Mizutani *et al.* (2003) but the pumping and the cold fluid cost are lower. So, the global cost is lower (11,572.56 vs. 19,641). The outlet temperature of the cold fluid is 335.73 K, higher than 316 K, the value obtained by Mizutani *et al.* (2003).

In Part C, the area cost is higher but pumping, cold fluid and auxiliary cooling service cost are lower and because of this combination, the global annual cost is lower than the

presented by Mizutani *et al.* (2003). The outlet cold fluid temperature is 338.66 K, higher than the value obtained by the authors and the outlet hot fluid temperature is 316 K, lower than the value obtained by Mizutani *et al.* (2003).
The PSO success rates 74%, 69% and 65% for Parts A, B and C, respectively.

4. Conclusions

In the present chapter two models for the optimal design of heat exchangers were presented, one based on Mathematical Programming and other one based on the PSO algorithm.
The first one (Ravagnani and Caballero, 2007a) is based on GDP and the optimisation model is a MINLP, following rigorously the Standards of TEMA. Bell-Delaware method was used to calculate the shell-side variables. The model was developed for turbulent flow on the shell side using a baffle cut of 25% but the model can consider other values of baffle cuts.
The model calculates the best shell and tube heat exchanger to a given set of temperatures, flow rates and fluids physical properties. The major contribution of this model is that all the calculated heat exchanger variables are in accordance with TEMA standards, shell diameter, outlet tube bundle diameter, tube arrangement, tube length, tube pitch, internal and external tube diameters, number of baffles, baffle spacing, number of tube passes, number of shells and number of tubes. It avoids heat exchanger parameters adjustment after the design task. The tube counting table proposed and the use of DGP makes the optimisation task not too hard, avoiding non linearities in the model. The problem was solved with GAMS, using the solver SBB. During the solution of the model, the major problems were found in the variables limits initialisation. Two examples were solved to test the model applicability. The objective function was the heat exchange area minimization and in area and pumping expenses in the annual cost minimization. In the studied examples comparisons were done to Shenoy (1995) and Mizutani *et al.* (2003). Having a larger field of TEMA heat exchanger possibilities, the present model achieved more realistic results than the results obtained in the literature. Besides, the task of heat exchanger parameters adjustment to the standard TEMA values is avoided with the proposed MINLP formulation proposition. The main objective of the model is to design the heat exchanger with the minimum cost including heat exchange area cost and pumping cost or just heat exchange area minimization, depending on data availability, rigorously following the Standards of TEMA and respecting shell and tube sides pressure drops and fouling limits. Given a set of fluids data (physical properties, pressure drop and fouling limits and flow rate and inlet and outlet temperatures) and area and pumping cost data the proposed methodology allows to design the shell and tube heat exchanger and calculates the mechanical variables for the tube and shell sides, tube inside diameter (d_{in}), tube outside diameter (d_{ex}), tube arrangement, tube pitch (pt), tube length (L), number of tube passes (np^t) and number of tubes (N^t), the external shell diameter (D^s), the tube bundle diameter ($Dotl$), the number of baffles (Nb), the baffles cut (lc) and the baffle spacing (ls). Also the thermal-hydraulic variables are calculated, heat duty (Q), heat exchange area (A), tube-side and shell-side film coefficients (h^t and h^s), dirty and clean overall heat transfer coefficients (Ud and Uc), pressure drops (ΔP^t and ΔP^s),

	Part A			Part B		Part C	
	Mizutani *et al.* (2003)	Ravagnani and Caballero (2007a)	Ravagnani *et al.* (2009)	Mizutani *et al.* (2003)	Ravagnani *et al.* (2009)	Mizutani *et al.* (2003)	Ravagnani *et al.* (2009)
Total Cost ($/year)	5,250	5,028.29	3,944.32	19,641	11,572.56	21,180	15,151.52
Área Cost ($/year)	2,826	3,495.36	3,200.46	3,023	4,563.18	2,943	4,000.38
Pumping ($/year)	2,424	1,532.93	743.86	1,638	1,355.61	2,868	1,103.176
Cold Fluid ($/year)	*	*	--	14,980	5,653.77	11,409	6,095.52
Aux. Cool. ($/year)	*	*	--	*	--	3,960	3,952.45
m_c (kg/s)	*	*	*	58		46	
T_c^{out} (K)	*		--	316	335.73	319	338.61
T_h^{out} (K)	*		--	*	--	316	315.66
Área (m^2)	202	264.63	250.51	227	386.42	217	365.63
D_s (mm)	0.687	1.067	0.8382	0.854	1.219	0.754	1.219
length (mm)	4.88	4.88	6.09	4.88	3.66	4.88	4.88
d_{out}^t (mm)	15.19	25.04	19.05	19.05	19.05	19.05	25.40
d_{in}^t (mm)	12.6	23.00	15.75	14.83	14.20	14.83	18.60
Tubes arrangement	Square	Square	Square	Square	Triangular	Triangular	Square
Baffle Cut	**	25%	25%	**	25%	**	25%
Baffle spacing (mm)	0.542	0.610	0.503	0.610	0.732	0.610	0.732
Baffles	8	7	11	7	4	7	5
No. of tubes	832	680	687	777	1766	746	940
Tube passes	2	8	4	4	8	4	8
No. of shell passes	**	1	1	**	3	**	2
ΔP^s (kPa)	7,494	4,431	4,398.82	7,719	5,097.04	5,814	2,818.69
ΔP^t (kPa)	22,676	23,312	7,109.17	18,335	15,095.91	42,955	17,467.39
h_s (kW/m^2°C)	1,829	3,240.48	5009.83	4,110	3,102.73	1,627	3,173.352
h_t (kW/m^2°C)	6,480	1,986.49	1322.21	2,632	1,495.49	6,577	1,523.59
U (W/m^2°C)	860	655.29	700.05	857	598.36	803	591.83
r_d (m^2°C/W)	**	3.46e-4	3.42e-4	**	3.40e-4	**	3.40e-4
Ft factor	0.812	0.812	0.812	0.750	0.797	0.750	0.801
Hot Fluid Allocation	Shell	Tube	Tube	Tube	Tube	Shell	Tube
v_t (m/s)	**	1.058	1.951	**	1.060	**	1.161
v_s (m/s)	**	0.500	0.566	**	0.508	**	0.507

* Not applicable
** Not available

Table 13. Results for Example 6

fouling factor (*rd*), log mean temperature difference (*LMTD*), the correction factor of LMTD (*Ft*) and the fluids location inside the heat exchanger.
The second model is based on the Particle Swarm Optimization (PSO) algorithm. The Bell-Delaware method is also used for the shell-side calculations as well as the counting table presented earlier for mechanical parameters is used in the model. Three cases from the literature cases were also studied. The objective function was composed by the area or by the sum of the area and pumping costs. In this case, three different situations were studied. In the first one all the fluids temperatures are known and, because of this, the heat load is also a known parameter. In the second situation, the outlet hot and cold fluids are unknown. In this way, the optimization model considers these new variables. All of the cases are complex non linear programming problems. Results shown that in all cases the values obtained for the objective function using the proposed PSO algorithm are better than the values presented in the literature. It can be explained because all the optimization models used in the literature that presented the best solutions in the cases studied are based on MINLP and they were solved using mathematical programming. When used for the detailed design of heat exchangers, MINLP (or disjunctive approaches) is fast, assures at least a local minimum and presents all the theoretical advantages of deterministic problems. The major drawback is that the resulting problems are highly nonlinear and non convex and therefore only a local solution is guarantee and a good initialization technique is mandatory which is not always possible. PSO have the great advantage that do not need any special structure in the model and tend to produce near global optimal solutions, although only in an 'infinite large' number of iterations. Using PSO it is possible to initially favor the global search (using an l-best strategy or using a low velocity to avoid premature convergence) and later the local search, so it is possible to account for the tradeoff local vs. global search.
Finely, considering the cases studied in the present chapter, it can be observed that all of the solutions obtained with MINLP were possibly trapped in local minima. By using the PSO algorithm, a meta-heuristic method, because of its random nature, the possibility of finding the global optima in this kind or non-linear problems is higher. The percentage of success is also higher, depending on the complexity of the problem. Computational time (about 18 minutes for all cases) is another problem and the user must work with the possibility of a trade off between the computational effort and the optimum value of the objective function. But for small-scale problems the PSO algorithm proposed in the present paper presents the best results without excessive computational effort.

5. References

Blackwell, W. W. and Haydu, L. (1981), Calculating the Correct LMDT in Shell-and-Tube Heat Exchangers, *Chemical Engineering*, 101-106.

Chen, J. J. (1987), Letter to the Editor: Comments on improvement on a replacement for the logarithmic mean, *Chemical Engineering Science*, 42: 2488-2489.

Hall, S. G., Ahmad, S. and Smith, R. (1990). Capital Cost Targets for Heat Exchanger Networks Comprising Mixed Materials of Construction, Pressure Ratings an Exchanger Types, *Computers and Chemical Engineering*, Vol. 14, No 3, pp. 319-335

Jegede, F. O., Polley, G. T. (1992). Optimum Heat Exchanger Design, *Transactions of the Institute of Chemical Engineering*, 70 (A2): 133-141.

Kennedy, J.; Eberhart, R. (2001). Swarm Intelligence. Academic Press, London.

Kern, D. Q. (1950). Process Heat Transfer, McGraw Hill.

Mizutani, F. T., Pessoa, F. L. P., Queiroz, E. M., Hauan, S. and Grossmann, I. E. (2003). Mathematical Programming Model for Heat Exchanger Network Synthesis Including Detailed Heat Eschanger Designs. 1. Shell-and-Tube Heat Exchanger Design, *Industrial Engineering Chemistry Research*, 42: 4009-4018.

Panjeh Shahi, M. H. (1992). Pressure Drop Consideration in Process Integration, *Ph. D. Thesis - UMIST* - U.K.

Polley, G. T., Panjeh Shah, M. H. M. and Jegede, F. O. (1990). Pressure Drop Considerations in the Retrofit of Heat Exchanger Networks, *Transactions of the Institute of Chemical Engineering*, 68: 211-220.

Polley, G. T., Panjeh Shah, M. H. M. (1991). Interfacing Heat Exchanger Network Synthesis and Detailed Heat Exchanger Design, *Transactions of the Institute of Chemical Engineering*, 69: 445-447.

Ravagnani, M. A. S. S. (1994). Projeto e Otimização de Redes de Trocadores de Calor, *Ph.D. Thesis*, FEQ-UNICAMP-Campinas - Brazil. (in portuguese).

Ravagnani, M. A. S. S., Silva, A. P. and Andrade, A. L. (2003). Detailed Equipment Design in Heat Exchanger Networks Synthesis and Optimization. *Applied Thermal Engineering*, 23: 141 - 151.

Ravagnani, M. A. S. S. and Caballero, J. A. (2007a). A MINLP model for the rigorous design of shell and tube heat exchangers using the TEMA standards. Trans. IChemE, Part A, *Chemical Engineering Research and Design*, 85(A10): 1 - 13.

Ravagnani, M. A. S. S. e Caballero, J. A. (2007b). Optimal heat exchanger network synthesis with the detailed heat transfer equipment design. *Computers & Chemical Engineering*. 31: 1432 - 1448.

Ravagnani, M. A. S. S., Silva, A. P., Biscaia Jr, E. C. e Caballero, J. A. (2009). Optimal Design of Shell-and-Tube Heat Exchangers Using Particle Swarm Optimization. *Industrial & Engineering Chemistry Research*. 48 (6): 2927-2935.

Serna, M. and Jiménez, A. (2004). An Efficient Method for the Design of Shell and Tube Heat Exchangers, *Heat Transfer Engineering*, 25 (2), 5-16.

Shenoy, U. V. (1995). Heat Exchanger Network Synthesis - Process Optimization by Energy and Resource Analysis, Gulf Publishing Company.

Smith, R. (2005). Chemical Process Design and Integration, Wiley.

Taborek, J. (1983). Shell-and-Tube Heat Exchangers, Section 3.3, *Heat Exchanger Design Handbook*, Hemisphere.

TEMA. (1988). Standards of the Tubular Heat Exchanger Manufacturers Association, 7th ed.; Tubular Exchanger Manufacturers Association: New York,

Vieira, R. C. and Biscaia Jr., E. C., Métodos Heurísticos De Otimização. Notas De Aula Da Escola Piloto Virtual Do Peq/Coppe/Ufrj. Available Under Consultation: biscaia@peq.coppe.ufrj.br, 2002.

8

Enhancement of Heat Transfer in the Bundles of Transversely-Finned Tubes

Pis'mennyi, E.N., Terekh, A.M. and Razumovskiy, V.G.
National Technical University of Ukraine "Kyiv Polytechnic Institute"
Ukraine

1. Introduction

The problem of improving heat transfer surfaces made in the form of the bundles of transversely finned tubes remains very pressing. Studying the works, devoted to this problem, and a number of patents, including patents of the USA, Great Britain, Germany, and Japan, revealed that the ways of increasing thermoaerodynamic efficiency of the transversely finned surfaces mainly involve a search for the most rational types of finning and arrangement of the finned tube bundles.

A great many works on developing intensified surfaces are associated with setting up conditions for the breakdown of thickened boundary layers on relatively high fins and for the organization of a developed vortex flow, if possible, over the entire surface. Such conditions are attained by corrugation of the transverse fins (Tolubinskiy & Lyogkiy, 1964), their perforation (Migai et al., 1992; Eckels & Rabas, 1985), cutting into short sections with ends bent to the opposite sides (Taranyan et al., 1972; Sparrow & Myrum, 1985) and with the use of the so-called segment finning.

Analyzing these developmental works and related investigations allows us to note the following. The fin corrugation leads to a noticeable augmentation of heat transfer, but the heat enhancement is accompanied in this case by a still more noticeable increase of the aerodynamic drag: by data (Tolubinskiy & Lyogkiy, 1964), the replacement of smooth fins on a single finned cylinder by corrugated fins at Re = 10^4 enhances heat transfer by 12-15% with a drag increase by 65-70%. This circumstance, in conjunction with a difficult manufacturability of the tubes with corrugated fins, renders their wide use problematic.

The literature offers ample coverage of the results for thermoaerodynamic characteristics of bundles of tubes with cut fins (Taranyan et al., 1972; Kuntysh & Iokhvedov, 1968; Antufiev & Gusyev, 1968; Iokhvedov et al., 1975; Antufiev, 1965) (Fig. 1). Such heat transfer surfaces are fabricated from the tubes with a typical helical finning by cutting the fins into short sections by a thin mill along the generatrix of a carrying cylinder or along a helical line at an angle of 45°. Cutting does not practically diminish the fin surface and, according to data of the above-mentioned works, enhances heat transfer by 12-36% depending on the fin parameters and the method of cutting fins. The effect of flow turbulization, produced in this case, is the more appreciable, the higher are the cut fins. However, in all cases an increase in aerodynamic drag markedly outstrips an increase in heat transfer which on the whole noticeably reduces the total effect of heat transfer enhancement. Besides, the production of tubes with cut fins requires additional technological operations, which, in conjunction with a high susceptibility to contamination of the heat transfer surfaces from such tubes and complexity of their cleaning, substantially limited their application.

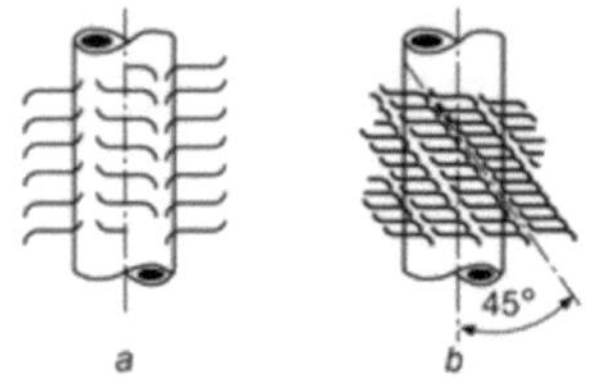

Fig. 1. Tubes with cut fins (Taranyan et al., 1972; Kuntysh & Iokhvedov, 1968; Iokhvedov et al., 1975)

More technologically advantageous, with similar characteristics of thermoaerodynamic efficiency, is a solution involving the application of flow-agitating notches on the ends of rolled-on fins (Kokorev et al., 1978; Kuntysh & Piir, 1991; Kuntysh, 1993), owing to which it found use in manufacturing heat transfer surfaces from aluminum tubes for some air cooling devices. However, this situation, apart from the limited range of application as to temperature conditions, retained essential operational drawbacks of the cut fins with bent edges, viz. an increased susceptibility to contamination and complexity of cleaning the interfin gaps, aggravated by their considerable blocking by the deformed fin edges.

Fairly well-known is the type of finning referred to in the literature as segment finning (Fig. 2). Judging from advertizing materials and from data (Weierman, 1976), such finning can produce an appreciable effect even in the case of some loss by heat transfer surface in comparison with continuous fins with the same height, thickness and pitch. Heat transfer is enhanced here not only because of a decrease in the boundary layer thickness as a result of a small width of an individual "segment" and the turbulization of flow on its separation from sharp edges of the fin but also because each individual element of the segment fin is, in essence, a straight rectangular fin with efficiency higher than of a disk fin. Estimates according to data (Weierman, 1976) manifest that replacing a typical helical finning by segment finning, with other conditions being equal, can decrease the number of tubes in a bundle by 18-20%. However, reviewing the structures of present-day heat exchangers, including power devices, indicates that, regardless of the above-stated assets that are combined with easier manufacturability (the segment finning is formed by welding a preliminarily notched steel strip to the tube by high-frequency currents), the considered type of the intensified heat transfer surface has not found wide application as yet. This fact is to some extent linked with a scanty investigation of thermoaerodynamic characteristics of the surfaces from tubes with segment finning. Insufficiently reliable, by our data, are design equations forming the basis for estimating thermoaerodynamic characteristics of the bundles of such tubes.

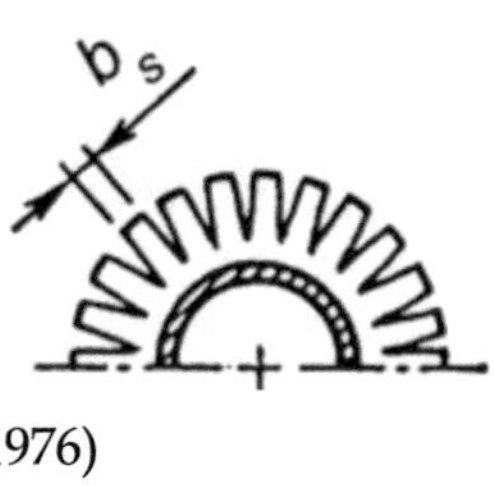

Fig. 2. Segmented fins (Weierman, 1976)

The enhancement method, described in (Fiebig et al., 1990), also involves perforation of a thin fin. Its essence lies in the formation, on a rectangular fin behind the carrying cylinder, of two delta wings representing bent parts of the perforated fin. The wings, inclined toward the incident flow, generate longitudinal vortices enhancing transfer in the near-wall region, which, with reference to actual heat exchangers, in the authors' opinion, can increase heat transfer by 20% and decrease operational expenditure by 10%. These estimates, relying on the exploration of experimental data for a single finned cylinder to a multirow heat exchanger, are too optimistic, considering the variation in the flow turbulence over the depth of a finned bundle.
Study (Kuntysh & Kuznetsov, 1992) attributes the improvement of mass and dimensional characteristics of the surfaces, made from circular tubes with a helical rolled-on finning, to the removal of a finned part lying in the wake region behind the carrying cylinder, where the heat transfer rate on the whole, as is well known, is relatively low. For the removal of the finned part to be possibly more adaptable to manufacture, the authors suggest that fins should be cut off on the chord along the plane parallel to the tube midsection (Fig. 3). According to data (Kuntysh & Kuznetsov, 1992), the heat transfer coefficients, related to a total surface of the tubes with a finning cutoff in the indicated fashion throughout height h, increase in comparison with the case of typical finned tubes by 1.23 times at $Re = 3 \cdot 10^3$ and by 1.3 times at $Re = 2.5 \cdot 10^4$. Here, the aerodynamic drag is practically unchangeable. However, due to the decrease in the area of the heat transfer surface, a total heat extraction diminishes by 13% and 23%, respectively. Nonetheless, the authors assert that, with other conditions being equal, up to 28% of the metal consumed for the finning fabrication can thus be saved. Overall, the above-considered way of perfecting transversely finned surfaces cannot be recognized as rational for obvious reasons.
Yet another trend for improving thermoaerodynamic characteristics of the tubular transversely finned surfaces that involves a change of their geometry is a search for new types of the arrangement of finned tube bundles. Main ideas of such developmental works have been mostly borrowed from studies (Yevenko & Anisin, 1976; Lokshin et al., 1982; Migai & Firsova, 1986) conducted with smooth tube bundles. Among them are:

- a so-called crossed or lattice arrangement of the tubes;
- arrangements intermediate between purely in-line and purely staggered that can be produced by changing the angle between the axes of longitudinal rows of ordinary in-line tube bundles and the velocity vector of the incident flow;
- zigzag arrangements formed by an alternating longitudinal displacement of the tubes in transverse rows of staggered bundles (Fig. 4); and
- bundles with an unequal number of tubes in transverse row.

A significant volume of investigations in these directions has been conducted in works (Kuntysh & Kuznetsov, 1992; Kuntysh et al., 1991; Kuntysh & Stenin, 1993; Stenin, 1994; Kuntysh et al., 1990).

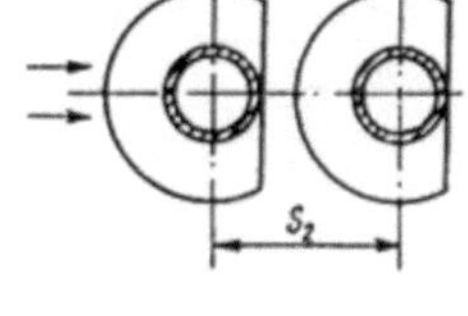

Fig. 3. Tubes with finning cut off in the rear part

When characterizing the results for thermoaerodynamic characteristics of crossed bundles it should be noted that, in conformity with data (Kuntysh & Kuznetsov, 1992), such arrangements do not enhance heat transfer in comparison with an ordinary layout of finned tubes, although the drag decreases somewhat, and should in all probability be regarded as inexpedient.

Studying the characteristics of intermediate in-line - staggered arrangements (Kuntysh & Stenin, 1993; Stenin, 1994) revealed the effect of heat transfer enhancement reaching 5% relative to the data for the original purely staggered bundle with a dense distribution of tubes, which is comparable with an error of the experiments of this kind. An appreciably greater effect can be attained, as study (Pis'mennyi, 1991) showed, using normal staggered arrangements with optimal pitch relationships.

The use of zigzag arrangements can be justified to some extent primarily because the frontal width of a bundle can be diminished (Fig. 4). Discrepancy of the data (Kuntysh & Kuznetsov, 1992) allows us to assume that there is no noticeable effect of heat transfer enhancement when the tubes in transverse rows of staggered bundles are displaced. The matter is that, in the above-mentioned study, in experiments with zigzag tube bundles with the fin factor ψ = 12.05 and with the drag equal to that of original ordinary staggered bundles heat transfer increased by 8-17% and experiments with zigzag tube bundles with the fin factor ψ = 17.5 indicated a decrease in the surface-average heat transfer. It is very doubtful that the recorded relatively insignificant variation in the parameters can lead to a substantial change in transfer in the bundles of transversely finned tubes. Obviously, the above effects are linked with methodical errors of the experiments.

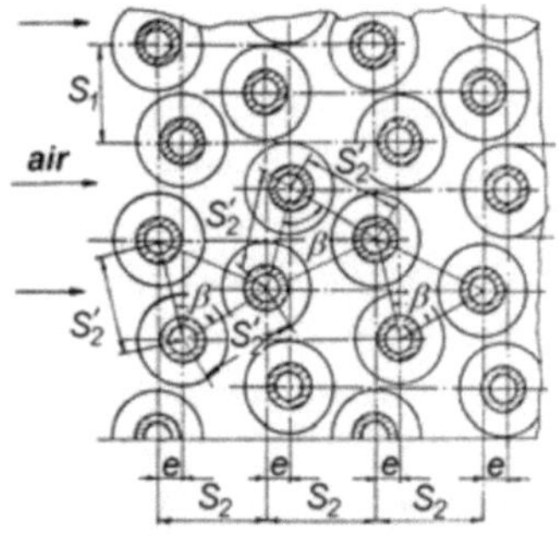

Fig. 4. Zigzag tube arrangement (Kuntysh & Kuznetsov, 1992)

Some studies (Kuntysh & Fedotova, 1983; Samie & Sparrow, 1986; Khavin, 1989) considered the possibility of enhancing heat transfer by inclining the finned tubes with respect to the direction of the incident flow. In this case, an additional turbulization of the flow occurs as a result of its separation from the inlet edges of fins whose planes have a positive attack angle. Experiments, performed in the region of Reynolds numbers Re = $5 \cdot 10^3$ - $5 \cdot 10^4$ with a single finned tube (Samie & Sparrow, 1986) and with staggered tube bundles (Kuntysh & Fedotova, 1983), showed an increase the surface in average heat transfer coefficients by 20-30% with an increase in the inclination angle by up to 40°. Here, the most intense rise in α is observed for the inclination angles raging from 0 to 30°, which, with a relatively monotonic increase in the aerodynamic drag, allows a selection of an optimal inclination angle of the tubes.

A certain reserve for increasing thermoaerodynamic efficiency of transversely finned tubes resides in converting to a noncircular shape of the cross section of the fin-carrying tube. Under definite conditions, specifically, with stringent limitations on the aerodynamic drag of a heat exchanger, it is reasonable to use shaped (generally plane-oval or elliptical) tubes in lieu of circular carrying tubes (Antufiev, 1966; Berman, 1965; Yudin & Fedorovich, 1992; Ilgarubis et al., 1987). Geometric characteristics of such tube bundles include additional parameters among which are the relation of longitudinal and lateral dimensions of the cylinder cross section and the attack angle of the profile with respect to the direction of the incident flow (Fig. 5). As works on heat transfer and aerodynamics of smooth and finned shaped cylinders (Antufiev, 1966; Yudin & Fedorovich, 1992; Ilgarubis et al., 1987) demonstrated, the search for optimal values of these parameters can be tied with the prospects for improving the developed surfaces.

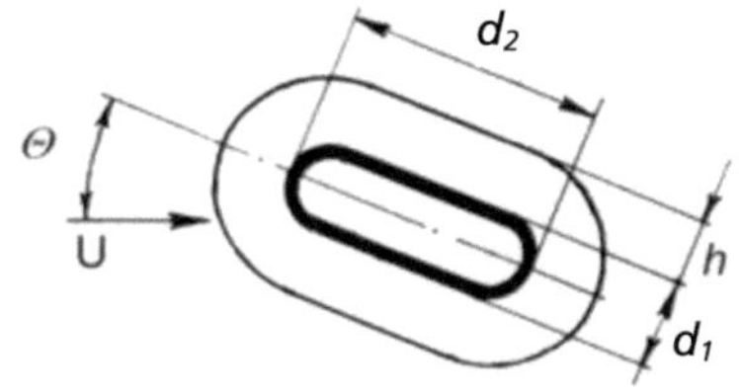

Fig. 5. Geometric characteristics of a plane-oval finned tube

In conclusion of the review part it should be noted that at the present time, in connection with a significant increase in the metal cost for large volumes of the production of the heat exchange equipment, it is considered expedient to use the ideas and designs leading to decrease in the specific amount of metal per structure only by a few percent, with other conditions being equal. Here, a good deal of attention is given to manufacturability of the developed surfaces: their elements should be fabricated with the aid of waste-free high-efficient technologies (like welding, rolling on and molding).

2. Physical substantiation of the proposed designs

In order to determine the ways of enhancement of local heat transfer in the bundles of transversely finned tubes an experimental research of the effects of fin tube geometry, of tube type, and of in-line and staggered arrangement on the distribution of heat transfer coefficients over the fin surface was conducted in NTUU "Kiev Polytechnic Institute".

In addition, the relationships governing local heat transfer incident to gas flows across bundles of tubes with radial and helically-wound fins are of interest in calculating the temperature distributions over the heating surfaces (especially at high loads), and in determination of the real fin efficiencies E. The latter depend strongly upon the distribution of heat transfer coefficients a over the fin surface, and are used in analytical engineering methods for converting from convection to reduced values of heat transfer coefficients.

The measurement of the distribution of a over a fin involves great procedural difficulties, so relatively little was done on this problem.

Here we present the results of the studies on the local heat transfer coefficients in tubes with radial fins of different geometries, operating in various bundle arrangements.

We worked with three tube models, described in Table 1. The model geometries conformed to those of industrial use. In addition, we aimed to investigate the widest possible range of ratio h/d which, from previous visual observations, we knew to have a large effect on the flow pattern over the fin. The tubes were assembled into staggered and in-line bundles (Table 2) and the measurements were performed in the first rows encountered by the gas flow, as well as in inner rows.

The local values of α were determined under steady conditions by means of foil-type heat flux rate transducers, known as the Gardon transducers (Gardon, 1960). Pickups of this type have a fairly high sensitivity, low inertia, linear dependence of the output signal on the sensed heat flux, and are relatively simple to construct. Six to seven such pickups, with thermocouples for determining the fin-surface temperature embedded in their casings, were spaced along the mid-height of copper fins of the tubes in which the heat flux was metered. The sensing elements of the pickups were 0.05-mm thick constantan foil disks of 6-mm diameter. The number of pickups per tube depended on the desired radial resolution of the measurements (i.e., over the height of the fin). Variations in azimuth were obtained by turning the entire heat-metering tube about its longitudinal axis in 5° to 10° increments. The three types of electrically-heated heat-metering tubes used in the experiments were identical in all respects except for dimensions.

Type of tube	Diameter of tube supporting the fins	Fin height d, mm	Fin pitch t, mm	Fin thickness S, mm	Relative fin height h/d	Surface area extension factor ψ
1	32.0	30.0	12.7	5.0	0.932	10.9
2	63.0	22.5	12.0	2.5	0.357	6.1
3	83.5	32.5	12.0	2.5	0.257	3.8

Table 1. Geometric parameters of the finned tubes

Assembly number	Tube type (see Table 1)	Arrangement	Relative transverse tube pitch σ_1	Relative longitudinal tube pitch σ_2	σ_1/σ_2
1	1	In-line	3.47	2.97	1.17
2	1	"	3.47	5.31	0.65
3	1	Staggered	3.47	2.56	1.30
4	1	"	3.47	5.31	0.65
5	1	"	4.31	2.65	1.62
6	1	"	4.31	5.31	0.81
7	2	"	3.33	1.30	2.56
8	2	"	3.33	1.76	1.89
9	2	"	4.09	1.30	3.15
10	2	"	4.09	1.76	2.32
11	3	"	3.02	1.30	2.32
12	3	"	3.91	1.30	3.01

Table 2. Geometric parameters of the bundles of finned tubes

The studies were performed on an air loop, described in detail in (Pis'mennyi & Lyogkiy, 1984). The values of a in the sensor locations were determined from expression

$$\alpha_i = \frac{q_i}{t_i - t_{fl}} \tag{1}$$

where q_i is the flux density in the pickup site; t_i is the fin surface temperature at this site and t_{fl} is the temperature of the flowing air. The values of q_i were determined from calibration curves $q_i = f(e_i)$, relating q_i to the value e_i of the electrical output of the pickup.

The measurements were performed at several fixed values of Re between $1{\cdot}10^4$ and $5{\cdot}10^4$. The reference dimension was diameter d of the tube supporting the fins, the reference velocity was the velocity U in the narrowest net free cross section of the bundle. The physical properties were taken at flow temperature t_{fl}. The results were worked up in the form $a_i/a_{av} = f(\varphi, P)$ at Re = const, where φ is the angle (in the plane of the fin) measured in the frontal (i.e., facing the air flow) generatrix of the tube, and $P = r - r_0/R - r_0$ is a dimensionless coordinate, measured along the radius of the tube that made this angle φ, the point $P = 0$ being the base of the fin ($r = r_0$), and the fin tip ($r = R$) having the coordinate $P = 1$. The local heat transfer coefficients a_i were related to the surface-averaged coefficient a_{av}, calculated as

$$\alpha_{av} = \frac{\sum_{i=1}^{N} a_i \Delta t_i H_i}{\Delta t_{av} \sum_{i=1}^{N} H_i} \tag{2}$$

where Δt_{av} is the surface-averaged temperature difference; Δt_i is the local temperature difference; H_i is the area of heat-transferring surface adjoining the ith heat flux rate pickup, and N is the number of such pickups. The values of a_{av} were also calculated from correlations for the average coefficients of heat transfer of transversely-finned tubes (Pis'mennyi, 1993). The differences in the values of a_i/a_{av} obtained by these two procedures were less than the experimental error, which we estimate to have been no more than 14.5%. Some of the results are plotted in the figures that follow, on which the distributions of a_i/a_{av} are shown in the form of five zones, depending on the level of this ratio.

The measurements that shed most light on the mechanisms of the processes occurring in the space between the fins and on the effect of the finned-tube geometry on the distribution of heat transfer coefficients over the fin surface are those in the first rows of bundles with fairly large relative longitudinal tube pitches σ_2 (Figs. 6 and 8). At these σ_2, the pitches cease to affect the heat transfer. These data show that the distribution of a over the fins surface is very uneven, with the highest a occurring in the root zones.

The shape of these distributions fits the model of the flow over a finned tube, which we have developed previously (Pis'mennyi, 1984, 1993) on the basis of experimental data.

The distinguishing feature of that model is that it reflects the existence of intensive secondary circulatory flows near the fin roots (Fig. 9). This circulation explains all the features of the flow over finned tubes and of the distributions of a over their surfaces not explained by traditional theory, namely the relatively high (and in some cases peak) a near fin roots, high a at the boundaries of the trailing vortical zones, etc.

The distributions of a_i/a_{av} plotted in Fig. 6 were obtained for a type *1* tube with relatively high fins ($h/d = 0.932$), for which the flow over the front (facing the flow or "windward") half of the tube circumference is as shown in Fig. 7a. The high values of a_i/a_{av} at the fin base result from the reattachment of secondary flows to its surface (zone A_2, Fig. 7a), while the minimum in the region of P = 0.3 to 0.35 occurs in the zone of separation A_3 where the secondary (recirculating) and primary flow collide. The static pressures, measured in the axial plane of the windward half circumference of the fin, suggest the existence of an adverse pressure gradient, which, of course, is the necessary condition for flow separation. The local peak observed at P = 0.7 is associated with reattachment of the flow that has separated from the sharp inlet edge of the fin which, in this case, is quite thick. The shapes of the distributions of a_i/a_{av} do not vary greatly as Re is varied from $2{\cdot}10^4$ to $5{\cdot}10^4$. However, the peak a_i/a_{av} is shifted from the root region of the fin (at Re = $2{\cdot}10^4$) to the region of its front edge (at Re = $5{\cdot}10^4$).

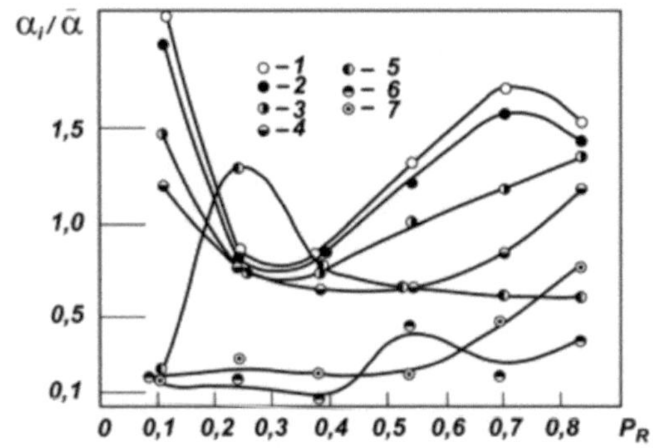

Fig. 6. Distribution of the relative heat transfer coefficient over the height of type *1* fin (h/d = 0.932) located in the first (along the gas flow) row. Re = $2{\cdot}10^4$. φ, °: 1) 0; 2) 30; 3) 60; 4) 90; 5) 120; 6) 150; 7) 180

The peaks of a_i/a_{av} below the middle section of the tube (curves for $\varphi > 90°$), observed in Fig. 6, stem from the spreading along the flow of the recirculation vortices from the fin root zones upon interaction with the layers that were sheared off from the supporting cylinder (tube). On the whole, the value of a_i/a_{av} in the rear part of the fin is lower because of formation (at high h/d) of a broad trailing vortex zone with relatively weak recirculation flow. For the same reason Fig. 6 does not exhibit any significant rise in a on the rear side of the tube itself. The slight rise in a on the rear edge of the fin at φ = 120 to 180° is produced by macroscale vortices, into which the sheared off layers (mentioned above) that separate from the tube are rolled up by the secondary flows. The large size of the rear vortex zone at high h/d is due to the significantly greater thickness of the boundary layer on the tall fin and, as a result, significant displacement of flow from the interfin space.

At low relative fin heights ($h/d \leq 0.357$), the flow pattern on the fin (Fig. 7b) and the distribution of a_i/a_{av} over its surface (Fig. 8) change. The effect of the tube itself on the flow pattern in the windward part of the fin increases, increasing the strength of the root vortex in region A_2 and forming of a local pressure trough beneath its center. As a result, an adverse pressure gradient is induced in the zone of the recirculating flow. This gradient causes separation of the boundary layer from the fin surface in zone A_3 and the formation, in addition to the small corner zone A_1 of still another circulation zone, namely A_3. The secondary flow is reattached in zone A_4 and then continues counter the main stream up to the windward edge of the fin. In this way the zone of secondary flow extends to the front

edge of the fin. The net result is that much of the fin area is virtually devoid of flow patterns predicted by the traditional theory.
The distribution of a_i/a_{av} over the fin surface fits the above flow pattern. In the windward half of the fin, the largest a_i/a_{av} occurs in the root region. As in the case of $h/d = 0.932$, this is caused by secondary flows. However, at the Re = $1{\cdot}10^4$ (Fig. 8a) there are two peaks in this region: the first is close to the base ($P_R < 0.15$) in the point of reattachment of the secondary flow to the fin surface, while the second is slightly farther from the base ($P_R = 0.33$). The second peak is induced by the transition from laminar to turbulent flow in the wall boundary layer in the accelerated flow region of zone A_2 (Fig. 7b). The second peak is not observed at higher Re (Fig. 8b) because the boundary-layer flow is now turbulent until the reattachment point. At all Re we observed beyond this point a reduction in a_i/a_{av}, which dips in the point of separation of the secondary flow from the fin surface (zone A_3, Fig. 7b) at approximately midheight of the fin ($P_R = 0.5$).

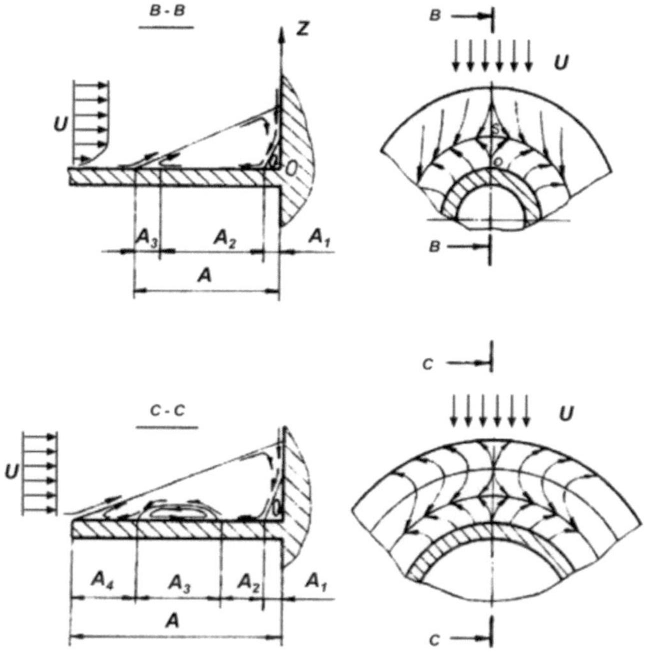

Fig. 7. Patterns of flow upstream of the windward part of the finned tube. a) At high h/d; b) at low h/d; A zone of secondary circulation flows; A_1 and A_3 zones of separation of the boundary layer from the surface; A_2 and A_4 reattachment regions

The a_i/a_{av} in the rear part of the fin are higher than in the case of high h/d, mainly because of the more intensive recirculation flows in the relatively shallower and wider trailing vortical zone. These flows create three-dimensional secondary flows in the root zone of the rear part of the fin ($\varphi > 120^o$ and $P_R < 0.2$), which are qualitatively similar to the recirculation flows in the front root zone A. Their superposition on the high flow turbulence in the rear one accounts for the high a_i/a_{av} in the rear of the tube at low h/d.
Note that at $h/d = 0.263$ maximum a occurs at the fin base at all Re. These results show up quite clearly in Figs. 9 and 10.

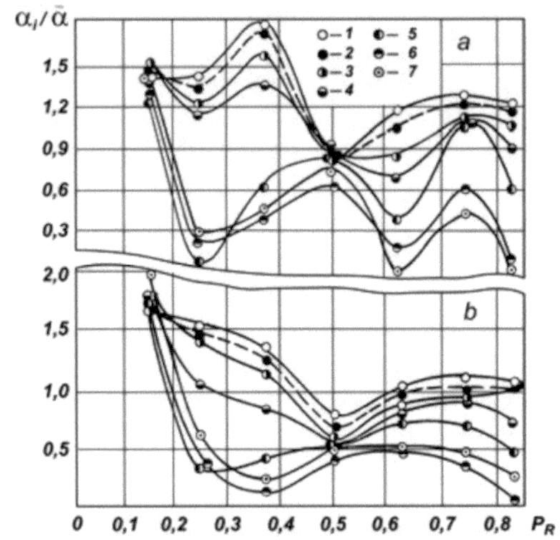

Fig. 8. Distribution of the relative heat transfer coefficient over the height of type 2 fin (h/d = 0.357) located in the first (in the direction of gas flow) row. Re: a) $1 \cdot 10^4$; b) $2 \cdot 10^4$. φ, °: 1) 0; 2) 30; 3) 60; 4) 90; 5) 120; 6) 150; 7) 180

Fig. 9. Distribution of the relative heat transfer coefficient over the surface of type 1 fin (h/d = 0.932) located in the frontal row of the bundle. Re = $2 \cdot 10^4$; the free-stream direction is from the top down. a_i/a_{av}: 1) 1.93 to 1.53; 2) 1.53 to 1.14; 3) 1.14 to 0.75; 4) 0.75 to 0.35; 5) 0.35 to 0.03

The above is also confirmed by the spike in turbulence level ε on the rear surface of the tube, as well as by the fact reported in (Neal & Hitchcock, 1966) that the variations in ε in the zones of φ = 0 and 180° at low h/d are similar. The local minima and maxima of the curves of $a_i/a_{av} = f(P_R)$ at φ > 90° (Fig. 8a) are probably caused by transitions in the recirculation boundary layer in the rear region of the fin, and also by propagation of secondary flows from the frontal zone of A toward the rear. With reduction in row pitch σ_2, the distribution of a over the fin surfaces of the inner-row tubes is increasingly affected by the type of tube arrangement within the bundle (Figs. 11 and 12). Thus, the value of a on the front parts ($\varphi \approx \pm 30°$) of fins of the inner tubes in in-line bundles decreases steeply along (Fig. 12a) because of the effect of vortical wake of the upstream tubes in the same rows (these values of a drop to those typical of the rear vortical zones of the upstream tubes).

Fig. 10. Distribution of the relative heat transfer coefficient over the surface of type *2* fin ($h/d = 0.357$) located in the frontal row of the bundle. Re = $2{\cdot}10^4$; the free-stream direction is from the top down. a_i/a_{av}: 1) 1.94 to 1.66; 2) 1.66 to 1.24; 3) 1.24 to 0.83; 4) 0.83 to 0.41; 5) 0.41 to 0.11

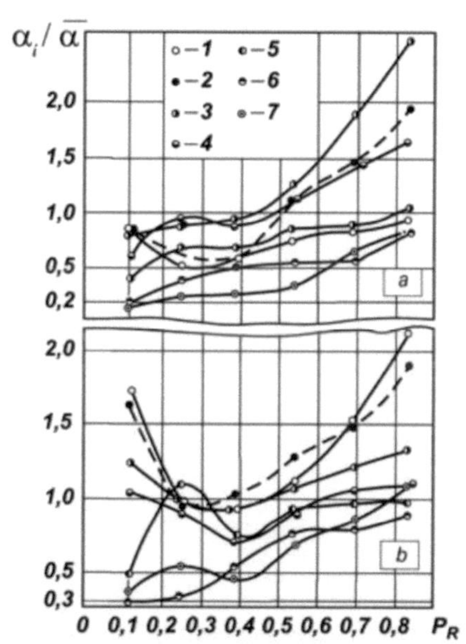

Fig. 11. Distribution of the relative heat transfer coefficient over the surface of type *1* fin ($h/d = 0.932$) in the 4th row of a six-row bundle. Re = $2{\cdot}10^4$; for legend see Fig. 7 or 9.
a) in-line bundle, $\sigma_1 = 3.47$, $\sigma_2 = 2.97$; b) staggered bundle, $\sigma_1 = 3.47$, $\sigma_2 = 2.66$

The drop in a increases at smaller σ_2, and is explainable by the decrease of the intensity of circulation in the wake with decreasing relative pitch L/d between interacting tubes (for in-line bundles $L/d = \sigma_2$). According to data from (Migay, 1978), the rear wake exhibits an approximately constant turbulence level ε. Thus, in the inner rows of in-line bundles, the highest a occur in fin zones with $\varphi \approx \pm$ 50 to 70°, where the fin is impacted by a flow outside its aerodynamic shadow.

In a staggered bundle with longitudinal and transverse pitches similar in those in an in-line bundle, the L/d is double that of the latter bundle ($L/d = 2\sigma_2$), so that at $\sigma_2 > 2$ the distributions of a on the fronts of inner-row tubes (Fig. 12b) stays approximately the same as on the first row, i.e., with a peak at $\varphi = 0°$. The uniformity of the distributions of a in these bundles improves with the forcing effect of adjoining tubes, which is maximum at $\sigma_1/\sigma_2 = 2\sqrt{3}$. In this case each tube operates as if it were surrounded by a circular deflector formed of six adjoining tubes.

As σ_2 in staggered bundles is decreased to $\sigma_2 < 1.5$ (which is possible at quite large σ_1 and relatively low h/d), the flow pattern begins to resemble that in in-line bundles. That is, the inner-row tubes operate in the near vortex wakes of upstream tubes, and the distributions of a over the fin circumference (Fig. 13) acquire the configuration exhibiting the low a in the front that is typical of in-line bundles. It remains to represent the experimental data on the local values of a in dimensionless form. In paper (Pis'mennyi, 1991), in which we described the surface-average values of a for bundles of transversely finned tubes, the high values of exponent $\overline{m}$ in the equation for the average heat transfer coefficients

$$N\overline{u} = C\,\mathrm{Re}^{\overline{m}} \tag{3}$$

which are typical of bundles with low L/d, were attributed to a direct correlation between the values of ε and m. Workup of data on local a for a type *1* finned tube (Table 1) located in an inner row of an in-line bundle with $\sigma_1 = 3.47$ and $\sigma_2 = 2.97$ confirmed the existence of this correlation.

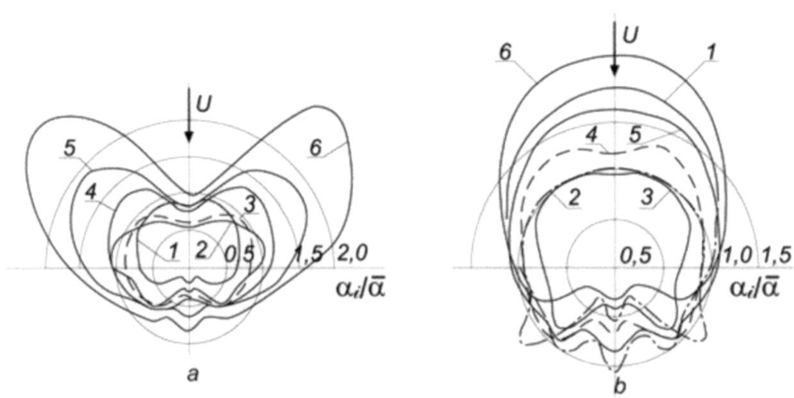

Fig. 12. Distribution of the relative heat transfer coefficient over the circumference of type *1* fin ($h/d = 0.932$) in the 4th row of a six-row bundle. Re = $2{\cdot}10^4$; a) in-line bundle, $\sigma_1 = 3.47$, $\sigma_2 = 2.97$; b) staggered bundle, $\sigma_1 = 3.47$, $\sigma_2 = 2.66$. *P*: 1) 0.117; 2) 0.247; 3) 0.393; 4) 0.540; 5) 0.697; 6) 0.833

The highest levels of ε in both the front and rear vortical wakes correlate with high values of m in the equation for the local heat transfer coefficient

$$\mathrm{Nu}_l = C_l \mathrm{Re}^m. \tag{4}$$

These results are listed in Table 3 in a form convenient for comparison with Figs. 11a and 12a, which present the distributions of α for this bundle. Averaging of local values over the surface of the finned tube yields $\bar{m}$ = 0.836, which is virtually identical to the value of $\bar{m}$ = 0.833, calculated from the correlation in (Pis'mennyi, 1993; Pis'mennyi & Terekh, 1991).

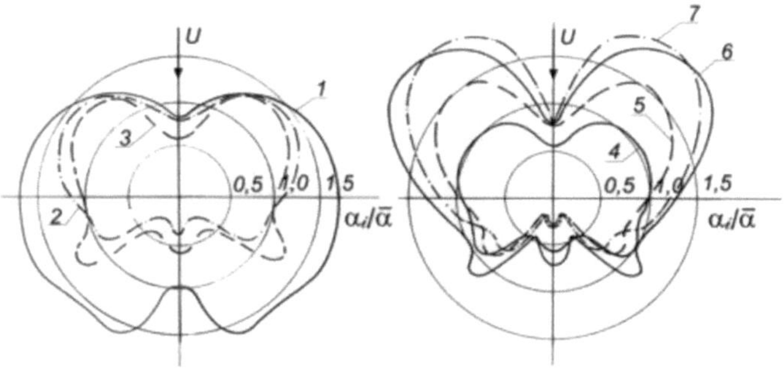

Fig. 13. Distribution of the relative heat transfer coefficient over the circumference of type 2 fin (h/d = 0.357) in the 5th row of a seven-row bundle with σ_1 = 3.33 and σ_2 = 1.30. Re = $2{\cdot}10^4$. P: 1) 0.13; 2) 0.22; 3) 0.34; 4) 0.47; 5) 0.60; 6) 0.72; 7) 0.81

P	Local values of $\bar{m}$						
	0°	30°	60°	90°	120°	150°	180°
0.117	0.86	0.80	1.00	1.07	1.30	1.22	1.30
0.247	0.88	0.82	0.80	0.77	0.88	0.85	1.04
0.393	0.73	0.72	0.56	0.72	0.71	0.95	1.15
0.540	0.77	0.65	0.58	0.64	0.82	0.99	1.10
0.697	0.78	0.58	0.57	0.64	0.90	1.10	1.05
0.893	0.91	0.78	0.62	0.69	0.94	1.06	0.89

Table 3. Values of $\bar{m}$ in the equation for the local heat transfer coefficient

We have thus gained deeper insight into the physics of the processes occurring in bundles of transversely finned tubes, improved our understanding of the temperature distributions in standard industrial tube bundles operating at high heat flux densities.

The developed heat transfer surfaces applied in large power plants have, as a rule, a staggered arrangement with large lateral S_1 and small longitudinal S_2 tube pitches, for which there are corresponding increased values of the parameter S_1/S_2 = 2.5 to 4.0. Large values of S_1 are dictated by a need for ensuring repairs of the heat exchange device. Besides, the bundles with large lateral pitch are less contaminated and more fitted for cleaning. On the other hand, relatively small values of the longitudinal pitch S_2 are dictated by a need for providing sufficient compactness of the heat exchange device as a whole.

As results for the flow (Pis'mennyi, 1991) and local heat transfer revealed, the arrangement parameters (S_1, S_2, and S_1/S_2) largely determine the flow past the bundles and the distribution of heat transfer rates over their surface.

Dimensions of the rear vortex zone are at a maximum in the bundles characterized by large values of parameter S_1/S_2. In such bundles, the neighboring tubes exert a slight reducing effect on the flow in interfin channels and, being displaced as the boundary layer at the fin thickens in the direction from the axis of the incident flow, the flow forms a wide rear zone (Fig. 14).

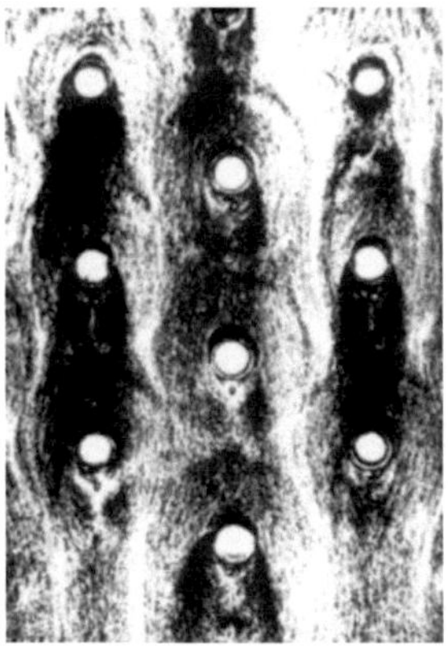

Fig. 14. Flow pattern in the finned tube bundle with $S_1/S_2 = 3.0$ (Re = $5.3{\cdot}10^4$) (Pis'mennyi, 1991)

In this case, the distribution of heat transfer rates over the finned tube surface is essentially uneven: in the forepart of a circular diagram of the relative heat transfer coefficients there is a crevasse associated with a superposition of the near vortex wake from the streamwise preceding tube (Fig. 13). The same pattern is observed also in the rear part of the tube. Thus, frontal and rear sections of the finned tubes, which are in the region of aerodynamic shadow in the discussed cases of large values of parameter S_1/S_2, show low-efficiency. In this case, the highest levels of the heat transfer rate are displaced into the lateral regions of tubes interacting with the flow outside the zone of the aerodynamic shadow.

In the typical case considered there are two ways of increasing thermoaerodynamic efficiency of the heat transfer surface:

- the first way is linked with constructive measures that make it possible to engage low-efficient sections of the finned tube surface in a high-rate heat transfer; and
- the second way involves the use of heat transfer surfaces not having a finned part that lies in the region of aerodynamic shadow and is, in fact, useless.

3. Bundles of the tubes with the fins bent to induce flow convergence

The first of the two ways is applied to the case of finned tubes with circular cross section. It is suggested that this be done by bending the fins to induce flow convergence (Fig. 15).

This method of a development of the idea of parallel bending of fins suggested at the Podol'sk Machine Building Plant (Russian Federation) (Ovchar et al., 1995) in order to reduce the transverse pitches of tubes in bundles and to improve the compactness of heat exchangers as a whole. Surfaces with fins bent to induce flow convergence can be made of ordinary tubes with welded or rolled on transverse fins, by deforming the latter, something that is achieved by passing the finned tube through a "draw plate" or another kind of bending device. In addition to parameters of bundles of ordinary finned tubes, the geometry of such surfaces is described by two additional quantities: the convergence angle γ and bending ratio b/h. For this reason the possibility of attaining the maximum enhancement of heat transfer when using the suggested method for tubes with specified values of d, h, t, and

δ, in addition to finding their optimal layout represented by ratios σ_1 and σ_2, involves finding the optimal values of γ and b/h. Special investigations were performed for determining the extent of the enhancement and the optimum values of the above parameters.

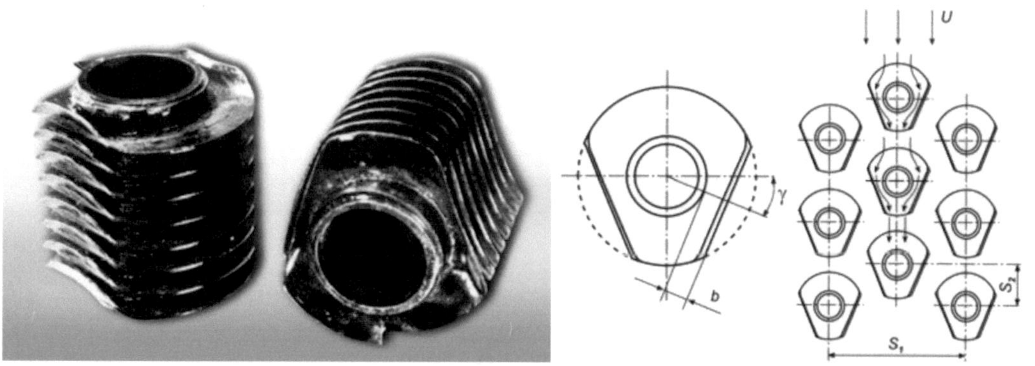

Fig. 15. Tubes with the fins bent to induce flow convergence

Studies of heat transfer, aerodynamic drag and specifics of flow over bundles of tubes with fins to bend in order to induce convergent flow were carried out using experimental methods, the most important features of which are:

- complete thermal simulation attained by electrically heating all the tubes in the bundle;
- determination, in the course of experiments, of surface-average convective heat transfer coefficients, by measuring the temperature distribution over the surface of the fin and of the wall of the finned tube.

The experiments were performed using steel tubes with welded-on transverse fins and the following geometric parameters: d = 42 mm, h = 15 mm, t = 8 mm, δ = 1.3 mm, ψ = 5.98, and b/h = 0.5. Tubes with these dimensions are extensively used in various heat exchangers, including units used in power equipment.

The effect of the value of γ on the thermoaerodynamic performance of finned-tube bundles was determined with the specially constructed bundles with $\gamma = 7°$, 14°, and 20°.

The value of b/h was selected with consideration of investigations of heat transfer and aerodynamic drag of the bundles of tubes with parallel bent fins, which showed that the value of b/h for tubes of these dimensions should be taken equal to 0.5. A further increase in this ratio causes a marked rise in drag while contributing virtually nothing to heat transfer enhancement.

Calorimetering tubes that served for measuring the temperature field of the fin and the tube were made of turned steel blanks in the form of two parts screwed together with one another. This provided access to the surface of the tube heightwise middle fin into which, as into the wall of the tube at its base, were lead-caulked in 18 copper-constantan thermocouples that used 0.1 mm diameter wires. The beads of the latter were, prior to this, welded in points with specified coordinates. The thermocouples were installed at a pre-bent fin. The fins were bent by pressing the tube in a specially constructed "draw plate" with a specified distance and angle between bending plains. The device was capable of producing fins with different values of γ.

The geometric parameters of the staggered tube bundles used in the experiments are listed in Table 4.

Location number	S_1, mm	S_2, mm	σ_1	σ_2	σ_1/σ_2	d_{eq}, mm
1	135	38	3.21	0.90	3.55	26.9
2	135	54	3.21	1.29	2.50	34.6
3	135	65	3.21	1.55	2.08	38.3
4	135	75	3.21	1.79	1.80	38.3
5	135	85	3.21	2.02	1.59	38.3
6	127	38	3.02	0.90	3.34	23.8
7	111	54	2.64	1.29	2.06	27.0
8	86	75	2.05	1.79	1.15	17.1
9	86	85	2.05	2.02	1.01	17.1

Table 4. Geometric parameters of the bundles of tubes with fins bent to induce flow convergence

A total of 24 staggered tube bundles were used in the experiments; the planes of the bent parts of the fins of all the tubes were oriented symmetrically relative to the direction of the free stream. The surface-average heat transfer of internal rows of tubes was investigated at Re between $3{\cdot}10^3$ and $6{\cdot}10^4$. The experimental data were approximated by power-law equations in the form

$$Nu = C_q{\cdot}Re^m. \quad (5)$$

Table 5 lists value of experimental constants m and C_q in equation (5) for the 24 bundles that were investigated. The extent of heat transfer enhancement was assessed by comparing our data with those for ordinary bundles (in which the fins were not bent).

Analysis of results shows that bending the fins enhances heat transfer in all the cases under study, but that its level, defined by the ratio of Nusselt numbers for the experimental and basic fins (Nu/Nu_b), depends highly on the value of γ and on the tube pitches (Fig. 16).

As expected, the highest values of Nu/Nu_b were obtained in bundles with large transverse and relatively small longitudinal pitches ($\sigma_1/\sigma_2 > 2$) when the conditions of washing the leading and trailing parts of basic finned tubes are highly unfavorable (Pis'mennyi, 1991).

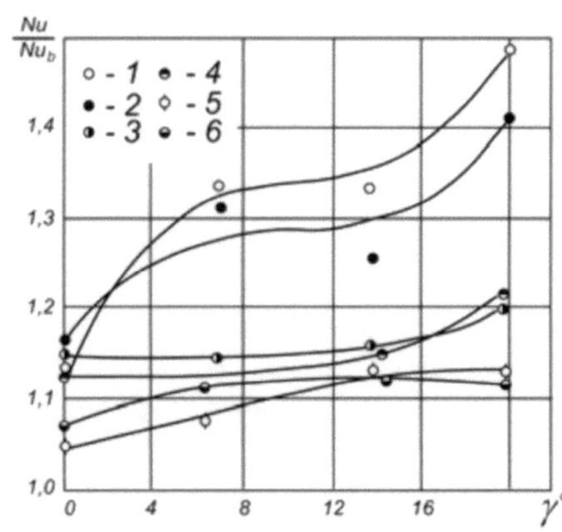

Fig. 16. Enhancement of heat transfer as a function of convergence angle γ at Re = $1.3{\cdot}10^4$. $\sigma_1 = 3.21$; σ_2: 1) 1.29; 2) 1.55; 3) 1.79; 4) 2.02; $\sigma_1 = 2.05$; σ_2: 5) 1.79; 6) 2.02

Location number	σ1	σ2	γ = 7°		γ = 14°		γ = 20°	
			m	C_q	*m*	C_q	*m*	C_q
1	3.21	0.90	-	-	0.69	0.135	0.71	0.112
2	3.21	1.29	0.64	0.270	0.66	0.232	0.71	0.158
3	3.21	1.55	o.67	0.204	0.70	0.150	0.68	0.196
4	3.21	1.79	0.66	0.187	0.69	0.151	0.68	0.162
5	3.21	2.02	0.61	0.276	0.69	0.148	0.67	0.185
6	3.02	0.90	-	-	0.77	0.068	0.74	0.084
7	2.64	1.29	-	-	0.73	0.132	0.76	0.111
8	2.05	1.79	0.71	0.105	0.71	0.107	0.71	0.107
9	2.05	2.02	0.66	0.155	0.71	0.102	0.72	0.094

Table 5. Experimental constants *m* and C_q in Eq. (5)

The bent tube segments in this case press the flow toward the trailing part of the finned tube, thus directing highly-intense secondary flows that are generated in the root region of the leading part of the tube (Pis'mennyi, 1984; Pis'mennyi & Terekh, 1993b) deeper into the space downstream of the tube. This, in the final analysis, decreases markedly the size of the trailing vertical zone, which is clearly seen by comparing Figs. 17a and b, obtained by visualizing the flow on the standard and bent fins of tubes of the same dimensions under otherwise same flow conditions. Significant segments of the trailing surfaces of the tube and fin then participate in high-rate heat transfer, thus increasing the overall surface-average heat transfer rate. This rate increases both because of reduction in the size of regions with low local velocities and by increasing the fraction of the surface of the finned tube that interacts with high-intensity secondary circulating flows, which are induced to come into contact with the peripheral lateral parts of the fin and also due to increasing the length of vortex filaments within a given area (Fig. 18).

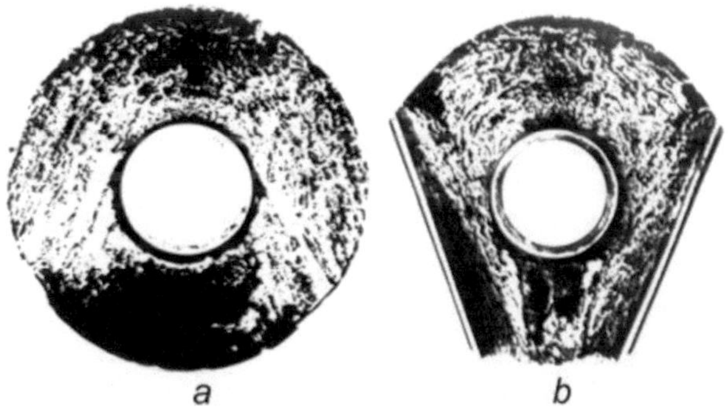

Fig. 17. Flow on the surface of an ordinary cylindrical (a) and bent (b) fins at Re = $2{\cdot}10^4$

The flow pattern in the wake of the finned tube also changes radically. The leading part of the further downstream tube interacts in this case with a relatively intensive jet that is discharged from the trailing convergent part of the tube-fins set (Fig. 15), rather than with the ordinarily encountered weak recirculation flow. This also increases the heat transfer coefficient, because of the increase in the local velocities and also because of intensification of secondary circulation flows at the fin root and increasing the region of their activity in the leading part of the finned tube (Fig. 18).

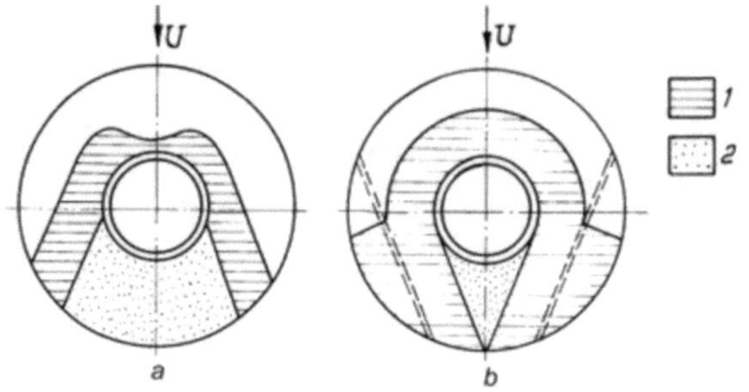

Fig. 18. Transformation of the dimensions of typical regions on the surface of a finned tube in the inward part of a bundle with $\sigma_1/\sigma_2 > 2$ with fin bent to provide for flow convergence. (a) an ordinary (basic) fin, and (b) bent fin. 1) region of intensive secondary circulating flows; 2) the trailing vortex zone

The level of perturbation of the wake flow which, as is known, controls, together with the local velocities, the rate of heat transfer remains rather high with the bent fins. This is promoted by turbulization of the flow after its separation from the outer surfaces of the perforated wall of the convergent "nozzle" that is formed by the bent parts of the fins (Fig. 15) and injection through gaps between their edges of a part of the flow from the spaces between the fins transverse to the free stream (Fig. 19).

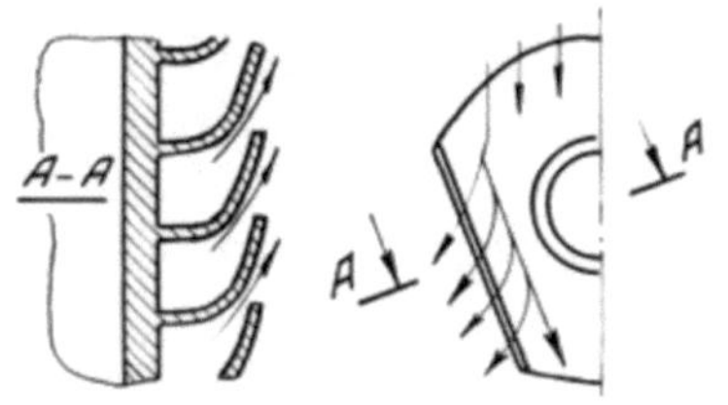

Fig. 19. Injection of flow into the space between the tubes through slots in the walls of the "convergent nozzle"

Taken together, all the above increases the surface-averaged heat transfer coefficient. Here exist optimal values of σ_2 which give, in case of $\sigma_1/\sigma_2 > 2$ under study, the greatest gain in the heat transfer coefficient. Thus, at $\sigma_1 = 3.21$ the value of Nu/Nu_b is highest at $\sigma_2 \approx 1.3$. The slight deterioration in the improvement at lower values of σ_2 is caused by increasing the mutual shading of tubes of the deeper-lying rows, which interferes with the supply of "fresh" flow from the spaces between the tubes to the convergent passages formed by the bent fins. A much greater reduction in the value of Nu/Nu_b is observed when the value of σ_2 is increased above the optimal. This is also caused by redistribution of the flow in the spaces between the tubes and the fins so as to reduce the flow rates within the latter.

The dominant effect of the relationship between the flow rates in the spaces between the fins and those between the tubes is also confirmed by the fact that reducing the values of σ_1 while maintaining the values of σ_1/σ_2 constant causes blockage of spaces between the tubes, over which a part of the flow was bypassed past the convergent passages formed by the fins (Fig. 20), which causes the flow rate through the latter to increase.

It is typical that the maximum gain in the rate of heat transfer is observed in layouts that also provide for the highest absolute values of the surface-average heat transfer coefficients.

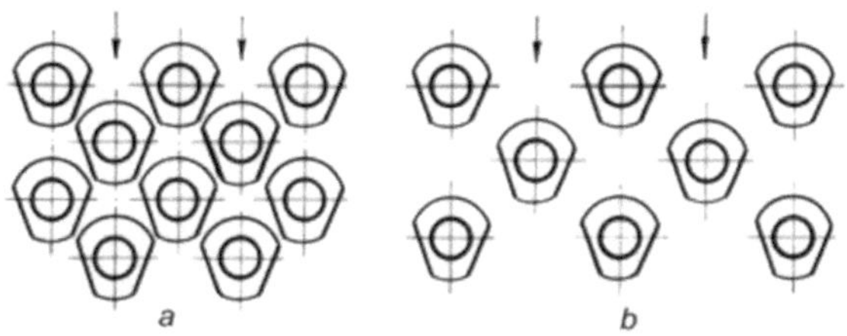

Fig. 20. Comparison of configurations of bundles with $\sigma_1 = 3.21$, $\sigma_2 = 1.55$ ($\sigma_1/\sigma_2 = 2.08$) (b) and with $\sigma_1 = 2.64$, $\sigma_2 = 1.29$ ($\sigma_1/\sigma_2 = 2.06$) (a)

As previously mentioned, the effect of γ on the rate of heat transfer is very clearly observed, but is much more complex than it would appear at first sight. This is seen from Fig. 16 which, in addition to data obtained in the present experimental study at γ between 7 and 20°, also presents experimental results on bundles of the same size with parallel bending of fins ($\gamma = 0°$). The effect of γ is most perceptible at the ranges between 0 to 7° and 14 to 20°. As noted, the effect of the fin bending ratio b/h on the heat transfer rate was investigated using tubes with parallel fin bending. Experiments performed over the range of b/h = 0.3 to 0.5 showed that Nu/Nu_b increases only slightly (up to 5%) with an increase in b/h. There are grounds to believe that this tendency prevails also when the fins are bent to provide flow convergence.

It was found in investigating the aerodynamic drag of bundles of tubes with flow-convergence inducing bending of fins that the experimental data at Re_{eq} between $3 \cdot 10^3$ and $6 \cdot 10^4$ are satisfactorily approximated by an expression such as

$$Eu_0 = C_r \cdot Re_{eq}^{-n}. \tag{6}$$

Table 6 lists the values of experimental constants n and C_r for the tube bundles under study. Bending of fins to provide for flow convergence was found to cause a marked rise in the aerodynamic drag as compared with bundles where the fins were not so bent over the entire range of pitches, pitch ratios and values of γ. The rise in drag can be represented by the ratio of Euler number for the bundle under study and for the base bundle Eu_0/Eu_0^b at Re_{eq}= const.

It is seen from Fig. 21 that the variation in $Eu_0/Eu_0^b = f(\gamma)$ is monotonous. The highest rise in drag (to 90-100%) is observed at γ = 20°. These data were compared with separately obtained results for tubes with parallel bent fins. It is remarkable that the rise in Eu_0/Eu_0^b as compared with the case of γ = 0° does not exceed 30%. This indicates that inducing convergence of flow in the spaces between the fins is only one of the reasons of the rise in drag in such bundles. Another factor is that bending of fins as such, even at $\gamma = 0°$, causes a transformation of the half-open spaces between the fins into narrow closed curved channels with wedge-shape cross sections (Fig. 19), the flow between which involves a marked energy loss, in particular because it is subjected to the decelerating effect of the walls over the entire perimeter of its cross section.

It follows from the analysis above that improving the flow pattern within the bundle may allow attaining a significant rise in the heat transfer rate without an excessive increase in drag. Depending on the fin-bending parameters, layout and Reynolds number for the tubes of the size under study the enhancement of heat transfer ranges from 15 to 77% at a respective rise in drag between 40 and 11% as compared with ordinary fins.

Location number	σ_1	σ_2	$\gamma = 7°$		$\gamma = 14°$		$\gamma = 20°$	
			n	C_r	n	C_r	n	C_r
1	3.21	0.90	-	-	0.11	0.744	0.15	1.271
2	3.21	1.29	0.15	1.410	0.15	1.473	0.16	1.775
3	3.21	1.55	0.14	1.319	0.17	1.717	0.17	1.857
4	3.21	1.79	0.13	0.943	0.15	1.280	0.16	1.485
5	3.21	2.02	0.13	0.859	0.14	1.080	0.17	1.615
6	3.02	0.90	-	-	0.13	1.070	0.14	1.263
7	2.64	1.29	-	-	0.14	1.626	0.13	1.553
8	2.05	1.79	0.13	1.065	0.14	1.256	0.13	1.176
9	2.05	2.02	0.13	1.073	0.14	1.173	0.14	1.241

Table 6. Experimental constants n and C_r in Eq. (6)

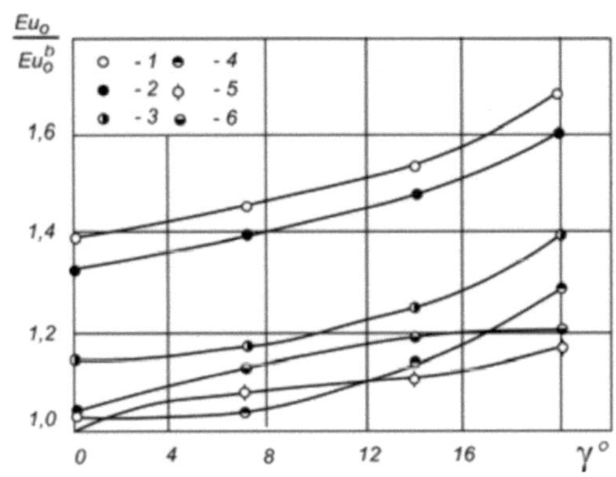

Fig. 21. Rise in aerodynamic drag as a function of γ at Re = 1.3·10^4. σ_1 = 3.21; σ_2: 1) = 1.29; 2) 1.55; 3) 1.79; 4) 2.02; σ_1 = 2.05; σ_2: 5) 1.79; 6) 2.02

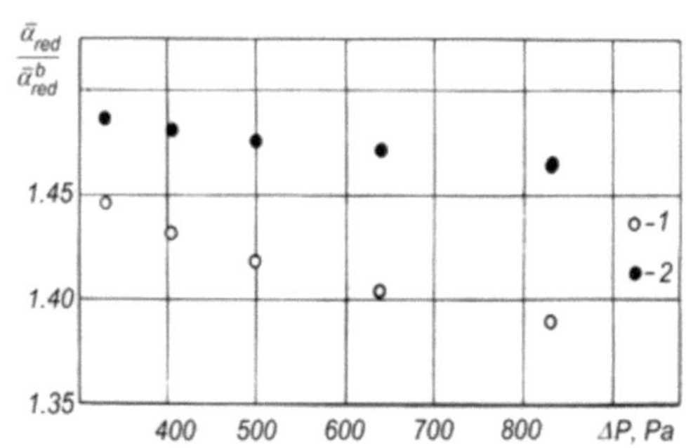

Fig. 22. Ratio of surface-averaged reduced heat transfer coefficients of the enhanced and basis bundles at the same values of drag and σ_2 = 1.29; σ_1: 1) 3.21; 2) 2.64

The effect of using a given method of enhancement of external heat transfer in finned-tube bundles can be uniquely estimated by comparing the reduced heat transfer coefficients of the ordinary and enhanced bundles at equal pressure drops ΔP. Estimates performed in this

manner show that the best performance is exhibited by bundles with $\sigma_2 = 1.29$ and $\sigma_1 = 3.21$ and 2.64 at $\gamma = 20^o$. Figure 22 is a plot of the ratio of surface-averaged reduced heat transfer coefficients $\overline{\alpha}_{red}$ of enhanced and basis bundles obtained at ΔP = idem.
The range of values of drag corresponds to Re between $8{\cdot}10^3$ and $13{\cdot}10^3$ which is most typical for power-equipment heat exchangers. It follows from the figure that the net gain in the external heat transfer of convergence-inducing bending of fins for $\sigma_1 = 3.21$ and $\sigma_2 = 1.29$ is from 38 to 44% and for the case of $\sigma_1 = 264$ and $\sigma_2 = 1.29$ it is at least 47%. Metal consumption of the device decreases correspondingly.

4. Surfaces of partially finned flattened oval tubes

The second of the ways for improving the thermoaerodynamic performance of transversely-finned heat transfer surfaces that involves removing ineffective parts of fins appears advisable in cases when configured (oval, flattened-oval, etc.) finned tubes are used in heat exchangers in order to reduce the aerodynamic drag. In such cases it is suggested to replace fully finned configured (for example, flattened-oval) tubes by partially finned ones, i.e., such in which parts of the cylindrical surface with a high curvature (the leading and trailing parts) are not finned (Fig. 23). This means that the suggested type of surface is missing a part of the fin area which "works" relatively poorly not only because it, as a rule, is located in the region of the aerodynamic shadow, but also because its efficiency factor E is lower than that for fins located on the flat lateral sides of the tube.

Fig. 23. Partially finned flattened oval tubes

The principal geometric parameters of the tubes (Table 7) were selected to be close to those of fully finned oval tubes, the heat transfer and aerodynamics of which were investigated in (Yudin & Fedorovich, 1992). This made it possible to compare their thermoaerodynamic performance and to evaluate the effect of replacing fully finned tubes by those with a partially finned surface.
The surface-average heat transfer was investigated by the traditional method of complete thermal modeling consisting in electric heating of all the tube bundles. The main quantity of interest were the reduced heat transfer coefficients a_{red}. The heat transfer coefficients a were computed from the reduced coefficients using the expression

$$\alpha_{red} = \alpha\,(E{\cdot}H_f/H_t + H_{ft}/H_t) \quad (7)$$

The fin efficiency factor E was calculated from a formula for a straight rectangular fin. For comparison of the heat transfer data with corresponding data for fully-finned tubes from the paper (Yudin & Fedorovich, 1992), which also presents reduced heat transfer coefficients,

the latter were also recalculated to their convective counterparts by means of equation (7). The values of *E* for the oval fin were then determined by averaging values calculated separately for segments with smaller and greater curvature over the surface.
It is sensible to compare heat transfer data for tubes with different fin patterns only when the convective heat transfer coefficients are referred to the surface of the tube, for which reason the experimental results were represented in the form

$$\mathrm{Nu_{c}} \cdot \psi = f(\mathrm{Re}) \tag{8}$$

Quantity	Designation	Partially finned tubes	Fully finned tubes [1]
Transverse dimension of finned tube	d_1, mm	15.0	14.0
Longitudinal dimension of finned tube	d_2, mm	38.0	36.0
Height of fins	h, mm	11.5	10.0
Fin pitch	t, mm	3.5	3.0
Fin thickness	δ, mm	0.8	0.5
Surface extension factor	Ψ	5.22	10.2
Aspect ratio of tube cross section	d_2/d_1	2.53	2.57
Relative height of fin	h/d_1	0.76	0.71
Relative fin pitch	t/d_1	0.23	0.21
Relative fin thickness	δ/t_1	0.05	0.04

Table 7. Geometric parameters of configured finned tubes

The data on aerodynamic drag were represented in the form of Euler numbers referred to a single transverse row of a bundle.
The experiments were performed with seven staggered and two in-line bundles, in which the flattened oval tubes were arranged with their major axis along the free-stream velocity vector (Fig. 24a and d).

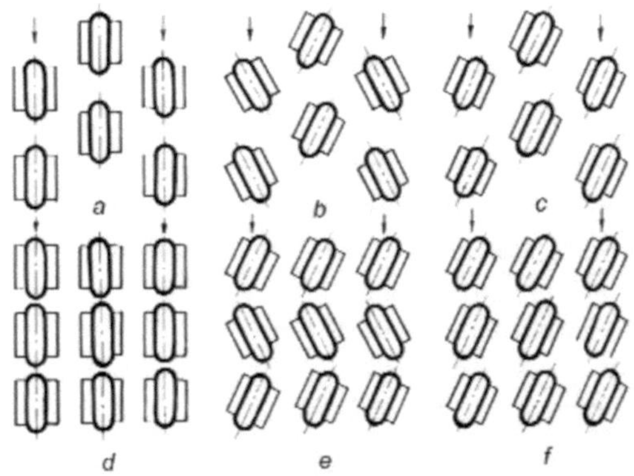

Fig. 24. Geometric arrangements of configured tubes within the bundles: (a) through (c) staggered bundles; (d) through (f) in-line bundles

[1] Yudin & Fedorovich, 1992

At the same time, configured tubes can be placed within bundles in a number of ways by varying the angle of attack of their profile Θ, and also by using different combinations of mutual arrangement of the tubes with nonzero angle Θ. In this manner we analyzed two principal versions of in-line and staggered arrangements: with successive alternation of the sign of angle Θ across the bundle (Fig. 24b and e) and without such alternation (Fig. 24c and f). This means that we investigated a total of 15 versions of bundle arrangements. Their geometric parameters are listed in Table 8.

The results on heat transfer (Fig. 25) show, in the first place, that replacing fully finned oval tubes (ψ = 10.2) with partially finned tubes (ψ = 5.22) does not reduce the heat flux from the tube bundle, all other conditions remaining equal. This holds for all the four bundle geometries (Nos. 1, 2, 4, and 5, Table 8) that had pitches which allowed comparison with data for bundles of fully finned tubes obtained in (Yudin & Fedorovich, 1992), which validates the physical assumptions for the modification of the tubes. Moreover, the heat flux removed from bundles of partially finned tubes is in these cases even slightly higher than from bundles of fully finned tubes. The mutual location of curves of $Nu_K \cdot \psi = f(Re)$ for all pairs of bundles being compared (curves for tubes with ψ = 10.2 lie lower and are shallower) allows the assumption that the reason for the lower heat transfer efficiency of the fully finned tubes is the existence of thermal contact resistance between the oval fins that have been placed on them and the tube wall, the role of which increases with increasing Re, as follows from paper (Kuntysh, 1993). On the other hand, partially finned tubes have a perfect thermal contact between the fins and the tube wall.

Arrange-ment number	Bundle geometry	S_1, mm	S_2, mm	S_1/d_1	S_2/d_1	S_1/S_2	Θ^o	Bundle geometry[2]
1	staggered	47.5	46.0	3.17	3.07	1.03	0	a
2	"	47.5	58.0	3.17	3.87	0.82	0	a
3	"	47.5	75.0	3.17	5.00	0.63	0	a
4	"	63.3	36.0	4.22	2.40	1.73	0	a
5	"	63.3	42.0	4.22	2.80	1.51	0	a
6	"	63.3	46.0	4.22	3.07	1.38	0	a
7	"	63.3	58.0	4.22	3.87	1.09	0	a
8	"	63.3	46.0	4.22	3.07	1.38	30	b
9	"	63.3	46.0	4.22	3.07	1.38	30	c
10	in-line	63.3	46.0	4.22	3.07	1.38	0	d
11	"	63.3	58.0	4.22	3.87	1.09	0	d
12	"	63.3	46.0	4.22	3.07	1.38	15	e
13	"	63.3	46.0	4.22	3.07	1.38	30	e
14	"	63.3	46.0	4.22	3.07	1.38	15	f
15	"	63.3	46.0	4.22	3.07	1.38	30	f

Table 8. Geometric parameters of the bundles of configured partially finned tubes

Investigations of the effect of bundle configuration showed that at the same pitches and Reynolds numbers in-line bundles have virtually one half of the drag of staggered bundles.

[2] as depicted in Fig. 24

The in-line geometry gives on the average 40 to 50% lower values of a as compared with the staggered bundle with the same values of S_1 and S_2, or which reason the effect of pitch at $\Theta = 0^o$ was investigated primarily with staggered bundles.

The heat transfer coefficient varied by 20 to 25% over the range of S_1/d_1 between 3.17 and 4.22, of S_2/d_1 from 2.4 to 5 and S_1/S_2 between 1.03 and 1.76: it increased with S_1/S_2 and with S_1/d_1 and decreased and stabilized with increasing S_2/d_1. The highest heat transfer coefficients were obtained with arrangement 4 ($S_1/d_1 = 4.22$ and $S_2/d_1 = 2.4$).

This allows the assumption that in certain cases in-line bundles of configured finned tubes may become preferable to staggered bundles.

The effect of S_1/d_1 and S_2/d_1 on the drag was investigated in staggered bundles. The data show that for the given pitches of bundles of partially finned tubes decreases with an increase in both these geometric ratios. As to the effect of angle of attack Θ, it was found (Fig. 26) that the drag increases markedly with increasing Θ both in staggered and in-line bundles. The drag is virtually independent on the mutual arrangements of the tubes at $\Theta \neq 0$ (Fig. 24). Still it appears that Θ has a somewhat more perceptible effect on the drag of in-line as compared with staggered bundles: in the first case increasing Θ from 0° to 30° at $Re_{fs} = 10^4$ increases Eu_0 by approximately 90%, whereas in the second - by approximately 70%. In addition, the shape of curves of $Eu_0 = f(Re_{fs})$ for the in-line bundles changes with Θ: in the case of Θ - 30° the curves become virtually self-similar ($n \approx 0$) over the entire range of Re under study, whereas at $\Theta = 0$ they have a perceptible slope (n = -0.16). On the other hand, for staggered bundles these curves are virtually equidistant both at $\Theta = 0^o$ and 30°.

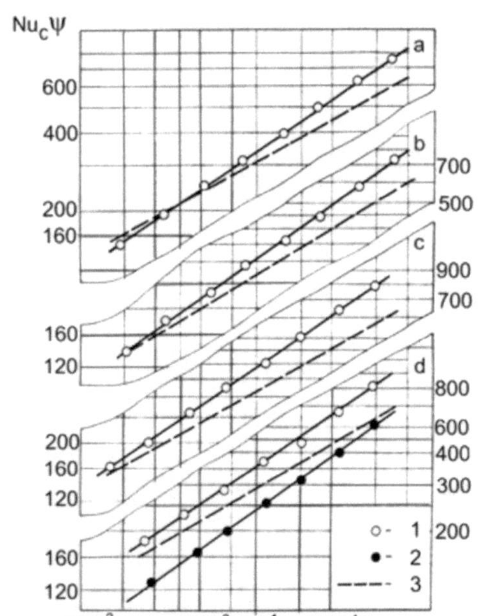

Fig. 25. Heat transfer from bundles of configured finned tubes at $\Theta = 0$. a) $S_1/d_1 = 3.17$, $S_2/d_1 = 3/07$; b) $S_1/d_1 = 3.17$, $S_2/d_1 = 3.87$; c) $S_1/d_1 = 4.22$, $S_2/d_1 = 2.40$; d) $S_1/d_1 = 4.22$, $S_2/d_1 = 2.80$; 1) staggered bundles of partially finned tubes; 2) in-line bundle of partially finned tubes; 3) staggered bundles of fully finned tubes (Yudin & Fedorovich, 1992)

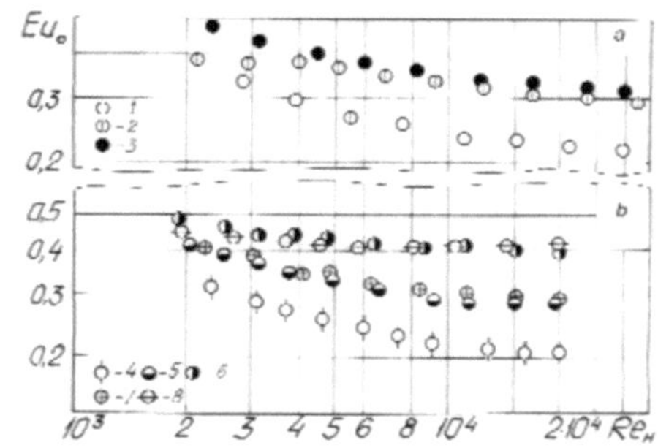

Fig. 26. Aerodynamic drag of bundles of partially finned tubes at $\Theta > 0$. a) staggered bundles; b) in-line bundles; 1) geometry 6a; 2) geometry 8b; 3) geometry 8c; 4) geometry 10d; 5) geometry 12e; 6) geometry 13e; 7) geometry 14f; 8) geometry 15f (the geometry numbers correspond to Table 8)

5. Conclusions

Thus, the use of tubes with fins bent to induce flow convergence makes it possible to markedly reduce the weight and size of heat exchangers under the same thermal effectiveness. In addition, the suggested type of enhanced finned surfaces is of interest also in the following aspects:

- tubes with fins of the suggested type can be manufactured employing standard technologies of rolling-on and welding-on of fins coupled with a relatively simple fin-bending operation, i.e., does not require extensive retooling and large additional expenditures; and
- bundles equipped with fins of the suggested type should exhibit better self-cleaning properties in dust laden flows than bundles using standard finned tube, since foulants usually accumulate in the aerodynamic shadow zone in the rear and front parts of the finned tube.

It is possible to use bundles of partially finned configured tubes which, in the first place, will allow a large saving of fin metal. On the assumption that heat fluxes removed from two bundles with similarly spaced fully and partially finned tubes with the same heights *h*, pitches *t*, fin thicknesses δ, shape and dimensions of the tubes are at least equal, then, if their aerodynamic drag values are also equal, replacing these by the others may save about half of the metal used for fins of fully finned tubes. This may amount to 20-30% of the total weight of the heat exchanger. The reasons why the heat flux density removed from these two types of tubes remains the same and maybe even increases somewhat in spite of the reduction in the heat-transmission area may be the following:

- the fins that are eliminated are parasitic, since they are usually located in the aerodynamic shadow;
- the fins placed on the flat lateral surfaces of flattened oval and similar tubes have efficiency *E* higher than oval fins;
- the technology of producing partially finned tubes allows providing for virtually ideal thermal contact between the fins ant the tube wall, which is not true of the currently employed technologies of producing fully finned oval tubes; and
- the elimination of the leading and trailing parts of the fins eliminates additional thermal resistance in the form of foulants that deposit between these fins.

6. Nomenclature

d - diameter of finned tube;
d_1 and d_2 - lateral and longitudinal dimensions of the cross section of the shaped tube, respectively;
E – fin efficiency factor;
h – fin height;
H – heat-transfer area;
P_R – dimensionless coordinate: $(r - r_0)/(R - r_0)$;
r_0 and *R* – radius of finned tube at the basis and end of the fin, respectively;
S_1 and S_2 – transverse and longitudinal tube pitches, respectively;
t – fin pitch;
ΔP – pressure drop;
U – flow velocity;
Ψ – fin factor;
Θ – inclination angle of a logitudinal axis of the shaped tube cross section to the velocity vector of the incident flow;
$\sigma_1 = S_1/d$ and $\sigma_2 = S_2/d$ - relative transverse and longitudinal tube pitches, respectively;
δ – fin thickness
Subscripts:
eq – equivalent;
fs – free stream;
ft – surface of non-finned part of the tube;
red – reduced

7. References

Tolubinskiy, V.I. & Lyogkiy, V.M. (1964). Heat Transfer Coefficients and Aerodynamic Drags of Single Finned Cylinders in Cross Air Flow (*in Russian*). *Voprosy Radioelektroniki*, Ser.1, *Electronika*, Issue 9, pp. 114-120

Migai, V.K., Bystrov, P.G., & Fedotov, V.V. (1992). Heat Transfer in the Bundles of Tubes with Lug-Type Finning in Cross Flow (*in Russian*). *Heavy Mechanical Engineering (Тяжелое машиностроение)*, No.7, pp. 8-10

Eckels, P.W. & Rabas, T.J. (1985). Heat Transfer and Pressure Drop Performance of Finned Tube Bundles. *Journal of Heat Transfer*, Vol.107, pp. 205-213

Taranyan, I.G., Iokhvedov, F.M., & Kuntysh V.B. (1972). Study of the Effect of Finning Parameters on Heat Transfer and Drag of Staggered Bundles of Tubes with Transverse Smooth and Integral Fins (*in Russian*). *Thermophysics of High Temperatures (Теплофизика высоких темпeратур)*, Vol.10, No.5, pp. 1049-1054

Kuntysh, V.B. & Iokhvedov, F.M. (1968). Heat Transfer and Aerodynamic Drag of the Bundles of Tubes with Slotted Fins (*in Russian*). *Refrigerating Engineering (Холодильная техника)*, No.6, pp. 14-18

Antufiev, V.M., & Gusyev, Ye.K. (1968). Enhancement of Heat Transfer of Finned Surfaces in Cross Flow (*in Russian*). *Heat Power Engineering (Теплоэнергетика)*, No.7, pp. 31-34

Iokhvedov, F.M., Taranyan, I.G., & Kuntysh, V.B. (1975). Heat Transfer and Aerodynamic Drag of Staggered Tube Bundles with Various Shapes of a Transverse Slotted Fin (*in Russian*). *Power Mechanical Engineering (Энергомашиностроение)*, No.11, pp. 23-26

Sparrow, E.M., & Myrum, T.A. (1985). Crossflow Heat Transfer for Tubes with Periodically Interrupted Annular Fins. *International Heat Mass Transfer*, Vol.28, No.2, pp. 509-512

Weierman, C. (1976). Correlations to Ease the Selection of Finned Tubes. *Oil and Gas Journal*, Vol.74, No.36, pp. 94-100

Antufiev, V.M. (1965). Study of Efficiency of Various Shapes of Finned Surfaces in Cross Flow (*in Russian*). *Heat Power Engineering*, No.1, pp. 81-86

Kokorev, V.I., Vishnevskiy, V.U., Semyonov, S.M. et al. (1978). Results of Studying Heat Transfer Tubes with Slotted Transverse Fins (*in Russian*). *Heat Power Engineering*, No.2, pp. 35-37

Kuntysh, V.B. & Piir, A.E. (1991). Enhancement of Heat Transfer of the Tube Bundles of Air Cooling Devices by Notching the Edges of Spiral Rolled-on Fins (*in Russian*). *Izvestiya VUZov. Energetika*, No.8, pp. 111-115

Kuntysh, V.B. (1993). Enhancement of Heat Transfer of Staggered Tube Bundles by Peripheral Notching of Spiral Fins (*in Russian*). *Izvestiya VUZov. Energetika*, No.5-6, pp. 111-117

Fiebig, M., Mitra, N., & Dong, Y. (1990). Simultaneous Heat Transfer Enhancement and Flow Loss Reduction of Fin-Tubes. *Heat Transfer-1990. Proceedings of 9th International conference*, Vol.4, pp. 51-55, 1990, (Jerusalem, August 19-24), New York

Kuntysh, V.B. & Kuznetsov, N.M. (1992). Thermal and Aerodynamic Calculations of Finned Air-Cooled Heat Exchangers (*in Russian*). *Energoatomizdat Press*, St. Petersburg Division, 280 p.

Yevenko, V.I. & Anisin, A.K. (1976). Improving the Efficiency of Heat Transfer of the Tube Bundles in Cross Flow (*in Russian*). *Heat Power Engineering (Теплоэнергетика)*, No.7, pp. 37-40

Lokshin, V.A., Fomina V.N., & Titova Ye.Ya. (1982). On One of the Methods of Enhancing Convective Heat Transfer in Smooth Tube Bundles in Cross Flow (*in Russian*). *Heat Power Engineering (Теплоэнергетика)*, No.11, pp. 17-18

Migai, V.K. & Firsova, E.V. (1986). Heat Transfer and Drag of Tube Bundles (*in Russian*). *Nauka Press*, 195 p.

Kuntysh, V.B., Stenin, N.N., & Krasnoshchyokin, L.F. (1991). Study of Thermoaerodynamic Characteristics of Staggered Bundles with Nontraditional Arrangement of Finned Tubes (*in Russian*). *Refrigerating Engineering (Холодильная техника)*, No.6, pp. 11-13

Kuntysh, V.B. & Stenin, N.N. (1993). Heat Transfer and Aerodynamic Drag of In-line – Staggered Finned Tube Bundles in Cross Flow (*in Russian*). *Heat Power Engineering (Теплоэнергетика)*, No.2, pp. 41-45

Stenin, N.N. (1994). The Development and Study of Promising Arrangements of Finned Tubes for Air-Cooled Heat Exchangers (*in Russian*). *Abstract of Candidate's Dissertation*. St. Petersburg, 21 p.

Kuntysh, V.B., Piir, A.E., & Gerasimenko A.N. (1990). Heat Transfer and Aerodynamic Drag of Staggered Bundles with a Variable Number of Tubes in a Row (*in Russian*). *Izvestiya VUZov, Energetika*, No.5, pp. 82-86

Pis'mennyi, E.N. (1991). Special Features of Flow and Heat Transfer in Staggered Bundles of Transversely Finned Tubes. *Journal of Engineering Physics*, Vol.60, No.6, pp. 676-681

Kuntysh, V.B. & Fedotova, L.M. (1983). Effect of the Attack Angle of the Air Flow on Heat Transfer and Drag of a Staggered Finned Tube Bundle (*in Russian*). *Izvestiya VUZov, Energetika*, No.4, pp. 93-96

Samie, F. Sparrow, E. (1986). Heat Transfer from a Finned Tube Oriented at an Angle to Flow. *Heat Transfer*, No.2, pp. 205-208

Khavin, A.A. (1989). The Effect of the Angle of Incidence of Flow on Thermoaerodynamic Characteristics of Finned Tube Bundles (*in Russian*). *Institute of Engineering Thermophysics (Институт технической теплофизики АН УССР), Dep. In VINITI*, No.6957-V89

Antufiev, V.M. (1966). Efficiency of Various Shapes of Convective Heating Surfaces (*in Russian*). *Energiya Press*, 184 p.

Berman, Ya.A. (1965). Study and Comparison of Finned Tubular Heat Transfer Surfaces in a Wide Range of Reynolds Numbers (*in Russian*). *Chemical and Oil Mechanical Engineering (Химическое и нефтяное машиностроение)*, No.10, pp. 21-26

Yudin, V.F. & Fedorovich, Ye.D. (1992). Heat Transfer of Oval-Shaped Finned Tube Bundles (*in Russian*). *The International Minsk Forum 1992, Convective Heat Transfer*, Minsk, Vol.1, Part 1, pp. 58-61

Ilgarubis, V.-A.S., Ulinskas, R.B., & Butkus, A.V. (1987). Drag and Average Heat Transfer of Compact Plane-Oval Finned Tube Bundles (*in Russian*). *Proceedings of the Academy of Sciences of Lithuanian SSR (Trudy Akad. Nauk LitSSR)*, Set B, Vol.158, pp. 49-55

Ota, T., Nishiyama, H., & Taoka, Y. (1984). Heat Transfer and Flow around an Elliptic Cylinder. *International Journal of Heat Mass Transfer,* Vol.27, No.10, pp. 1771-1776

Pis'mennyi, E.N. & Lyogkiy, V.M. (1984). Toward the Calculation of Heat Transfer of Multi - Row Staggered Bundles of Tubes with Transverse Finning. *Thermal Engineering,* No.31 (6), pp. 349-352

Pis'mennyi, E.N. (1984). Study of Flow on the Surface of Fins on Cross Finned Tubes. *Journal of Engineering Physics*, Vol.47, No.1, pp. 761-765

Pis'mennyi, E.N. & Terekh, A.M. (1993a). A Generalized Method for Calculating Convective Heat Transfer with Cross Flow over Tube Banks Having External Annular and Coil-Tape Finning. *Thermal Engineering*, Vol. 40, No.5, pp. 394-398

Pis'mennyi, E.N. & Terekh, A.M. (1993b). Local Heat Transfer in Bundles of Transversely Finned Tubes. *Heat Transfer Research,*, Vol.25, No.6, pp. 825-835

Skrinska, A.J., Žukauskas, A.A., & Štašiulevičius, J.K. (1964). An Experimental Study of the Local Coefficients of Heat Transfer from Helically-Finned Tubes (*in Russian*). *Proceedings of the Academy of Sciences of Lithuanian SSR (Trudy Akad. Nauk LitSSR),* Set B, Vol.4 (39), pp. 213-218

Žukauskas, A.A., Ulinskas, R.V., & Zinevičius, F.V. (1984). Local Parameters of Heat Transfer and Flow over Bundles of Staggered Finned Tubes (*in Russian*). *Ibid.*, Vol.2 (141), pp. 46-55

Neal, S.B.H.C. & Hitchcock, J.A. (1966). A Study of the Heat Transfer Processes in Bundles of Finned Tubes in Cross Flow Using a Large Model Technique. *Proceedings of the 3rd International Heat Transfer Conference (Chicago)*, pp. 290-293

Lyogkiy, B.M., Zholudov, Ya.S., & Gerashchenko, O.A. (1976). Local Heat Transfer in Crossflow over a Single Circular Tube with External Circular Fins (*in Russian*). *Journal of Engineering Physics (Инженерно-физический журнал)*, Vol.30, No.2, pp. 274-280

Krückels, W. & Kottke, V. (1970). Untersuchung über die Verteilung des Wärmeübergangs an Rippen und Rippenrohr-Modellen. *Chemie-Ing. Technik*, Bd. 42, No.6, S. 355-362

Gardon, R. (1960). A Transducer for the Measurement of Heat Flow Rate. *Journal of Heat Transfer,* Vol.82, No.4, pp. 396-398

Migay, V.K. (1978). Calculation of Heat Transfer in Bundles of Staggered Tubes Operating in Crossflow (*in Russian*). *Heat Power Engineering (Теплоэнергетика)*, No.2, pp. 31-34

Pis'mennyi, E.N. & Terekh, A.M. (1991). Heat Transfer in Bundles of Transversely Finned Tubes with Small Numbers of Tube Rows (*in Russian*). *Industrial Heat Engineering (Промышленная теплотехника)*, Vol.13, No.3, pp. 55-60

Ovchar, V.G. et al. (1995). Certain Aspects of Improving the Performance of Steam Boilers and Thermal Electric Power Plants (*in Russian*). *Heat Power Engineering (Теплоэнергетика)*, No.8, pp. 2-8

Kuntysh, V.B. (1993). Investigation of Heat Transfer and Its Enhancement in Tube Bundles of Air-Cooled Heat Exchangers (*in Russian*). *Abstract of Doctor's Dissertation.* St. Petersburg, 45 p.

9

On the Optimal Allocation of the Heat Exchangers of Irreversible Power Cycles

G. Aragón-González, A. León-Galicia and J. R. Morales-Gómez
PDPA, Universidad Autónoma Metropolitana-Azcapotzalco
México

1. Introduction

Thermal engines are designed to produce mechanical power, while transferring heat from an available hot temperature source to a cold temperature reservoir (generally the environment). The thermal engine will operate in an irreversible power cycle, very often with an ideal gas as the working substance. Several power cycles have been devised from the fundamental one proposed by Carnot, such as the Brayton, Stirling, Diesel and Otto, among others. These ideal cycles have generated an equal number of thermal engines, fashioned after them. The real thermal engines incorporate a number of internal and external irreversibilities, which in turn decrease the heat conversion into mechanical power.

A standard model is shown in Fig. 1 (Aragón-González et al., 2003), for an irreversible Carnot engine. The temperatures of the hot and cold heat reservoirs are, respectively, T_H and T_L. But there are thermal resistances between the working fluid and the heat reservoirs; for that reason the temperatures of the working fluid are T_1 and T_2, for the hot and cold isothermal processes, respectively, with $T_1 < T_H$ and $T_L < T_2$. There is also a heat loss $\dot{Q}_{leak}$ from the hot reservoir to the cold reservoir and there are other internal irreversibilities (such as dissipative processes inside the working fluid). This Carnot-like model was chosen because of its simplicity to account for three main irreversibilities above, which usually are present in real heat engines.

On the other hand, the effectiveness of heat exchangers (ratio of actual heat transfer rate to maximum possible heat transfer rate), influence over the power cycle thermal efficiency. For a given transfer rate requirement, and certain temperature difference, well-designed heat exchangers mean smaller transfer surfaces, lesser entropy production and smaller thermal resistances between the working fluid and the heat reservoirs. At the end all this accounts for larger power output from the thermal engine.

Former work has been made to investigate the influence of finite-rate heat transfer, together with other major irreversibilities, on the performance of thermal engines. There are several parameters involved in the performance and optimization of an irreversible power cycle; for instance, the isentropic temperature ratio, the allocation ratio of the heat exchangers and the cost and effectiveness ratio of these exchangers (Lewins, 2000; Aragón-González et al., 2008 and references there included). The allocation of the heat exchangers refers to the distribution of the total available area for heat transfer, between the hot and the cold sides of an irreversible power cycle. The irreversible Carnot cycle has been optimized with respect to the allocation ratio of the heat exchangers (Bejan, 1988; Aragón-González et al., 2009).

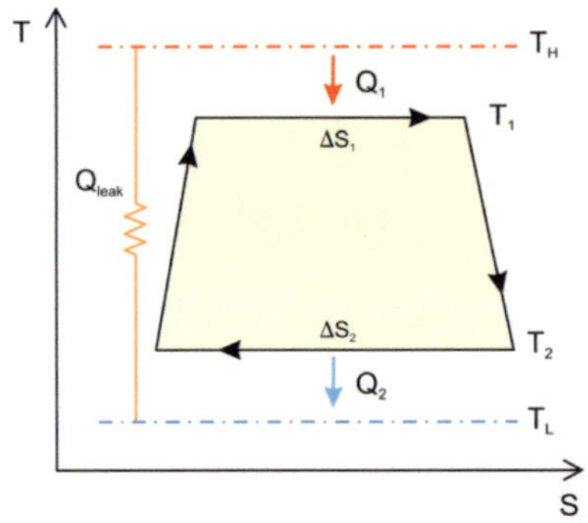

Fig. 1. A Carnot cycle with heat leak, finite rate heat transfer and internal dissipations of the working fluid.

1.1 Heat exchangers modelling in power cycles

Any heat exchanger solves a typical problem, to get energy from one fluid mass to another. A simple or composite wall of some kind divides the two flows and provides an element of thermal resistance between them. There is an enormous variety of configurations, but most commercial exchangers reduce to one of three basic types: a) the simple parallel or counterflow configuration; b) the shell-and-tube configuration; and c) the cross-flow configuration (Lienhard IV & Lienhard V, 2004). The heat transfer between the reservoirs and the hot and cold sides is usually modeled with single-pass counterflow exchangers; Fig. 2. It is supposed a linear relation with temperature differences (non radiative heat transfer), finite one-dimensional temperature gradients and absence of frictional flow losses. For common well-designed heat exchangers these approximations capture the essential physics of the problem (Kays & London, 1998).

Counterflow heat exchangers offer the highest effectiveness and lesser entropy production, because they have lower temperature gradient. It is well-known they are the best array for single-pass heat exchanging. It has also been shown they offer an important possibility, to achieve the heating or cooling strategy that minimizes entropy production (Andresen, B. & Gordon J. M., 1992). For the counterflow heat exchanger in Fig. 2, the heat transfer rate is

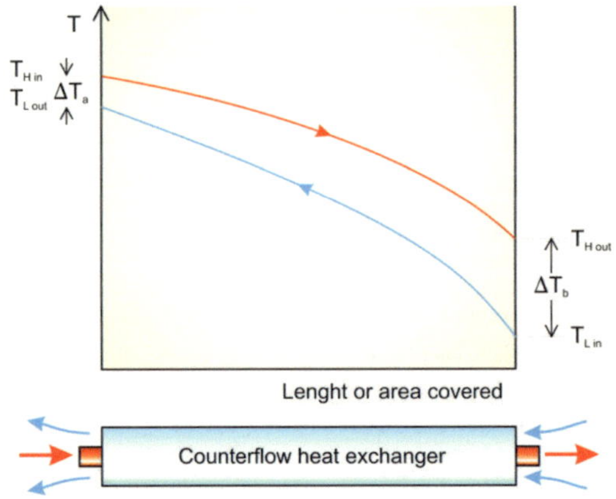

Fig. 2. Temperature variation through single-pass counterflow heat exchanger, with high and low temperature streams.

(Lienhard IV & Lienhard V, 2004):

$$q = UA\,\Delta T_{mean} \tag{1}$$

where U (W/m²K) is the overall heat transfer coefficient, A (m²) is the heat transfer surface and ΔT_{mean} is the logarithmic mean temperature difference, LMTD (K) (see Fig. 2).

$$LMTD = \Delta T_{mean} = \frac{\Delta T_a - \Delta T_b}{\ln \frac{\Delta T_a}{\Delta T_b}} \tag{2}$$

For an isothermal process exchanging heat with a constant temperature reservoir, as it happens in the hot and cold sides of the irreversible Carnot cycle in Fig. 1, it appears the logarithmic mean temperature difference is indeterminate (since $\Delta T_a = \Delta T_b$). But applying L'Hospital's rule it is easily shown:

$$LMTD = \Delta T_a = \Delta T_b\,. \tag{3}$$

For the Brayton cycle (Fig. 3) with external and internal irreversibilities which has been optimized with respect to the total inventory of the heat transfer units (Aragón-González G. et al., 2005), the hot and cold sides of the cycle have:

$$LMTD_H = \frac{T_3 - T_{2s}}{\ln \frac{T_H - T_{2s}}{T_H - T_3}} \quad \text{and} \quad LMTD_L = \frac{T_{4s} - T_1}{\ln \frac{T_{4s} - T_L}{T_1 - T_L}} \tag{4}$$

The design of a single-pass counterflow heat exchanger can be greatly simplified, with the help of the effectiveness-NTU method (Kays and London, 1998). The heat exchanger effectiveness (ε) is defined as the ratio of actual heat transfer rate to maximum possible heat transfer rate from one stream to the other; in mathematical terms (Kays & London, 1998):

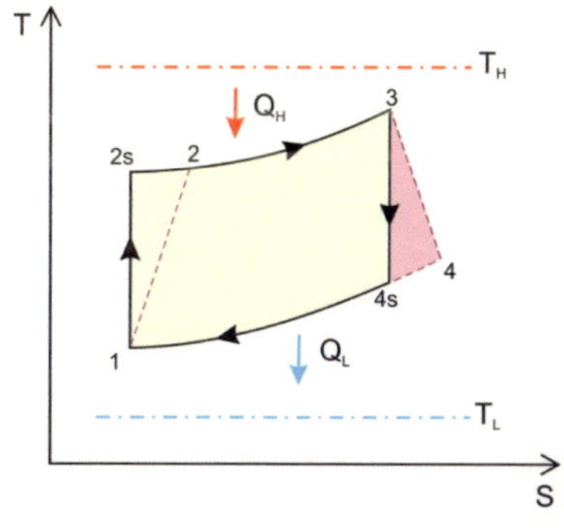

Fig. 3. A Brayton cycle with internal and external irreversiblities.

$$\varepsilon = \frac{C_H\left(T_{Hin} - T_{Hout}\right)}{C_{min}\left(T_{Hin} - T_{Lin}\right)} = \frac{C_L\left(T_{Lout} - T_{Lin}\right)}{C_{min}\left(T_{Hin} - T_{Lin}\right)} \tag{5}$$

it follows that:

$$q_{actual} = \varepsilon C_{min} (T_{Hin} - T_{Lin}) \tag{6}$$

The number of transfer units (NTU) was originally defined as (Nusselt, 1930):

$$NTU = \frac{UA}{C_{min}}; \tag{7}$$

where C_{min} is the smaller of $C_L = (\dot{m}\, c_p)_L$ and $C_H = (\dot{m}\, c_p)_H$, both in (W/K); with $\dot{m}$ the mass flow of each stream and c_p its constant-pressure specific heat. This dimensionless group is a comparison of the heat rate capacity of the heat exchanger with the heat capacity rate of the flow. Solving for ε gives:

$$\varepsilon = \frac{1 - e^{-\left(1 - \frac{C_{min}}{Cmax}\right)NTU}}{1 - \frac{C_{min}}{C_{max}} e^{-\left(1 - \frac{C_{min}}{Cmax}\right)NTU}} \tag{8}$$

Equation (8) is shown in graphical form in Fig. 4. Entering with the ratio C_{min}/C_{max} and NTU = UA/C_{min} the heat exchanger effectiveness ε can be read, and with equation (6) the actual heat transfer rate is obtained.
When one stream temperature is constant, as it happens with both temperature reservoirs in the hot and cold sides of the irreversible Carnot and Brayton cycles, the capacity rate ratio C_{min}/C_{max} is equal to zero. This heat exchanging mode is called "single stream heat exchanger", and the equation (8) reduces to:

$$\varepsilon_{singlestream} = 1 - e^{-NTU} \tag{9}$$

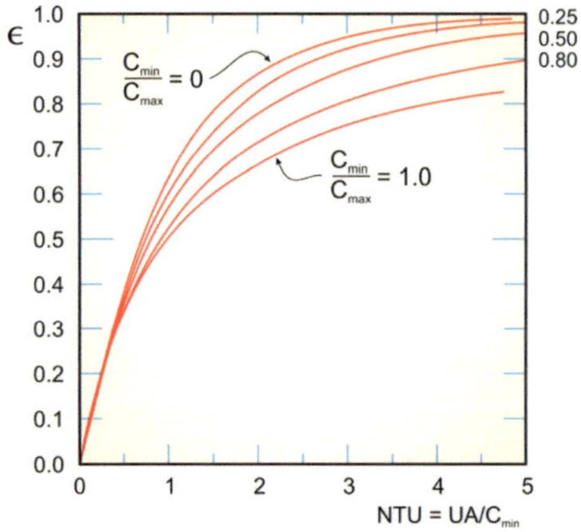

Fig. 4. Effectiveness of counterflow heat exchangers is a function of NTU and C_{min}/C_{max}.

The following sections will be dedicated to the optimal allocation of counterflow heat exchangers which are coupled in the hot-cold sides of irreversible Carnot-like and Brayton-like cycles (Fig. 1 and Fig. 3, respectively).

2. The optimal allocation of the heat exchangers for a Carnot-like cycle

The optimal allocation of the heat exchangers of irreversible power cycles was first analyzed for A. Bejan (Bejan, 1988). He optimized the power for the endoreversible Carnot cycle and found that the allocation (size) of the heat exchangers is balanced. Furthermore, Bejan also found for the model of Carnot the optimal isentropic temperature ratio $x = T_2/T_1$ by a double maximization of the power. He obtained the optimal ratio: $x_{mp}=\sqrt{\mu}$; $\mu = T_L/T_H$; which corresponded to the efficiency to maximum power proposed previously for Novikov-Chambadal-Curzon-Ahlborn (Bejan, 1996 and Hoffman et al., 1997):

$$\eta_{CNCA} = 1 - \sqrt{\mu} \tag{10}$$

The equation (10) was also found including the time as an additional constraint (see Aragón et al., 2006; and references there included). Recently in (Aragón-González et al., 2009), the model of the Fig. 1 has been optimized with respect to x and to the allocation ratio φ of the heat exchangers of the hot and cold side for different operation regimes (power, efficiency, power efficient, ecological function and criterion $\dot{\Omega}(x,\varphi)$). Formerly, the maximum power and efficiency have been obtained in (Chen, 1994; Yan, 1995; and Aragón et al., 2003). The maximum ecological function has been analyzed in general form in (Arias-Hernández et al., 2003). In general, these optimizations were performed with respect to only one characteristic parameter: x including sometimes also time (Aragón-González et al., 2006)). However, (Lewins, 2000) has considered the optimization of the power generation with respect to other parameters: the allocation, cost and effectiveness of the heat exchangers of the hot and cold sides (Aragón-González et al., 2008; see also the reviews of Durmayas et al., 1997; Hoffman et al., 2003). Also, effects of heat transfer laws or when a property is independent of the heat transfer law for this Carnot model, have been discussed in several works (Arias-Hernández et al., 2003; Chen et al., 2010; and references there included), and so on.
Moreover, the optimization of other objective functions has been analyzed: $\dot{\Omega}$ criterion (Sanchez-Salas et al., 2002), ecological coefficient of performance (ECOP), (Ust et al., 2005), efficient power (Yilmaz, 2006), and so on.
In what follows, Carnot-like model shown in Fig. 2 will be considered, it satisfies the following conditions (Aragón-González (2009)): The working fluid flows through the system in stationary state. There is thermal resistance between the working fluid and the heat reservoirs. There is a heat leak rate from the hot reservoir to the cold reservoir. In real power cycle leaks are unavoidable. There are many features of an actual power cycle which fall under that kind of irreversibility, such as the heat lost through the walls of a boiler, a combustion chamber, or a heat exchanger and heat flow through the cylinder walls of an internal combustion engine, and so on. Besides thermal resistance and heat leak, there are the internal irreversibilities. For many devices, such as gas turbines, automotive engines, and thermoelectric generator, there are other loss mechanisms, i.e. friction or generators losses, and so on, which play an important role, but are hard to model in detail. Some authors use the compressor (pump) and turbine isentropic efficiencies to model the internal loss in the gas turbines or steam plants. Others, in Carnot-like models, use simply one constant greater than one to describe the internal losses. This constant is associated with the entropy produced inside the power cycle. Specifically, this constant makes the Claussius inequality to become equality:

$$\frac{\dot{Q}_2}{T_2} - I\frac{\dot{Q}_1}{T_1} = 0 \tag{11}$$

where $\dot{Q}_i$ (i = 1, 2) are the heat transfer rates and $I = \Delta S_2/\Delta S_1 \geq 1$ (Chen, 1994). The heat transfer rates $\dot{Q}_H$, $\dot{Q}_L$ transferred from the hot-cold reservoirs are given by (Bejan, 1988):

$$\dot{Q}_H = \dot{Q}_1 + \dot{Q}; \quad \dot{Q}_L = \dot{Q}_2 + \dot{Q} \tag{12}$$

where the heat leak rate $\dot{Q}$ is positive and $\dot{Q}_1$, $\dot{Q}_2$ are the finite heat transfer rates, between the reservoirs T_H, T_L and the working substance. By the First Law and combining equations (11) and (12), the power P, heat transfer rate $\dot{Q}_H$ and thermal efficiency are given by:

$$\begin{aligned} P &= \dot{Q}_H - \dot{Q}_L = \dot{Q}_1 - \dot{Q}_2 = \dot{Q}(1 - Ix); \\ \dot{Q}_H &= \dot{Q}_1 + \dot{Q} = \frac{P}{1 - Ix} + \dot{Q} \\ \eta &= \frac{P}{f(x)P + \dot{Q}} \end{aligned} \tag{13}$$

where $x = \frac{T_2}{T_1}$ is the internal isentropic temperature ratio and $f(x) = \frac{1}{1 - Ix}$ is always positive. The entropy-generation rate and the entropy-generation rate multiplied by the temperature of the cold side gives a function Σ (equations (13)):

$$\begin{aligned} S_{gen} &= \frac{\dot{Q}_L}{T_L} - \frac{\dot{Q}_H}{T_H} > 0 \\ \Sigma &= T_L S_{gen} = T_L\left(\frac{\dot{Q}_H - P}{T_L} - \frac{\dot{Q}_H}{T_H}\right) = \dot{Q}_H(1 - \mu) - P \\ \Sigma &= g(x)P + \dot{Q}(1 - \mu) \end{aligned} \tag{14}$$

where $g(x) = f(x)(xI - \mu)$ is also positive. The ecological function (Arias-Hernández et al., 2003), if T_L is considered as the environmental temperature, and the efficient power (Yilmaz, 2006) defined as power times efficiency, are given by

$$E = P - \Sigma = (1\text{-}g(x))P - \dot{Q}\,(1 - \eta) \tag{15}$$

$$P_\eta = \eta P$$

and g(x) should be less than one (Arias-Hernández et al. (2003)). This is fulfilled if and only if E > 0 (see conditions on it in subsection 2.3). Finally, the $\dot{\Omega}$ criterion states a compromise between energy benefits and losses for a specific job and for the Carnot model discussed herein, it is expressed as (Sanchez-Salas et al., 2002):

$$\dot{\Omega} = \frac{2\eta - \eta_{max}}{\eta} P \tag{16}$$

where η_{max} is a constant.

2.1 The fundamental optimal relations of the allocation and effectiveness of the heat exchangers

The relevance of the optimization partial criterion obtained in (Aragón-González et al. (2009)) is that can also be applied to any parameter z different from x, and to any objective function that is an algebraic combination of the power and/or efficiency (as long as the objective function has physical meaning and satisfies the equations (20) and (21) below). In particular, for all the objective functions $P(x,z)$, $\eta(x,z)$, $E(x,z)$, $P_\eta(x,z)$, $\dot{\Omega}(x,z)$ and also for other characteristic parameters (not only these presented in (Aragón-González et al. (2009)). In what follows, let z be any characteristic parameter of the power plant different to x and the following operation regimes will be considered:

$$G(x,z)=P(x,z),\ \eta(x,z),\ E(x,z),\ P_\eta(x,z),\dot{\Omega}(x,z) \tag{17}$$

(power, efficiency, ecological function, efficient power, and $\dot{\Omega}$ criterion, respectively). Assuming, the parameter z can be any characteristic parameter of the cycle different to x. Thus, if $z \neq x$ and z_{mp} is the point in which the power P achieves a maximum value, then:

$$\left.\frac{\partial P}{\partial z}\right|_{z_{mp}} = 0 \text{ and } \left.\frac{\partial^2 P}{\partial z^2}\right|_{z_{mp}} < 0 \tag{18}$$

and, from the third equation of (13):

$$\frac{\partial \eta}{\partial z} = \frac{\dot{Q}\left(\frac{\partial P}{\partial z}\right)}{\left[f(x)P + \dot{Q}\right]^2} \tag{19}$$

since $\dot{Q}$ does not depend of the variable z. Thus,

$$\left.\frac{\partial \eta}{\partial z}\right|_{z_{me}} = 0 \Leftrightarrow \left.\frac{\partial P}{\partial z}\right|_{z_{mp}} = 0 \tag{20}$$

where z_{me} is the point in which the efficiency η achieves its maximum value. This implies that their critical values are the same $z_{mp} = z_{me}$ (necessary condition). The sufficiency condition is obtained by:

$$\left.\frac{\partial^2 \eta}{\partial Z^2}\right|_{z_{mp}=z_{me}} = \frac{\dot{Q}\left(\left.\frac{\partial^2 P}{\partial z^2}\right|_{z_{mp}=z_{me}}\right)}{\left[f(x)P + \dot{Q}\right]^2} < 0 \tag{21}$$

The optimization described by the equations (18)-(21) can be applied to the operation regimes given by equations (17) (the operation regime Σ (equation (14)) does not have a global minimum as was shown in (Aragón-González et al., 2009)). Thus, if z_{mec}, $z_{mP\eta}$, $z_{m\Omega}$ are the values in which the objective functions $E(x,z)$, $P_\eta(x,z)$, $\dot{\Omega}(x,z)$ reach their maximum value, then: $z_{mp} = z_{me} = z_{mec} = z_{mp\eta} = z_{m\Omega}$. Furthermore, the optimization performed, with respect to x, is invariant to the law of heat transfer no matter the operation regime G(x,z).

As an illustration, only two design rules corresponding to z will be considered. The first rule is when the constrained internal conductance, which is applied to the allocation of the heat exchangers from hot and cold sides with the same overall heat transfer coefficient U by unit of area A in both ends (see equations (1) and (3)). Thus,

$$\begin{aligned} \alpha + \beta &= \gamma \\ \tfrac{\alpha}{U} + \tfrac{\beta}{U} &= A \end{aligned} \tag{22}$$

where γ is a constant, α, β are the thermal conductances on the hot and cold sides, respectively, and in parametrizing as:

$$\varphi = \frac{\alpha}{UA}; \quad 1 - \varphi = \frac{\beta}{UA} \tag{23}$$

The second rule corresponds to total constrained area. Now, the total area A is fixed, but when distributed it has different overall heat transfer coefficients (different effectiveness) on hot and cold sides (see equations (1) and (3)). Then,

$$A = A_H + A_L = \frac{\alpha}{U_H} + \frac{\beta}{U_L} \tag{24}$$

where A_H and A_L are heat transfer areas on hot and cold sides, and U_H and U_L are overall heat transfer coefficients on the hot and cold sides, respectively. In parametrizing again:

$$\varphi^* = \frac{\alpha}{U_H A}; \quad 1 - \varphi^* = \frac{\beta}{U_L A} \tag{25}$$

The following criterion can be established:

Criterion 1. If x is fixed, z ≠ x is a characteristic parameter arbitrary of the irreversible Carnot cycle, the law of heat transfer is any, including the heat leak, and the objective functions are $G(x,z) = P(x,z), \eta(x,z), E(x,z), P_\eta(x,z), \dot{\Omega}(x,z)$. Then, the objective function G(x,z) reaches his maximum in: $z_{mG}=z_{mP}=z_{me}=z_{mec}=z_{mP\eta}=z_{m\Omega}$. In particular, if $z=\varphi$ or φ^*, the thermal conductances, overall heat transfer coefficients and areas are given either by the equations (22) or (25), and G(x,z) represent any operation regime given by the equation (17), then, $z_{mG} = \varphi_{mG}$ or φ^*_{mG}, and are given by:

$$\begin{aligned} \varphi_{mG} &= \frac{1}{1+\sqrt{I}} \\ \text{or} \quad \varphi^*_{mG} &= \frac{\sqrt{R_U}}{\sqrt{I}+\sqrt{R_U}} \end{aligned} \tag{26}$$

where $R_U = \frac{U_L}{U_H}$. Consequently, the optimal area ratio of the heat exchangers is given by:

$$\frac{A_L}{A_H} = \sqrt{I} \tag{27}$$

or the optimal distribution of the heat exchangers areas is:

$$A_H^* = \frac{A}{1+\sqrt{I\frac{U_H}{U_L}}};\ A_L^* = \frac{A}{1+\sqrt{\frac{U_L}{IU_H}}} \tag{28}$$

Indeed, it is enough to choose as objective function G(x,z), the power and the transfer heat law by conduction, since they are the algebraically simplest. For the first design rule (equation (23)), the dimensionless power output, $p = \frac{P}{UAT_H}$ is:

$$p = \frac{(1-Ix)\left(1-\frac{\mu}{x}\right)}{\frac{1}{\varphi}+\frac{I}{1-\varphi}} \tag{29}$$

or for the second design rule, the dimensionless power $p^* = \frac{P}{AU_H T_H}$ is given by:

$$p^* = \frac{(1-Ix)\left(1-\frac{\mu}{x}\right)}{\frac{1}{\varphi^*}+\frac{I}{(1-\varphi^*)R_U}} \tag{30}$$

In optimizing p or p* with respect to φ or φ* (respectively), the equations (26) are derived and combining the equations (23) or (25) and (26), the equations (27) and (28) are obtained; from these equations, the Eq. (26) or Eq. (27) are derived.

2.2 Efficiencies to maximum G(x, φ)

In this section, the efficiencies to maximum G(x, φ) are calculated by the substitution of the optimal value φ_{mG} given by the first equation (26) that doesn't depend on the aforesaid operation regimes G and neither on the transfer heat law. The power as objective function and the transfer heat law by conduction are newly chosen. The optimal values of x will be adapted or extended from the current literature for each objective function. The efficiencies to maximum G, for numerical values given, will be compared.

Now, if all heat transfer rates are assumed to be linear in temperature differences, from the equation (29), the dimensionless power is: $p = \frac{(1-Ix)\left(1-\frac{\mu}{x}\right)}{\left(1+\sqrt{I}\right)^2}$.

The optimization of p, η, $e = \frac{E}{UAT_H}$ with respect to x has been discussed in (Aragón-González et al., 2000; 2003; 2008), and are given by:

$$x_{mP} = \sqrt{\frac{\mu}{I}};$$

$$x_{me} = \frac{I\mu + \left(1+\sqrt{I}\right)\sqrt{\mu L(1-\mu)(C-I\mu)}}{CI}; \tag{31}$$

$$x_{mec} = \sqrt{\frac{\mu(1+\mu)}{2I}};$$

For the objective functions $p_\eta = \frac{P_\eta}{UAT_H}$, $\dot{\omega} = \frac{\dot{\Omega}}{UAT_H}$, they have been calculated only numerically in (Yilmaz, 2006; Sanchez-Salas et al., 2002). However, they can be calculated analytically. For instance, the optimal value $x_{mp\eta}$ for maximum efficient power p_η is obtained by:

$$p_\eta = \frac{p^2}{f(x)p + L(1-\mu)}; \quad \frac{\partial p_\eta}{\partial x} = 0 \tag{32}$$

and from the equation (32), the following cubic equation is obtained:

$$\left[\frac{\partial p}{\partial x}\left(2\left(L(1-\mu) + \frac{p}{1-Ix}\right)(1-Ix)^2 - p(1-Ix)\right) - p^2 I\right]_{x_{mp\eta}} = 0 \tag{33}$$

then, the root with physical meaning is chosen; the solution is too large to be presented here. This value of $x_{p\eta}$ extends to one presented in (Yilmaz, 2006). Similarly, a closed form for $x_{m\Omega}$ which extends to one presented in (Sanchez-Salas et al., 2002) is equally calculated, which is:

$$x_{m\Omega} = \sqrt{\frac{Ix_{me}^2 + \mu}{2I}} \tag{34}$$

Using equations (13), (30), (32) and (33) the efficiencies η_{mG} (η_{mp}, η_{max}, η_{mec}, $\eta_{mp\eta}$ and $\eta_{m\Omega}$) to maximum $G = P, \eta, E, P_\eta$ and $\dot{\Omega}$ are given by:

$$\eta_{mG} = \frac{1 - Ix_{mG}}{1 + \frac{L(1-\mu)\left(\sqrt{I}+1\right)^2}{1 - \frac{\mu}{x_{mG}}}} \tag{35}$$

where $L = K/UA$, being K the thermal conductance of the heat leak $\dot{Q} = K(T_H - T_L)$. For numerical values of (Aragón-González et al., 2009): $I = 1.235$ and $L = 0.01$ the efficiencies η_{mG} can be contrasted. Fig. 5 shows the behavior between, η_{mE}, $\eta_{mp\eta}$, $\eta_{m\Omega}$ with respect to η_{mp} and η_{max}, versus μ; it can be seen that the following inequality is satisfied: $\eta_{mp} \leq \eta_{mec}$, and $\eta_{p\eta}$ $\eta_{m\Omega} \leq \eta_{max}$.

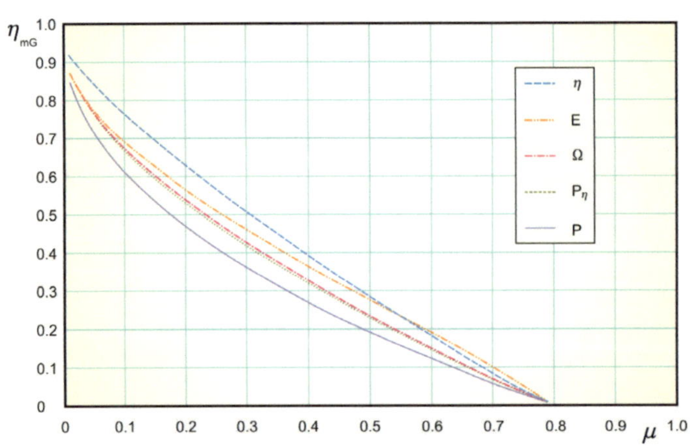

Fig. 5. Behaviors of η_{mp}, η_{max}, η_{mE}, $\eta_{mp\eta}$, $\eta_{m\Omega}$, with respect to μ if $I = 1.235$ and $L = 0.01$.

A completely analogous analysis can be performed by the substitution of the second equation of (26) (or any other optimal parameter, v. gr. costs per unit heat transfer) in the equation (30) for the power (or some other equation corresponding to the power including costs as was proposed in (Aragón-González et al., 2008)) for the same transfer heat law. Also, this can be made applying the same objective function (power) and only changing the heat transfer law. But this is not covered by this chapter's scope.

2.3 The ecological coefficient of performance (ECOP)

In (Ust et al. (2005)) was indicated that the ecological function may take negative values, due to the loss rate of availability term, $T_0 S_{gen}$ (T_0 is the environment temperature), can be greater than the actual power output. In general, the ecological function takes only positive values for efficiencies greater than the half of the Carnot efficiency η_C. Indeed, from the equation (15) for Σ the following is obtained:

$$\begin{aligned}\Sigma &= P\left(\frac{\eta_C - \eta}{\eta}\right) > 0\\ E &= \left(\frac{2\eta - \eta_C}{\eta}\right)P > 0 \Leftrightarrow \eta > \frac{\eta_C}{2}\end{aligned} \tag{36}$$

where $T_0 = T_L$ has been taken.

Then, a new ecological objective function was proposed in (Ust et al. (2005)), in order to identify the effect of loss rate of availability on power output. This objective function has always positive values just like the power, efficiency and efficient power, and is an algebraic expression only of the efficiency η. The objective function was called the ecological coefficient of performance (ECOP) and is defined as the power output per unit loss rate of availability, i.e. $ECOP = \frac{P}{\Sigma} > 0$; where $T_L = T_0$ have been supposed. A performance analysis upon the Carnot-like model, using the ECOP criterion as objective function, was carried out by Ust et al. Their optimization was performed only for x and taking some numerical optimal values of φ^* (Figure 3 (d) of Ust et al., 2005). It was found (see Figure 4 of Ust et al. 2005) that the optimal conditions to maximum ECOP and efficiency η coincides, although their functional forms are different. This can be shown applying the optimization described in the subsection 2.1 (equations (18) to (21)) and in a form more general. Indeed, let $z = x$, φ or φ^* (or any other parameter); from the first equation of (36), the ECOP can be wrote:

$$ECOP = \frac{\eta}{\eta_C - \eta} > 0 \tag{37}$$

and as

$$\begin{aligned}\left.\frac{\partial ECOP}{\partial z}\right|_{z_{mEcop}} &= \left.\frac{\frac{\partial \eta}{\partial z} ECOP}{(\eta_C - \eta)}\right|_{z_{mEcop}} = 0 \Leftrightarrow \left.\frac{\partial \eta}{\partial z}\right|_{z_{me}} = 0\\ \left.\frac{\partial^2 ECOP}{\partial z^2}\right|_{z_{mEcop}} &= \frac{\left.\frac{\partial^2 \eta}{\partial z^2}\right|_{z_{mEcop}} (1 + ECOP)}{(\eta_C - \eta)} < 0\end{aligned} \tag{38}$$

Furthermore, the following criterion can be established and can be shown in simpler form:
Criterion 2. The ECOP criterion reaches its maximum if $\eta = \eta_{max}$. Moreover, the optimal conditions are the same for the ECOP and η objective functions, independently of the heat

transfer law. In particular, the second equation of (31) is fulfilled. Indeed, it is enough to write (37) as:

$$ECOP = \frac{1}{\frac{\eta_C}{\eta} - 1}; \qquad \frac{1}{\frac{\eta_C}{\eta} - 1} \leq \frac{1}{\frac{\eta_C}{\eta_{max}} - 1} \tag{39}$$

since $\eta \leq \eta_{max}$, i.e. the maximum is reached if $\eta = \eta_{max}$ and as none heat transfer explicit law has been used, then, it is satisfied for any the heat transfer law and for any characteristic parameter. And clearly the second equation of (31) is fulfilled. From the Criterion 2 follows that the optimal conditions are the same to maximum ECOP and efficiency η. Thus the work of (Ust et al., 2005) has been extended to any characteristic parameter and to any heat transfer law. Nevertheless, the optimal conditions coincide for ECOP and η independently of the heat transfer law. They contain different thermodynamic meanings: the efficiency gives information about the necessary fuel consumption in order to produce a certain power level whereas ECOP gives information about the entropy generation (loss rate of availability). If the parameters are x and φ^*, a discussion is presented in (Ust et al., 2005). Also, (Ust et al., 2005) have noted that the actual power equals to the theoretical power output minus the loss rate of availability, i.e. $P = P_{actual} - T_0 S_{gen}$ (cfr. with (Sonntag et al., 2003) for more details). They have concluded that the loss rate of availability has been considered twice in the ecological function. It will need interpretation in order to understand what it means thermodynamically.

3. The optimal allocation of the heat exchangers for a Brayton-like cycle

A. Bejan (Bejan, 1988) optimized the power for the endoreversible Brayton cycle and found that the allocation (size) of the heat exchangers is balanced. The Brayton endoreversible model discussed for him corresponds to the cycle 1-2s-3-4s of the Fig. 3. Formerly, H. Leff (Leff, 1987)) was focused on the reversible Brayton cycle and obtained that the efficiency to maximum work corresponds to the CNCA efficiency (equation (10) with one $\mu^* = T_1/T_3$, see Fig. 3. In (Wu et al; 1991) a non-isentropic Brayton model was analyzed and found that the isentropic temperatures ratio (pressure ratio), that maximizes the work, is the same as a CNCA-like model (Aragon-González et al., 2000; 2003). In (Swanson, 1991) the endoreversible model was optimized by log-mean temperature difference for the heat exchangers in hot and cold sides and assumed that it was internally a Carnot cycle. In (Chen et. al., 2001) the numerical optimization for density power and distribution of a heat exchangers for the endoreversible Brayton cycle is presented. Other optimizations of Brayton-like cycles can be found in the following reviews (Durmayas et al. 1997; Hoffman et al. 2003). Recent optimizations of Brayton-like models were made in (Herrera et al., 2006; Lewins, 2005; Ust, 2006; Wang et al., 2008).

For the isentropic Brayton cycle (1-2s-3-4s-1) its efficiency is given by (Fig. 3):

$$\eta = 1 - x \tag{40}$$

where $x = \varepsilon^{1-\frac{1}{\gamma}}$, with $\varepsilon = p_{2s}/p_1$ the pressure ratio (maximum pressure divided by minimum pressure) and $\gamma = c_p/c_v$, with c_p and c_v being the constant-pressure and the constant-volume specific heats. Furthermore, the following temperature relations are satisfied:

$$T_{2s} = \frac{T_1}{x}; \; T_{4s} = T_3 x \tag{41}$$

where x is given by the equation (40). If a non-isentropic Brayton cycle, without external irreversibilities (see 1-2-3-4 cycle in Fig. 3) is considered, with isentropic efficiencies of the turbine and compressor η_1 and η_2, respectively, and from here the following temperature relations are obtained (Aragón-González et al., 2000):

$$\begin{gathered} \eta_1 = \frac{T_3 - T_4}{T_3 - T_{4s}}; \; \eta_2 = \frac{T_{2s} - T_1}{T_2 - T_1}; \\ T_2 = T_1\left(1 + \frac{1 - x}{\eta_2 x}\right); \; T_4 = T_3(1 - \eta_1(1 - x)) \end{gathered} \tag{42}$$

Now, if we consider the irreversible Brayton cycle of the Fig. 3, the temperature reservoirs are given by the constant temperatures T_H and T_L. In this cycle, two single-pass counterflow heat exchangers are coupled to the cold-hot side reservoirs (Fig. 2 and Fig. 3). The heat transfer between the reservoirs and the working substance can be calculated by the log mean temperature difference LMTD (equation (2)). The heat transfer balances for the hot-side are (equations (1) and (6)):

$$Q_H = U_H A_H LMTD_H = mc_p(T_3 - T_2); \; Q_L = U_L A_L LMTD_H = mc_p(T_4 - T_1) \tag{43}$$

where $LMTD_{H.L}$ are given by the equations (4). The number of transfer units NTU for both sides are (equation (7)):

$$N_H = \frac{U_H A_H}{mc_p} = \frac{T_3 - T_2}{LMTD_H}; \quad N_L = \frac{U_L A_L}{mc_p} = \frac{T_4 - T_1}{LMTD_L} \tag{44}$$

Then, its effectiveness (equation (9)):

$$\varepsilon_H = 1 - e^{-N_H} = \frac{T_3 - T_2}{T_H - T_2}; \quad \varepsilon_L = 1 - e^{-N_L} = \frac{T_4 - T_1}{T_4 - T_L} \tag{45}$$

As the heat exchangers are counterflow, the heat conductance of the hot-side (cold side) is $U_H A_H$ ($U_L A_L$) and the thermal capacity rate (mass and specific heat product) of the working substance is C_W. The heat transfer balances results to be:

$$Q_H = C_W \varepsilon_H (T_H - T_2) = C_W(T_3 - T_2); \quad Q_L = C_W \varepsilon_L (T_4 - T_L) = C_W(T_4 - T_1) \tag{46}$$

The temperature reservoirs T_H and T_L are fixed. The expressions for the temperatures T_2 and T_4, including the isentropic efficiencies η_1 and η_2, the effectiveness ε_H and ε_L and $\mu = T_L/T_H$ are obtained combining equations (41), (42), and (45):

$$T_2 = \frac{\left[\varepsilon_L \mu x^{-1} + \varepsilon_H(1 - \varepsilon_L)\right]\left(\frac{1-x}{\eta_2} + x\right)}{\left[\varepsilon_L + \varepsilon_H(1 - \varepsilon_L)\right]} T_H, \quad T_4 = \frac{\left[\varepsilon_H x + \varepsilon_L \mu(1 - \varepsilon_H)\right]\left(\frac{1}{x} - \frac{(1-x)\eta_1}{x}\right)}{\left[\varepsilon_L + \varepsilon_H(1 - \varepsilon_L)\right]} T_H \tag{47}$$

And, the dimensionless expressions, $q = Q / C_W T_H$, for the hot-cold sides are:

$$q_H = \varepsilon_H\left(1 - \frac{T_2}{T_H}\right); \quad q_L = \varepsilon_L\left(\frac{T_4}{T_H} - \mu\right) \tag{48}$$

From the first law of the Thermodynamic, the dimensionless work $w = W/C_W T_H$ of the cycle is given by:

$$w = \varepsilon_H\left[1 - \frac{T_2}{T_H}\right] - \varepsilon_L\left[\frac{T_4}{T_H} - \mu\right] \tag{49}$$

and substituting the equations (47), the following analytical relation is obtained:

$$w = \varepsilon_H\left[1 - \frac{\varepsilon_L \mu x^{-1} + \varepsilon_H(1-\varepsilon_L)}{\varepsilon_L + \varepsilon_H(1-\varepsilon_L)}\left(\frac{1-x}{\eta_2} + x\right)\right] - \varepsilon_L\left[\frac{\varepsilon_H x + \varepsilon_L \mu(1-\varepsilon_H)}{\varepsilon_L + \varepsilon_H(1-\varepsilon_L)}\left(\frac{1}{x} - \frac{(1-x)\eta_1}{x}\right) - \mu\right] \tag{50}$$

This relation will be focused on the analysis of the optimal operating states. There are three limiting cases: isentropic [$\varepsilon_H = \varepsilon_L = \eta_1 = \eta_2 = 1$]; non-isentropic [$\varepsilon_H = \varepsilon_L = 1$, $0 < \eta_1, \eta_2 < 1$]; and endoreversible [$\eta_1 = \eta_2 = 1$, $0 < \varepsilon_H, \varepsilon_L < 1$]. Nevertheless, only the endoreversible cycle is relevant for the allocation of the heat exchangers (see subsection 3.2). However, conditions for regeneration for the non-isentropic cycle are analyzed in the following subsection.

3.1 Conditions for regeneration of a non-isentropic Brayton cyle for two operation regimes

J. D. Lewins (Lewins, 2005) has recognized that the extreme temperatures are subject to limits: a) the environmental temperature and; b) in function of the limits on the adiabatic flame or for metallurgical reasons. The thermal efficiency η (see equation (40)) is maximized without losses, if the pressure ratio ε_p grows up to the point that the compressor output temperature reaches its upper limit. These results show that there is no heat transferred in the hot side and as a consequence the work is zero. The limit occurs when the inlet temperature of the compressor equals the inlet temperature of the turbine; as a result no heat is added in the heater/combustor; then, the work vanishes if $\varepsilon_p = 1$. Therefore at some intermediate point the work reaches a maximum and this point is located close to the economical optimum. In such condition, the outlet temperature of the compressor and the outlet temperature of the turbine are equal ($T_{2s} = T_{4s}$; see Fig. 3). If this condition is not fulfilled ($T_{2s} \neq T_{4s}$), it is advisable to couple a heat regeneration in order to improve the efficiency of the system if $T_{2s} < T_{4s}$ (Lewins, 2005). A similar condition is presented when internal irreversibilities due to the isentropic efficiencies of the turbine (η_1) and compressor (η_2) are taken into account (non-isentropic cycle): $T_2 < T_4$ (see Fig. 3 and equation (20) of (Zhang et al., 2006)).
The isentropic cycle corresponds to a Brayton cycle with two coupled reversible counterflow heat exchangers (1-2s-3-4s in Fig. 3). The supposition of heat being reversibly exchanged (in a balanced counterflow heat exchanger), is an equivalent idealization to the supposed heat transfer at constant temperature between the working substance of a Carnot (or Stirling) isentropic cycle, and a reservoir of infinite heat capacity. In this cycle $C_W T_H = mc_p T_3$, $T_H = T_3$, $T_H = T_3$ and $T_L = T_1$, then,

$$w = (1-x) - \left(\frac{1}{x} - 1\right)\mu^*; \quad q_H = 1 - x\mu^*; \quad q_L = x - \mu^* \tag{51}$$

For maximum work:

$$x_{mw} = \sqrt{\mu^*}\ ; \quad \eta_{CNCA} = 1 - \sqrt{\mu^*} \tag{52}$$

where $\mu^* = \frac{T_1}{T_3}$ and η_{CNCA} corresponds to the CNCA efficiency (equation (10)). Furthermore, in condition of maximum work:

$$\frac{T_1}{T_{2s}} = x_{mw}\ ; \quad \frac{T_1}{T_{4s}} = \frac{T_1}{T_3}\frac{T_3}{T_{4s}} = \frac{x_{mw}^2}{x_{mw}} = x_{mw} \tag{53}$$

so $T_{2s} = T_{4s}$. In other conditions of operation, when $T_{2s} < T_{4s}$, a regenerator can be coupled to improve the efficiency of the cycle. An example of a regenerative cycle is provided in (Sontagg et al., 2003).

On the other hand, the efficiency of the isentropic cycle can be maximized by the following criterion (Aragón-González et al., 2003).

Criterion 3. Let $\eta = \frac{w}{q_H} = 1 - \frac{q_L}{q_H}$. Suppose that $\frac{\partial^2 q_H}{\partial x^2} < 0$ and $\frac{\partial^2 q_L}{\partial x^2} = 0$, for some x. Then, the maximum efficiency η_{max} is given by:

$$\eta_{max} = \frac{\frac{\partial w}{\partial x}\Big|_{x=x_{me}}}{\frac{\partial q_H}{\partial x}\Big|_{x=x_{me}}} = 1 - \frac{\frac{\partial q_L}{\partial x}\Big|_{x=x_{me}}}{\frac{\partial q_H}{\partial x}\Big|_{x=x_{me}}} \tag{54}$$

where x_{me} is the value for which the efficiency reaches its maximum.

Criterion 3 hypothesis are clearly satisfied: $\frac{\partial^2 q_H}{\partial x^2} < 0$ and $\frac{\partial^2 q_L}{\partial x^2} = 0$ for some x (Fig. 6). Thus, the maximum efficiency η_{max} is given by the equation (54):

$$1 - \frac{\frac{1}{\mu}}{\frac{1}{x_{me}^2}} = 1 - \frac{x_{me}^2}{\mu} \tag{55}$$

In solving, $x_{me} = \mu$ and $\eta_{max} = 1 - \mu$ which corresponds to the Carnot efficiency; the other root, $x_{me} = 0$, is ignored. And the work is null for $x_{me} = \mu$; as a consequence the added heat is also null (Fig. 6). Now regeneration conditions for the non-isentropic cycle will be established.

Again $C_W T_H = mc_p T_3$, $T_H = T_3$ and $T_L = T_1$ (cycle 1-2-3-4 in Fig. 3) and T_2 and T_4 are given by the equations (42). Thus, using equations (42) and the structure of the work in the equation (51), the work w and the heat q_H are:

$$\begin{aligned} w &= \eta_1(1 - x) - \frac{1}{\eta_2}\left(\frac{1}{x} - 1\right)\mu^*; \\ q_H &= \left[1 - \left(1 + \frac{(1 - x)}{\eta_2 x}\right)\mu^*\right] \end{aligned} \tag{56}$$

Maximizing,

$$T_{4s} = IT_{2s};\ x_{NI} = \sqrt{I\mu^*} \text{ and } \eta_{NI} = 1 - \frac{I\eta_2(1 - \mu^*) + \sqrt{I\mu^*} - 1}{\sqrt{I}\left(\sqrt{I}\eta_2(1 - \mu^*) + \sqrt{\mu^*}\left(\sqrt{I\mu^*} - 1\right)\right)} \tag{57}$$

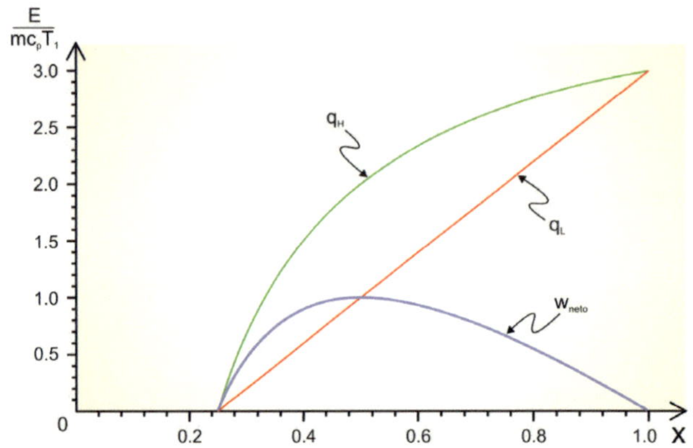

Fig. 6. Heat and work qualitative behavior for μ=0.25.

where $I = 1/\eta_1\eta_2$ and η_{NI} is the efficiency to maximum work of the non-isentropic cycle. Furthermore, the hypotheses from the Criterion 3 are fulfilled (the qualitative behavior of w and q_H is preserved, Fig. 6). In solving the resulting cubic equation, the maximum efficiency, its extreme value and the inequality that satisfies are obtained:

$$\eta_{max} = 1 - \frac{\eta_2}{\mu\eta_1}\left(\frac{\eta_1\mu + \sqrt{\eta_1\mu(1-\mu)((1-\eta_2)\mu + \eta_2(1-\eta_1)}}{\mu(1-\eta_2) + \eta_2}\right)^2$$
$$x_{me} = \frac{\eta_1\mu + \sqrt{\eta_1\mu(1-\mu)(1-\eta_2)\mu + \eta_2(1-\eta_1)}}{\eta_1(\mu(1-\eta_2) + \eta_2)} \tag{58}$$
$$I\mu \le x_{me} \le x_{mw}$$

Now, following (Zhang et al., 2006), in a Brayton cycle a regenerator is used only when the temperature of the exhaust working substance, leaving the turbine, is higher than the exit temperature in the compressor ($T_4 > T_2$). Otherwise, heat will flow in the reverse direction decreasing the efficiency of the cycle. This point can be directly seen when $T_4 < T_2$, because the regenerative rate is smaller than zero and consequently the regenerator does not have a positive role. From equations (42) the following relation is obtained:

$$T_4 = T_3(1 - \eta_1(1-x)) > T_1\left(1 + \frac{1}{\eta_2}\left(\frac{1}{x} - 1\right)\right) = T_2 \tag{59}$$

which corresponds to a temperature criterion which is equivalent to the first inequality of:

$$x > x_{min} = \frac{-\beta + \sqrt{\beta^2 + 4I\mu}}{2}; \ \beta = \left(\tfrac{1}{\eta_1} - 1\right) + \left(I - \tfrac{1}{\eta_1}\right)\mu > 0 \tag{60}$$

Indeed, from the equation (59):

$$x^2 + \beta x - I\mu > 0$$
$$x > x_{min} = \frac{-\beta + \sqrt{\beta^2 + 4I\mu}}{2} > 0 \tag{61}$$

the inequality is fulfilled since $\sqrt{\beta^2 + 4I\mu} > \beta$. The other root is clearly ignored. Therefore, if $x \le x_{min}$, a regenerator cannot be used. Thus, the first inequality of (60) is fulfilled.

Criterion 3. If the cycle operates either to maximum work or efficiency, a counterflow heat exchanger (regenerator) between the turbine and compressor outlet is a good option to improve the cycle. For other operating regimes is enough that the inequality (61) be fulfilled. When the operating regime is at maximum efficiency the inequality of (61) is fulfilled. Indeed,

$$
\begin{aligned}
&x_{me} > x_{min}\\
&x_{me} - x_{min} = \eta_1\mu + \beta + \left(\frac{\sqrt{\beta^2 + 4I\mu}}{2} - \frac{\sqrt{\eta_1\mu(1-\mu)((1-\eta_2)\mu + \eta_2(1-\eta_1))}}{\eta_1(\mu(1-\eta_2) + \eta_2)}\right)\\
&\left(\frac{\sqrt{\beta^2 + 4I\mu}}{2}\right)^2 - \left(\frac{\sqrt{\eta_1\mu(1-\mu)((1-\eta_2)\mu + \eta_2(1-\eta_1))}}{\eta_1(\mu(1-\eta_2) + \eta_2)}\right)^2 = \dots\\
&\eta_1^2\left((\eta_2(1-\mu)+\mu)\left(\beta^2(\eta_2(1-\mu)+\mu) + 4I\mu^2\right) + 4\eta_2\mu(1-\mu)\right)>0\\
&\frac{\sqrt{\beta^2+4I\mu}}{2} > \frac{\sqrt{\eta_1\mu(1-\mu)((1-\eta_2)\mu+\eta_2(1-\eta_1))}}{\eta_1(\mu(1-\eta_2)+\eta_2)}
\end{aligned}
\tag{62}
$$

where the following elementary inequality has been applied: If $a, b > 0$, then $a < b \Leftrightarrow a^2 < b^2$. If the operating regime is at maximum work, the proof is completely similar to the equations (62). An example of a non-isentropic regenerative cycle is provided in (Aragón-González et al., 2010).

3.2. Optimal analytical expressions

If the total number of transfer units of both heat exchangers is N, then, the following parameterization of the total inventory of heat transfer (Bejan, 1988) can be included in the equation (50):

$$
N_H + N_L = N; \quad N_H = yN \text{ and } N_L = (1-y)N \tag{63}
$$

For any heat exchanger $N = \frac{UA}{C}$, where U is the overall heat-transfer coefficient, A the heat-transfer surface and C the thermal capacity. The number of transfer units in the hot-side and cold-side, N_H and N_L, are indicative of both heat exchangers sizes. And their respective effectiveness is given by (equation (9)):

$$
\varepsilon_H = 1 - e^{-yN}; \quad \varepsilon_L = 1 - e^{-(1-y)N} \tag{64}
$$

Then, the work w (equation (50)) depends only upon the characteristics parameters x and y. Applying the extreme conditions: $\frac{\partial w}{\partial x} = 0$; $\frac{\partial w}{\partial y} = 0$, the following coupled optimal analytical expressions for x and y, are obtained:

$$
\begin{aligned}
x_{NE} &= \sqrt{\frac{(z_1 - z)(Cz - B)}{(z-1)(Az_1 - Bz)}\mu};\\
y_{NE} &= \frac{1}{2} + \frac{1}{2N}\ln\left(\frac{Ax - B\mu}{Bx - C\mu}\right)
\end{aligned}
\tag{65}
$$

where $z_1 = e^N$; $z = e^{yN}$; $A = \eta_1\eta_2 e^N + 1 - \eta_2$; $B = e^N(\eta_1\eta_2 + 1 - \eta_2)$ and $C = e^N - \eta_2 + \eta_1\eta_2$.
The equations (65) for x_{NE} and y_{NE} cannot be uncoupled. A qualitative analysis and its asymptotic behavior of the coupled analytical expressions for x_{NE} and y_{NE} (equations (65)) have been performed (Aragón-González (2005)) in order to establish the bounds for x_{NE} and y_{NE} and to see their behaviour in the limit cases. Thus the following bounds for x_{NE} and y_{NE} were found:

$$0 < x_{NI} \le x_{NE} < 1; \quad 0 < y_{NE} < \tfrac{1}{2} \tag{66}$$

where x_{NI} is given by the equation (57). The inequality (66) is satisfied because of $1 < I \le \frac{(z_1 - z)(Cz - B)}{(z-1)(Az_1 - Bz)}$. If I = 1 ($\eta_1 = \eta_2 = 100\%$), the following values are obtained: $x_{NE} = x_{CNCA} = \sqrt{\mu}$; $y_{NE} = y_E = ½$ which corresponds to the endoreversible cycle. In this case necessarily: $\varepsilon_H = \varepsilon_L = 1$. Thus, the equations (65) are one generalization of the endoreversible case [$\eta_1 = \eta_2 = 1$, $0 < \varepsilon_H, \varepsilon_L < 1$]. The optimal allocation (size) of the heat exchangers has the following asymptotic behavior: $\lim_{N\to\infty} y_{NE} = 1/2$; $\lim_{\eta_1,\eta_2\to 1} y_{NE} = 1/2$. Also, x_{NE} has the following asymptotic behavior: $\lim_{N\to\infty} x_{NE} = x_{NI}$; $\lim_{N\to\infty} \eta_{NE} = \eta_{NI}$. Thus, the non-isentropic [$\varepsilon_H = \varepsilon_L = 1$, $0 < \eta_1, \eta_2 < 1$] and endoreversible [$\eta_1 = \eta_2 = 1$, $0 < \varepsilon_H, \varepsilon_L < 1$] cycles are particular cases of the cycle herein presented. A relevant conclusion is that the allocation always is unbalanced ($y_{NE} < ½$).
Combining the equations (65), the following equation as function only of z, is obtained:

$$\sqrt{\mu}\left(\frac{Bz_1 - Cz^2}{Az_1 - z^2B}\right) = \sqrt{\frac{(z_1 - z)(Cz - B)}{(z-1)(Az_1 - Bz)}} \tag{67}$$

which gives a polynomial of degree 6 which cannot be solved in closed form. The variable *z* relates (in exponential form) to the allocation (unbalanced, $\varepsilon_H < \varepsilon_L$) and the total number of transfer units N of both heat exchangers. To obtain a closed form for the effectiveness ε_H, ε_L, the equation (67) can be approximated by:

$$\sqrt{\mu}\left(\frac{Bz_1 - Cz^2}{Az_1 - z^2B}\right) = \left(\frac{1}{2} + \frac{1}{2}H\right) \tag{68}$$

with $H = \frac{(z_1 - z)(Cz - B)}{(z-1)(Az_1 - Bz)}$; and using the linear approximation: $\sqrt{H} = 1 + \frac{1}{2}(H-1) + O\left((H-1)^2\right)$.

It is remarkable that the non-isentropic and endoreversible limit cases are not affected by the approximation and remain invariant within the framework of the model herein presented. Thus, this approximation maintains and combines the optimal operation conditions of these limit cases and, moreover, they are extended. The equation (68) is a polynomial of degree 4 and it can be solved in closed form for z with respect to parameters: μ or N, for realistic values for the isentropic efficiencies (Bejan (1996)) of turbine and compressor: $\eta_1 = \eta_2 = 0.8$ or 0.9, but it is too large to be included here. Fig. 7 shows the values of z (z_{mp}) with respect to μ. Using the same numerical values, Fig. 8 shows that the efficiency to maximum work η_{NE}, with respect to μ, can be well approached by the efficiency of the non-isentropic cycle η_{NI} (equation (57)) for a realistic value of N = 3 and isentropic efficiencies of 90%. The behavior of y_{NE} with respect to the total number of transfer units N of both heat exchangers, with the same numerical values for the isentropic efficiencies of turbine and compressor and μ = 0.3, are presented in Fig. 9. When the number of heat transfer units, N, is between 2 to 5, the allocation for the heat exchangers y_{NE} is approximately 2 - 8% or 1 - 3%, less than its asymptotic value or ½, respectively.

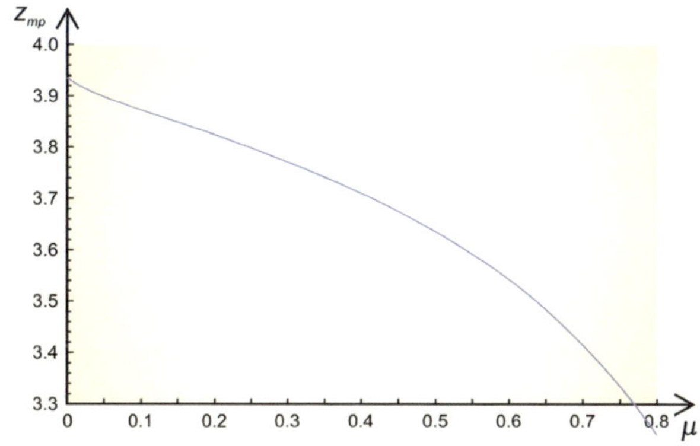

Fig. 7. Behaviour of $z(z_{mp})$ versus μ, if $\eta_1 = \eta_2 = 0.8$ and N=3.

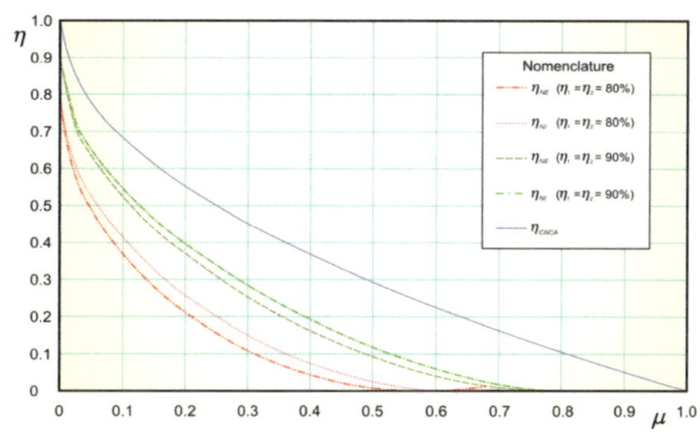

Fig. 8. Behaviour of η_{NE}, η_{NI} and η_{CNCA} versus μ, if $\eta_1 = \eta_2 = 0.8$ or 0.9 and N=3.

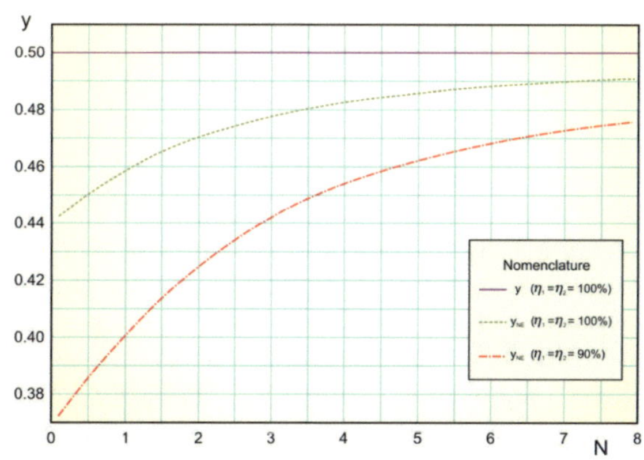

Fig. 9. Behavior of y_{NE} versus N, when $\eta_1 = \eta_2 = 0.8$ or 0.9 and $\mu = 0.3$.

This result shows that the size of the heat exchanger in the hot side decreases. Now, if the Carnot efficiency is 70% the efficiency η_{NE} is approximately 25 - 30% or 10 - 15%, when the

number of heat transfer units N is between 2 and 5 and the isentropic efficiencies are $\eta_1 = \eta_2 = 0.9$ or 0.8 respectively, as is shown in Fig. 9.
Now, if $\eta_1 = \eta_2 = 0.8$ (I = 1.5625) ; $y_{NE} = 0.45$ then $N \cong 3.5$ (see Fig. 9) and for the equations (64): $\varepsilon_H = 0.74076$ and $\varepsilon_L = 0.80795$. Thus, one cannot assume that the effectiveness are the same: $\varepsilon_H = \varepsilon_L < 1$; whilst I > 1. Current literature on the Brayton-like cycles, that have taken the same less than one effectiveness and with internal irreversibilities, should be reviewed. To conclude, $\varepsilon_H = \varepsilon_L$ if and only if the allocation is balanced (y = ½) and the unique thermodynamic possibility is: optimal allocation balanced ($y_{NE} = y_E = ½$); that is $\varepsilon_H = \varepsilon_L$. And $\varepsilon_H < \varepsilon_L$ if and only if I>1 there is internal irreversibilities.

4. Conclusions

Relevant information about the optimal allocation of the heat exchangers in power cycles has been described in this work. For both Carnot-like and Brayton cycles, this allocation is unbalanced. The expressions for the Carnot model herein presented are given by the Criterion 1 which is a strong contribution to the problem (following the spirit of Carnot's work): to seek invariant optimal relations for different operation regimes of Carnot-like models, independently from the heat transfer law. The equations (26)-(28) have the above characteristics. Nevertheless, the optimal isentropic temperatures ratio depends of the heat transfer law and of the operation regime of the engine as was shown in the subsection 2.2 (Fig. 5). Moreover, the equations (26) can be satisfied for other objective functions and other characteristic parameter: For instance, algebraic combination of power and/or efficiency and costs per unit heat transfer; as long as these objective functions and parameters have thermodynamic sense. Of course, the objective function must satisfy similar conditions to the equations (20) and (21). But this was not covered by this chapter's scope.
The study performed for the Brayton model combined and extended the optimal operation conditions of endoreversible and non-isentropic cycles since this model provides more realistic values for efficiency to maximum work and optimal allocation (size) for the heat exchangers than the values corresponding to the non-isentropic or the endoreversible operations. A relevant conclusion is that the allocation always is unbalanced ($y_{NE} < ½$). Furthermore, the following correlation can be applied between the effectiveness of the exchanger heat of the hot and cold sides:

$$\varepsilon_H = \frac{1 - \frac{1}{z_{NE}}}{1 - z_{NE} e^{-N}} \varepsilon_L \tag{69}$$

where z_{NE} is calculated by the equation (68) and shown in Fig. 7, which can be used in the current literature on the Brayton-like cycles. In subsection 3.1 the problem of when to fit a regenerator in a non-isentropic Brayton cycle was presented and criterion 3 was established. On the other hand, the qualitative and asymptotic analysis proposed showed that the non-isentropic and endoreversible Brayton cycles are limit cases of the model of irreversible Brayton cycle presented which leads to maintain the performance conditions of these limit cases according to their asymptotic behavior. Therefore, the non-isentropic and endoreversible Brayton cycles were not affected by our analytical approximation and remained invariant within the framework of the model herein presented. Moreover, the optimal analytical expressions for the optimal isentropic temperatures ratio, optimal allocation (size) for the heat exchangers, efficiency to maximum work and maximum work obtained can be more useful than those we found in the existing literature.

Finally, further work could comprise the analysis of the allocation of heat exchangers for a combined (Brayton and Carnot) cycle with the characteristics and integrating the methodologies herein presented.

5. References

Andresen, B. & Gordon J. M. Optimal heating and cooling strategies for heat exchanger design. J. Appl. Phys. 71, (January 1992) pp. 76-79, ISSN: 0021 8979.

Aragón-González G., Canales-Palma A. & León-Galicia A. (2000). Maximum irreversible work and efficiency in power cycles. *J. Phys. D: Appl. Phys.* Vol. 33, (October 2000) pp. 1403-1410, ISSN: 1361-6463.

Aragón-González G., Canales-Palma A., León-Galicia A. & Musharrafie-Martínez, M. (2003) A criterion to maximize the irreversible efficiency in heat engines,. *J. Phys. D: Appl. Phys.* Vol. 36, (Janaury 2003) pp. 280-287, ISSN: 1361-6463.

Aragón-González G., Canales-Palma A., León-Galicia A. & Musharrafie-Martínez, M. (2005). The fundamental optimal relations and the bounds of the allocation of heat exchangers and efficiency for a non-endoreversible Brayton cycle. *Rev. Mex. Fis.* Vol. 51 No. 1, (January 2005), pp. 32-37, ISSN: 0035-00IX.

Aragón-González G., Canales-Palma A., León-Galicia A. & Morales-Gómez, J. R. (2006). Optimization of an irreversible Carnot engine in finite time and finite size, Rev. Mex. Fis. Vol. 52 No. 4, (April 2006), pp. 309-314, ISSN: 0035-00IX.

Aragón-González G., Canales-Palma A., León-Galicia A. & Morales-Gómez, J. R. (2008). Maximum Power, Ecological Function and Eficiency of an Irreversible Carnot Cycle. A Cost and Effectiveness Optimization. *Braz. J. of Phys.* Vol. 38 No. 4, (April 2008), pp. 543-550, ISSN: 0103-9733.

Aragón-González G., Canales-Palma A., León-Galicia A. & Rivera-Camacho, J. M. (2009). The fundamental optimal relations of the al location, cost and effectiveness of the heat exchangers of a Carnot-like power plant. *Journal of Physics A: Mathematical and Theoretical.* Vol. 42, No. 42, (September 2009), pp. 1-13 (425205), ISSN: 1751-8113.

Aragón-González G., Canales-Palma A., León-Galicia A. & Morales-Gómez, J. R. (2010). A regenerator can fit into an internally irreversible Brayton cycle when operating in maximum work. *Memorias del V Congreso Internacional de Ingeniería Física,* ISBN: 978-607-477-279-1, México D.F., May 2010.

Arias-Hernández, L. A., Ares de Parga, G. and Angulo-Brown, F. (2003). On Some Nonendoreversible Engine Models with Nonlinear Heat Transfer Laws. *Open Sys. & Information Dyn.* Vol. 10, (March 2003), pp. 351-75, ISSN: 1230-1612.

Bejan, A. (1988). Theory of heat transfer-irreversible power plants. *Int. J. Heat Mass Transfer.* Vol. 31, (October 1988), pp. 1211-1219, ISSN: 0017-9310.

Bejan, A. (1995) Theory of heat transfer-irreversible power plants II. The optimal allocation of heat exchange equipment. *Int. J. Heat Mass Transfer.* Vol. 38 No. 3, (February 1995), pp. 433-44, ISSN: 0017-9310.

Bejan, A. (1996). *Entropy generation minimization,* CRC Press, ISBN 978-0849396519, Boca Raton, Fl.

Chen, J. (1994). The maximum power output and maximum efficiency of an irreversible Carnot heat engine. *J. Phys. D: Appl. Phys.* Vol. 27, (November 1994), pp. 1144-1149, ISSN: 1361-6463.

Chen L., Cheng J., Sun F., Sun F. & Wu, C. (2001). Optimum distribution of heat exchangers inventory for power density optimization of an endoreversible closed Brayton cycle. *J. Phys. D: Appl Phys.* Vol. 34, (Janaury 2010), pp. 422-427, ISSN 1361-6463.

Chen, L., Song, H., Sun, F. (2010). Endoreversible radiative heat engine configuration for maximum efficiency. *Appl. Math. Modelling,* Vol. 34 (August 2010), pp. 1710–1720, ISSN: 0307-904X.

Durmayaz, A. Sogut, O. S., Sahin, B. and Yavuz, H. (2004). Optimization of thermal systems based on finite time thermodynamics and thermoeconomics. *Progr. Energ. and Combus. Sci.* Vol. 30, (January 2004), pp. 175-217, ISSN: 0360-1285.

Herrera, C. A., Sandoval J. A. & Rosillo, M. E. (2006). Power and entropy generation of an extended irreversible Brayton cycle: optimal parameters and performance. J. *Phys. D: Appl. Phys.* Vol. 39. (July 2006) pp. 3414-3424, ISSN: 1361-6463.

Hoffman, K. H.,. Burzler, J. M and Shuberth, S. (1997). Endoreversible Thermodynamics. *J. Non-Equilib. Thermodyn.* Vol. 22 No. 4, (April 1997), pp. 311-55, ISSN: 1437-4358.

Kays, W. M. & London, A. L. (1998). *Compact heat exchangers* (Third edition), McGraw-Hill, ISBN: 9780070334182, New York.

Leff, H. S.(1987). Thermal efficiency at maximum work output: New results for old engines. *Am..J. Phys.* Vol. 55(February 1987), pp. 602-610, ISSN: 0894-9115.

Lewins, J. D. (2000). The endo-reversible thermal engine: a cost and effectiveness optimization. *Int. J. Mech. Engr. Educ.,* Vol. 28, No. 1, (January 2000), pp. 41-46, ISSN: 0306-4190

Lewins, J. D. (2005). A unified approach to reheat in gas and steam turbine cycles. *Proc. Inst. Mech. Engr. Part C: J. Mechanical Engineering Science.* Vol. 219, No 2 (November 2000), (March 2005), pp. 539-552., ISSN: 0263-7154.

Lienhard IV, J. H. & Lienhard V, J. H. (2011) *A Heat Transfer Textbook* (Fourth edition), Phlogiston Press, ISBN: 0-486-47931-5, Cambridge, Massachusetts

Nusselt, W. Eine Neue Formel für den Wärmedurchgang im Kreuzstrom. *Tech. Mech. Thermo-Dynam.* Vol. 1, No. 12, (December 1930), pp. 417–422.

Sanchez Salas, N., Velasco, S. and Calvo Hernández, A. (2002). Unified working regime of irreversible Carnot-like heat engines with nonlinear heat transfer laws. *Energ. Convers. Manage.* Vol. 43 (September 2002), pp. 2341—48, ISSN: 0196-8904.

Sontagg, R. E., Borgnankke, C. & Van Wylen, G. J. (2003), *Fundamentals Of Thermodynamics* (Sixth edition), John Wyley and Sons, Inc., ISBN: 0-471-15232-3, New York

Swanson L. W (1991). Thermodynamic optimization of irreversible power cycles wit constant external reservoir temperatures. *ASME J. of Eng. for Gas Turbines Power* Vol. 113 No. 4, (May 1991), pp. 505-510, ISSN: 0742-4795.

Ust Y., Sahin B. and Kodal A. (2005). Ecological coefficient of performance (ECOP) optimization for generalized irreversible Carnot heat engines. *J. of the Energ. Inst.* Vol. 78 No. 3 (January 2005), 145-151, ISSN: 1743-9671.

Wang L.G., Chen L., F. R. Sun & Wu, C. (2008). Performance optimisation of open cycle intercooled gas turbine power with pressure drop irreversibilities. *J. of the Energ. Inst.* Vol. 81 No. 1, (January 2008), pp. 31-37, ISSN: 1743-9671.

Wu C. & Kiang R. L. (1991). Power performance of a non-isentropic Brayton cycle. *ASME J. of Eng. for Gas Turbines Power* Vol. 113 No.4, (April 1991), pp. 501-504, ISSN: 0742-4795.

Yan Z. and L. Chen . The fundamental optimal relation and the bounds of power output and efficiency for and irreversible Carnot engine. J. Phys. A: Math. Gen. Vol. 28, (December 1995) pp. 6167-75, ISSN: 1751-8113.

Yilmaz T. (2006). A new performance criterion for heat engines:efficient power. *J. Energy Inst.,* Vol. 79, (January 2006), pp. 38—41, ISSN: 1743-9671.

Zhang, Y., Ou, C., Lin, B., and Chen, J. (2006). The Regenerative Criteria of an Irreversible Brayton Heat Engine and its General Optimum Performance Characteristics. *J. Energy Resour. Technol.* Vol, 128 No. 3, (2006), pp. 216-222, ISSN: 0195-0738.

Part 3

Gas Flow and Oxidation

10

Gas-Solid Flow Applications for Powder Handling in Industrial Furnaces Operations

Paulo Douglas Santos de Vasconcelos[1]
and André Luiz Amarante Mesquita[2]
[1]Albras Alumínio Brasileiro S/A
[2]Federal University of Pará
Brazil

1. Introduction

Gas-solid flow occurs in many industrial furnaces operations. The majority of chemical engineering units operations, such as drying, separation, adsorption, pneumatic conveying, fluidization and filtration involve gas-solid flow.

Poor powder handling in an industrial furnace operation may result in a bad furnace performance, causing errors in the mass balance, erosion caused by particles impacts in the pipelines, attrition and elutriation of fines overloading the bag houses. The lack of a good gas-solid flow rate measurement can cause economic and environmental problem due to airborne.

The chapter is focused on the applications of powder handling related with furnaces of the aluminum smelters processes such as anode baking furnace and electrolytic furnace (cell) to produce primary aluminum.

The anode baking furnace illustrated in figure 1 is composed by sections made up of six cells separated by partitions flue walls through which the furnace is fired to bake the anodes. The cell is about four meters deep and accommodates four layers of three anode blocks, around which petroleum coke is packed to avoid air oxidation and facilitate the heat transfer. During the baking process, the gases released are exhausted to the fume treatment center

(a) (b)

Fig. 1. a) Anode baking furnace building overview; b) Petroleum coke being unpacked from anode coverage by vacuum suction.

(FTC) where the gases are adsorbed in a dilute pneumatic conveyor and in an alumina fluidized bed. The handling of alumina is made via a dense phase conveyor.
The baked anode is the positive pole of the electrolytic furnace (cell) which uses 18 of them by cell. The pot room and the overhead multipurpose crane are illustrated in figure 2.

(a) (b)

Fig. 2. a) Aluminum smelter pot room, b) Overhead crane being fed of alumina from a day bin by a standard air slide.

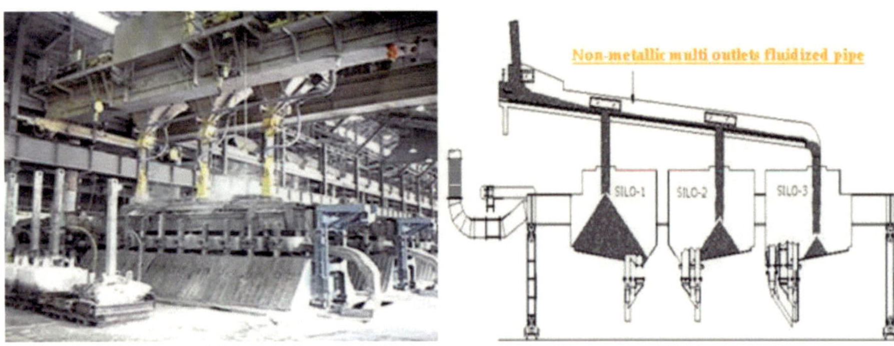

(a) (b)

Fig. 3. a) Electrolytic furnace being fed of alumina by the overhead crane; b) Sketch of electrolytic furnace being continuous fed of alumina by a special fluidized pipeline.

The old aluminum smelters feed their electrolytic cells with the overhead cranes as can be seen in figures 2 and 3. This task is very hard to the operators and causes spillage of alumina to the pot room workplace. This nuisance problem is being solved by the development of a special multi-outlets nonmetallic fluidized pipeline.
The fundamentals of powder pneumatic conveying and fluidization will be discussed in this chapter, such as the definition of a pneumatic conveying in dilute and dense phase, the fluidized bed regime map as illustrated in figure 5 and finally the air fluidized conveyor.
Firstly, petro coke and alumina used as raw materials in the primary aluminum process is characterized using sieve analyses (granulometry size distribution). Then, bulk and real density are determined in the laboratory analyses; with these powder physical properties, they can be classified in four types using the Geldart's diagram as illustrated in figure 4.

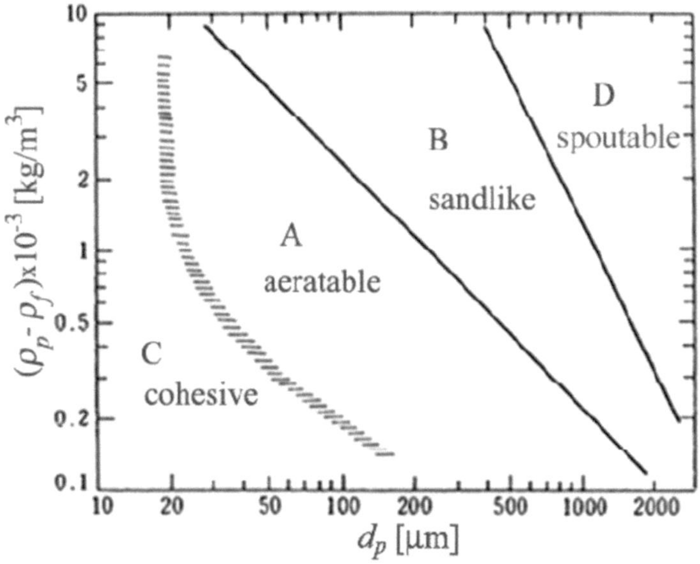

Fig. 4. Powder classification diagram for fluidization by air - source: (Geldart, 1972).

The majority of powders used in the aluminum smelters belong to groups A and B considering Geldart's criteria.

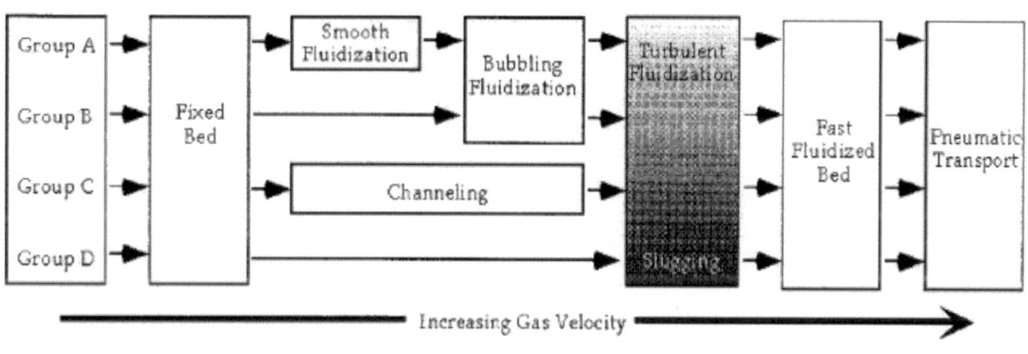

Fig. 5. Flow regime map for various powders.

This figure 5 summarized the fluidized bed hydrodynamics related with powders classified according to Geldart's criteria.
Once the velocities associated with each mode of operation are determined, the pressure drop of the regime is calculated so that the gas-solid flow is predicted using the modeling and software adequated to optimize the industrial installation.
The pipeline and air fluidized conveyors feeding devices are also discussed in this chapter.
Finally two cases studies applied in the baking furnace of pneumatic powder conveying in dilute phase are shown as a result of a master degree dissertation. Another case study is the development of an equation to predict the mass solid flow rate of the air-fluidized conveyor as a result of thesis of doctorate. The equation has design proposal and it was used in the design of a fluidized bed to treat the gases from the bake furnace and to continuously alumina pot feeding the electrolyte furnaces to produce primary aluminum.

2. Fundamentals of pneumatic conveying of solids

Pneumatic conveying of solids is an engineering unit operation that involves the movement of millions of particles suspend by draft in dilute phase or in a block of bulk solids in dense phase inside a pipeline. Figure 6 illustrates a pneumatic conveying of solids with the essential components, like the air mover, feeding device, pipeline and bag house.

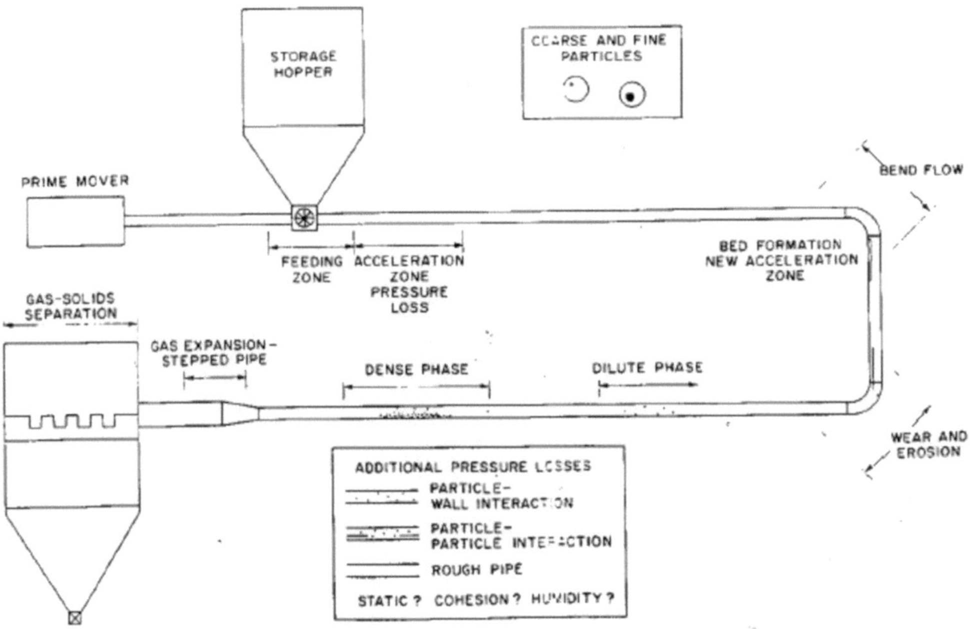

Fig. 6. Typical pneumatic conveyor layout – source: (Klinzing *et al*, 1997).

A good criterion showed in table 1 to decide if the transport of solids in air will be in dilute phase or in dense phase is the mass load ratio (μ) calculated by the equation 1.

$$\mu = \dot{G} / \dot{v}\,\rho_g \tag{1}$$

Where $\dot{G}$, $\dot{v}$ and ρ_g are the solid mass flow rate, gas (air) volume flow rate and gas (air) density.

Mode of transport of solids	Solids –to – air ratio (μ)
Dilute phase	0 - 15
Dense phase	> 15

Table 1. Systems' classification concerning solids-to-air ratio - source: (Klinzing et al, 1997).

Figure 7 illustrates a variety of solids modes of transportation and the states diagrams showing the log of the pipeline pressure drop versus the log of the air velocity inside the pipeline. From figure 7 it is concluded that in dilute phase the pneumatic conveyor has high air velocity, low mass load ratio, and low pressure drop in the pipeline. In dense phase mode the conveyor operates with high mass load rate, low air velocity but high pressure drop in the pipeline. The engineer responsible for the project has to analyze which is the best solution for each case study.

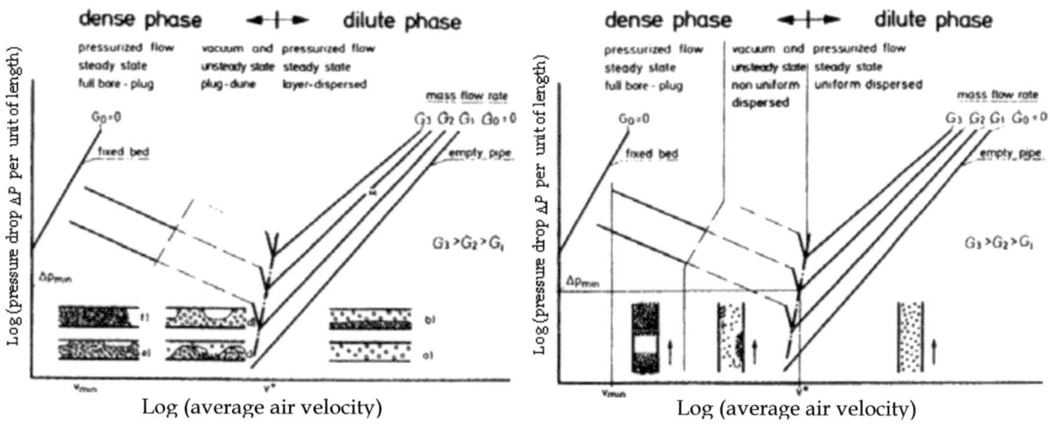

Fig. 7. Conveying conditions showing the changes in solids loading; left: State diagram horizontal flow, right: State diagram vertical flow - source: (Klinzing et al, 1997).

2.1 Pressure drop calculation in the pipeline

The equations given here are based on the hypothesis that the gas-solid flow is in dilute phase. Some assumptions such as: transients in the flow (Basset forces) are not considered nor the pressure gradient around the particles (this is considered negligible in relation to the drag, gravitational and friction forces).
The pressure drop due to particle acceleration is not considered.
The flow is considering incompressible, omnidimensional and the concentration of solids particles is uniform. The physical properties of the two phases are temperature dependent.
The mass flow rate for each phase can be expressed by the following equations:

$$\dot{m}_g = \rho_g V_{mc} \varepsilon_{mc} A \tag{2}$$

$$\dot{G} = \rho_s V_s \varepsilon_s A \tag{3}$$

Where the suffixes g and s denote gas and solid, respectively, V_{mc} is the minimum air velocity in dilute phase in relation to the pipe cross section A, ε_s and ε_{mc} are the volume fraction occupied by the solid and gas inside the pipeline as follows:

$$\varepsilon_s = \frac{A_s}{A} = \frac{4\dot{G}}{\rho_s \pi D^2 V_s} \tag{4}$$

D is the pipe diameter and V_s is the particle velocity. The void or porosity at the minimum air velocity ε_{mc}, in other words is the fraction of volume occupied by the gas (air).

$$\varepsilon_{mc} = \frac{A_g}{A} = 1 - \varepsilon_s \tag{5}$$

Velocity of the particle V_s and the particle terminal velocity V_t calculation:
In this chapter it is considered the models of (Yang, 1978) for the pressure drop calculation.

$$V_s = \frac{V_{mc}}{\varepsilon_{mc}} - V_t\sqrt{1 + \frac{2f_s V_s^2}{gD}\varepsilon_{mc}^{4.7}} \tag{6}$$

$$V_t = \frac{g d_p^2 (\rho_s - \rho_g)}{18\mu_g}, \quad K < 3.3 \tag{7}$$

$$V_t = \frac{0.153 g^{0.71} d_p^{1.14} (\rho_s - \rho_g)^{0.71}}{\rho_g^{0.29} \mu_g^{0.43}}, \quad 3.3 < K < 43.6 \tag{8}$$

$$V_t = 1.74\left[\frac{g d_p (\rho_s - \rho_g)}{\mu_g}\right]^{1/2}, \quad 43.6 < K < 2360 \tag{9}$$

K is a factor that determines the range of validation for the drag coefficient expressions, when the particle Reynolds number is unknown, and given by:

$$K = d_p\left[\frac{g\rho_g(\rho_s - \rho_g)}{\mu_g^2}\right]^{1/3} \tag{10}$$

Where μ_g is the gas dynamic viscosity, g acceleration due gravity, ρ_g gas (air) density, d_p the particle diameter, f_s is the solid friction factor.
The total pressure drop, ΔP_T for gas-solid flow is calculated with the contribution of the static pressure ΔP_E, and friction loss ΔP_F for both phases:

$$\Delta P_T = (\Delta P_E + \Delta P_F)_s + (\Delta P_E + \Delta P_F)_g \tag{11}$$

$$(\Delta P_E)_s = \rho_s \varepsilon_s L g \tag{12}$$

$$(\Delta P_E)_g = \rho_g \varepsilon_g L g \tag{13}$$

The contribution due to the friction factor given by the Darcy equation:

$$(\Delta P_F)_s = \frac{2 f_s \rho_s V_p^2 L}{D} \tag{14}$$

$$(\Delta P_F)_g = \frac{2 f_g \rho_g V_g^2 L}{D} \tag{15}$$

Where, f_s and f_g are the friction factor for the solid and gas (air), respectively. The friction factor for the gas is calculated by the Colebrook equation.

$$\frac{1}{\sqrt{4f_g}} = 1.74 - 2\log\left[2\xi + \frac{18.7}{\mathrm{Re}\sqrt{4f_g}}\right] \tag{16}$$

$$\mathrm{R}_e = \frac{\rho_g V_{mc} D}{\mu_g} \tag{17}$$

R_e is the Reynolds number, ξ is the relative roughness of the pipe, this equation is implicit for f_g. The friction factors due to solids in vertical and horizontal flow are obtained by the model of (Yang, 1978).

$$f_{sv} = 0.00315\frac{1-\varepsilon_{mc}}{\varepsilon_{mc}^3}\left[\frac{(1-\varepsilon_{mc})V_t}{\frac{V_{mc}}{\varepsilon_{mc}} - V_s}\right]^{-0.979} \tag{18}$$

$$f_{sh} = 0.0293\frac{1-\varepsilon_{mc}}{\varepsilon_{mc}^3}\left[\frac{(1-\varepsilon_{mc})\frac{V_g}{\varepsilon_{mc}}}{\sqrt{gD}}\right]^{-1.15} \tag{19}$$

3. Fundamentals of powder fluidization

Fluidization is an engineering unit operation that occurs when a fluid (liquid or gas) ascend trough a bed of particles, and that particles get a velocity of minimum fluidization V_{mf} enough to suspend the particles, but without carry them in the ascending flow. Since this moment the powder behaves like a liquid at boiling point, that is the reason for term "fluidization". Figure 8 gives a good understanding of the minimum fluidization velocity.

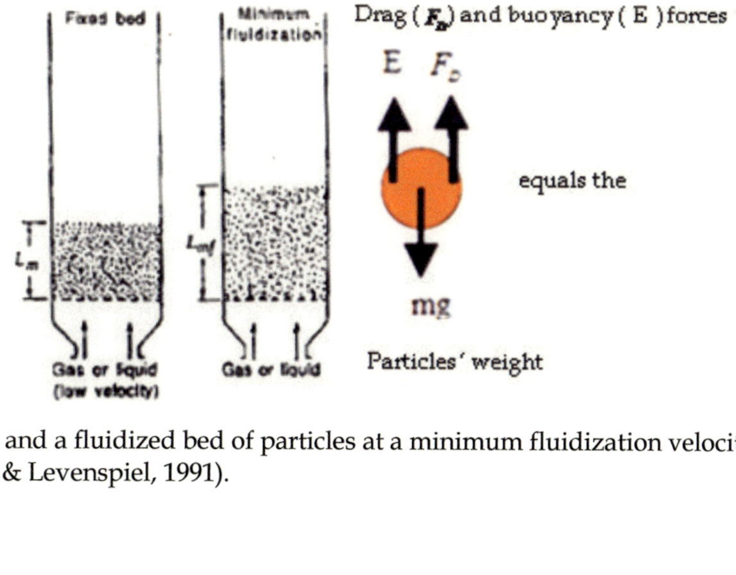

Fig. 8. Fixed and a fluidized bed of particles at a minimum fluidization velocity - adapted from (Kunii & Levenspiel, 1991).

3.1 Minimum fluidization velocity calculation

In this chapter it will be point out the beginning (fluidized beds) and the ending (pneumatic transport) of the flow regime map illustrated in figure 5.
The minimum fluidization velocity will be calculated by the (Ergun, 1952) equation 20 and it's experimental value obtained using a permeameter as showed in figure 19.

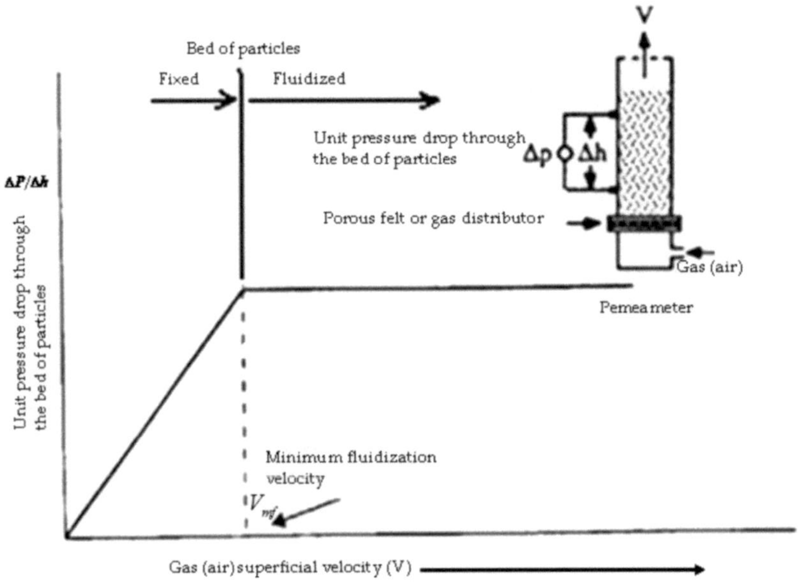

Fig. 9. Pressure drop through a bed of particles versus superficial air velocity – source: (Mills, 1990).

$$(1-\varepsilon_{mf})(\rho_s-\rho_g)g = 150\frac{(1-\varepsilon_{mf})^3}{\varepsilon_{mf}^{\ 3}}\frac{\mu_g V_{mf}}{(\varphi_s d_p)^2} + 1.75\frac{(1-\varepsilon_{mf})}{\varepsilon_{mf}^3}\frac{\rho_g V_{mf}^2}{\varphi_s d_p} = AV_{mf} + BV_{mf}^2 \quad (20)$$

Calculating C by equation 21 we get an equation of second power, which the positive solution is calculated by equation 22.

$$C = (1-\varepsilon_{mf})(\rho_s-\rho_g)g \quad (21)$$

$$V_{mf} = \frac{-A+\sqrt{A^2+4BC}}{2B} \quad (22)$$

Where A and B are the viscous and the inertial factors of the Ergun equation, C is the weight per unit volume of the bed of particles.
Fluidization is related with small velocities, the factor B is negligible and the Ergun equation can be simplified with an error less than 1% by the equation 23.

$$V_{mf} = \frac{C}{A} \Rightarrow V_{mf} = \frac{(\rho_s-\rho_g)g\varepsilon^3{}_{mf}(\phi_s d_p)^2}{150(1-\varepsilon_{mf})\mu_g} \quad (23)$$

For an incipient fluidization, when the weight of particles equals the drag force, it is a good attempt to consider the porosity at the minimum fluidization velocity ε_{mf} equals the porosity ε of the fixed bed. The porosity of the fixed bed is calculated by the equation 24.

$$\varepsilon = 1 - \frac{\rho_{bnv}}{\rho_s} \tag{24}$$

$$\rho_{bnv} = \frac{M_s}{V_{total}} \tag{25}$$

Where ρ_{bnv} is the non-vibrated bulk density, ρ_s is the solid real density calculated in a laboratory by a pycnometer, M_s is the total mass of particles weighted in a electronic scale, V_{total} is the total volume of particles and voids in the sample previously weighted in a electronic scale, d_p is the particle mean diameter obtained by sieve analysis in a laboratory, ϕ_s is the particle sphericity, that can be estimated by equation 26 with d_p in (m).

$$\frac{1-\varepsilon}{\phi_s} = 0.255Log(d_p) + 1.85 \tag{26}$$

Other important velocity in pneumatic transport and fluidization is the particle terminal velocity V_t that is calculated by equations 7 to 10.

4. Air-fluidized conveyors

4.1 Pile powder flow - simple model

Assuming that the block of alumina is made of non-cohesive material with uniform porosity ε, and is inclined at the moment of analysis in an angle α to the horizontal plane, this elemental block has a constant width Δz.

Also assuming that the flow is isothermal in the (y) direction, a simple force balance requires that:

Gravitational force component = Drag force + Frictional force due to apparent viscosity (27)

$$\rho_s.(1-\varepsilon)gsen\alpha(\Delta x\Delta y\Delta z) = \Delta P(\Delta x\Delta z) + \tau(\Delta y\Delta z) \tag{28}$$

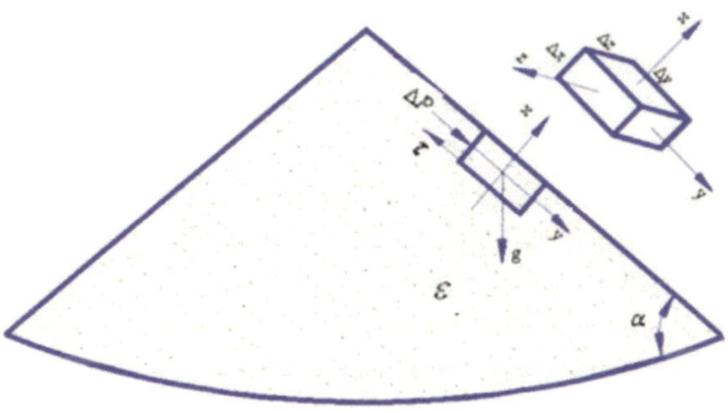

Fig. 10. Force balance acting on an elemental pecked bed block of alumina flowing to repose.

Assuming that the particles of the block are ready to slip over each other, the internal friction between the particles is obtained from equation 29 – source: (Schulze, 2007).

$$\tau_{xy} = \tan\varphi_i.\sigma_x \tag{29}$$

Assuming that the cohesion between particles is negligible, τ_{xy} is the shear stress in the plane parallel to the plane (y-z), ϕ_i is the angle of internal friction of the powder, and σ_x the normal stress in the direction (x) is:

$$\sigma_x = \frac{\rho_s(1-\varepsilon)g\cos\alpha(\Delta x\Delta y\Delta z)}{\Delta y\Delta z} \tag{30}$$

Rearranging the equations 28 to 30 is obtained the pressure drop ΔP of flow in the (y) direction by the equation 31.

$$\frac{\Delta P}{\Delta y} = \rho_s(1-\varepsilon)g(sen\alpha - \tan\varphi_i\cos\alpha) \tag{31}$$

The pile of alumina powder reaches the equilibrium between gravity force and Interparticles forces with an angle of repose β, so in this moment the angle alpha of pile turns beta ($\alpha \rightarrow \beta$).

The friction coefficient between the particles of the pile is the tangent of the particles internal friction angle, given by equation 32.

$$\mu = \tan\phi_i \tag{32}$$

4.2 Engineering model for design proposal to air fluidized conveyors

The proposed model for gas-solid flow is based in the figure 11 and is fitted by the following considerations:

1. The height of the moving fluidized bed in the (y) and (z) directions is constant;
2. The flow of the fluidizing air and that of moving fluidized bed are both steady and full developed in the mean flow;
3. The flow of the moving fluidized bed is in the (y) direction;
4. The slip velocity between the particles and the air in the (y) direction will be negligible;
5. The direction of the fluidizing air will have components in the (x) and (y) directions both in the inlet and outlet of the moving fluidized bed;
6. The pressure of fluidization is constant for every conveyor inclination;
7. The gas-solid flow is considered isothermal and irrotacional;
8. It will not be considered the mass and heat transfer between the particles and the air;
9. The shear stress τ_{xy} of particles will vary in the (x) direction and will be maximum in the bottom and minimum in the top of the elemental block of alumina;
10. The electrostatic and van der Walls forces in the proposed model it will not be considered;
11. The porosity of the moving fluidized bed varies with the fluidizing air velocity but will be considered isotropic;
12. The shear stress of the particles is a function of the moving fluidized bed porosity;

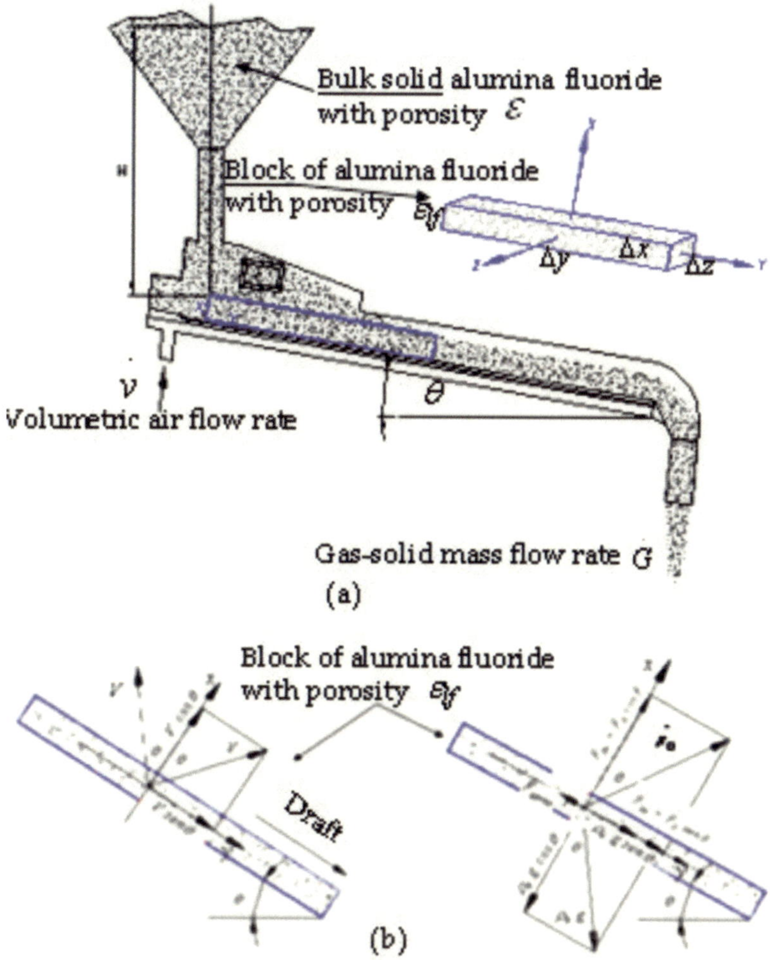

Fig. 11. (a) Elemental block of alumina in an air-fluidized conveyor inclined of an angle θ, (b) in the left: elemental block alumina being fluidized by the components of superficial air velocity V, in the right: balance of forces per unit of volume acting on the elemental block of alumina.

13. The moving fluidized bed has a rheological behavior similar to a liquid;
14. The stress in the side walls of the air fluidized conveyor is considered negligible due to air lubrication,
15. The friction coefficient between particles and the fluidizing felt in the bottom will be minimized by the fluidizing air velocity, and will follow the model of (Kozin & Baskakov, 1996) for angle of repose in the fluidized state, but adjusted with k factor;
16. There will not be rate of mass accumulation inside the air fluidized conveyor see equation 33;
17. It will not have rate of momentum accumulation inside of the air fluidized conveyor see equation 12;
18. The drag force of the particles inside the moving fluidized bed will follow the Ergun (1952) equation.

19. The elemental block of alumina will be considered continuum;
20. The maximum solid mass flow rate will be determined by the model proposed by (Jones, 1965) for powder discharge from small silos – powder liquid-like behavior;
21. For a full opened gate valve of the feed bin the mass flow rate of the assemble bin-air fluidized conveyor will follow the rhythm of the air fluidized conveyor;
22. The model will not consider the contribution of the height of the bulk solids H in the feeding bin.

4.2.1 Equation of continuity

It was assumed in the above considerations that there will not be rate of mass accumulation inside the air fluidized conveyor according to equation 33.

$$\begin{pmatrix}\text{Rate of mass}\\ \text{accumulation}\end{pmatrix}=\begin{pmatrix}\text{Rate of mass}\\ \text{in}\end{pmatrix}-\begin{pmatrix}\text{Rate of mass}\\ \text{out}\end{pmatrix}=0 \quad (33)$$

Equation 33 points out that the mass of gas and solid mixture entering the air fluidized conveyor must be the same as in the outlet of the air fluidized conveyor during the full-developed steady flow regime. So, for the two phases of the flow we have.
Gas phase:

$$\frac{\partial}{\partial x}\varepsilon_{lf}\rho_g V_x + \frac{\partial}{\partial y}\varepsilon_{lf}\rho_g V_y = 0 \quad (34)$$

Where: V_x and V_y are the component of the superficial air velocity in the (x) and (y) directions, ρ_g is the gas or air density, ε_{lf} is the moving fluidized bed porosity.
Solid phase:

$$\frac{\partial}{\partial y}(1-\varepsilon_{lf})\rho_s V_s = 0 \quad (35)$$

Where: V_s is the velocity of the moving fluidized bed in the (y) direction as it was supposed in the model considerations, ρ_s is the solid or particle density.

4.2.2 The momentum equation

Let's consider the elemental block of alumina in figure 11 with porosity ε_{lf} flowing in a fluid with constant density ρ_g and bulk density ρ_b of the mixture gas-solid and air superficial velocities V_x and V_y in the (x) and (y) directions and the particles flowing in the (y) direction without slipping the air in this direction with velocity V_s.
As it was supposed there will not be rate of momentum accumulation inside of the air fluidized conveyor according to equation 36.

$$0=\begin{pmatrix}\text{Rate of}\\ \text{momentum in}\end{pmatrix}-\begin{pmatrix}\text{Rate of}\\ \text{momentum out}\end{pmatrix}+\begin{pmatrix}\text{Sum of forces}\\ \text{acting on system}\end{pmatrix} \quad (36)$$

The equation 36 is in truth is a force balance equation; the terms used in this equation are as follows:

Rate of momentum in across surface at (x) in the beginning of the elemental block
(Moment transport due to powder apparent viscosity); $\tau_{xy}(\Delta y \Delta z)|x$
Rate of momentum in across surface at (x+ Δx) in the end of the elemental block
(Moment transport due to powder apparent viscosity); $\tau_{xy}(\Delta y \Delta z)|x+\Delta x$
Rate of momentum in across surface at (y=0)
(Momentum due to gas-solid motion) $(\Delta x \Delta z V_y)(\rho_b V_y)|y$
Rate of momentum in across surface at (y + Δy)
(Momentum due to gas-solid motion) $(\Delta x \Delta z V_y)(\rho_b V_y)|y+\Delta y$
Pressure force on the elemental block at (x) in the (x) direction $P(\Delta y \Delta z)|x$
Pressure force on the elemental block at (x+ Δx) in the(x) direction $P(\Delta y \Delta z)|x+\Delta x$
Pressure force on the elemental block at (y=0) in the (y) direction $P(\Delta x \Delta z)|y$
Pressure force on the elemental block at (y+ Δy) in the (y) direction $P(\Delta x \Delta z)|y+\Delta y$
Gravity force acting on the mixture gas-solid in the (x) direction $(\Delta x \Delta y \Delta z)\rho_b g \cos\theta$
Gravity force acting on the mixture gas-solid in the (y) direction $(\Delta x \Delta y \Delta z)\rho_b g sen\theta$
Drag force acting on the mixture gas-solid in the (x) direction F_{Dx}
Drag force acting on the mixture gas-solid in the y) direction F_{Dy}
Friction force due to gravity and fluid during block motion in the(y) direction $F_{friction}$
Substituting the terms relating with (x) direction in the moment balance equation 36 results:

$$P(\Delta y \Delta z)|x - P(\Delta y \Delta z)|x+\Delta x - F_{Dx} + (\Delta x \Delta y \Delta z)\rho_b g \cos\theta = 0 \tag{37}$$

Equation 11 is now divided by the elemental volume $\Delta x \Delta y \Delta z$ and, if Δx is allowed to be infinitely small $\Delta x \to 0$, it is obtained:

$$\frac{\partial P}{\partial x} = \rho_b g \cos\theta - \bar{F}_{Dx} \tag{38}$$

Where $\bar{F}_{Dx}$ is the drag force per unit of volume of the elemental block in (x) direction.
Substituting the terms relating with (y) direction in the moment balance equation 10 results equation 39:

$$\tau_{xy}(\Delta y \Delta z)|x - \tau_{xy}(\Delta y \Delta z)|x+\Delta x + (\Delta x \Delta z V_y)(\rho_b V_y)|y - (\Delta x \Delta z V_y)(\rho_b V_y)|y+\Delta y +$$

$$+ P(\Delta x \Delta z)|y - P(\Delta x \Delta z)|y+\Delta y + (\Delta x \Delta y \Delta z)\rho_b g sen\theta + F_{Dy} - F_{friction} = 0 \tag{39}$$

Equation 37 is now divided by the elemental volume $\Delta x \Delta y \Delta z$ and, if Δx and Δy are allowed to be infinitely small $\Delta x \to 0$ and $\Delta y \to 0$ it is obtained:

$$-\frac{\partial \tau_{xy}}{\partial x} - \rho_b V_y \frac{\partial V_y}{\partial y} - \frac{\partial P}{\partial y} + \bar{F}_{Dy} - \bar{F}_{friction} + \rho_b g sen\theta = 0 \tag{40}$$

One of the considerations of the proposed model is that the particle velocity doesn't vary in the (y) direction, so the second term of equation 40 is zero. Rearranging the equation 40, it is obtained:

$$\frac{\partial P}{\partial y} = \bar{F}_{Dy} - \frac{\partial \tau_{xy}}{\partial x} - \bar{F}_{friction} + \rho_b g sen\theta \tag{41}$$

Where $\bar{F_{Dy}}$, the drag force per unit of volume of the elemental block in the (y) direction is calculated concerning the 18th consideration of the proposed model as follows:
$\bar{F_{Dy}}$ is calculated by the Ergun equation 42

$$\bar{F_{Dy}} = 150\frac{(1-\varepsilon_{lf})^2}{\varepsilon_{lf}^3}\frac{\mu_g V_y}{(\varphi_s d_p)^2} + 1,75\frac{(1-\varepsilon_{lf})}{\varepsilon_{lf}^3}\frac{\rho_g V_y^2}{\varepsilon_{lf}^3(\varphi_s d_p)}, \varepsilon_{lf} < 0,79 \tag{42}$$

$\bar{F}_{friction}$ is the force due to gravity and drag forces of the moving fluidized bed in the (y) direction.

$$\bar{F}_{friction} = \mu_a \frac{\partial P}{\partial x} \tag{43}$$

Substituting the equation 38 in equation 43 we get.

$$\bar{F}_{friction} = \mu_a(\rho_b g \cos\theta - \bar{F}_{Dx}) \tag{44}$$

$$\bar{F_{Dx}} = AV_x + BV_x^2 \tag{45}$$

$$\mu_a = \tan\left[(1-0.1\frac{V}{V_{mf}})^2 \phi_i\right] \tag{46}$$

$$\bar{F_{Dy}} = AV_y + BV_y^2 \tag{47}$$

$$A = 150\frac{(1-\varepsilon_{lf})^2}{\varepsilon_{lf}^{\ 3}}\frac{\mu_g}{(\varphi_s.d_p)^2} \tag{48}$$

$$B = 1.75\frac{(1-\varepsilon)}{\varepsilon_{lf}^{\ 3}}\frac{\rho_g}{\varphi_s d_p} \tag{49}$$

$$V_x = V\cos\theta \tag{50}$$

$$V_y = Vsen\theta \tag{51}$$

$$\bar{F_{Dx}} = AV\cos\theta + BV^2\cos^2\theta \tag{52}$$

$$\bar{F_{Dy}} = AVsen\theta + BV^2 sen^2\theta \tag{53}$$

A and B are the viscous and inertial factors of Ergun's equation, these factors are calculated by the equations 48 and 49, μ_a is the coefficient of friction between the particles and the bottom of the air fluidized conveyor at fluidized state.

If we reanalyze the equation 28 it is possible to compare the fluidized powder flowing rather like a liquid, which is sheared by the powder apparent viscosity and the friction due to inertia force. Therefore, by rearranging equation 28, we can obtain the maximum shear stress in the bottom of the block of height h and length of Δy by equation 54.

$$-\tau_{0xy} = -\rho_b g sen\theta h + \mu_a \rho_b g \cos\theta \frac{h^2}{L} \tag{54}$$

In the plane parallel with the bottom arising shear stress as a function of the deepness of the block in the (x) direction calculated by equation 55.

$$\tau_{xy} = -\rho_b g \cos\theta \left[(2\mu_a \frac{h}{L} - \tan\theta)(h-x) - \mu_a \frac{(h-x)^2}{L} \right] - \tau_{0xy} \tag{55}$$

Equation 55 describes the shear stress distribution and is according to the considerations of the proposed model. In others words, in the top of the block (x equals 0) the shear stress is minimum and maximum in the bottom of the block (x equals h).
Deriving equation 55, it is found the shear stress gradient by equation 56.

$$-\frac{\partial \tau_{xy}}{\partial x} = -\rho_b g \left[2\frac{\mu_a}{L} x \cos\theta - sen\theta \right] \tag{56}$$

Algebraically rearranging the equations 43 to 53 and introducing the results in equation 41, we get:

$$\frac{\partial P}{\partial y} = AVK_1 + BV^2 K_2 + \rho_b g K_3 \tag{57}$$

$$K_1 = sen\theta + \mu_a \cos\theta \tag{58}$$

$$K_2 = sen^2\theta(1-\mu_a) + \mu_a \tag{59}$$

$$k_3 = 2sen\theta - \mu_a \cos\theta(2\frac{x}{L} + 1) \tag{60}$$

Equation 34 represents a weight of bulk solids per unit of volume of the elemental block of alumina as can be seen in figure 11. The elemental block of alumina flows in the (y) direction with porosity less than 0.8 in an air-fluidized conveyor inclined at an angle θ in relation to the horizontal plane, with (x equals h – height of the bed) and a length of L meters long.
Rearranging 57 as weight of bulk solids by its total volume we get equation 61.

$$\frac{\Delta mg}{\Delta V_{total}} = AVK_1 + BV^2 K_2 + \rho_b g K_3 \tag{61}$$

Dividing the left side of equation 61 by the time (t) and integrating it, we get the mass solid flow rate of the air-fluidized conveyor in equation 62.

$$\dot{G} = \frac{AVK_1 + BV^2K_2 + \rho_b g K_3}{g}\dot{v} \tag{62}$$

Where Δm and ΔV_{total} are the mass and total volume of the elemental block of alumina, $\dot{m}$ is the mass solid flow rate predicted to the air-fluidized conveyor, $\dot{v}$ is the volumetric flow rate of the fluidizing air, V is the superficial air velocity, V_{mf} is the minimum fluidization velocity.

$$V = \frac{\dot{v}}{\Delta z \Delta y} \tag{63}$$

$$V_s = \frac{\dot{G}}{\rho_b A} \tag{64}$$

Where $\Delta z = b$, $\Delta y = L$ and A are the width, length and cross section of the air fluidized conveyor.

According to (Jones, 1965), the flow of bulk solids from a small silo (bin) or from an orifice in the bottom of a fluidized bed, behaves like a liquid flowing and is calculated by the energy balance equation of Bernoulli as follows in the equation 65 adjusted by a reduction of 50% of the liquid flow from in the same condition.

$$\dot{G} = 0{,}5\rho_b A_o \sqrt{2gH} \tag{65}$$

$$\rho_b = (1-\varepsilon)\rho_s \tag{66}$$

Where $\dot{m}$ is the solid mass flow rate, ρ_b is the solid bulk density, A_o is the area of the bin's discharge orifice (normally the orifice has a gate valve to control the mass solid flow rate) ε is the material packed or non-aerated porosity, g is the acceleration due to gravity and H is the height of bulk solid material in the bin - see figure 11.

5. Pipeline feeding devices

There are many feeding device used in pneumatic conveying of solids. Low pressure devices are used in dilute phase conveyor such as venture feeder and rotary airlock valve as is illustrated in figure 12a. For pneumatic conveying in dense phase blow vessel is used a as is illustrated in figure 12b.

The feeding device must guarantee a good sealing between the pipeline and the bag house, hopper or silo used to collect or store the powder that will be pneumatically conveyed.

6. Petro coke handling at Albras anode bake furnace

Albras increased its productive capacity of 350,000 t/y to 390,000 t/y in the year 2000. To reach this production it was necessary to increase the bake furnaces capacity and, in consequence to increase the furnaces suction system capacity, which at that time didn't comply with efficiency the current demand of the plant.

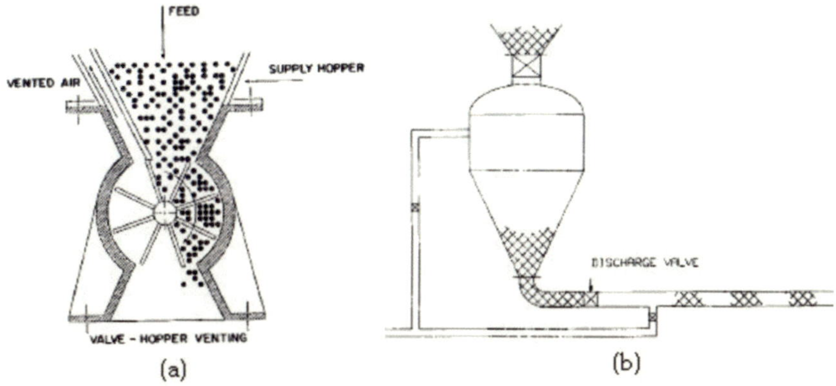

Fig. 12. Feeding devices used in pneumatic conveying of powders - source: (Klinzing *et al*, 1997).

The Carbon Plant Engineering Group accepts the challenge to upgrade the original system to increase the coke suction capacity of 15 t/h to 80 t/h. So software in FORTRAN was developed using the equations 1 to 19 to calculate the pressure drop in the circuit showed in figure 13, and to specify a new exhauster, redesign the bag house, the telescope tube and a new nozzle.
Following, the steps of the project, from the coke characterization and the circuit topology.

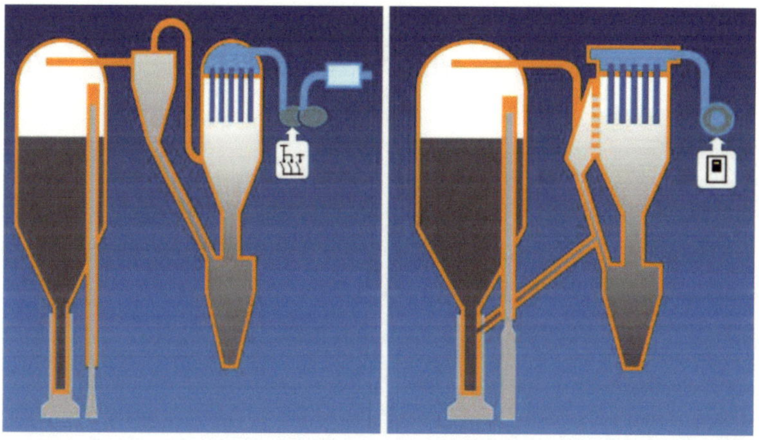

Fig. 13. Bake furnace suction system - old system in the left, new system in the right - source: (Vasconcelos, 2000).

The system capacity was increased from 15 t/h to 110 t/h of coke suction capacity. Figure 14 shows the nozzle details of the old and new systems.

7. Petro coke handling at Albras carbon plant

In smelters that use prebaked anodes, operations such as butt cleaning, butt and anode reject crushing and grinding, handling of coke packing material in the bake furnace, floor sweeping and discharge of dust from bake furnace cranes cause significant problems with the high generation of carbon dust and consequent environmental pollution.

Fig. 14. Albras bake furnace coke suction system before and after modifications - source: Albras Alumínio Brasileiro SA.

To control the dust pollution is a very difficult task. Another problem is how to convey and store the dust collected.
This case study presents the problem existing in the carbon plant of Albras, describing the pneumatic conveyor system developed by the Carbon Plant Engineering department, and the storage of the dust collected in the carbon plant.
Using the same software developed by the team of professor Mesquita of Federal Univerty of Pará to collect the carbon dust from the bake furnace and rodding shop and pneumatic convey to a silo in the paste plant at Albras aluminum smelter.
Figure 15 presents the originals problems relating the emission of carbon dust during the carbon plant process.

Fig. 15. Petro coke handling problem at Albras, in left bake furnace, in center and in the right bag house dust discharge at the rodding shop - source: Albras Alumínio Brasileiro SA.

Figure 16 shows the situation of coke handling at the bake furnaces during the multipurpose overhead crane coke dust discharge. The ultra fine carbon dust is generated during the pack and unpacked coke used to cover the anodes to avoid air-oxidation and facilitate the heat transfer to anode during the baking process.
Figure 17 shows the computer screen of a pneumatic conveying in dilute phase developed to collect the carbon dust in the carbon plant reducing the pollution in the workplace. The dust

collect is send to a cement plant reducing the consumption of charcoal in the cement's process.

Fig. 16. Dust discharging at Albras bake furnace, implemented solution in the left side, in the center discharge of dust in big bags, free falling of dust in truck in the right - source: Albras Alumínio Brasileiro SA.

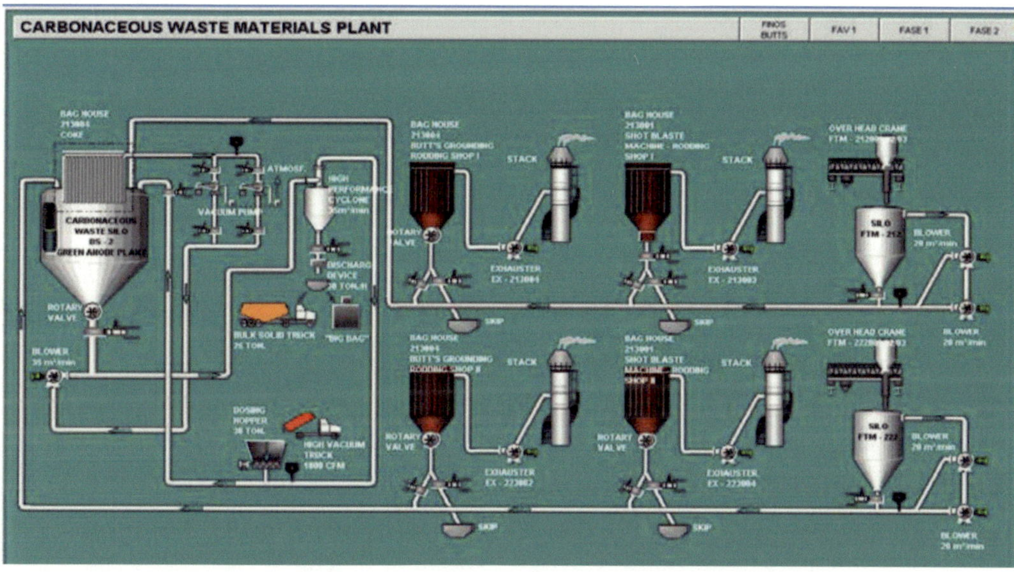

Fig. 17. Computer screen of a pneumatic conveying system in dilute phase at Albras aluminum smelter - source: (Vasconcelos & Mesquita, 2003).

8. Air fluidized conveyor

It was developed a non-conventional air slide called air fluidized conveyor to be of low weight, non-electrical conductor, heat resistant, easy to install, maintain and also operates at a very low cost compared with the conventional air slides. Figure 18 shows in the left a conventional air slide with rectangular shape, with one inlet and one outlet and in the right the round air fluidized conveyor with possibility to have multiples outlets.

Fig. 18. The Albras aluminum smelter air fluidized conveyor and a conventional air slide in the left.

8.1 Predict and experimental results of the air fluidized conveyor for fluoride alumina

The properties calculated and obtained from experiments with alumina fluoride used at Albras aluminum smelter are summarized in table 2.

Material property	value
Specific gravity	3.5
Non- aerated/vibrated bulk density - kg/m^3	1000
Aerated bulk density at ($0.5\,V_{mf}$) - kg/m^3	999.66
Aerated bulk density at ($0.75\,V_{mf}$) - kg/m^3	999.66
Aerated bulk density at ($0.875\,V_{mf}$) - kg/m^3	999.66
Aerated bulk density at ($1.0\,V_{mf}$) - kg/m^3	990.86
Aerated bulk density at ($1.5\,V_{mf}$) - kg/m^3	868.47
Aerated bulk density at ($2.0\,V_{mf}$) - kg/m^3	786.86
Aerated bulk density at ($2.5\,V_{mf}$) - kg/m^3	726.77
Minimum fluidization velocity by Ergun equation (cm/s)	1.83
Minimum fluidization velocity - experimental (cm/s)	1.77
Mean particle diameter - μm	99.44
Non- aerated angle of repose - °	35
Non- aerated angle of internal friction - °	70
Normal packed porosity (-)	0.71428
Geldart classification according figure 4 – group	B

Table 2. Properties of the alumina fluoride.

Figure 19 shows the pictures of the permeameters used to determine experimentally the minimum fluidization velocity of alumina fluoride.

Fig. 19. Permeameters used at Albras laboratory to survey the minimum fluidization velocity of the powders used in the primary aluminum industry - source: Albras Alumínio Brasileiro SA.

8.2 Predict and experimental results of the air fluidized conveyor for alumina fluoride

Two air-fluidized conveyors using the equation 62 were developed as result of a thesis for doctorate. The results for the conveyor with diameter of 3 inches and 1.5 m long showed in figure 20 are summarized in table 3.

Fig. 20. Air-fluidized conveyor of 1.5 m long with three outlets.

Mass gas-solid flow rate for alumina fluoride (t/h) - air fluidized conveyor of (3"/1.5m)							Inclination (°)
0	0	0	0	1.973	4.400	6.965	(-1)
0	0	0	0	2.145	4.616	7.222	(-0.5)
0	0	0	0.109	2.317	4.831	7.478	0
0	0	0	0.359	2.660	5.261	7.990	1
0	0	0	0.608	3.002	5.689	8.499	2
0	0	0	0.858	3.344	6.116	9.005	3
40	60	70	80	120	160	200	Air flow rate (LPM)
0.5	0.75	0.875	1	1.5	2	2.5	V/V

Table 3. Predicted solid mass flow rate of a 3"-1.5 m air-fluidized conveyor based on equation 62.

The experimental results for the air-fluidized conveyor showed in figure 20 are summarized in table 4.

Mass gas-solid flow rate for alumina fluoride (t/h) - air fluidized conveyor of (3"/1.5m)							Inclination (°)
0	0	0	2.148	4.709	6.173	7.978	(-1)
0	0	0	2.825	5.871	7.633	8.649	(-0.5)
0	0	0.944	3.563	6.448	8.099	8.851	0
0	1.293	3.529	4.952	7.787	8.917	9.127	1
0	2.778	4.261	6.002	8.478	10.523	10.309	2
0	3.192	4.829	6.208	9.764	11.497	13.195	3
40	60	70	80	120	160	200	Air flow rate (LPM)
0.5	0.75	0.875	1	1.5	2	2.5	V/V_{mf}

Table 4. Experimental results from the tests runs at Albras Aluminum smelter laboratory.

Figure 21 shows the other air-fluidized conveyor of 3 inches diameter and 9.3 m long designed using equation 62, which will be used as prototype to feed continuously the electrolyte furnace with alumina fluoride.

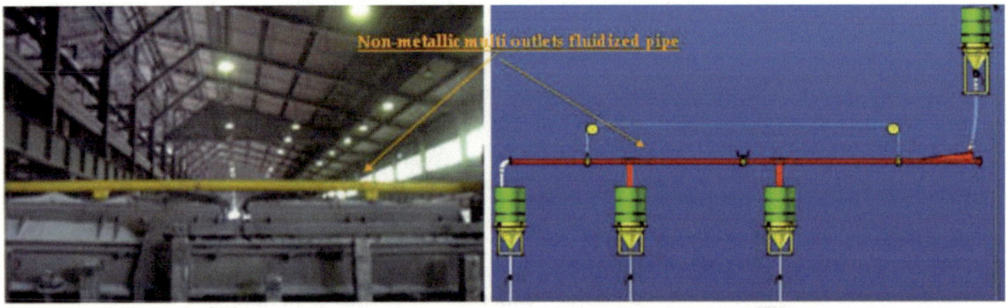

(a) (b)

Fig. 21. a) The nonmetallic fluidized pipe during tests in electrolytic aluminum cell; b) Sketch of the nonmetallic fluidized pipe for performance test at the fluidization laboratory.

The equation 62 predicts a mass solid flow rate of 7.29 t/h for that conveyor, but observed was a mass solid flow rate of 6.6 t/h at 1.5 V_{mf} and a downward inclination of 0.5° was used during the test run depicted in figure 22.

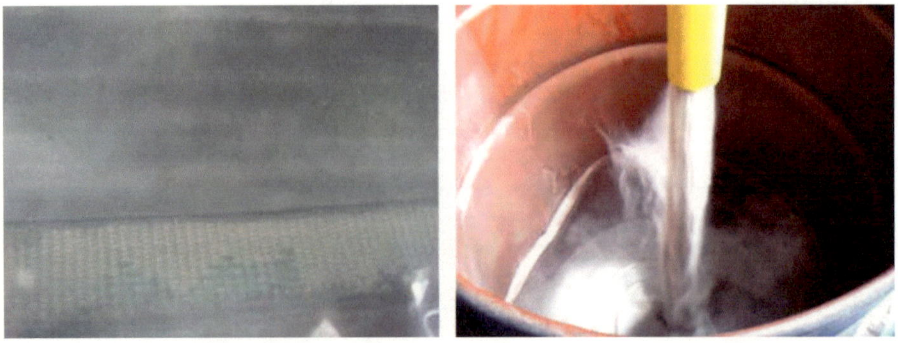

Fig. 22. Test rig to measure the mass solid flow rate of a, 9.3 m long 3 inches diameter air-fluidized conveyor at Albras aluminum smelter.

9. Conclusion

The objective of this chapter is to contribute with readers responsible for the design and operation of industrial furnaces.
Focused on the project of powder handling at high velocity, such as the two cases studies concerning pneumatic conveying in dilute phase applied at Albras aluminum smelter. The last case study regarding powder handling at very low velocity such is illustrated in figure 5 is used in several industrial applications and the intention in this case is to help project engineers to design air slides of low energy consumption. Based on the desired solid mass flow rate of the process using equation 62 is possible to design the conveyor, knowing the rheology of the powder that will be conveyed. In the application of Albras aluminum smelter the experiments results for the small conveyor the values obtained in the experiments was higher than that predict for horizontal and upward inclination in velocities less than the minimum fluidization velocity, because the equation doesn't take in to account the height of material in the feeding bin according (Jones, 1965) equation. In the case of the larger conveyor we have better results, because the conveyor is fed by a fluidized hose as can be seen in figure 21b. So in the next steps of the research it will be necessary to include the column H of the feeding bin in equation 62.

10. Acknowledgment

The authors would like to thanks the LORD GOD for this opportunity, Albras Alumínio Brasileiro SA for the authorization to public this chapter, the Federal University of Pará for my doctorate in fluidization engineering and to inTech - Open Access Publisher for the virtuous circle created to share knowledge between readers and authors.

11. References

Ergun, S. Fluid Flow through Packed Columns, Chem. Engrg. Progress, Vol. 48, No. 2, pp. 89 - 94 (1952).
Geldart, D. Types of Gas Fluidization Powder Technology, 7, 285 - 292 (1972 - 1973).
Jones, D. R. M. Liquid analogies for Fluidized Beds, Ph.D. Thesis, Cambridge, 1965.
Klinzing, G. E.; Marcus, R. D.; Risk, F. & Leung, L. S. Pneumatic Conveying of Solids - A Theoretical and Practical Approach, second edition, Chapman Hall. (1997).
Kozin, V. E.; Baskakov, A. & Vuzov, P., Izv., Neft 1 Gas 91 (2) (1996).
Kunii, D. & Levenspiel O. Fluidization Engineering, second edition, Butterworth-Heinemann, Boston (1991).
Mills, D. Pneumatic Conveying Design Guide, Butterworths, London, (1990).
Schulze, D. Powder and Bulk Solids, Behavior, Characterization, Storages and Flow, Spriger Heidelberg, New York (2007).
Vasconcelos, P.D. Improvements in the Albras Bake Furnaces Packing and Unpacking System - Light Metals 2000, pp. 493 - 497.
Vasconcelos, P.D & Mesquita, A. L. Exhaustion Pneumatic Conveyor and Storage of Carbonaceous Waste Materials - Light Metals 2003, pp. 583-588.

Yang, W. C. A mathematical definition of choking phenomenon and a mathematical model for predicting choking velocity and choking voidage, AIChE J., Vol. 21, 1013 (1978).

11

Equivalent Oxidation Exposure - Time for Low Temperature Spontaneous Combustion of Coal

Kyuro Sasaki and Yuichi Sugai
Department of Earth Resources Engineering, Kyushu University
Japan

1. Introduction

Coal is a combustible material applicable to a variety of oxidation scenarios with conditions ranging from atmospheric temperature to ignition temperature. One of the most frequent and serious causes of coal fires is self-heating or spontaneous combustion. Opening an underground coal seam to mine ventilation air, such as long-wall gob and goaf areas and coal stockpiles, creates a risk of spontaneous combustion or self-heating. Careful management and handling of coal stocks are required to prevent fires. Furthermore, the spontaneous combustion of coal also creates a problem for transportations on sea or land.

Generally, the self-heating of coal has been explained using the imbalance between the heat transfer rate from a boundary surface to the atmosphere and heat generation via oxidation reaction in the stock. The oxidation reaction depends on temperature and the concentrations of unreacted and reacted oxygen. When carbon monoxide exceeds a range of 100 to 200 ppm in the air around the coal and its temperature exceeds 50 to 55°C, the coal is in a pre-stage of spontaneous combustion. Thus, comprehensive studies of the mechanisms and processes of oxidation and temperature increase at low temperature (less than 50 to 55°C) have been investigated for long years.

Measurement of the heat generation rate using crushed coal samples versus constant temperature have been reported to evaluate its potential for spontaneous combustion. Miyakoshi et al.(1984) proposed an equation guiding heat generation in crushed coal via oxygen adsorption based on a micro calorimeter. Kaji et al. (1987) measured heat generation rate and oxygen consumption rate of three types of crushed coal at constant temperatures. They presented an equation to estimate heat generation rate against elapsed time. However, their time was defined under a constant temperature of coal, thus it is not able to be applied for the process with changing temperature of coal.

According to our observations of surface coal mines, the spontaneous combustion of coal initiates at coal seam surfaces as "hot spots," which have temperatures ranging from around 400 to 600 °C. Generally, the hot spot has a root located at a deeper zone from the outside surface of the coal seam or stock that is exposed to air. When the hot spot is observed on the surface, it is smoldering because of the low oxygen concentration. The heat generation rate from coal in the high temperature range (over 60°C) follows the Arrhenius equation, which is based on a chemical reaction rate that accelerates self-heating. Brooks and Glasser (1986) presented a simplified model of the spontaneous combustion of coal stock using the Arrhenius equation to estimate heat generation rate. They used a natural convection model to serve as a reactant transport mechanism. Carresl & Saghafil (1998) have presented a numerical model to predict spoil pile self heating that is due mainly to the interaction of coal and carbonaceous spoil materials with oxygen and water. The effects of the moisture content in the coal on the heat generation rate and temperature are not considered in this chapter. However, Sasaki et al. (1992) presented some physical modeling of these effects on coal temperature.

Yuan and Smith (2007) presented CFD modeling of spontaneous heating in long-wall gob areas and reported that the heat has a corresponding critical velocity. However, when the Arrhenius equation is used for a small coal lump, the calculation does not show a return to atmospheric temperatures. This can be seen from the data shown in Fig. 1. The reason, that the results cannot be applied to small amounts of coal stock, may be a type of ageing effect.

Nordon (1979) proposed this as a possible explanation using the Elovich equation that has been used in adsorption kinetics based on the adsorption capacity. He also presented a model for the self-heating reaction of coal and identified two steady-state temperature conditions one less than and one over 17°C. He also commented that the transport processes of diffusion and convection take the mobile reactant, oxygen, from the boundary to the distributed reaction where heat energy is released, and then convey the latter back to the boundary. However, his concept is difficult to apply to numerical models.

In this chapter, a model is presented for spontaneous combustions of coal seam and coal stock. It is based on time difference between thermal diffusion and oxygen diffusion. Furthermore, the concept of "Equivalent Oxidation Exposure Time (EOE time)" is presented. Also, we compared the aging time to the oxidation quantity to verify the mechanism presented. Numerical simulations matching both the thermal behaviors of large stocks and small lumps of coal were performed.

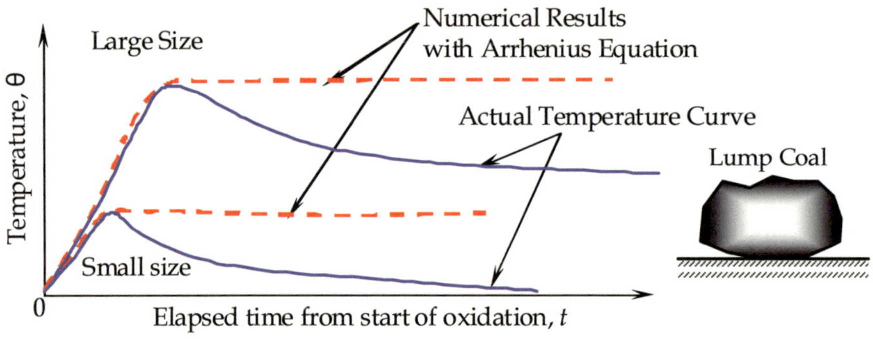

Fig. 1. Difference of temperature change between a numerical simulation result by Arrhenius equation and actual process for small and large amounts of coal stock

2. Mechanism of temperature rise in a large amount of coal stock

Coal exposed to air is oxidized via adsorbed oxygen in temperature ranges. It has a different time dependence than that expressed by the Arrhenius equation, which guides this behavior in the high temperature range. The adsorption rate of oxygen decreases with increasing time for a constant temperature, because coal has a limit of oxygen consumption.

A schematic showing the process of spontaneous combustion is shown in Fig. 2. Assume a coal stock has all but its bottom surface exposed to air of oxygen concentration, C_0 and and temperature, θ_0. Oxidation heat is generated in the coal is started from outside surface of the stock, because oxygen is supplied from the atmosphere. Some heat is lost to the atmosphere, but some also diffuse to inward to the center of the stock. The outer part of the stock returns to the atmospheric temperature, θ_0, after enough time. However, the oxygen concentration of the inside stock is kept at a relatively low concentration, because oxygen does diffuse to the inner zone via the oxidation zone. When coal at the center of the stock is preheated slowly without oxygen, a high temperature spot at the center is generated.

The oxidation and heat generation zone gradually moves from the stock surface to the center while shrinking and rising in temperature. Finally a hot spot is formed at the center (see Fig. 2 (a) to (c)). Oxygen diffuses to center region after formation of the hot spot. This time delay of oxygen diffusion allows the coal temperature to rise exponentially in the center by long preheating and inducing smaller EOE time (see 3.3). Thus, of the greater the volume in the coal stock, the more delay between preheating and oxygen diffusion.

After formation of the hot spot in the center, the coal begins to burn slowly without flames and projects toward the outer surface through paths with relatively high effective diffusivity, which has greater oxygen concentration than the surrounding coal. Finally, the hot spot appears on the outside surface of the stock, which marks the start of spontaneous.

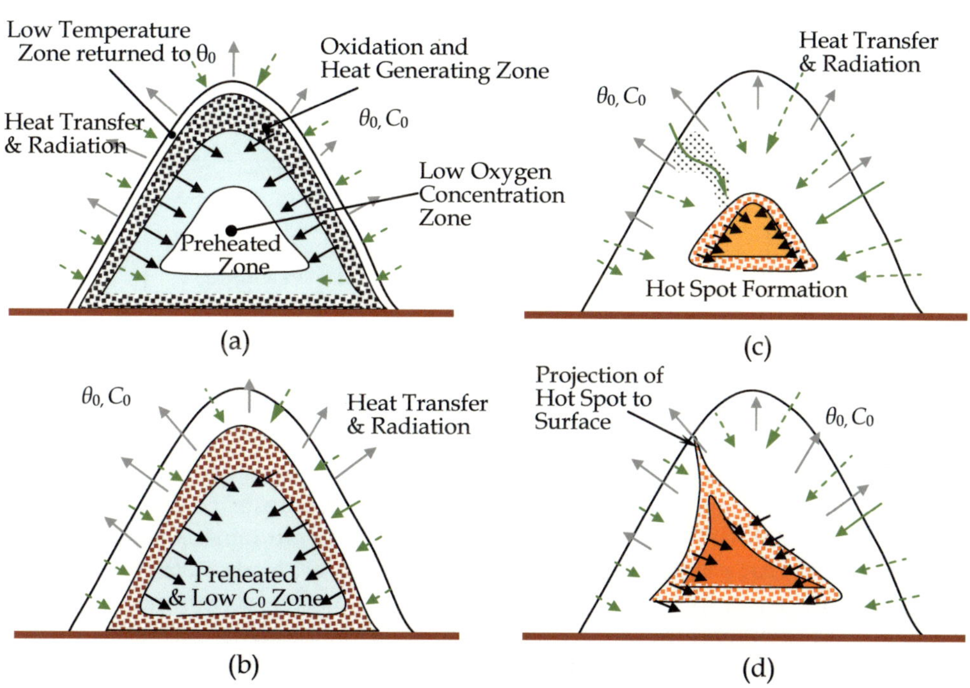

Fig. 2. Schematic process showing spontaneous combustion of large amount of coal stock, (a), (b) and (c): Hot spot forming process with accumulating heat and shrinking zone of oxidation and preheating zone, (d): Projection growth of hot spot toward to stock surface through high permeable path

3. EOE time and heat generation rate of coal

3.1 Heat generation rate from coal

In the present model, coal oxidation reaction includes physical adsorption and chemical adsorption via oxygen reaction at low temperatures. Measurements of the heat generation rate at the early stages of the process that show an exponential decrease have been reported by many experiments, such Kaji et al. (1987), shown in Fig. 3, and Miyakoshi et al.(1984). Based on their measurement results, the heat generation rate per unit mass of coal at temperature θ (°C), q (W/g or kW/kg), can be expressed with a function of elapsed time after being first exposed to air, τ (s):

$$q = C \cdot A\exp(-\gamma\tau) \tag{1}$$

where, A (kW/kg) is heat generating constant, C is molar fraction of oxygen, and γ (s^{-1}) is the decay power constant. The initial order of heat generating rate of coal for exposing air is $q(0) \approx 0.01$ to 0.001 kW/kg.

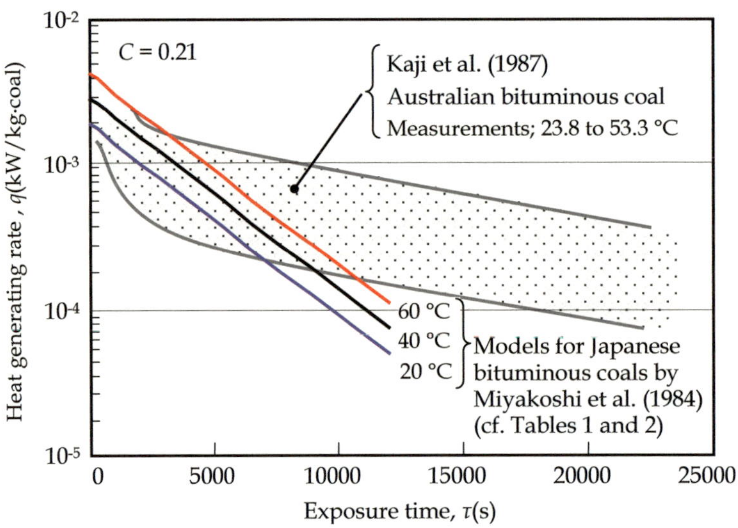

Fig. 3. Models of heat generating rate of coal vs. exposure time for constant temperatures

3.2 Arrhenius equation for coal oxidation

Kaji et al. (1987) measured rates of oxygen consumption due to coal oxidation in the temperature range 20 to 170 °C using coals ranging from sub-bituminous to anthracite coal. They reported that heat generated per unit mole of oxygen at steady state is h = 314 to 377 (kJ/mole), and their results of the Arrhenius plots, the oxygen consumption rate versus inverse of absolute temperature T^{-1} (K^{-1}), shows the Arrhenius equation. Thus, the higher the coal temperature; the faster the oxidation or adsorption rate is given. When the heat generation rate is proportional to oxygen consumption rate, the heat generated, A, can be estimated using the following equation,

$$A = A_0 \cdot \exp\left(-\frac{E}{RT}\right) \tag{2}$$

where, A_0 (kW/kg) is pre-exponential factor for A, E (J/mole) is the activation energy, R is gas constant (J/mol/K), and T (=273+θ) (K) is absolute temperature. Kaji et al.(1987) has reported that the coals have almost the same activation energy of around E=50 kJ/mole for temperature range of 20 to 170 °C. On the other hand, Miyakoshi et al. (1984) reported as $E \approx$ 20 kJ/mole for Japanese coals in temperature range lower than 50 °C based on measurements of oxygen adsorption heat using with a micro-calorimeter.

The activation energy of fresh coal is expected as much lower than that of exposed coal in the air, because fresh coal adsorbs oxygen physically at an initial stage of self-heating. Average activation energy and decay power constant, presented by Miyakoshi et al. for Japanese bituminous coals (see Tables 1 and 2), were used for present numerical simulations.

3.3 Equivalent oxidation exposure time

The heat generating rate, q, is expressed as a function of θ, C, and τ. Equations (1) and (2) can be used to calculate q for a constant temperature. However, they are not applicable for the calculation of the normal coal temperature change versus elapsed time. Its concept is partly similar to Elovich equation, but it provide a scheme to estimate q follows change of temperature of coal and EOE time.

For an example, assume a coal lump is placed in an environment in which C = 0.1 and θ=45°C, for elapsed time; τ=1 h, and then is stored in other one of C =0.2 and θ= 70°C for another 1 h period. It is not possible to reconstruct this situation by adding the former and later times with different oxidation rates. A new model of the elapsed time that considers the aging degree of the coal is required to overcome this difficulty. The cumulative generated heat of the coal, Q_m' (J/g) from elapsed time 0 to t, is defined as,

$$Q'_m = \int_0^t q'(\theta', C', t')dt' \tag{3}$$

where, the actual heat generation rate, $q'(\theta', C', t')$, θ' and C' are changing with the elapsed time, t'. However, the cumulative heat, Q_m, for constant θ and C, can be derived using Equations (1) and (2) from time 0 to τ^*:

$$Q'_m = \int_0^{\tau^*} q(\theta, C, t')dt' = \frac{CA}{\gamma}\left(1 - \exp(-\gamma\tau^*)\right) = Q_m \tag{4}$$

If the amounts of accumulated heat, Q'_m and Q_m, defined in Equations (3) and (4), are equal, τ^* in Eq. (4) expresses the aging time of the coal for constant temperature; $\theta = \theta'(t)$ and constant concentration; C=$C'(t)$, for the actual elapsed time (t'=t). In this paper, τ^* is defined as the EOE time (see Fig. 4). It is calculated based on a summation of generated heat $q'(\theta', C', t')\cdot\Delta t'$ over a numerical calculation interval time, $\Delta t'$. It is expressed by the following single-calculation equation:

$$\therefore \tau^* = -\frac{1}{\gamma}\cdot\ln\left[1 - \frac{\gamma}{CA}\left(\sum_i q'_i \cdot \Delta t'_i\right)\right] \tag{5}$$

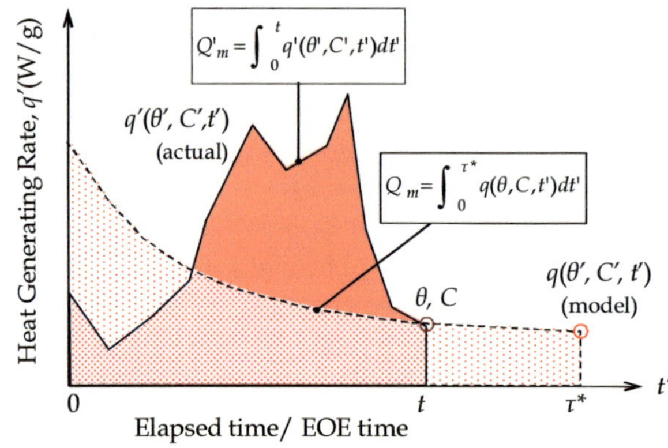

Fig. 4. Schematic definition of EOE-time of coal to estimate heat generating rate by matching total heat generations

The most important characteristic of the EOE is that if a part of coal releasing heat to its surrounding, its EOE time is increased. It means that receiving heat makes smaller EOE time due to temperature increasing. Using the EOE time, the actual heat generating rate of coal at t can be obtained by substituting τ^* instead of τ into Eq. (1).

$$q(t) = C(t) \cdot A_0 \cdot \exp\left(-\frac{E/R}{273+\theta(t)}\right)\exp(-\gamma\tau^*) \tag{6}$$

Assuming the reaction heat of unit volume of oxygen is ΔH, the oxygen consumption rate, v', and the accumulated consuming oxygen, V', are given by:

$$v(t) = \frac{q}{\Delta H} \tag{7}$$

$$V(t) = \int_0^{\tau^*} v'(\theta',C',t')dt' \tag{8}$$

The reaction heat of unit volume of oxygen was evaluated as $\Delta H \approx 16$ (J/cm^3O_2) based on the experimental results of heat generation rate by Kaji et al.(1987) and Miyakoshi et al.(1984) shown in Fig. 3. The oxygen consumption rate is used in the oxygen diffusion equation for its concentration.

3.4 Thermal conduction and diffusivity of coal stock consisting porous media

For a case of coal stock, thermal characteristics are required for a porous media consisting lump coals and air. Thermal conductivity of a porous media is dependent on the porosity or void fraction, ε, and specific internal surface area in it. Kunii & Smith (1960) have presented the equations predicting an effect of porosity and thermal conductivities of solid and fluid on the heat transfer properties. They have presented effective thermal conductivity of porous media versus porosity, ε. In present model of a coal stock, effective thermal conductivity can be estimated by a following equation revised from the Kunii & Smith's equation by omitting term of the thermal radiation effects due to low temperature range,

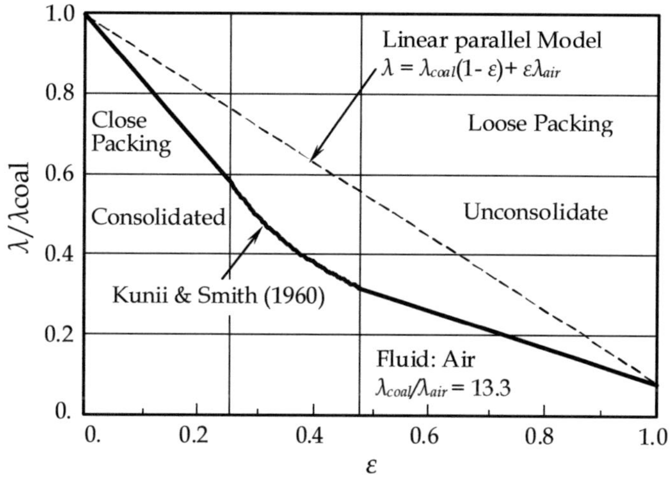

Fig. 5. Effective thermal conductivity vs. porosity of coal stock in the air evaluated by Kunii & Smith's equation(1960) (a case for $\lambda_{coal}/\lambda_{air} = 13.3$)

$$\frac{\lambda}{\lambda_{coal}} = \varepsilon\frac{\lambda_{air}}{\lambda_{coal}} + \frac{1-\varepsilon}{\Phi\frac{\lambda_{coal}}{\lambda_{air}} + \frac{2}{3}} \tag{9}$$

where λ is effective thermal conductivity of coal stock, λ_{coal} is thermal conductivity of lump coal, λ_{air} is thermal conductivity of air, and Φ is a ratio of effective thickness of air film over coal lump diameter given against ε that is classified into three regime of $\varepsilon \geq 0.476$ (loose packing/unconsolidated), $0.476 > \varepsilon \geq 0.260$ and $\varepsilon < 0.260$ (close packing/consolidated) adapted by Kunii & Smith (1960). Volumetric heat capacity, ρC_p (J/kg), and thermal diffusivity of the coal stock, α (m²/s) are derived from following equations,

$$\rho C_P = \varepsilon\rho_{air}C_{P\,air} + (1-\varepsilon)\rho_{coal}C_{P\,coal} \tag{10}$$

$$A = \frac{\lambda}{\rho C_p} \tag{11}$$

Suppose $\lambda_{coal}/\lambda_{air} = 13.3$ or , $\lambda_{coal} = 0.36$ W/m/°C for a typical thermal conductivity of coal, the effective thermal conductivity, λ, calculated by Eq. (9) is shown in Fig. 5 with thermal conductivity of linear parallel model; $\lambda = \lambda_{coal}(1-\varepsilon)+\varepsilon\lambda_{air}$. The effective thermal conductivity is lower than that of the linear parallel model, because air in coal stock gives a thermal resistance around coal lumps because of low thermal conductivity of air.

Coal Density	Specific Heat of Coal	Thermal diffusivity of Coal	Diffusion Coefficient
ρ_{coal}	C_p	a	D
1291 kg/m³	1210 J/kg/°C	6.8×10^{-8} m²/s	7.1×10^{-6} m²/s

Table 1. Thermal properties of coal for present numerical simulations

Decay power constant	Pre-exponential factor	Activation energy
γ	A_0	E
3.0×10^{-4} s^{-1}	2.9×10^{4} W/kg/kg	2.0×10^{4} J/mol

Table 2. Heat generating properties of coal for present numerical simulations

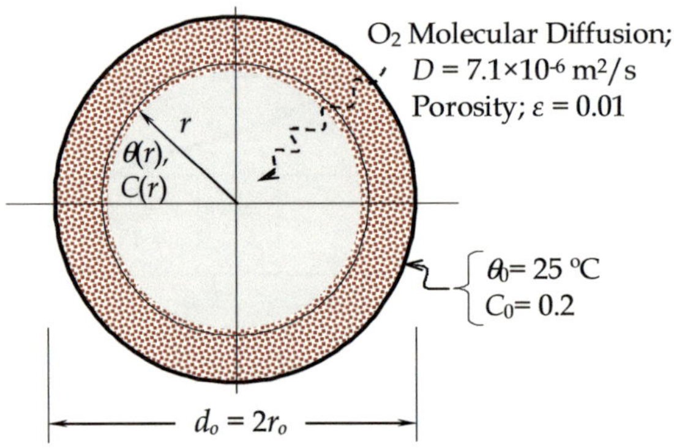

Fig. 6. Model definition of a sphere lump coal exposed to atmospheric air

4. Numerical simulation results and discussion

In this section, numerical simulations for three kinds of coal stock model carried out by authors are introduced to show the effectiveness of the EOE time simulating self-heating process of the coal stocks. Those were done using the finite difference method to solve the equations on heat transfer and oxygen advection and diffusion. Number of blocks used in the simulation was 100 for one-dimensional model and 10000 for two-dimensional model. Time interval of the numerical simulations was adapted as 20s to satisfy enough accuracy.

4.1 Sphere lump coal exposed to atmospheric air

The simulations on coal lump were carried out for simple one-dimensional sphere model as shown in Fig. 6. Its outer surface is open to air with constant temperature and constant O_2 concentration. Thus, oxygen is provided by molecular diffusion expressed as;

$$\frac{\partial\theta}{\partial t} = a\left(\frac{2}{r}\frac{\partial\theta}{\partial r} + \frac{\partial^2\theta}{\partial r^2}\right) + \frac{q}{C_p} \tag{12}$$

$$\varepsilon\frac{\partial C}{\partial t} = D\left(\frac{2}{r}\frac{\partial C}{\partial r} + \frac{\partial^2 C}{\partial r^2}\right) - \rho v \tag{13}$$

$$\left.\begin{matrix}\theta = \theta_0 \\ C = C_0\end{matrix}\right\} at \quad r = r_o \tag{14}$$

The thermal and heat generating properties of the coal seam used in the simulations are listed in Tables 1 and 2. Gas permeability, K, and diffusion coefficient, D, of lump coal and close packing of crushed coal were measured, and the correlated equations have been presented by Sasaki et al.(1987). The boundary conditions of temperature and oxygen concentration at the outer surface were fixed with constants expressed by Eq.(14).

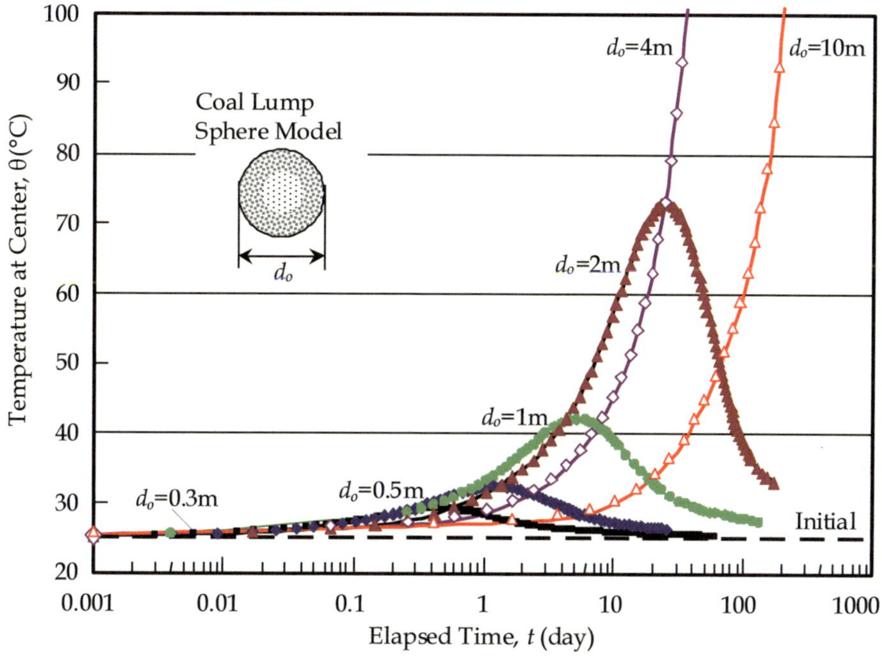

Fig. 7. Temperature at sphere center vs. elapsed time for different diameter

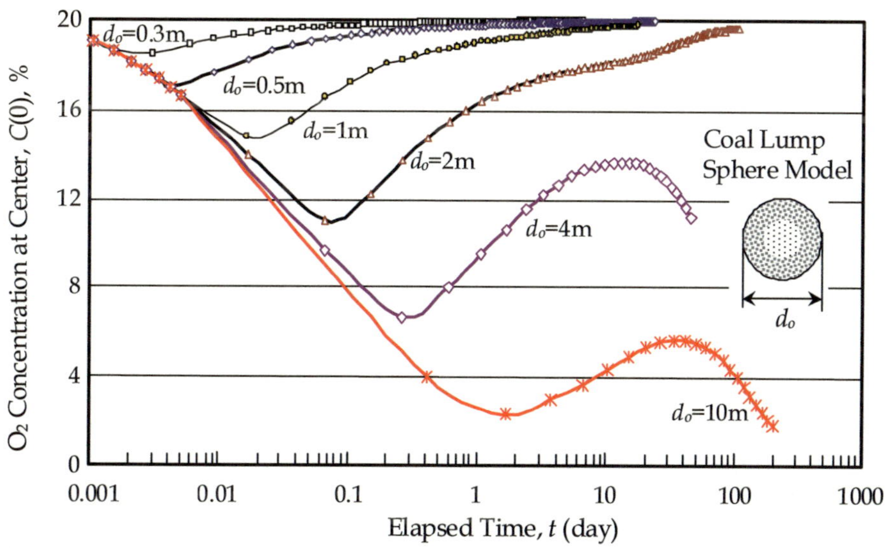

Fig. 8. Oxygen concentration at sphere center vs. elapsed time for different diameter

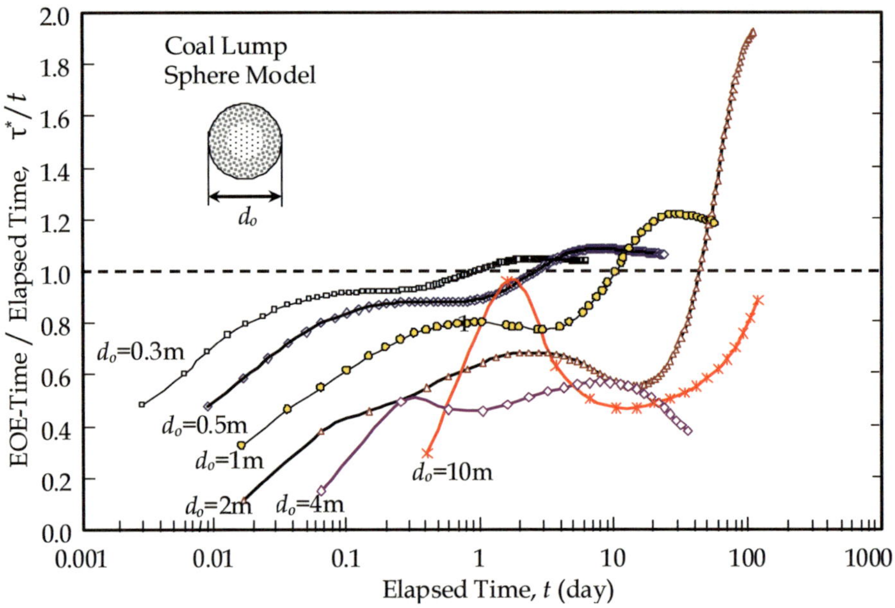

Fig. 9. Ratio of EOE time over elapsed time at sphere center of coal lump vs. elapsed time for different diameter

Figures 7 to 9 show the numerical simulation results on temperature, oxygen concentration and ratio of EOE time over elapsed time at the sphere center versus elapsed time for different diameter d_0= 0.3 to 10m . The temperature at center of the stock is increased with elapsed time, but the cases of $d_0 \leq 2$ m show the temperature return to atmospheric and initial temperature θ_0 =25 °C. This is because that the EOE time increased by heat transfer to surrounding air makes reducing heat generation rate of coal lump even if its location is at the center. However, the case of $d_0 \geq 4$ m, coal at the sphere center receiving enough heat in low oxygen concentration before oxygen diffuses into the center, and lower EOE time induces higher heat generation than that of $d_0 \leq 2$m before ignition and combustion of coal. The critical diameter is evaluated roughly as d_0= 3m for present model, it depends on the activation energy, E and the decay power constant, γ, of the coal.

4.2 A model of coal seam remained at goaf area in underground mines

In Fig. 10, a simple one-dimensional numerical model for a coal seam remained at goaf area that is cavity area behind a longwall working in underground coal mines. It is expected to expose to relatively high temperature air of 45 °C. Its faces are open to air with ventilation pressure difference in the goaf area, Δp = 10 mmH_2O= 98Pa. Thus, oxygen is provided by not only molecular diffusion, but also permeable airflow between two faces. Therefore, oxygen in the air diffuses from both ends and adsorbs in micro pores of the coal seam as it diffuses toward the center of the seam.

The thermal and heat generating properties of the coal seam used in the simulations are listed in Tables 1 and 2. In this study, the effects of the moisture content in the coal on the heat generation rate and temperature are not considered. However, Sasaki et al. (1992) presented some physical modeling of these effects on coal temperature.

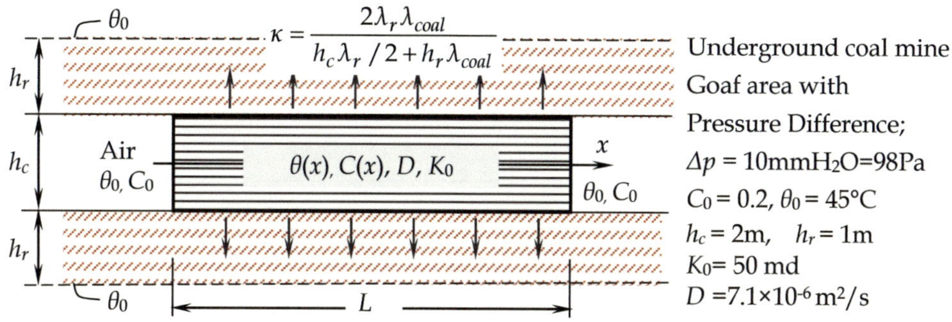

Fig. 10. Numerical simulation model of a coal seam using one dimensional model

$$\frac{\partial\theta}{\partial t} = a\frac{\partial^2\theta}{\partial x^2} - \frac{\kappa}{pC_p h_c}(\theta - \theta_0) + \frac{q}{C_q} \tag{15}$$

$$\varepsilon\frac{\partial C}{\partial t} = D\frac{\partial^2 C}{\partial x^2} - U\frac{\partial C}{\partial x} - \rho v \tag{16}$$

$$U = K_0\frac{\Delta p}{\mu_{air}L} \tag{17}$$

$$\left.\begin{array}{l}\theta = \theta_0 \\ C = C_0\end{array}\right\} at \quad x = 0, L \tag{18}$$

The coal seam is L= 5.0 m in length, has effective diffusion coefficient of D=7.1×10^{-6} m^2/s, and permeability of K_0= 10 to 100 md ≈ 10^{-15} to 10^{-14} m^2.

The results showing the temperature distribution are shown in Fig. 11. The zone with rising temperature and high oxygen consumption gradually moves toward the center and its maximum temperature also increases. The present results are similar to the simulation results presented by Nordon (1979), but temperature of the outer layer near the boundary surface decreases with the decreasing heat generation rate. This drop in the heat generation rate is due to increasing EOE time in the outer layer. As shown in Fig. 12, the larger the permeability, the larger the EOE time of the outer layer of the stock.

The thermal and heat generating properties of the coal seam used in the simulations are listed in Tables 1 and 2. In this study, the effects of the moisture content in the coal on the heat generation rate and temperature are not considered. However, Sasaki et al. (1992) presented some physical modeling of these effects on coal temperature.

The results showing the temperature distribution are shown in Fig. 11. The zone with rising temperature and high oxygen consumption gradually moves toward the center and its maximum temperature also increases. The present results are similar to the simulation results presented by Nordon (1979), but temperature of the outer layer near the boundary surface decreases with the decreasing heat generation rate. This drop in the heat generation rate is due to increasing EOE time in the outer layer. As shown in Fig. 12, the larger the permeability, the larger the EOE time of the outer layer of the stock.

Oxygen in the ventilation air diffuses from both faces of the coal seam toward its center region under the lowest oxygen concentration due to absorption at outer regions as shown in Fig. 13. It takes much longer time to make higher oxygen concentration in the center region.

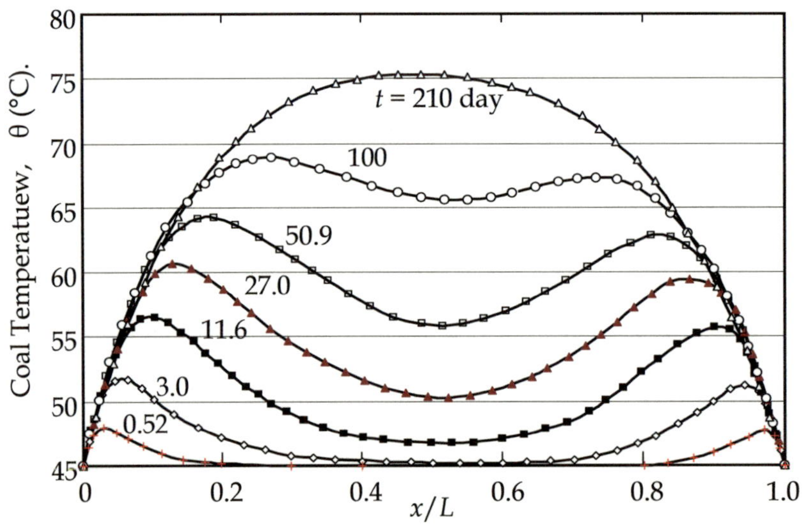

Fig. 11. Transition of temperature distribution of coal seam (L=5m, Initial temperature, $\theta(0) = \theta(L)$ = 45°C, outer oxygen concentration, $C(0) = C(L)$ =0.20, permeability K_0=50 md = 4.9×10^{-14} m^2, D=7.1×10^{-6} m^2/s)

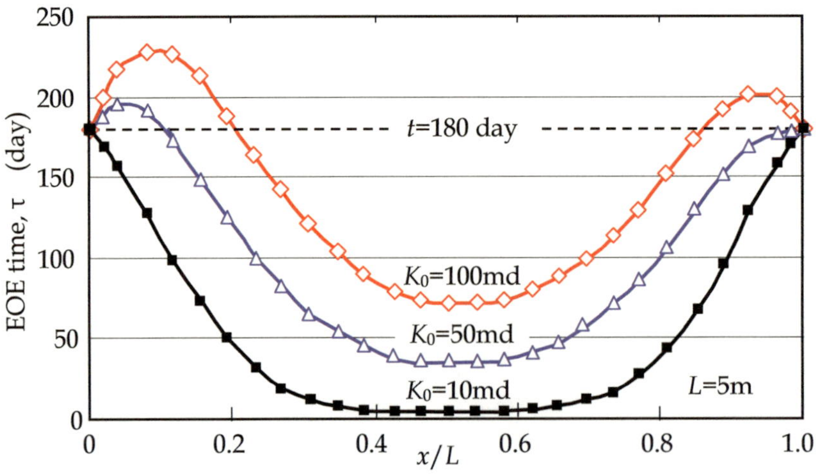

Fig. 12. Simulation results of EOE-time for different permeability of coal seam (L=5m, $\theta(0) = \theta(L)$ =45°C, t = 180 days, outer oxygen concentration, $C(0) = C(L)$ =0.20)

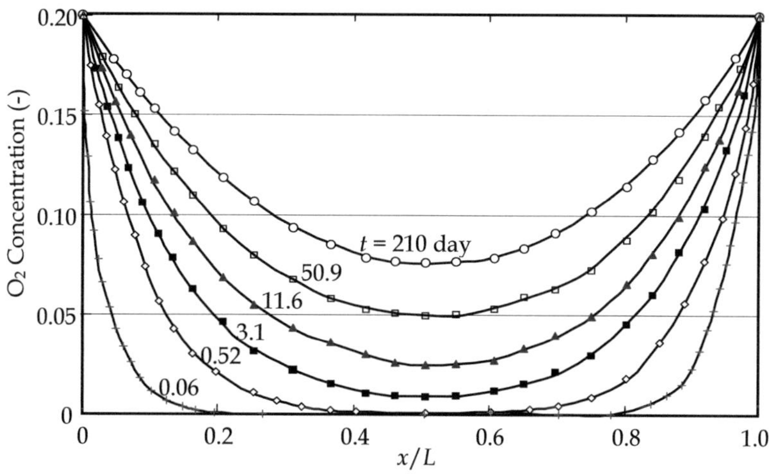

Fig. 13. Distributions of oxygen concentration in coal seam (L=5m, $\theta(0)$ = 45°C, oxygen concentration, $C(0) = C(L)$ =0.20, permeability K_0=50 md = 4.9×10^{-14} m^2, D=7.1×10^{-6} m^2/s)

4.3 Two-dimensional coal stock in considering internal natural convection flow

Nield & Bejan (1999) have presented numerical models and applications for convection flows in a porous media. Spontaneous combustion in a coal seam, that is consolidated porous media, has been modelled and analyzed by numerical simulations. The simulation was performed using the finite difference method to solve the equations of heat transfer, oxygen diffusion and permeable flow via ventilation pressure difference.

The EOE time has been applied to numerical simulations on spontaneous self-heating of coal stocks. The numerical simulations of the coal stock were performed while accounting for the natural convection flow and heat transfer in the stock as a porous media with two dimensional simulation model, which shown in Fig. 14. The right-hand region of width W and height H (x=0 to W, z = 0 to H) was calculated using its symmetry about x = 0.

By solving equations of stream function, ψ, and boundary condition at outer surface (x=L and z=H), the natural convection flow velocities (u, w) in horizontal(x) and vertical(z) directions (x, z) are expressed by numerical analysis with two dimensional equations for heat transfer and oxygen diffusion for the stock that is expanded from equations in one dimension described in former sections 4.1 and 4.2 (see Nield & Bejan, 1999). The numerical simulations were done using with 900 to 1800 blocks for the coal stock models.

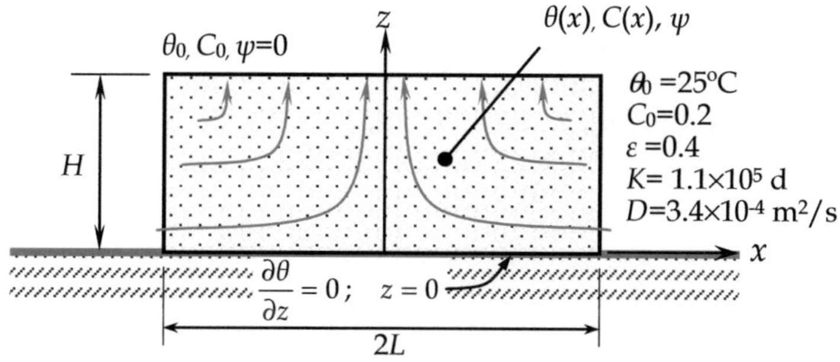

Fig. 14. Two-dimensional coal stockyard model with internal natural convection flow

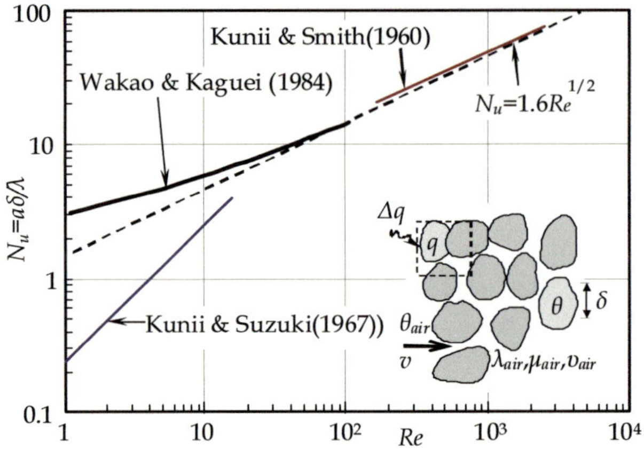

Fig. 15. Nusselt number, N_u vs. Reynolds number, $Re=v\delta/\upsilon_{air}$, in porous media consisting lump coals and air

$$\frac{\partial^2\Psi}{\partial x^2}+\frac{\partial^2\Psi}{\partial z^2}=\frac{K\quad\rho_{air}}{\mu_{air}}g\beta\frac{\partial\theta}{\partial x} \tag{19}$$

$$u=\frac{\partial\Psi}{\partial x};\quad w=-\frac{\partial\Psi}{\partial z} \tag{20}$$

where K is permeability of coal stock, β is thermal expansion coefficient of air, μ_{air} is air viscosity and g is acceleration of gravity. The boundary conditions at x=0, W and z=0, H are,

$$\begin{aligned}\Psi=0 \text{ at } x=0; \quad \frac{\partial\Psi}{\partial x}=0 \text{ at } z=W\\ \Psi=0 \text{ at } z=0; \quad \frac{\partial\Psi}{\partial z}=0 \text{ at } z=H\end{aligned} \tag{21}$$

A model on heat transfer rate between lump coals or coal matrix and airflow is needed to simulate internal temperature distribution in the stock. Wakao & Kaguei(1982) reviewed the effective heat transfer coefficient, a, for unconsolidated porous media. Expressions of Nusselt number, $N_u(=a\delta/\lambda)$, have been presented by for the interstitial heat transfer coefficients in porous media as shown in Fig. 15. From the figure, N_u is roughly proportional to square root of the Reynolds number, $R_e^{1/2}$, and it matches fairly well with equations presented by Kunii & Smith(1960), Kunii & Suzuki and Wakao & Kaguei(1982). In present numerical simulations, an approximated equation on heat transfer per unit volume, Δq ;

$$\Delta q=a(\theta_{air}-\theta)\xi=1.6R_e^{1/2}\frac{(\theta_{air}-\theta)\xi}{\delta}\lambda_{air} \tag{22}$$

where θ_{air} is air flow temperature, θ is lump coal temperature and ξ is internal surface area in the unit volume of the coal stock. Δq is used to calculate natural convection air flow temperature θ_{air} and lump coal temperature θ with heat generation rate of coal lump as q+Δq.

The coal stock W in width and H in height was simulated with the stock conditions; C_0=0.2, θ_0=30°C, K= 1.1×10^5 d and D=3.4×10^{-4} m^2/s. The stock bottom at z = 0 is set as adiabatic and impermeable boundary. Natural convection airflow in the stock is observed in Fig. 16 as that flow comes from side walls toward the center of the stock. It controls temperature rise, cooling, and oxygen supply. High temperature region that was generated at center and upper in the stock after t≈100 h. But the natural convection flow and distribution of oxygen concentration are complicated with rapid changing in early stage; t≈ 0 to 100 h of self-heating of coal stock (see Fig. 17). The region is also downstream of the convective airflow with low oxygen concentration but high temperature. The convection flow becomes faster with rising internal temperature. The mechanisms controlling the temperature rise are complex and affected by the EOE time. The temperature and convective flow velocity are affected each other, and coal temperature determines not only the heat generation rate by supplying oxygen, but also the cooling or heating rate proportional to temperature difference between air and coal lumps.

A comparison of simulation results for different aspect ratios; W/H = 1 and 2 is shown in Fig. 18. It is interesting that center region of longer ratio W/H= 2 shows relatively lower temperature compared with outer region. The reason is the internal natural convection flow from side walls is coming up to upper surface before closing to center region. Thus, the temperature distribution of right region is similar even if the aspect ratio is different.

Figure 19 shows the maximum temperature in the stock, θ_{max}, versus the elapsed time, t, for different aspect ratios W/H = 1 and 2. It rises to a temperature between 47 and 52°C in less than t=100 h, then holds this temperature during t=30 to 300 hours. Finally, the temperature decreases with time, because of the increasing the EOE time by releasing heat to the atmosphere. The natural convection airflow provides oxygen, but suppresses the maximum temperature in the stock by cooling effect and makes heat transfer increasing with the temperature difference to air temperature in atmosphere.

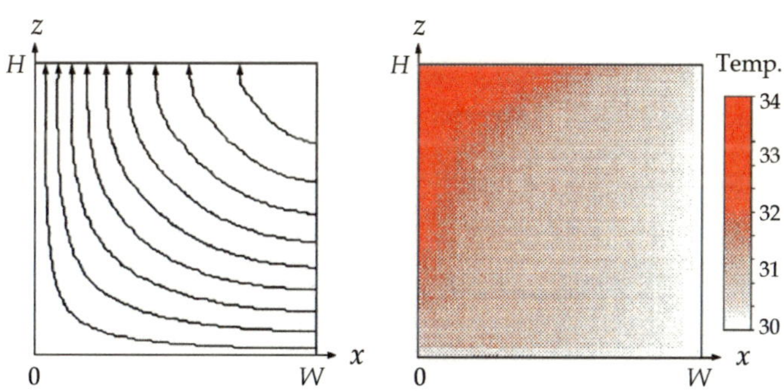

a) Flow stream line in right half b) Temperature distribution in right half,

Fig. 16. Streamlines of internal natural convection flow and temperature distribution at t = 500h in coal stock (2W=10m, H=5m, C_0=0.2, θ_0=30°C, K= 1.1×10^5 d, D=3.4×10^{-4} m^2/s)

Finally the simulation was done to get matching with a monitored temperature at a coal stockyard carried out by Coal Mining Research Center, Japan (CMRCJ, 1983). As shown in Fig. 20, a model of a coal stockyard is 30m in width and 5m in height with trapezoid shape. On the other hand, the simulation model is just rectangle shape consists same thermal and flow characteristics of the coal stock defined in Fig. 14. The temperature at the point in coal stockyard was compared. It shows fairly well matching with the monitored temperature data to corresponding position.

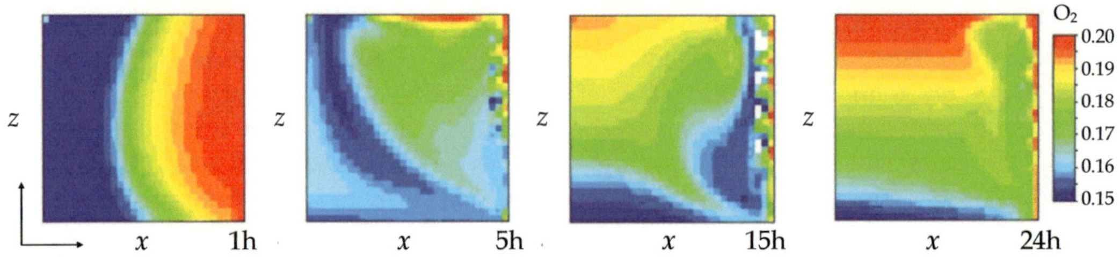

Fig. 17. Change of oxygen concentration distribution of two-dimensional coal stock (right half, $2W$=10m, H=5m, K= 1.1×10^5 d, D=3.4×10^{-4} m^2/s) with internal natural convection flow

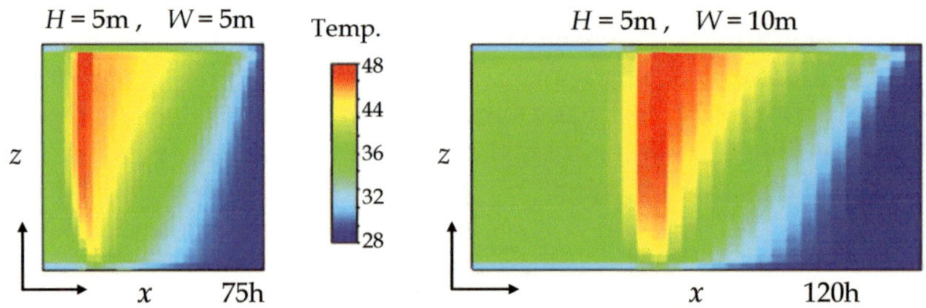

Fig. 18. Effect of aspect ratio, H/W, on temperature distribution in two-dimensional coal stock with internal natural convection flow ($2W$=10m, H=5m, K= 1.1×10^5 d, D=3.4×10^{-4} m^2/s)

An important result of the numerical simulations is that the oxidation of coal in the low temperature range reduces heat generating rate, because it provides cooling air accelerates increasing the EOE time. This could be used to ensure that coal stocks are kept within a safety level that prevents spontaneous combustion. Turnover of coal stocks at regular intervals works by increasing EOE time and releasing heat from center region of the stock.

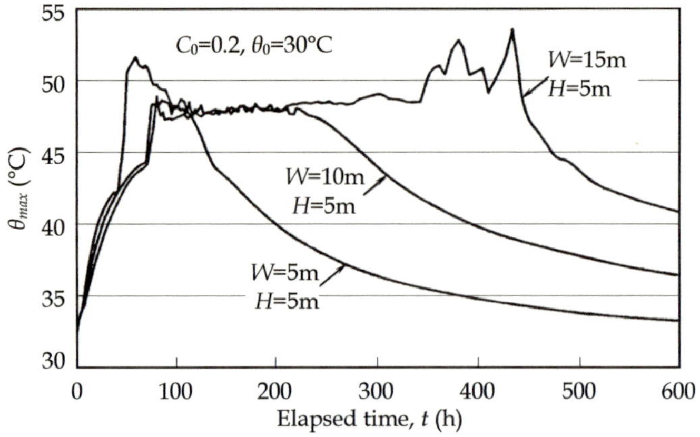

Fig. 19. Numerical simulation results of the maximum temperature transition in coal stock with three different aspect ratios

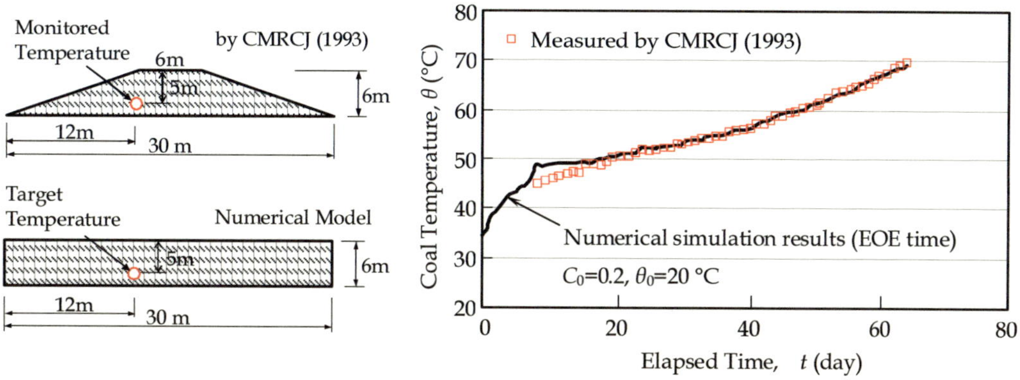

Fig. 20. Comparison of Numerical simulation results on coal temperature in a stockpile with monitored values by the Coal Mining Research Center, Japan (CMRCJ)

5. Summary

In this chapter, a thermal mechanism of spontaneous combustion of coal seams and stocks in low temperature has been described. It has been discussed that the reason to enhance self-heating of coal stocks is the time delay between preheating from thermal diffusion and oxygen provided via diffusion. Especially, preheating without supplying oxygen makes a situation with high risk of spontaneous combustion. Another important mechanism discussed is the formation of a hot spot through the shrinking of the heated oxidation zone from the outer layer toward into the center region of the coal stock.

Heat generating of coal via the oxidation at low temperature includes complex functions of temperature and elapsed time. Thus, numerical models using only the Arrhenius equation to express heat generating of coal bases on its temperature are not able to simulate actual heat and mass transfer phenomena.

The concept of equivalent oxidation exposure time (EOE time) has been introduced to express time decay of coal oxidation. We have used this concept to simulate the heat generation while considering coal aging; that is, the ratio of the cumulative amount of oxidation to its oxidation capacity. This concept allowed us to consider these factors using simple calculation procedures following temperature changing of the coal. The physical model agrees with some experimental measurements with decay rate of heat generating from coal during exposure to oxygen. It has been successfully applied to simulate the temperature of rump coal, coal seam and coal stock, which are exposed to ventilated or atmospheric air. We used the finite difference method to solve the equations of thermal diffusion, heat transfer, and oxygen diffusion in these models. For the case of the coal stock, natural convection flow was also considered. The results showed that natural convection flow provides oxygen, but suppresses the maximum temperature of the stock by convective heat flow moving to the atmosphere.

Low temperature oxidation of coal with cooling accelerates the increase of the EOE time and reduces heat generation rate to an inherently safe level. Turnovers of lump coals in the stock at regular intervals are expected to prevent spontaneous combustion effectively by increasing the EOE time; not only by releasing heat in the center of the stock.

6. Acknowledgment

The authors would like to thank Prof. Emeritus Dr. Miyakoshi and Mr. K. Sakamoto (Tone Co., LTD.) for their efforts on present study. We are also grateful to Prof. Ivana Lorkovic for encouraging me to write this chapter.

7. Nomenclature

a	=	effective heat transfer coefficient [W/m²/°C]
A	=	heat generating constant [kW/kg] or [W/g]
A_0	=	pre-exponential factor for A [kW/kg] or [W/g]
C	=	oxygen concentration or molar fraction [-]
C_0	=	oxygen concentration in atmosphere [-]
Cp	=	average specific heat of coal stock [J/kg/°C]
Cp_{air}	=	specific heat of air [J/kg/°C]
Cp_{coal}	=	specific heat of lump coal/coal seam [J/kg/°C]
d_o	=	outer diameter of coal lump/stock [m]
D	=	effective diffusion coefficient [m²/s]
E	=	activation energy [J/mole]
g	=	acceleration of gravity [m/s²]
H	=	height of two-dimensional coal stock [m]
h_c	=	height of coal seam [m]
h_r	=	height of upper and lower rock sediment layers [m]
K	=	permeability of coal stock [d] or [m²] (1 d= 9.8^{-13}m²/s)
K_0	=	permeability of coal seam [md] or [m²] (1md= 9.8×10^{-16}m²/s)
L	=	width of coal seam [m]
N_u	=	Nusselt number [-]
q	=	heat generation rate of coal [W/g] or [kW/kg]
Q_m	=	cumulative generated heat of coal [J/g] or [kJ/kg]
Q_m'	=	cumulative generated heat of coal [J/g] or [kJ/kg]
R	=	gas constant [J/K/mol]
Re	=	Reynolds number [-]
r	=	radius from center in coal lump/stock [m]
r_o	=	outer radius of coal lump/stock [m]
t	=	elapsed time [s]
T	=	absolute temperature of coal [K]
U	=	permeable flow velocity in a coal seam due to pressure difference [m/s]
u	=	natural convection flow velocity in x direction [m/s]
V	=	accumulated consuming volume of oxygen [cm³ O_2/g] or [m³O_2/s ·kg]
v	=	oxygen volume consumption rate [cm³O_2/s ·g] or [m³O_2/s ·kg]
W	=	half width of two-dimensional coal stock [m]
w	=	natural convection flow velocity in z direction [m/s]
x	=	horizontal axis [m]
z	=	vertical axis [m]
a	=	effective thermal diffusivity of coal seam or coal stock [m²/s]
β	=	thermal expansion coefficient of air [-]
γ	=	decay power constant [1/s]

ΔH	=	reaction heat of unit volume of oxygen [J/cm^{3}O$_2$]
Δq	=	heat transfer per unit volume of coal stock [kW/m^3] or [W/cm^3]
δ	=	average diameter of lump coals in coal stock [m]
ε	=	porosity or void fraction [-]
Φ	=	ratio of effective thickness over coal lump diameter [-]
κ	=	coefficient of heat transmission from coal seam [m^2/s]
λ	=	effective thermal conductivity of porous media [W/(m°C)]
λ_{air}	=	thermal conductivity of air [W/(m°C)]
λ_{coal}	=	thermal conductivity of lump coal [W/(m°C)]
λ_r	=	thermal conductivity of rock [W/(m°C)]
μ	=	air viscosity [Pas]
ξ	=	internal surface area in unit volume of coal stock [m^3/m^3] or [cm^3/cm^3]
τ	=	exposure time [s]
τ^*	=	equivalent oxidation exposure (EOE) time [s]
υ_{air}	=	dynamic viscosity coefficient of air [m^2/s]
ρ_{air}	=	air density [kg/m^3]
ρ_{coal}	=	density of lump coal or coal seam [kg/m^3]
θ	=	temperature of coal lump or coal stock [°C]
θ_0	=	air or atmospheric temperature and initial coal temperature [°C]
θ_{air}	=	temperature of internal natural convection air flow in coal stock [°C]
ψ	=	stream function of two dimensional natural convection flow [m^2/s]

8. References

Brooks, K. & Glasser, D. (1986). A Simplified Model of Spontaneous Combustion in Coal Stockpiles, *Fuel*, Vol. 65 Issue 8, pp. 1035-1041, August 1986, Pages 1035-1041, DOI 10.1016/0016-2361(86)90164-X

Brooks, K., Bradshaw, S. & Glasser, D. (1988). Experiment study of model compound oxidation on spontaneous combustion of coal, Chemical Engineering Science, Vol. 43, Issue 8, pp. 2139-2145, DOI 10.1016/0009-2509(88)87095-7

Carresl, J.N. & Saghafil, A. (1998). Predicting Spontaneous Combustion in Spoil Piles from Open Cut Coal Mines, Proceedings of Underground Coal Operators' Conference, Paper 234, February 18- 20 1998, WoUongong, http://ro.uow.edu.au/coal/234

Ho, Y. S. (2006). Review of Second-order Models for Adsorption Systems, Journal of Hazardous Materials, B136 (2006), pp. 681–689, ISSN 0304-3894

Koyata, K., Ono, T., Miyagawa, M., Orimoto, M., Koike, Y. & Ota, S. (1983). Development of the Coal Spontaneous Combustion Predicting System-Structure and Performance of Predicting System-, CRIEPI Report 283088, Central Research Institute of Electric Power Industry, Aug. 1983(in Japanese), ISSN 1340-6078

Kevin Brooks, Steven Bradshaw, David Glasser (1988) Experiment study of model compound oxidation on spontaneous combustion of coal, Chemical Engineering Science, Vol. 43, Issue 8, pp. 2139-2145, DOI 10.1016/0009-2509(88)87095-7

Kaji, R., Hishinuma, Y. & Nakamura, Y. (1987). Low Temperature Oxidation of Coals-A Calorimetric Study, Fuel, Vol. 66, Issue 2, February 1987, pp. 154-157, DOI 10.1016/0016-2361(87)90233-X

Kunii, D. & Smith, J.M. (1960). Heat Transfer Characteristics of Porous Rocks, A.1.Ch.E. Journal, Vol. 6-1 (March 1960), pp.71-78, DOI 10.1002/aic.690060115

Kunii, D. & Suzuki, M. (1967). Particle-to-Fluid Heat and Mass Transfer in Packed beds of Fine Particles, International Journal of Heat Mass Transfer, Vol. 10, pp.845-852, 1967, ISSN 0017-9310

Li Zeng-hua, Wang Ya-li, Song Na, Yang Yong-liang, Yang Yu-jing (2009). Experiment study of model compound oxidation on spontaneous combustion of coal, Procedia Earth and Planetary Science 1, pp.123–129, 2009, ISSN 1878-5220

Miyakoshi, H., Isobe, T. & Otsuka, K. (1984). Relationship between Oxygen Adsorption and Physico-chemical Properties of Coal, Journal of MMIJ, Vol. 100-1161, pp.1057-1062, 1984 (in Japanese), ISSN 0369−4194

Nield, D.A. & Bejan, A. (1999). *Convection in Porous Media*, Springer-Verlag, New, York, 1999. ISBN10 0-387-29096-6

Nordon, P. A. (1979). A Model for the Self-Heating Reaction of Coal and Char, *Fuel*, Vol. 58, pp.456-464, 1979, DOI 10.1016/0016-2361(79)90088-7

Sasaki, K., Miyakoshi, H. & Otsuka, K. (1987). Correlation between Gas Permeability and Macropore Structure of Coal, Journal of MMIJ, Vol.103-1198, pp.847-852, December 1987 (in Japanese), ISSN 03694194

Sasaki, K., Miyakoshi, H., Saitoh, A. & Chiba, T. (1992). Water Vapour Adsorption of Coal and Numerical Simulation Related Its Effects on Spontaneous Combustion in a Low Temperature Range, Journal of MMIJ, Vol.108-6, pp.479-486, June 1992 (in Japanese), ISSN 03694194

Wakao, N. and Kaguei, S. (1982). Heat and mass Transfer in Packed Beds, Gordon and Breach Science Publisher, New York and London, 1982. ISBN10 0677058608

Yuan, L. and Smith, A. (2007). Computational Fluid Dynamics Modeling of Spontaneous Heating in Longwall Gob Areas, *Proceedings of 2007 SME Annual Meeting and Exhibit(Denver)*, SMM, pp.1-7(NIOSHTIC-2 No. 20031657), February 25-28, 2007, Salt Lake City, Utah

Part 4

Heat Analysis

12

Integral Transform Method Versus Green Function Method in Electron, Hadron or Laser Beam - Water Phantom Interaction

Mihai Oane, Natalia Serban and Ion N. Mihailescu
National Institute for Laser, Plasma and Radiation Physics, Bucharest
Romania

1. Introduction

Many practical applications require the detailed study of the thermal behavior of different systems. The difficulties arise when these systems are inhomogeneous with respect to the parameters involved in the heat diffusion process. Currently, the heat diffusion equation has no analytical solution in this case. There exists however a wide range of methods to approximate the solution of the heat diffusion equation in inhomogeneous systems, starting from the numerical methods and ending with the exact analytical solution for a few particular cases, each of them presenting specific advantages and disadvantages [1-6].

We assume that the source term in the heat equation has the form: $f(x,y,z,t) = f(x,y,z)\cdot[h(t)-h(t-t_0)]$ (where h is the step function) and the thermal conductivity can be expressed as: $k(x,y,z) = k(x)$. We developed under these assumptions a computing method for solving the diffusion equation describing the heat propagation in inhomogeneous materials.

The procedure for solving the heat diffusion equation in inhomogeneous systems, with a prescribed accuracy in respect to the thermal conductivity, is outlined. To this aim, the thermal conductivity was considered a discontinuous function having a linear value in each layer. This approach is nonetheless valid in many cases like thin films or welding technologies. In these cases there exists nevertheless an interface that ensures the continuity of the thermal conductivity function.

We believe that the programs of simulation which are used in the present paper can be help - full for medical staff.

2. One-dimensional mathematical model

In this section we introduce the procedure for approximating the exact solution of the heat diffusion equation (1) with respect to the thermal conductivity in inhomogeneous media. Our approach is one dimensional, but it can be rather easily extended to multi-dimensional equations.

We consider a parallelepiped volume of dimensions a, b and c. The equation describing the heat diffusion is :

$$\frac{\partial}{\partial x}\left(k(x)\frac{\partial T(x,t)}{\partial x}\right)-\rho c\frac{\partial T(x,t)}{\partial t}=-f(x,t),a\leq x\leq b \tag{1}$$

Here: ρ is the mass density; c is the heat capacity; k is the thermal conductivity; T is the local temperature and t is current time. In each sub-domain $[x_i,x_{i+1}]$, the thermal conductivity is considered to be a linear function of the coordinate x, $k_i(x)=k(x_i)+m_i(x-x_i)$, where m_i is a real constant number.

When: $x\in[x_i,x_{i+1}]$, we have:

$$\frac{\partial}{\partial x}\left(k_i(x)\frac{\partial T_i}{\partial x}\right)-\rho_i c_i\frac{\partial T_i}{\partial t}=-f_i(x,t), \tag{2}$$

with the boundary conditions :

$$T_{i-1,j,k,l}(x,y,z,t)\big|_{x=x_i}=T_{i,j,k,l}(x,y,z,t)\big|_{x=x_i}, \tag{3}$$

$$T_{i+1,j,k,l}(x,y,z,t)\big|_{x=x_{i+1}}=T_{i,j,k,l}(x,y,z,t)\big|_{x=x_{i+1}},$$

$$k_{i-1}\cdot T_{i-1,j,k,l}'(x,y,z,t)\big|_{x=x_i}=k_i\cdot T_{i,j,k,l}'(x,y,z,t)\big|_{x=x_i},$$

$$k_{i+1}\cdot T_{i+1,j,k,l}'(x,y,z,t)\big|_{x=x_{i+1}}=k_i\cdot T_{i,j,k,l}'(x,y,z,t)\big|_{x=x_{i+1}}.$$

At the margins of the sample, we assume:

$$k_0\cdot T_{0,j,k,l}'(x,y,z,t)\big|_{x=x_0}=h_0\cdot T_{0,j,k,l}(x,y,z,t)\big|_{x=x_0},$$

$$k_i\cdot T_{i,j,k,l}'(x,y,z,t)\big|_{y=y_2/2}=-h_i\cdot T_{i,j,k,l}(x,y,z,t)\big|_{y=y_2/2},$$

$$k_i\cdot T_{i,j,k,l}'(x,y,z,t)\big|_{y=-y_2/2}=h_i\cdot T_{i,j,k,l}(x,y,z,t)\big|_{y=-y_2/2},$$

$$k_i\cdot T_{i,j,k,l}'(x,y,z,t)\big|_{z=z_3/2}=-h_i\cdot T_{i,j,k,l}(x,y,z,t)\big|_{z=z_3/2},$$

$$k_i\cdot T_{i,j,k,l}'(x,y,z,t)\big|_{z=-z_3/2}=h_i\cdot T_{i,j,k,l}(x,y,z,t)\big|_{z=-z_3/2},$$

$$k_n\cdot T_{n,j,k,l}'(x,y,z,t)\big|_{x=x_n}=-h_n\cdot T_{n,j,k,l}(x,y,z,t)\big|_{x=x_n},$$

and we have: $x_0=0$, $x_n=a$, $y_2=b$, $z_3=c$.

We next applied the integral operator $K_{ij}(x,\lambda_j)=\frac{1}{C(\lambda_j)}\int_{x_i}^{x_{i+1}}\overline{K}_{ij}(x,\lambda_i)dx$ to equation (2) with the kernel $\overline{K}_{ij}(x,\lambda_i)$.The kernel of this operator was derived by solving the partial differential equation :

$$\frac{\partial}{\partial x}\left(k_i(x)\frac{\partial \overline{K}_{ij}(x)}{\partial x}\right)+\lambda_j^2\overline{K}_{ij}(x)=0 \tag{4}$$

Equation (4) provides a series of positive eigenvalues $\lambda_j,\ j\in N$ and eigenfunctions $\overline{K}_{ij}(x,\lambda_i)$ of the differential operator: $\frac{\partial}{\partial x}\left(k(x)\frac{\partial}{\partial x}\right)$, [7]. The eigenfunctions were necessary for solving equation (2).

The solution of equation (4) has the form:

$$\overline{K}_{ij}(x,\lambda_i)=A_iJ_0\left(\frac{2\lambda_j\sqrt{\mu_i(k(x_i)+m_ix)}}{|m_i|}\right)+B_iY_0\left(\frac{2\lambda_j\sqrt{\mu_i(k(x_i)+m_ix)}}{|m_i|}\right),j\in N \tag{5}$$

Where: $\mu_i=\rho_i c_i$ and J_0 and Y_0 are the Bessel and Weber functions respectively.
After the application of the integral operator $K_i(x)$ equation (4) becomes:

$$-\lambda_j^2\overline{u}_i(\lambda_j,t)-\rho_i c_i\frac{\partial \overline{T}_i(\lambda_j,t)}{\partial t}=-\overline{f}_i(\lambda_j,t) \tag{6}$$

where:

$$\overline{u}_i(\lambda_j,t)=\frac{1}{C(\lambda_j)}\int_{x_i}^{x_{i+1}}T(x,t)\overline{K}_i(x,\lambda_j)dx$$

$$\overline{f}(\lambda_j,t)=\frac{1}{C(\lambda_j)}\int_{x_i}^{x_{i+1}}f(x,t)\overline{K}_j(x,\lambda_j)dx$$

and $C(\lambda_j)$ is a normalization factor.

Here we have [7]:

$$C(\lambda_j)=\sum_{i=0}^{n-1}\int_{x_i}^{x_{i+1}}\overline{K}_{ij}^2(x_i,\lambda_j)dx\,. \tag{7}$$

In the same manner, one can apply the functions: $\overline{K}_k(\mu_k,y)$ and $\overline{K}_l(\varepsilon_l,z)$, which satisfy the equations:

$$\begin{aligned}&\frac{\partial^2\overline{K_k}(\mu_k,y)}{\partial y^2}+\mu_k^2\overline{K_k}(\mu_k,y)=0\\&\frac{\partial^2\overline{K_l}(\varepsilon_l,z)}{\partial z^2}+\varepsilon_l^2\overline{K_l}(\varepsilon_l,z)=0\end{aligned} \tag{8}$$

This next gives: $\overline{K}_l(\varepsilon_l,z)=\cos(\varepsilon_l\cdot z)+(h/k\varepsilon_l)\cdot\sin(\varepsilon_l z)$, with, k being the thermal conductivity and h the heat transfer coefficient .

Then, the following equation is inferred:

$$\lambda_j^2\hat{T}_i(\lambda_j,\mu_k,\varepsilon_l,t)+\mu_k^2\hat{T}_i(\lambda_j,\mu_k,\varepsilon_l,t)+\varepsilon_l^2\hat{T}_i(\lambda_j,\mu_k,\varepsilon_l,t)+ \\ +\frac{\partial\hat{T}_i(\lambda_j,\mu_k,\varepsilon_l,t)}{\partial t}=\frac{\hat{f}_i(x,y,z,t)}{\rho_i c_i} \tag{9}$$

Where from it follows:

$$\hat{T}_i(\lambda_j,\mu_k,\varepsilon_l,t)=\frac{1}{C(\lambda_j)C(\mu_k)C(\varepsilon_l)}\times \\ \times\int_{x_i}^{x_{i+1}}\int_{-y_2/2}^{y_2/2}\int_{-z_3/2}^{z_3/2}T(x,y,z,t)\cdot\bar{K}_{ij}(\lambda_j,x)\cdot\bar{K}_k(\mu_k,y)\cdot\bar{K}_l(\varepsilon_l,z)dxdydz \tag{10}$$

In order to eliminate the time parameter t, we apply the direct and inverse Laplace transform to equation (9).
If we have, like in most cases: $f(x,y,z,t)=f(x,y,z)\cdot[h(t)-h(t-t_0)]$, one can get the solution:

$$T_i(x,y,z,t)=\sum_{j=1}^{\infty}\sum_{k=1}^{\infty}\sum_{l=1}^{\infty}\frac{1}{(\lambda_j^2+\varepsilon_l^2+\mu_k^2)}[1-e^{-\gamma\cdot(\lambda_j^2+\varepsilon_l^2+\mu_k^2)t}-(1-e^{-\gamma\cdot(\lambda_j^2+\varepsilon_l^2+\mu_k^2)(t-t_0)})h(t-t_0)]\times \\ g(\lambda_j,\mu_k,\varepsilon_l)\times\bar{K}_{ij}(\lambda_j,x)\times\bar{K}_k(\mu_k,y)\times\bar{K}_l(\varepsilon_l,z) \tag{11}$$

where:

$$g(\lambda_j,\mu_k,\varepsilon_l)=\frac{1}{C(\lambda_j)C(\mu_k)C(\varepsilon_l)}\times \\ \sum_{i=0}^{n-1}\int_{x_i}^{x_{i+1}}\int_{-y_2/2}^{y_2/2}\int_{-z_3/2}^{z_3/2}f_i(x,y,z,t)\cdot\bar{K}_{ij}(\lambda_j,x)\cdot\bar{K}_k(\mu_k,y)\cdot\bar{K}_l(\varepsilon_l,z)dxdydz \tag{12}$$

γ stands here for the thermal diffusivity. We point out that our semi-analytical solution becomes analytical, if we observe the, after 10 iterations, the solution becomes convergent (we have values of temperature less than 10^{-2}K for: i>10;j>10 and k>10). Under these conditions equation (10) becomes:

$$T_i(x,y,z,t)=\sum_{j=1}^{10}\sum_{k=1}^{10}\sum_{l=1}^{10}\frac{1}{(\lambda_j^2+\varepsilon_l^2+\mu_k^2)}[1-e^{-\gamma\cdot(\lambda_j^2+\varepsilon_l^2+\mu_k^2)t}-(1-e^{-\gamma\cdot(\lambda_j^2+\varepsilon_l^2+\mu_k^2)(t-t_0)})h(t-t_0)]\times \\ g(\lambda_j,\mu_k,\varepsilon_l)\times\bar{K}_{ij}(\lambda_j,x)\times\bar{K}_k(\mu_k,y)\times\bar{K}_l(\varepsilon_l,z) \tag{13}$$

3. Application of the theory: Laser-assisted hadron and electron beams therapy

It is well known [8] that the hadrons therapy (e.g. with protons) is much more suitable and efficient compared to electrons therapy, because the absorption curve is (in this case) a Dirac function. We can solve easily the heat equation for this case, and we obtain the temperature field in Fig.1. - (The Dirac absorption function is at 4 cm from the surface). Here dT is the temperature variation ($dT=T_f-T_i$), rather than the absolute temperature.

T_f and T_i are the final and initial temperatures respectively.

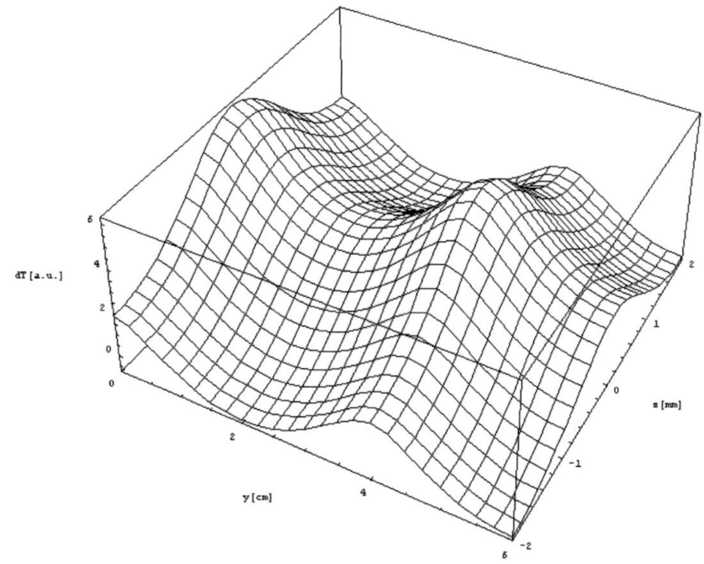

Fig. 1. Thermal field distribution in case of 1 MeV proton beam irradiation of a water phantom, for 120 sec.

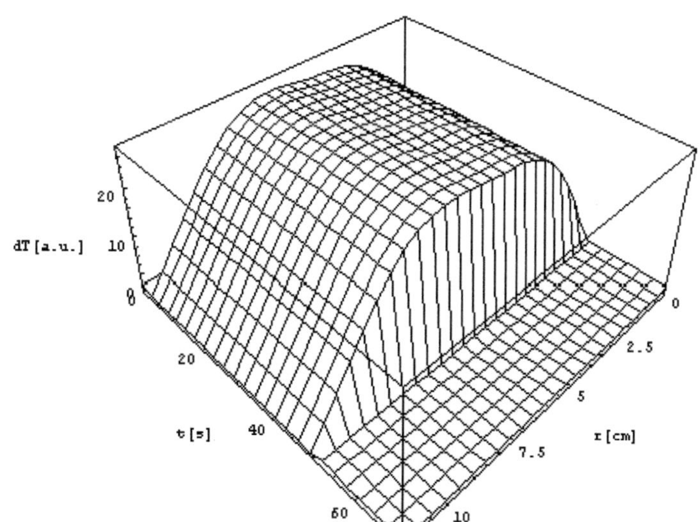

Fig. 2. Thermal field in water submitted to cw CO_2 laser irradiation for 50 sec.

The power of the cw CO_2 laser beam was P= 1W.
It is known from experience [8] that proton therapy is more efficient in the "presence" of a laser beam. We plotted in figure 2 the thermal gradient in water produced by cw CO_2 laser irradiation for 50 sec. (P = 1W). In Fig. 3 we presented the temperature field in water produced by an electron beam, when the "steady - state" is achieved. The white color corresponds to an increase of temperature, and the black color represents a decrease of temperature. We have use sub-domains of 0.25 cm. The thickness of the water phantom was 0.25 cm, and was contained in a plastic cube with a mass density close to 1 g/cm^3. Figure 3 was obtained using eq. (13).

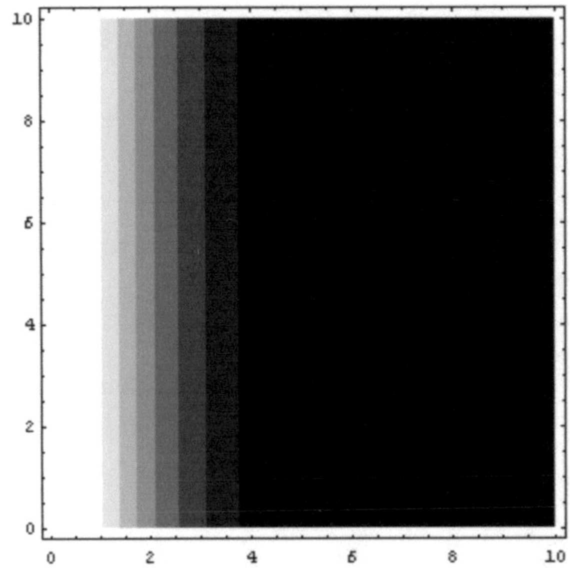

Fig. 3. Temperature field in water produce by an electron beam, when the "steady- state" is achieved.

The white color corresponds to the temperature increase while, the black color represents the temperature decrease. We have used sub-domains of 0.25 cm length.

4. The green function method

We start from the heat equation:

$$\left\{\frac{\partial}{\partial x}[K_x(T)\frac{\partial T}{\partial X}]+\frac{\partial}{\partial y}[K_y(T)\frac{\partial T}{\partial y}]+\frac{\partial}{\partial z}[K_z(T)\frac{\partial T}{\partial z}]\right\}=-S(x,y,z) \tag{14}$$

where S(x, y, z) is proportional with the absorbed dose. We consider [9], the case of a 10 MeV electron beam interactions with water. We have:

$$S(x,y,z)=K(y,z).D_{10}(x) \tag{15}$$

where according to experimental data from our laboratory:

$$D_{10}(x)=83.2337+18.6522\cdot x^2+15.1080\cdot x^3-4.1417\cdot x^4+0.3506\cdot x^5 \tag{16}$$

Here x stands for the direction of electron propagation. We will consider the radiation (electron beam) normal to water surface.

From the standard theory of Green function applied to multi-layer structures, we have:

$$K_{\perp} = \frac{l_1 + l_2 + \dots + l_n}{l_1/k_1 + l_2/k_2 + \dots + l_n/k_n} = \frac{l}{l_1/k_1 + l_2/k_2 + \dots + l_n/k_n} \tag{17}$$

where l_i is the length and k_i is the thermal conductivity of the *i-th* layer.

We introduce the area of the layer A_i:

$$K_{||} = \frac{K_1 A_1 + K_2 A_2 + \dots + K_n A_n}{A_1 + A_2 + \dots + A_n} = \frac{K_1 A_1 + K_2 A_2 + \dots + K_n A_n}{A} \tag{18}$$

We define the "linear" temperature

$$\theta(T) = \theta(T_0) + (1/K(T_0)) \int_{T_0}^{T} K(T')dT' \tag{19}$$

and we can write:

$$\theta_{\perp} = \frac{P[1 - R_{\perp}(T)]}{(\alpha)^{1/2}(\pi)^{3/2} K(T_0)} \int_0^{\infty} f_{\perp}(\xi) d\xi \tag{20}$$

where:

$$\alpha = \frac{K_{\perp}}{K_{||}}$$

and:

$$K_{||} = K(T) \tag{21}$$

The function $f_{\perp}$ is given by:

$$f_{\perp}(\xi) = \frac{\exp - [[X^2(\xi^2+1)] + [Y^2/(\xi^2+1)] + (Z^2/\alpha\xi^2)]}{\{(\xi^2+1)\}} \tag{22}$$

We plotted in Fig.4 the analytical results obtained with the Green function method.

The white color corresponds to temperature increase, and the black color represents a decrease of temperature. We have used sub-domains of 0.25 cm length. Figs. 3 and 4 allow for a direct comparison between the temperature fields in water computed with the integral transform technique and Green function method under identical conditions.

5. The thermal fields when we have multiple sources irradiations

We consider a parallelepiped sample with dimensions a, b, and c. The sample is irradiated by three laser beams which propagate along the Cartesian coordinate axes. The model is also valid for electron or hadrons beam irradiations.

Let us considering the following relations:

$$A(x,y,z,t) = A_1(x,y,z,t) + A_2(x,y,z,t) + A_3(x,y,z,t) \tag{23}$$

Therefore:

$$T(x,y,z,t) = T_1(x,y,z,t) + T_2(x,y,z,t) + T_3(x,y,z,t) \tag{24}$$

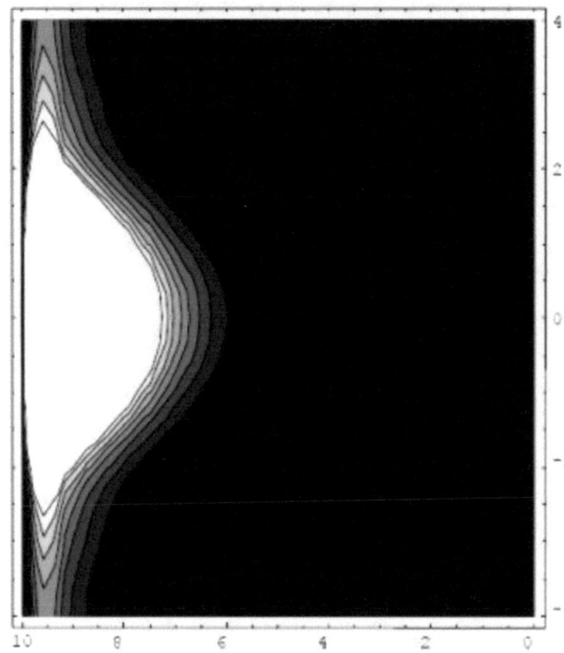

Fig. 4. The temperature field in water produced by a 10 MeV electron beam, when the "steady- state" is achieved.

We suppose that for the heat transfer coefficients: $h_1 = h_2 = h_3 = h_4 = h_5 = h_6 = h$. If we consider a linear heat transfer at the sample surface (the "radiation" boundary condition [11]), we have:
for the first laser beam , direction of propagation along x axis:

$$\left[\frac{\partial K_x}{\partial x} - \frac{h}{K}K_x\right]_{x=-\frac{a}{2}} = 0 \ ; \quad \left[\frac{\partial K_x}{\partial x} + \frac{h}{K}K_x\right]_{x=\frac{a}{2}} = 0 \ ; \quad \left[\frac{\partial K_y}{\partial y} + \frac{h}{K}K_y\right]_{y=\frac{b}{2}} = 0 \ ;$$
$$\left[\frac{\partial K_y}{\partial y} + \frac{h}{K}K_y\right]_{y=-\frac{b}{2}} = 0 \ ; \quad \left[\frac{\partial K_z}{\partial z} + \frac{h}{K}K_z\right]_{z=\frac{c}{2}} = 0 \ ; \quad \left[\frac{\partial K_z}{\partial z} + \frac{h}{K}K_z\right]_{z=-\frac{c}{2}} = 0 \tag{25}$$

for the second laser beam, direction of propagation along y axis:

$$\left[\frac{\partial L_x}{\partial x}+\frac{h}{K}L_x\right]_{x=-\frac{a}{2}}=0\ ;\quad \left[\frac{\partial L_x}{\partial x}+\frac{h}{K}L_x\right]_{x=\frac{a}{2}}=0\ ;\quad \left[\frac{\partial L_y}{\partial y}+\frac{h}{K}L_y\right]_{y=\frac{b}{2}}=0\ ;$$
$$\left[\frac{\partial L_y}{\partial y}-\frac{h}{K}L_y\right]_{y=-\frac{b}{2}}=0\ ;\quad \left[\frac{\partial L_z}{\partial z}+\frac{h}{K}L_z\right]_{z=\frac{c}{2}}=0\ ;\quad \left[\frac{\partial L_z}{\partial z}+\frac{h}{K}L_z\right]_{z=-\frac{c}{2}}=0 \tag{26}$$

for the third laser beam, direction of propagation along z axis:

$$\left[\frac{\partial M_x}{\partial x}+\frac{h}{K}M_x\right]_{x=-\frac{a}{2}}=0\ ;\quad \left[\frac{\partial M_x}{\partial x}+\frac{h}{K}M_x\right]_{x=\frac{a}{2}}=0\ ;\quad \left[\frac{\partial M_y}{\partial y}+\frac{h}{K}M_y\right]_{y=\frac{b}{2}}=0\ ;$$
$$\left[\frac{\partial M_y}{\partial y}+\frac{h}{K}M_y\right]_{y=-\frac{b}{2}}=0\ ;\quad \left[\frac{\partial M_z}{\partial z}+\frac{h}{K}M_z\right]_{z=\frac{c}{2}}=0\ ;\quad \left[\frac{\partial M_z}{\partial z}-\frac{h}{K}M_z\right]_{z=-\frac{c}{2}}=0 \tag{27}$$

The solution of the heat equation subjected to boundary conditions (25), (26) and (27) is:

$$\begin{aligned}\Delta T(x,y,z,t)=&\sum_{m,n}\left\{\sum_{i=1}^{\infty}\sum_{j=1}^{\infty}\sum_{o=1}^{\infty}a(\alpha_i,\beta_j,\chi_o)b(\alpha_i,\beta_j,\chi_o,t)K_x(\alpha_i,x)K_y(\beta_j,y)K_z(\chi_o,z)\right\}\\
&+\sum_{v,\varphi}\left\{\sum_{p=1}^{\infty}\sum_{r=1}^{\infty}\sum_{s=1}^{\infty}c(\delta_p,\varepsilon_r,\eta_s)d(\delta_v,\varepsilon_r,\eta_s,t)L_x(\delta_v,x)L_y(\varepsilon_r,y)L_z(\eta_s,z)\right\}\\
&+\sum_{r,\varphi}\left\{\sum_{t=1}^{\infty}\sum_{v=1}^{\infty}\sum_{w=1}^{\infty}e(\mu_t,\sigma_v,\omega_w)f(\mu_t,\sigma_v,\omega_w,t)M_x(\mu_t,x)M_y(\sigma_v,y)M_z(\omega_w,z)\right\}\end{aligned} \tag{28}$$

We have:

$$I_{mn}(x,y)=I_{0mn}\left[H_m\left(\tfrac{\sqrt{2}x}{w}\right)H_n\left(\tfrac{\sqrt{2}y}{w}\right)\times\exp\left[-\left(\tfrac{x^2+y^2}{w^2}\right)\right]\right]^2 \tag{29}$$

Here w is the width of the laser beam.

$$\begin{aligned}a(\alpha_i,\beta_j,\chi_o)=&\frac{\sum_{m,n}I_{0x,m,n}}{KC_iC_jC_o}\int_{-\frac{a}{2}}^{+\frac{a}{2}}(\alpha_i e^{-\alpha_i x}(1-r_S)K_x(\alpha_i,x)+\\
&+r_S\cdot\delta_S(x))dx\int_{-\frac{b}{2}}^{+\frac{b}{2}}\int_{-\frac{c}{2}}^{+\frac{c}{2}}I_{m,n}K_y(\beta_j,y)K_z(\chi_o,z)dydz\end{aligned} \tag{30}$$

where:

$$b(\alpha_i,\beta_j,\chi_o,t)=\frac{1}{\alpha_i^2+\beta_j^2+\chi_o^2}[1-e^{-\gamma_{ijo}^2 t}-(1-e^{-\gamma_{ijo}^2(t-t_o)})h(t-t_o)] \tag{31}$$

and

$$\gamma_{ijo}^2 = \gamma(\alpha_i^2 + \beta_j^2 + \chi_o^2) .$$

We have:

$$K_x(\alpha_i, x) = \cos(\alpha_i \cdot x) + (h / k\lambda_i) \cdot \sin(\alpha_i \cdot x) \tag{32}$$

The other formulas can be easily obtains by "rotations" of the indices. t -is the time and t_o the exposure time.

We have: r_S is the parameter which take care of the surface absorption and which make sense only for one photon absorption.

Here: $\alpha_i, \alpha_i^2, \delta_p, \delta_p^2, \mu_t, \mu_t^2$ are the eigenvalues corresponding to the eigenfunctions: $K_x, K_y, K_z, P_x, P_y, P_z, L_x, L_y, L_z, T_x, T_y, T_z, M_x, M_y, M_z, N_x, N_y, N_z$ [7].

$C_i C_j, C_o, C_p, C_r, C_s, C_t, C_v$ and C_w as well similarly formulas for two photon absorption.

$h(t - t_0)$ is the step function [7].

We can generalize formula (28) taking into account the one and two absorption coefficient. In this case we have the following solution:

$$\begin{aligned}
\Delta T(x,y,z,t) = & \sum_{m,n} \left\{ \sum_{i=1}^{\infty} \sum_{j=1}^{\infty} \sum_{o=1}^{\infty} a(\alpha_i, \theta_j, \chi_o) b(\alpha_i, \theta_j, \chi_o, t) K_x(\alpha_i, x) K_y(\theta_j, y) K_z(\chi_o, z) \right\} \\
& + \sum_{m,n} \left\{ \sum_{i=1}^{\infty} \sum_{j=1}^{\infty} \sum_{o=1}^{\infty} \beta(\alpha_i^2, \theta_j^2, \chi_o^2) b(\alpha_i^2, \theta_j^2, \chi_o^2, t) P_x(\alpha_i^2, x) P_y(\theta_j^2, y) P_z(\chi_o^2, z) \right\} \\
& + \sum_{v,q} \left\{ \sum_{p=1}^{\infty} \sum_{r=1}^{\infty} \sum_{s=1}^{\infty} c(\delta_p, \varepsilon_r, \eta_s) d(\delta_v, \varepsilon_r, \eta_s, t) L_x(\delta_v, x) L_y(\varepsilon_r, y) L_z(\eta_s, z) \right\} \\
& + \sum_{v,q} \left\{ \sum_{p=1}^{\infty} \sum_{r=1}^{\infty} \sum_{s=1}^{\infty} \pi(\delta_p^2, \varepsilon_r^2, \eta_s^2) d(\delta_v^2, \varepsilon_r^2, \eta_s^2, t) T_x(\delta_v^2, x) T_y(\varepsilon_r^2, y) T_z(\eta_s^2, z) \right\} \\
& + \sum_{r,\varphi} \left\{ \sum_{t=1}^{\infty} \sum_{v=1}^{\infty} \sum_{w=1}^{\infty} e(\mu_t, \sigma_v, \omega_w) f(\mu_t, \sigma_v, \omega_w, t) M_x(\mu_t, x) M_y(\sigma_v, y) M_z(\omega_w, z) \right\} \\
& + \sum_{r,\varphi} \left\{ \sum_{t=1}^{\infty} \sum_{v=1}^{\infty} \sum_{w=1}^{\infty} \psi(\mu_t^2, \sigma_v^2, \omega_w^2) f(\mu_t^2, \sigma_v^2, \omega_w^2, t) N_x(\mu_t^2, x) N_y(\sigma_v^2, y) N_z(\omega_w^2, z) \right\}
\end{aligned} \tag{33}$$

In formula (33) the upper index 2 means that the corresponding values are connected with two photon absorption. The eigenfunctions and the eigenvalues for two absorption phenomena can be calculated in the same way like in the case of one photon absorption with the only change that we have another absorption formula. It make no sense to take into account three or more photons absorption phenomena because in this situations the cross sections are very small.

In the next pages we will present three simulations, using the developed "multiple beam irradiation".

The different characteristics of dielectrics under one laser beam irradiation have been very well studied in literature. We will take the case of a ZnSe sample (all characteristics of the material can be found in reference [11]).

The sample is a cube with the dimension about 2 cm.

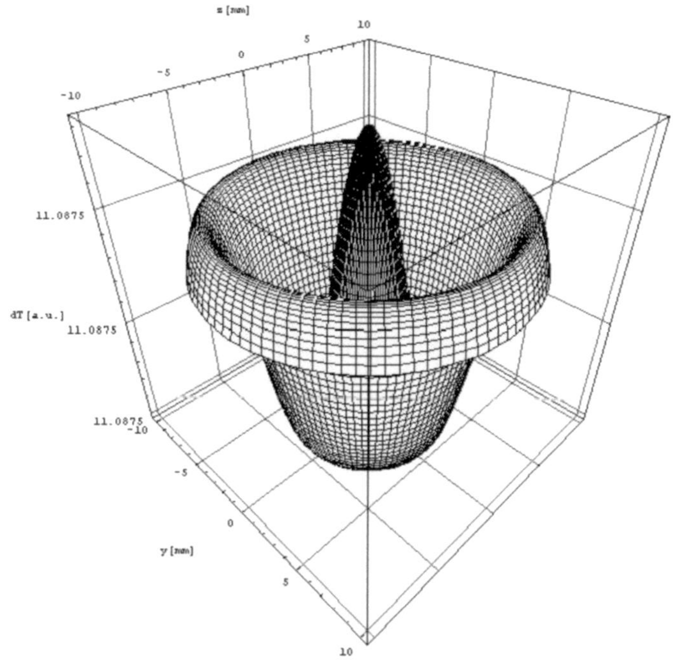

Fig. 5. Temperature field in the plane x=0, during a 100s irradiation with a 10 W CO_2 laser beam.

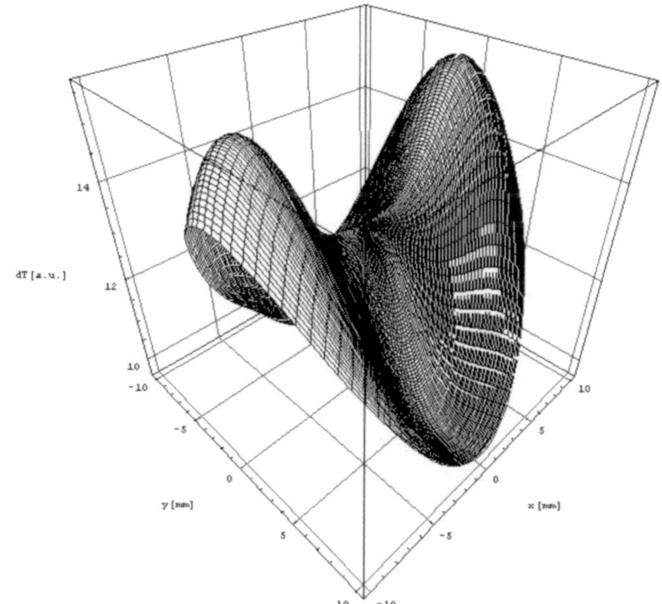

Fig. 6. Temperature field plotted during 100s irradiation with a 50 W CO_2 laser beam, operating in the TEM_{03}.

Our study indicates that for a sample under one, two or three laser irradiation, the heat equation has an exact semi-analytical solution. In fact it can be considered an analytical solution because the eigenvalues with index higher than 10 does not contribute to the solution of heat equation. This solution it is not simply the sum of solutions from three one-dimensional heat equations, because $T_1(x,y,z,t), T_2(x,y,z,t)$ and $T_3(x,y,z,t)$ are coupled via boundary conditions. Our model can be easily generalized for the cases when: $h_1 \neq h_2 \neq h_3 \neq h_4 \neq h_5 \neq h_6$ or $\alpha_x \neq \alpha_y \neq \alpha_z$ .The model could be applied to any laser-solid system whose interaction can be described by Beer law.
The integral transform technique has proved once again it's "power" in resolving heat equation problems [14-17].

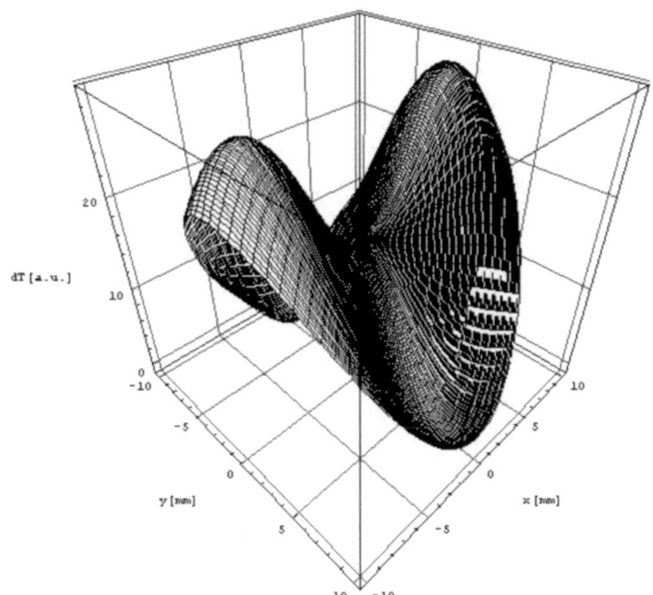

Fig. 7. Temperature field when the sample is irradiated simultaneously with the two laser beams, mentioned above (Fig.5 and Fig.6)

6. Discussions and conclusions

We developed a method for solving the heat diffusion equation- based on dividing the whole domain into small intervals, the length of each depending on the required accuracy of the final solution. The theory is applicable to laser, electrons and hadrons beams interaction with human tissues (which are simulated by a water phantom). In each of the obtained intervals the thermal conductivity function is approximated by a linear function. This function is introduced in the heat equation associated to each interval. At the interface between intervals, the continuity of temperature function and its first derivative are ensured, these conditions providing the values for the coefficients obtained in the final solution.

In order to solve such a system, the formalism of the integral operators with respect to the space and temporal dimensions was applied and the initial system becomes an algebraic one [7]. After solving the system, inverse transformations were applied and the final solution for each interval was obtained as a series of Bessel and Weber functions depending on the space coordinate.

We had thus developed a semi-analytical model for describing the beam - inhomogeneous medium interaction. It can be applied to beam-target interaction where the temperature variation is not very large. This experimental restraint is required because the model does not take into consideration the variation of the thermal parameters with the temperature.

From a practical point of view, the eigenvalues can be obtained from the boundary conditions. Also the constants A_i, B_i can be obtained easily from the same boundary conditions.

Here follows a few examples of the model applications: electron beam-water phantom interaction, proton-water phantom interaction, laser-optical components interaction and, in general, laser-solid media interaction (with the condition that the absorption coefficient keeps small).

We also made simulations using the Green function method. The results represented in Figures 3 and 4 are similar, with the exception of the edges temperature, where we believe that the Green function method is more close to the reality. In fact, the Green function method takes more into account that at the edges of the sample the heat transfer coefficients are higher and in consequence the temperatures get lower.

In previous papers different models (which were in fact particular cases of the present model) were applied to describe the interaction between a multi-mode cw CO_2 laser beam with multi-layered structures (of the type thin films substrate) [10] or with optical components [11].

The actual strength of the model is that it can take into account any form of the beam spatial distribution and any stationary type of interaction. That was the starting point for developing the semi-classical heat equation solution, which included the multi-photons laser-sample interaction [12]. The particular case $m_i = 0$ (i.e. when $k_i(x)=k(x_i)$) was analyzed in Ref. - [13].

The "power" of integral transform technique was emphasized in references [14-17]; both in classic and quantum physics.

Finally: a remark about figures 1-4. We mention that: dT (x, y, z, t) is in general proportional with *S* (x, y, z, t). This is not always true, but in our case is valid because the small values of the heat transfer coefficient. For a comprehensive discussion of the importance of heat transfer coefficient, see Appendix B in reference [11].

Our model offers a first simple approximation of the temperature field in (electron, proton, laser) beam (liquid, solid) target interaction.

The model can also describe the thermal fields for three different beams (electron, proton and laser), which act simultaneously onto a sample along the three Cartesian coordinates axes.

Figure 3 is illustrative for the strength of our model. The simulations performed using sub-domains of 0.25 cm were indeed in good agreement with the solutions given by the Green function method.

7. Acknowledgement

This work was supported by the UEFISCSU (Romania) under the project IDEI, 511/ 2009.

8. References

[1] A. P. Kubyshkin, M. P. Matrosov, and A. A. Karabutov, *Opt. Eng., 3214, 1996,*

[2] M. D. Dramicanin, Z. D. Ristovski, V. Djokovic, and S. Galovic: *Appl. Phys. Lett. 73, 321, 1998,*

[3] J. Opsal and A. Rosencwaig: *J. Appl. Phys. 53, 6, 4240, 1982,*

[4] Z. Bozóki , A. Miklós, and D. Bicanic , *Appl. Phys. Lett. 64, 11, 1362, 1994,*

[5] S. Bhattacharyya, A. Pal, and A. S. Gupta: *J. Heat Mass Transfer, 34, 41, 1998,*

[6] T. Aldoss , T. S. Chen , and B .F. Armly : *Int. J. Heat Mass Transfer, 36, 471, 1993,*

[7] N.S. Koshlyakov, M. M. Smirnov, and E. B. Gliner: *"Differential Equations of Mathematical Physics', Amsterdam: North-Holland Publishing Company, 1964,*

[8] R. Piana, *Oncology News International, Vol.17, No. 4, 1 April, 2008,*

[9] K. R. Hogstram and P. R. Almond, *Physics in Medicine and Biology, Vol.51, R 455, 2006,*

[10] M. Oane, I. Morjan, and R. Medianu, *Optics and Laser Technology, 36, 677, 2004,*

[11] M. Oane and D. Sporea, *Infrared Physics & Technology, 42, 31, 2001,*

[12] M. Oane and D. Apostol, *Optics and Laser Technology, 36, 219, 2004,*

[13] M. Oane, S. L. Tsao, and F. Scarlat, *Optics and Laser Technology, 39, 179, 2007,*

[14] M. Oane, A. Peled, Fl. Scarlat, I. N. Mihailescu, A. Scarisoreanu, and G. Georgescu, *Infrared Physics & Technology 51,242, 2008,*

[15] M. Oane, A. Peled, Fl. Scarlat, I.N. Mihailescu, and G. Georgescu, *Infrared Physics & Technology 51, 348, 2008,*

[16] M. Oane, Fl. Scarlat, and I. N. Mihailescu, *Infrared Physics & Technology, 51, 344, 2008,*

[17] M. Oane, *Lasers in Engineering, Vol.20, No.5-6, 329, 2010.*

13

Micro Capillary Pumped Loop for Electronic Cooling

Seok-Hwan Moon and Gunn Hwang
Electronics and Telecommunications Research Institute (ETRI)
Korea

1. Introduction

Electronic devices have been minimized, but their performance is becoming better and better. Their heat flux has been significantly increased and has already exceeded about 100 W/cm^2 recently. The insufficient dissipating of the heat flux may lead to performance decrease or failure of the electronic device and components. Heat flux in laptop computers has not been questioned; therefore, only a heat sink has been applied on cooling them. Recently, however, a more powerful cooling solution is sought for high heat flux. Solid materials with high thermal conductivity have been mainly used in low heat flux applications, whereas small-sized heat pipes have been utilized in high heat flux applications. The use of small-sized heat pipes in electronic devices like laptop computers has only been developed recently. For example, the use of heat pipes with diameter of 3–4 became common in laptop and desktop computers only during the early 2000s. Recently, as electronic devices have started to become smaller and thinner, heat pipes with diameter of 3–4 mm have been pressed to fit the form factor to them. However, a lot of problems were encountered in their thermal performance, thus micro heat pipes (MHPs) were developed to solve them. Specifically, a flat plate micro heat pipe (FPMHP) with diameter of less than 1.5 mm was developed by Moon (Moon et al., 2002). FPMHPs are being used mainly in display panel BLU applications and are being prepared to be used in the LED headlight of vehicles. However, in spite of their thermal performance and broad applications, FPMHPs may still show degradation in thermal performance in the case of thinner applications.
If we consider phase-change cooling devices like heat pipes that have thermal conductivity that is 500 times larger than copper rods for small-sized and thin electronic devices, there is a need to develop new cooling methods suitable for them.
A thin flat plate type micro capillary pumped loop (CPL) with thickness of less than 2 mm was developed by Moon as a trial product. The proposed micro CPL has two-staged grooves in the evaporator, instead of poles, for preventing the backflow of the vapor bubbles; this is a simpler structure compared to that of a micro CPL with poles. A large vapor space from the evaporator to the condenser was also constructed in the middle plate to allow for the reduction of the flow resistance of the vapor. The micro CPL was fabricated using MEMS technology and was composed of lower, middle, and upper substrates. The lower substrate was composed of silicon, while the middle and upper substrates were made from Pyrex glass for visualization. Through a preliminary test, it was verified that there was no leakage at the adhesion interface between the lower and the middle or upper substrates

and at the bonding interface between the lower substrate and the fill tube. Although the experimental studies for the micro CPL have been poor to date, we obtained reasonable experimental results in this study. The performance test result showed a heat transfer rate of 8.5 W for the micro CPL, and we could observe the operating characteristics of circulating or evaporating and condensing by visualization. Pure distilled water was used as working fluid.

2. Cooling methods for small-sized electronic devices

A cooling module with 3–4 mm diameter heat pipe, combined with Al heat sink, is mainly used for desktop PCs belonging to large-sized personal devices. However, a pressed 3–4 mm diameter heat pipe is used for laptop PCs due to its limited inner space (Moon et al., 2001). A decrease in heat transfer rate may occur in the case of pressed flat heat pipe due to the reduction of the inner space for the flow path of the working fluid and the deformation of the specific wick structure for capillary pressure (Kim et al., 2001). Thus, the thickness of the heat pipe is limited for its normal performance. Studies on flat plate heat pipes or MHPs that are less than 2 mm thick have been conducted (Moon et al., 1999; Moon et al., 2002). A heat pipe of small thickness or diameter is not suitable for high heat transfer rate application, but for small-sized mobile devices as cooling solution. In the case of an MHP, it is not easy for the wick for liquid flow path to be inserted into it due to its small size. Therefore, grooves were fabricated on the MHP envelope or the sharp corners of the polygonal structure by reforming itself act as the wick. Tubular type MHPs with circular or polygonal cross section are suitable for small-sized application, whereas flat plate type MHPs are suitable for display application. Cooling solutions that may be considered for small-sized electronic and telecommunication devices are as follows. Materials with high thermal conductivity like copper and aluminum are cooling solutions that can be used easily. Such materials are widely used as cooling solutions in the fields of electronic packaging module and system levels. A liquid cooling using micro channels is suitable for high heat flux application due to its high heat transfer rate. However, because it has constraint in the form factor, it is not suitable for mobile application. Furthermore, overall, the spray cooling method and thermo-electric cooling (TEC) may be considered for a specific application.

3. Micro Heat Pipes (MHPs)

3.1 Characteristics of the MHP

Because MHPs have limited inner space compared to medium-size heat pipes, inserting additional wick for liquid flow path is not easy. Therefore, MHPs are characterized to have capillary structure on their wall. MHPs have small vaporizing amount with latent heat due to their small size, therefore they are not suitable for high heat flux application. The existence of non-condensable gas, albeit in very small amount, can lead to the decrease in performance; therefore, high-quality fabricating process is needed.

Cotter (Cotter, 1984) was first to propose the concept of an MHP for the purpose of cooling electronic devices. The MHP was so small a heat pipe that the mean radius of curvature of the liquid-vapor interface is comparable to the hydraulic radius of the flow channel. Typically, MHPs have a convex but cusped cross section with hydraulic diameters of 10–500 μm and lengths of 10–50 mm (Faghri, 1995). Since the initial conceptualization by Cotter in 1984, numerous analytical and experimental investigations have been reported so far.

According to the previous investigations reported by Cotter (Cotter, 1984), Babin and Peterson (Babin et al., 1990), and Gener (Gerner, 1989), the maximum heat transport capacity of the MHP with 0.01–0.5 mm hydraulic radius was 0.03–0.5 W and corresponded to 10 W/cm^2 in the heat flux based on the surface area of the evaporator. Wu and Peterson (Wu et al., 1991) also reported that Q_{max} = 4–5 W or q''_{lim} = 5 W/cm^2 for a flat MHP with D_h = 1 mm. Thus, Faghri (Faghri, 1995) mentioned that it was necessary to improve the maximum heat transport capacity in order to cover the thermal load (> 10 W/cm^2) encountered in most main ICs of current computers and to widely apply the MHP to computers as a cooling device.
The maximum heat transfer rate of the MHP is increased as the operating temperature is increased, which is similar to the medium-sized heat pipe. Fig. 1 shows the maximum heat transfer rate according to the operating temperature variation for the curved rectangular MHP fabricated through this study, which has 30% of working fluid compared with the total volume. It is a characteristic of the MHP to have liquid block at the condenser during its operation due to its small inner fluid flow space. The liquid block induces a decrease in its performance since vapor cannot reach the condenser area, which is being occupied by the liquid block, and thus cannot accomplish phase-change heat transfer in there. Thus, optimum design of the working fluid amount is important to minimize the volume of the liquid block. Fig. 2 shows the capillary radius distribution over axial direction of the MHP in conditions of 50 °C of operating temperature. The capillary radius can be calculated by equation (1):

$$P_c = P_v - P_l = \sigma\left(\frac{1}{r_{c1}} + \frac{1}{r_{c2}}\right) \tag{1}$$

Where r_{c1} and r_{c2} are the principal capillary radii of the meniscus. For the MHP designed in this study, the capillary radius can be considered as $r_{c1} \approx r_{c(x)}$ and $r_{c2} \approx \infty$ since capillary radius variation over axial direction occurred, but the capillary radius variation through radial direction nearly not occurred. Fig. 3 shows the vapor and liquid pressure distributions over axial direction of the MHP in the condition of 50 °C of operating temperature.

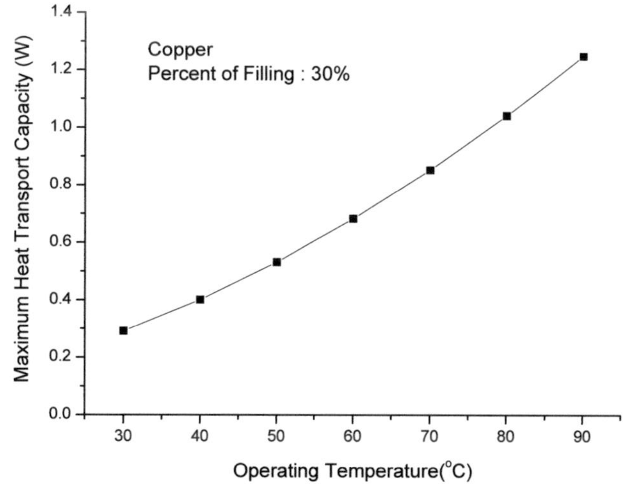

Fig. 1. Maximum heat transport capacity according to operating temperatures

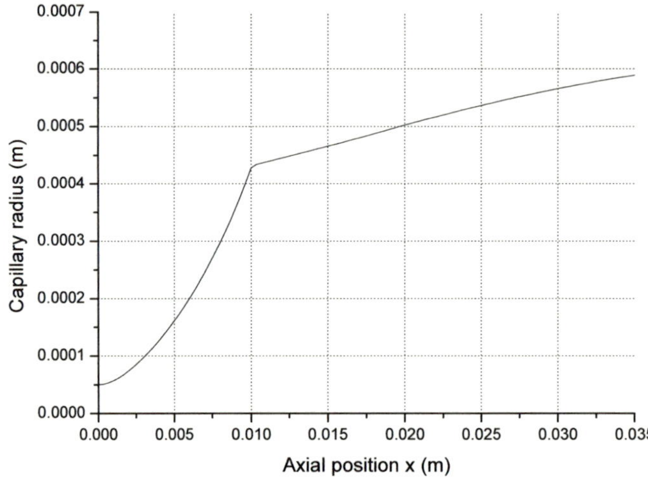

Fig. 2. Capillary radius variation over axial position

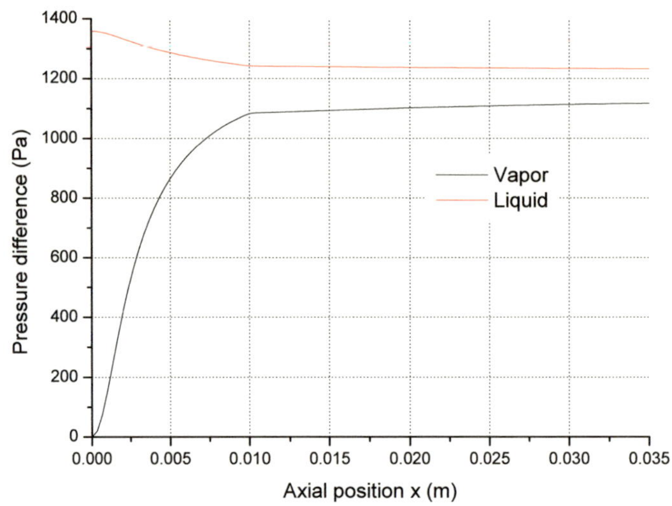

Fig. 3. Pressure distributions of vapor and liquid over axial position

3.2 Fabrication of the MHP

The size of the MHP developed in this study is smaller than that of the miniature heat pipe with diameter of 3-4 mm. The MHP could be manufactured using current mechanical technology of the simple manufacturing process. Therefore, this manufacturing process of the MHP has a good productivity. In fact, the microstructures of the MHP may be manufactured with etching process (Gerner, 1989), but such process has disadvantages in terms of productivity and cost. The container of the MHP was manufactured with the drawing process, and Fig. 1 shows its cross section. This MHP does not have additional wick installed on the inner wall of the general heat pipe, but has sharp corners made with structural deformation of its wall, which serves as wick.

The condensed liquid at the condenser of the MHP returns to the evaporator through the capillary force of the liquid. It is important for the MHP to have sharp edges in the corners to enable the working liquid to return from the condenser to the evaporator. The rectangular MHP with curved sides (Fig. 4a) has advantages for thermal performance, such as larger inner space for vapor flow and an additional corner for liquid path, than those of the triangular MHP (Fig. 4b). Because the edge angle at the corner of the triangular MHP is sharper than that of the rectangular MHP, the capillary force of the working fluid at the triangular MHP is larger than that at the rectangular MHP. The specification of the MHP is shown in Table 1.

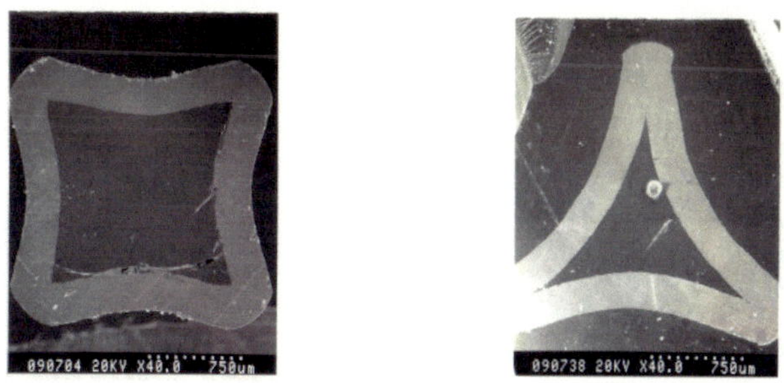

(a) Curved rectangular (b) Curved triangular

Fig. 4. Cross section of MHP

	Triangular MHP	Curved rectangular MHP
Total length (mm)	50/100	
Evaporator length	10 mm	
Adiabatic section length	15 mm	
Condenser length	25 mm	
Working fluid	Pure water	
Fill ratio of working fluid	20%	
Number of corners	3	4
Container material	Oxygen-free copper	
Container manufacturing method	Drawing	

Table 1. Experimental specification of MHP

FPMHPs with cross section of rectangular, modified rectangular, and triangular types were newly developed in the present study. The container of the FPMHPs was manufactured by the extrusion process, which could effectively form a sharp edge and significantly enhance the productivity of FPMHPs. Moreover, mass production of FPMHPs is possible through this method. The FPMHPs have three types of cross section – rectangular, modified rectangular, and triangular. Fig. 5 shows the cross section of the FPMHPs; Table 2 presents the dimensions of the FPMHPs.

	Thickness	Width	Total length
Rectangular type	2 mm	12 mm	50 mm
Modified rectangular type	1.5 mm	14 mm	50 mm
Triangular type	1.5 mm	12 mm	50 mm

Table 2. Dimensions of the FPMHPs

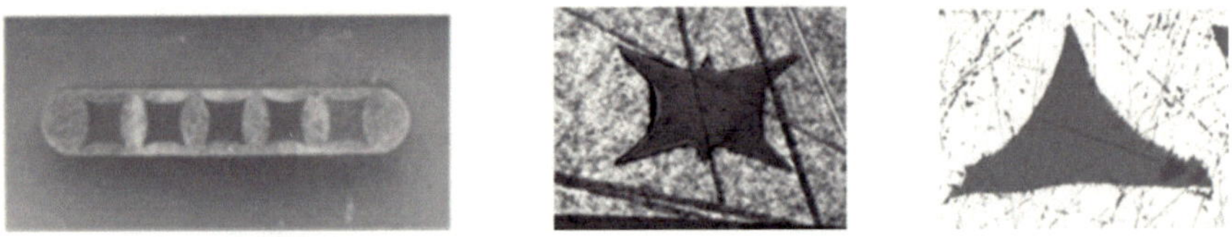

(a) Rectangular (b) Modified rectangular (c) Triangular

Fig. 5. Cross sections of the FPMHPs

3.3 Thermal performance of the MHP

The testing apparatus for the thermal performance of the MHP was composed of an MHP, a vacuum chamber unit, a constant temperature bath for cooling the MHP, a data acquisition system, and a DC power supplying unit, as shown in Fig. 6. The evaporator of the MHP was heated using the electric resistance heater and DC power supply unit. The wire with 0.36 mm diameter and 10 Ω/m resistance per meter, as a heater, was wound around the copper block with an interval of 0.5 mm for supplying the constant thermal load, which was attached on the outer wall of the evaporator. The condenser of the MHP is cooled by the water jacket with circulating water. Thermal grease (0.74 W/m °C) was filled between the pipe wall and the water jacket in order to minimize the thermal contact resistance.

The vacuum chamber (10^{-2}–10^{-3} torr) made of acryl was used to minimize the heat loss to the environment, as shown in Fig. 6. This vacuum chamber with a cylindrical body could be set rotationally for the inclination angle tests. To measure the wall temperature of the MHP, K-type thermocouples (Φ0.08 mm) were bonded by soldering at two points at the evaporator wall, one point at the adiabatic section, and two points at the condenser. The locations of the thermocouples are shown as X in Fig. 6 for the MHP with a total length of 50 mm. In the case of the 100 mm length, the distance between the locations of the thermocouples would be two times longer than that for the case of 50 mm length. The measured temperatures were recorded by using the data acquisition system.

The MHP is made of oxygen-free copper with wall thickness of between 0.27 and 0.28 mm. Pure water was used for the working fluid, which has relatively large surface tension and latent heat characteristic from 30–160 °C of the operating temperature. The filling ratio of the working fluid was 20% to the inner volume of the MHP. The liquid blocking region where the vapor could not be reached may be created at the condenser, the end of the MHP with 1–2 mm of equivalent diameter. The heat transfer by phase change of the working fluid could not be accomplished in the liquid blocking region. The temperature at the end of the condenser is 5–20 °C lower than that of the area adjacent to the condenser, and the heat transfer rate of the MHP is decreased due to the liquid blocking. Therefore, the inactivated amount of the working fluid due to liquid blocking should be considered at the design step of the MHP for calculating the filling ratio of the working fluid. The generation mechanism for the liquid blocking has not been reported in detail at any papers and still remained as an undeveloped field.

The experimental test was performed to investigate the thermal performance of the MHP. The operating temperature of the MHP identical with the temperature at the adiabatic section of the MHP was considered for the four cases - 60 °C, 70 °C, 80 °C, and 90 °C. The temperature and the amount of coolant circulating between the water jacket and the constant temperature bath were controlled carefully to maintain the conditions of the constant operating temperature of the MHP. The thermal load supplied to the MHP was increased stepwise by 0.5 W from 0.5 W. The wall temperature of the MHP was recorded at the steady state by each thermal load step. The measurement was stopped when the wall temperature of the evaporator of the MHP rapidly increased due to dry-out. The wall temperatures of the evaporator, the adiabatic section, and the condenser of the MHP were averaged in each zone. This test was measured in the chamber with vacuum condition of 10-2–10^{-3} torr. The results of the present study included errors in measurement, i.e., the tolerance in the heat supply (±0.05 V for voltage, ±0.01 A for current) and that in the temperature measurement (±0.1 °C). The performance testing apparatus of the FPMHP is similar to that of the circular type MHP.

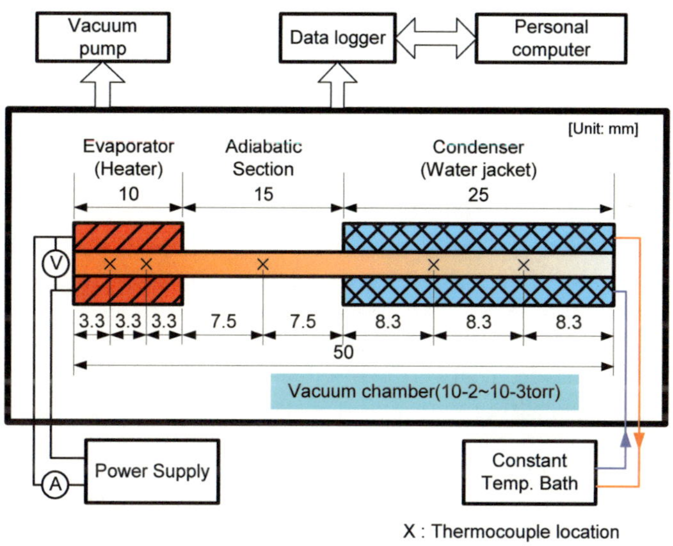

Fig. 6. Experimental apparatus

A heat pipe can transport a large amount of heat with a slight temperature difference between the evaporator and the condenser. In general, one of the test procedures to check whether a non-condensable gas exists in the heat pipe or a heat pipe is operated well just after the end of the manufacturing process is to measure the temperature difference between the evaporator and the condenser. In that case, however, we should remember that each heat pipe has a temperature difference as one's own. In the case of a small-sized heat pipe like the one in the present study, high-precision technologies are needed in the manufacturing process because the presence of non-condensable gases or contaminants, albeit small in amount, can be detrimental to the heat pipe performance.

Fig. 7 shows the temperature distribution by the axial length of 50 mm. The tested MHP has a curved triangular cross section and a 20% filling ratio to the inner total volume. The heat was dissipated only at the condenser with the conduction heat transfer. The temperatures were averaged over 60 seconds after steady state to reduce minor temperature fluctuation error. As shown in Fig. 7, the wall temperatures of the MHP are increased as the thermal load is increased. This means that the thermal equilibrium, which is the isothermal property of the MHP from the evaporator to the condenser, is well accomplished. The temperature differences between the evaporator and the condenser were 4.3-9.8 °C over the thermal loads of 0.5–4 W. However, the temperature difference of 9.8 °C between the evaporator and the condenser in the thermal load of 1 W is higher than that in other thermal loads. This is because the amount of latent heat to be transported toward the condenser is small by insufficient vaporization at the evaporator, and the thermal resistance is high by a relatively thick liquid film under low thermal load near 1 W.

Fig. 8 shows the temperature distribution by the axial length at the operating temperature of 90 °C, which equals the temperature at the adiabatic section. The tested MHP is the same as the one in Fig. 7. It is seen that the temperature difference between the evaporator and the condenser is increased as the thermal load is increased at the constant operating temperature of 90 °C. This can be explained by the fact that vapor flow velocity is increased as the thermal load is increased. Therefore, the friction force on the vapor-liquid interface and the pressure drop in the liquid flow are increased. Because the space for the vapor flow in the MHP is narrower than that in the conventional heat pipe, the pressure drop caused by the friction on the vapor-liquid interface may largely affect the MHP performance.

Fig. 9 shows the effect of the inclination angle on the thermal performance of the triangular MHP. In the figure, the negative inclination angle indicates a top heating mode in which the evaporator is located higher than the condenser, and conversely, the positive inclination angle means a bottom heating mode in which the evaporator is located lower than the condenser. As shown in Fig. 9, the effect of the inclination angle on the thermal performance is small. The thermal performance of the MHP was almost the same for the tilting mode from the horizontal mode to the top heating mode with -90 degrees. However, there was a decrease in the thermal performance of the MHP as the inclination of the MHP was rotated from the bottom heating mode to the top heating mode. That is, the thermal performance of the triangular MHP was very stable in the bottom heating mode, and it is seen that the capillary force of the working fluid was enough to flow from the condenser to the evaporator. The triangular MHP has the limiting power of 4.51 W at the top heating mode of -90 degrees.

Fig. 10 shows the overall heat transfer coefficient according to the total length of the triangular MHP. The considered lengths of the MHP were 50 mm and 100 mm. In the case of the MHP with a small-sized equivalent diameter smaller than 2 mm, the effect of the pipe length on the thermal performance of the MHP could be large. This is due to the fact that the pressure losses by friction at the vapor-liquid interface and the capillary limitation for returning the condensed liquid are significantly dominant as the pipe length is increased. As shown in Fig. 6, the thermal performance of the triangular MHP tends to be increased according to the decrease in the pipe length. In the case of the triangular MHP, the overall heat transfer coefficient was enhanced about 92% when the total length was decreased from 100 mm to 50 mm for the thermal load of 3 W. In the future, more detailed experimental results for the effect of the pipe length on the thermal performance of the MHP will be studied.

A dry-out state occurs when the temperature at the evaporator's lowest end for the bottom heating mode is abruptly increased compared to other temperatures at the evaporator. The

thermal load just prior to the state in which heat transfer by phase change can no longer be conducted due to a dry-out in the evaporator is defined as the heat transfer limit.
Fig. 11 shows the thermal resistances and heat transfer limits for the triangular MHP and the rectangular MHP. The thermal resistance can be calculated by equation (2):

$$R = \frac{T_e - T_c}{Q} \tag{2}$$

Where T_e and T_c are the wall temperatures at the evaporator and the condenser of the MHP, respectively, and Q(W) is a thermal load at the evaporator.
The tested MHP has a fill ratio of 20% to the internal total volume of the MHP. The operating temperature is not constant, but increases as the thermal load is increased. The heat dissipating at the condenser of the MHP was accomplished by circulating 20 °C water, which was controlled by a constant temperature bath. As shown in Fig. 11, the heat transfer limit of the triangular MHP is 1.6 times larger than that of the rectangular MHP. The heat transfer limits were 4.5 W and 7 W for the rectangular MHP and the triangular MHP, respectively. This is because the corners for the rectangular MHP are not developed sharply compared to that for the triangular MHP and the capillary force needed for returning the condensed liquid to the evaporator cannot be obtained sufficiently.
The property of the rectangular MHP having one additional corner than the triangular MHP may make an advantage in its thermal performance. However, because a radius of curvature at the corner is not sufficiently small to retain capillary pressure, the performance of the rectangular MHP cannot be superior to that of the triangular MHP. The performance of the MHP is largely restricted by the capillary limit.
The factor that mainly affects the capillary limit is the radius of curvature at the corner. The radius of curvature (r) is a function of a corner aperture angle (ϕ), a contact angle (α), and a location of the meniscus contact point in the corner (χ), as shown in equation (3). Except for the two factors of α and χ, which is constant as a heat pipe type, the operating condition is the most important factor to the radius of curvature. The smaller the corner aperture angle is, the smaller the radius of curvature is. Then, high capillary pumping pressure and high maximum heat transfer limit can be obtained under this condition.

$$r = f(\phi, \alpha, \chi) \tag{3}$$

The corner aperture angle of 60–70 degrees in the rectangular MHP of the present study is larger than that of the angle 30–40 degrees in the analytical result of Zaghdoudi (Zaghdoudi et al., 1997); therefore, the small radius of curvature cannot be obtained.
Fig. 12 presents the experimental thermal resistance and heat transfer limit of the triangular MHP for various operating temperatures. The tested MHP has a fill ratio of 20% and was performed for the various operating temperature of 60 °C, 70 °C, 80 °C, and 90 °C. As shown in Fig. 8, the heat transfer limit is a function of the operating temperature; it increases as the operating temperature is increased. The heat transfer limits were 6.18, 7.59, 8.01, and 10 W for the operating temperatures of 60 °C, 70 °C, 80 °C, and 90 °C, respectively.
Fig. 13 presents the experimental results of the present study and those of Moon (Moon et al., 1999) for the heat transfer limit. The MHP tested in Moon et al. had a curved triangular cross section and a stainless steel as container material. Pure water was used as working fluid. As shown in Fig. 13, the heat transfer limit of the present study is 1.7–2.1 times larger than that of

Moon et al. over the operating temperature of 60–80 °C. This result shows that large capillary limit was obtained in the present study compared to that in Moon (Moon et al., 1999). High productivity and simple manufacturing process were considered, and enhanced performance was obtained compared to that of Moon et al. for the future applications.

Figs. 14 and 15 show the performance test results of temperature distributions along with the length of the FPMHP. In both figures, it can be seen that the temperatures of the condenser instantly follow the temperature of the evaporator for overall thermal loads. This phenomenon means that the FPMHPs have good constant temperature characteristics as a heat pipe. The temperature differences between the evaporator and the condenser are 2.5–6.4 °C for the FPMHP with fill ratio of 25% and 2.2–11.9 °C for the one with fill ratio of 15%, respectively.

When the temperature of the evaporator was considered within the temperature of 120 °C, Table 3 shows the heat transfer rate of the FPMHP with fill ratio of 20%. In the table, the heat transfer rate of the rectangular FPMHP is higher than that of other FPMHPs. The heat transfer rate of the modified rectangular FPMHP is as high as that of the rectangular FPMHP is due to the capillary force of the former being higher than that of the latter. Meanwhile, the heat transfer rate of the triangular FPMHP is lower than that of other FPMHPs due to its characteristic of having small space for vapor flow.

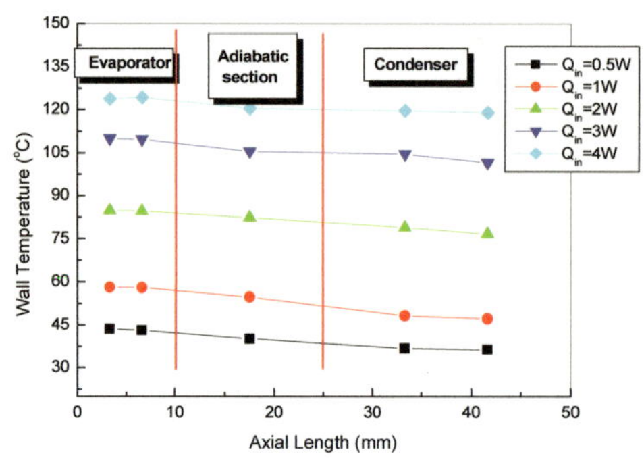

Fig. 7. Wall temperature distribution along the longitudinal axis by thermal loads

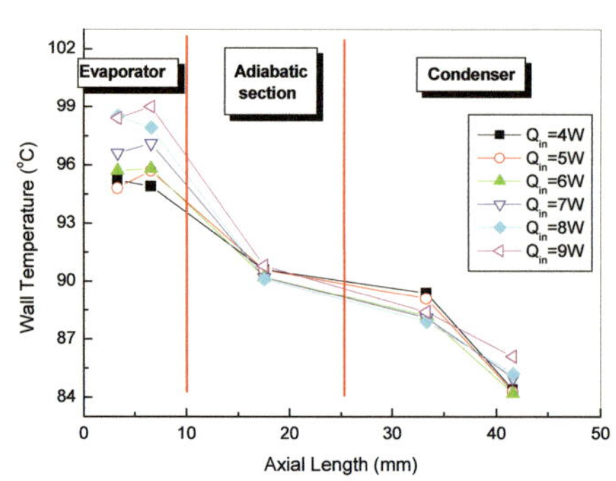

Fig. 8. Wall temperature distribution along the longitudinal axis of the MHP at Tv = 90 °C

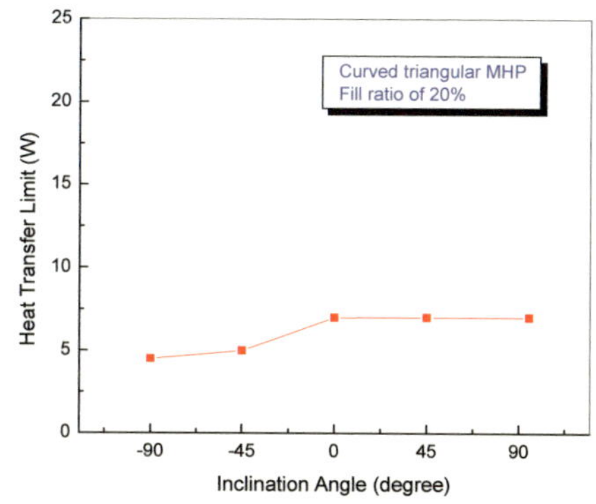

Fig. 9. Thermal performance by inclination angle

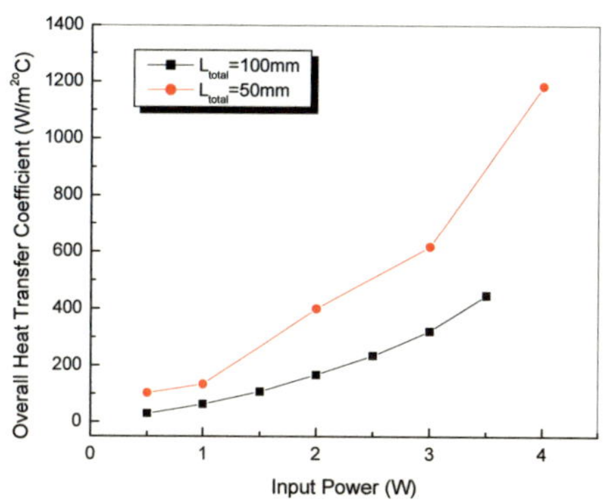

Fig. 10. Overall heat transfer coefficient by total length

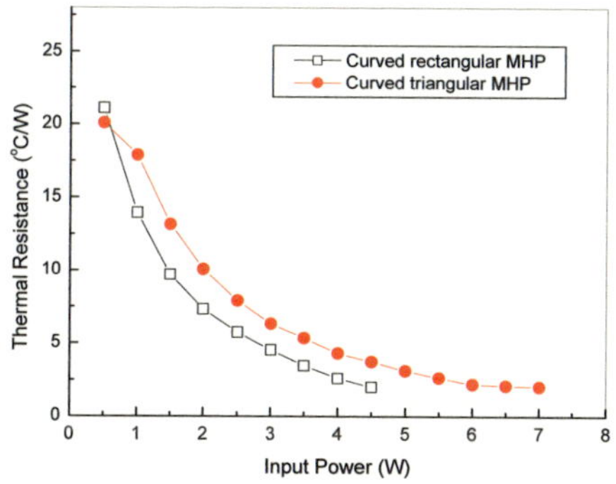

Fig. 11. Performance comparison by cross-section type

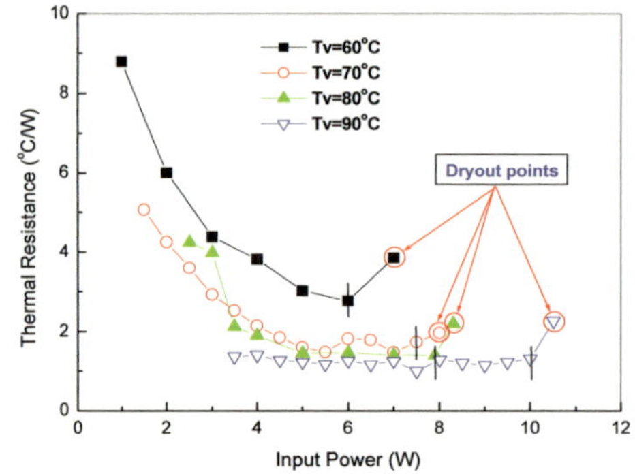

Fig. 12. Thermal performance by the operating temperature

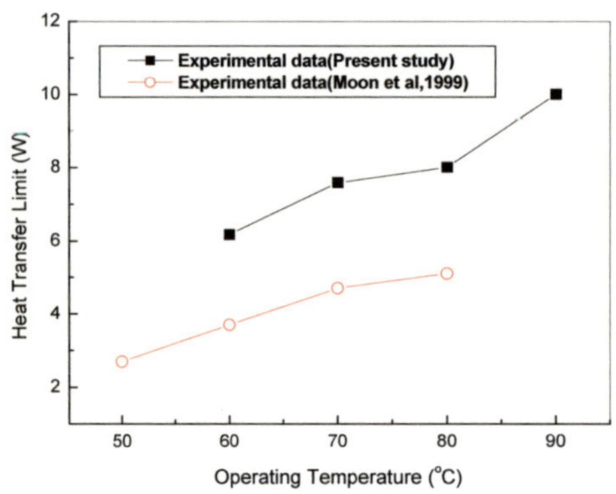

Fig. 13. Experimental results comparison between the present study and Moon et al.

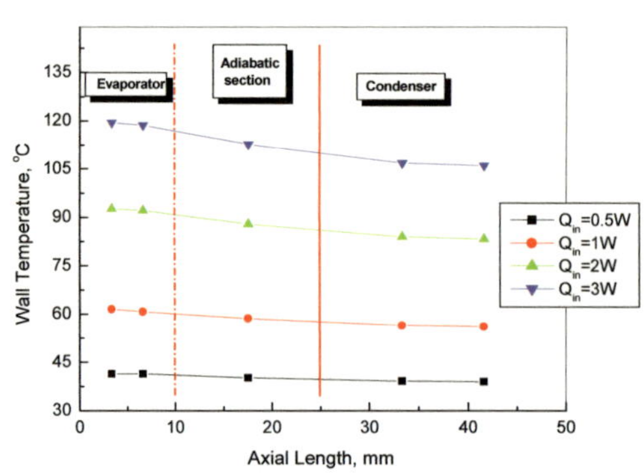

Fig. 14. Temperature distributions along FPMHP with fill ratio of 25% with the axial length for rectangular

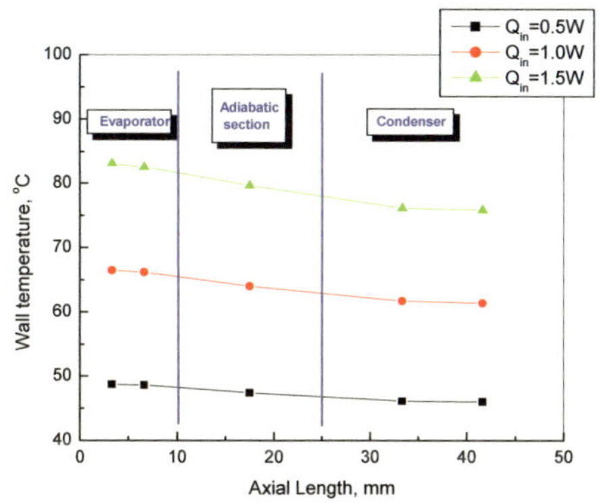

Fig. 15. Temperature distributions along with the axial length for rectangular FPMHP with fill ratio of 15%

	Heat transfer rate
Rectangular	13.66 W
Modified rectangular	13 W
Triangular	8 W

Table 3. Heat transfer rate of each FPMHP with fill ratio of 20%

3.4 Why micro CPL is needed in the future

Vapor and liquid have counter flow pattern in the heat pipe. Therefore, pressure drop on the vapor-liquid interface is created, which then leads to a decrease in the heat transfer performance of the heat pipe. Fig. 16 shows the effect of shear force on the vapor-liquid interface to the maximum heat transport capacity. The result shows that the case wherein no shear force is considered is about twice the maximum heat transport capacity of that which considers shear force. From the result, we can realize that the shear force on the vapor-liquid interface significantly affects the performance of the small-sized heat pipe. The micro CPL could be considered as an alternative solution. It has separated vapor and liquid flow path compared to the MHP. Therefore, when there is no shear force on the vapor-liquid interface, then pressure drop by shear force cannot occur. However, the successful fabrication of small-sized micro CPL with flat plate shape shows that a normal operating characteristic is not easy.

4. Micro CPL

4.1 Design of the micro CPL

Considering the tendency of portable electronic and communication devices to become thinner and thinner, flat plate type cooling devices offer great convenience to be applied to such devices compared with circular type cooling devices. In the present, the flat heat pipe pressed from a circular shape has been widely used in notebook PCs, sub-notebook PCs,

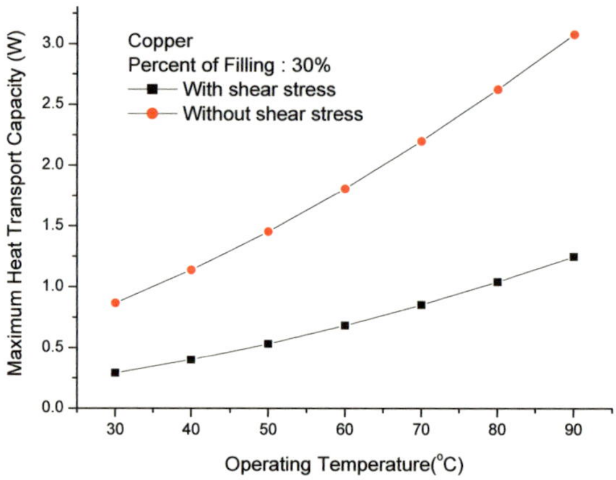

Fig. 16. Effects of shear force on maximum heat transport capacity

game machines, etc. However, the limit in thickness reduces the heat transfer capacity of the pressed circular type heat pipe. In addition, the thermal resistance of such pipe was greatly increased with the pressed thickness limit of 2 mm, caused by the tendency of the central part of the circular type heat pipe to be depressed when it is pressed. Another reason is the fact that it is not easy to secure sufficient space for vapor flow in a pressed heat pipe with the wick. Therefore, thin flat plate type cooling devices are needed for the thermal management of electronic packaging with limited inner space.

Fig. 17 shows the micro CPL designed in the present study. The micro CPL was designed to have flat plate shape and was composed of three layers - a bottom silicon layer, a middle glass layer, and a top glass layer. The structural characteristics of the micro CPL are as follows. Contrary to the structure of a conventional heat pipe, the micro CPL is fabricated by stacking several plates. The plates with evaporator, condenser, vapour, and liquid path are fabricated by each fabrication process and then integrated in order to construct one envelope for the micro CPL. By designing the evaporator in the bottom silicon plate to have two-step structures, the large vapor flow space in it could be secured in the limited inner space of the micro CPL. By designing the upper surface of the two-step structure to interconnect with the liquid line vertically, the backflow of the vapor bubbles could be protected. The liquid flow path at the condenser was designed to have zigzag pattern and cross section, which is reduced gradually to prevent the flow pressure drop from changing abruptly in the path and to condense the vapor sufficiently. Meanwhile, the grooves in the upper glass plate were designed to have relatively large width to achieve dispersion effect of the condensed liquid on it. The bottom silicon layer was constructed out of evaporator, condenser, reservoir, and heat blocks that are placed between the evaporator and the condenser. The heat blocks reduce heat conduction via the wall and calculate the pure heat transfer rate by the vapor-liquid phase change. The fine grooves of the two-step structure were constructed in parallel at the evaporator. The middle glass layer was composed of the vapor line and vapor flow space at the part corresponding to the evaporator. The liquid lines constructed in the middle glass layer meet vertically with the upper surface of the two-step grooves at the evaporator.

The operating principle of the micro CPL is slightly different from that of the heat pipe. First, the vapor created by the input power at the evaporator is transferred toward the condenser via the vapor line that is placed in the center of the micro CPL and has larger space than the liquid line. The vapor is condensed at the condenser by releasing the heat to the environment and the condensed liquid is returned toward the evaporator via the two liquid lines that are placed at both sides of the micro CPL. The vapor and the liquid circulate continuously. The working fluid was supplied through a hole located at the top of the reservoir. Fig. 18 shows the operating principle of the micro CPL, and Table 4 shows the specification of the one designed in the present study. The micro CPL is 1.5 mm thick with each plate having thickness of 0.5 mm. Considering thermal performance, the micro CPL should be thick; however, considering its application, it should be thin. In this study, the micro CPL was designed to be thin, but its thickness was determined by considering a minimum demand condition in the fabrication process. When the length of the small heat pipe exceeded 100 mm, its thermal performance significantly decreased. Hence, the total length of 50-70 mm was considered within 100 mm. Micro CPLs that are 50-70 mm long basically have the same structure, except that the length of their evaporator is 10 mm and 15 mm, respectively.

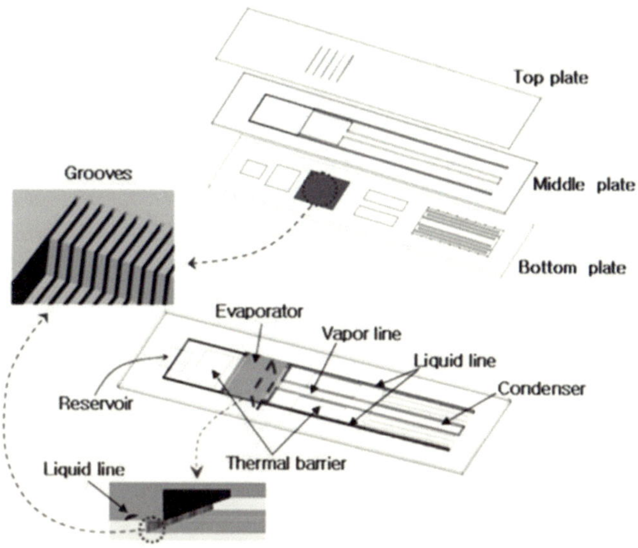

Fig. 17. Detailed structure of the flat plate type micro CPL

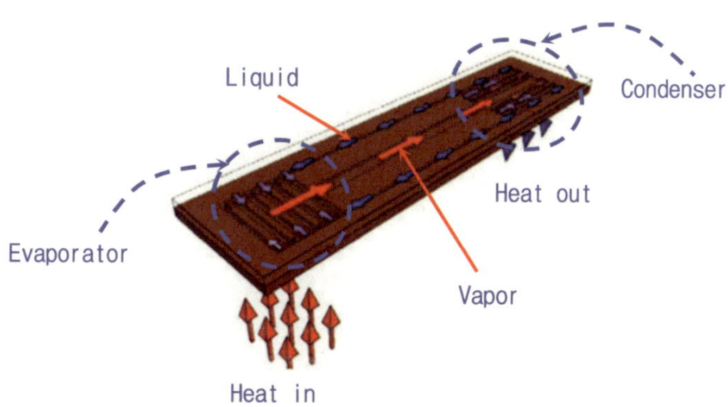

Fig. 18. An operating principle of the flat plate type micro CPL

Section	50mm length	70mm length
Evaporator Area	9*10mm	
Groove Width/Height	20*60μm	
Groove Number	167	
Vapor Line Height/Width	500*2000μm	
Vapor Line Length	20mm	40 mm
Liquid Line Height/Width	500*250μm	
Liquid Line Length	35mm	55mm
Condenser Area	10*20mm	

Table 4. Specifications of the flat plate type micro CPL

4.2 Fabrication of the micro CPL using MEMS

Silicon was used for the bottom plate of the micro CPL, while glass was used for the middle and top plates. The silicon bottom plate, which has an evaporator with two-step grooves, a condenser, and a reservoir, was fabricated through the deep reactive ion etch (DRIE) process. The glass middle plate, which consists of the vapor space at the evaporator, the vapor line, and the feed line for supplying the working fluid into the envelope, was fabricated by wet etching and sanding processes. Meanwhile, the glass top plate with grooves for distributing the liquid at the evaporator was fabricated by the wet etching process. Fig. 19 illustrates the fabrication process flow.

The detailed fabrication processes are as follows. First, the 3 μm oxide, as the masking layer, was deposited on the silicon bottom wafer through wet thermal oxidation. After the PR patterning and dry etching of the oxide layer, the oxide of 8,000Å – as the second masking layer – was deposited and patterned. Afterwards, first-step grooves (Fig. 19(a)) were etched to the depth of 60 μm by the first DRIE process. After removing the remaining PR, the second-step grooves were etched to the depth of 200 μm by the second DRIE process. The poly-silicon of 5,000 Å – as the masking layer – was deposited on the glass middle wafer to fabricate the feed line with relatively small cross-section area. After the PR patterning and dry etching of the poly-silicon layer, the grooves were etched to the depth of 60 μm by wet etching. After removing the remaining poly-silicon, the dry film resist (DFR) film was attached on the glass wafer for use as masking layer. The vapor line penetrating the glass wafer was fabricated by the sanding process (Figs. 19(b) and (c)). The grooves were then etched to the depth of 60 μm by wet etching at the evaporator in the top wafer, which is similar to the fabrication process of the middle wafer. After the grooves were fabricated, the DFR film was attached on the top glass wafer to fabricate the through hole. After patterning the DFR film, the through hole was then fabricated through the sanding process. The connection of the fill tube on the top of the glass wafer was one of the most difficult fabricating processes in the present study. The copper material was used for the fill tube like that of the general heat pipe. A fragment silicon plate on which the metal layer of circular bands shape was deposited was used for bonding with the fill tube. The metal layer has a combination of Ti/Cu/Ni. Ni of 5 μm was deposited through electroplating to guarantee that the fill tube and the metal layer are bonded firmly. The fragment silicon plate was bonded on the top glass wafer of the micro CPL by anodic bonding process. The bottom silicon wafer was bonded with the middle glass wafer by the anodic bonding process as well. The top glass wafer was bonded on the middle and bottom wafers by direct bonding process. Then the envelope for the micro CPL was completed (Fig. 20(a)). Filling the working fluid into the envelope was accomplished by the conventional method of filling after degassing. Pure water was used as working fluid. The working fluid was filled until all grooves and the liquid lines in the micro CPL were saturated; the amount of working fluid was weighed and recorded about 0.0854 g.

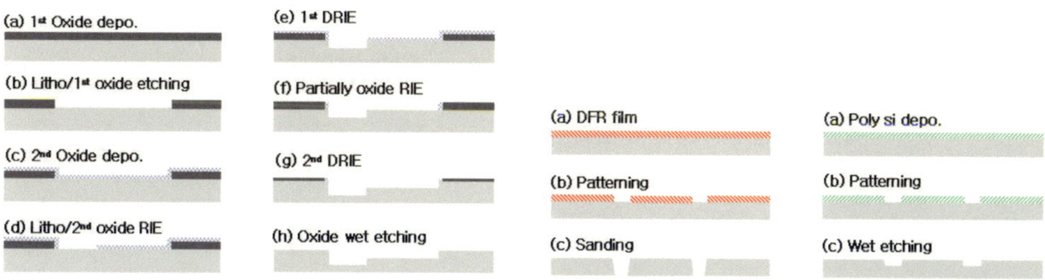

(a) (b) (c)

Fig. 19. The fabrication process of the flat plate type micro CPL

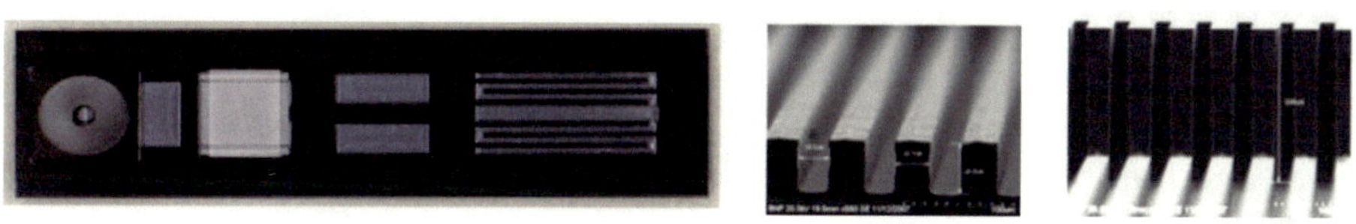

(a) (b) (c)

Fig. 20. Fabricated micro CPL: (a) top view of the micro CPL; (b) SEM images of grooves; and (c) SEM images of two-step grooves

4.3 Thermal performance of the micro CPL

Fig. 21 shows the experimental apparatus used to investigate the isothermal characteristics and the heat transfer rate of the fabricated micro CPL. The apparatus was composed of a vacuum chamber in which the micro CPL was placed, a data acquisition system, and a constant temperature water bath. Heat was supplied into the micro CPL electrically by DC power supply (HPS 60100 model). A heater fabricated with Ni-Cr (98.8 Ω/m) wire with diameter of 0.08 mm was attached on the backside of the evaporator of the bottom silicon plate. The T-type thermocouples were used in order to measure the wall temperature; they were attached by epoxy to two locations on the evaporator, one location on the vapor and the liquid line, and two locations on the condenser. The heat dissipated to the environment was controlled by a liquid cooling system using water jacket that was attached on the backside of the condenser of the bottom silicon plate. The temperature and the flow rate of the cooling liquid could be controlled by the constant temperature water bath. The performance test for the fabricated micro CPL was done in a vacuum chamber in order to minimize the heat loss to the environment. The vacuum chamber was maintained to the degree of a vacuum of $2*10^{-1}$ torr during the test by continuous vacuum pumping. The thermal load was supplied to the evaporator by steps of 1 or 0.5 W, starting with an initial power level of 1 W. The wall temperatures of the micro CPL in each power level were recorded when the temperatures of the wall reached a steady state. Meanwhile, visual inspection was conducted through the top glass plate in order to understand the operating characteristics of the micro CPL. The experimental results of the present study might include error rate in the measurement, i.e., the error rate in the measurement of the heat supply (±0.05 V for voltage, ±0.01 A for current) and temperature (±0.1 °C).

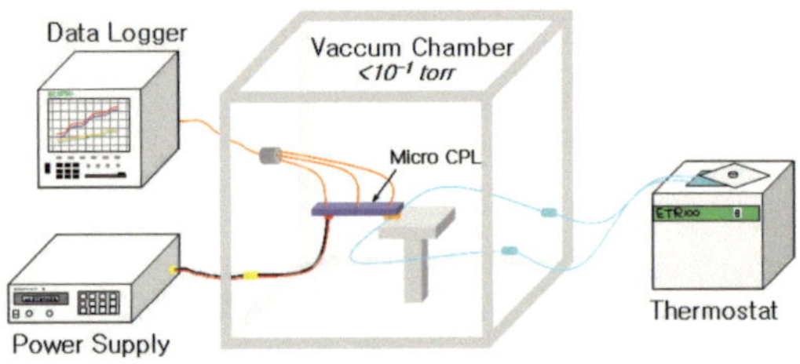

(a)

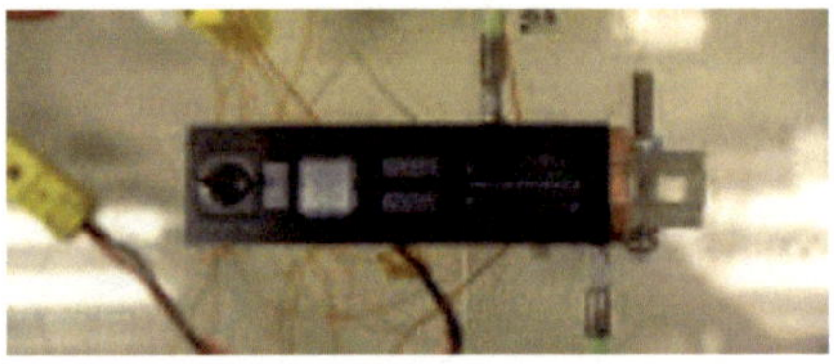

(b)

Fig. 21. Experimental apparatus: (a) performance test system and (b) top view of the micro CPL set-up

Considering its application in portable electronic devices, the micro CPL in the present study was designed to have flat plate shape. Nevertheless, the flow pressure drop in the fluid flow path of the micro CPL may be increased because its evaporator, vapor and liquid lines, and condenser were placed in the same plane as the flat plate with thickness of 1.5 mm. The fluid flow path with reduced cross section has significant effects on the decrease in the thermal performance of small heat pipes, which depend on the flow pressure drop. For a normal operation of the micro CPL, as shown in equation (4), the total pressure drop through the loop must be less than the capillary pressure limit created at the evaporator.

$$\Delta P_c \geq \Delta P_e + \Delta P_v + \Delta P_{cn} + \Delta P_l \tag{4}$$

Where ΔP_e, ΔP_v, ΔP_{cn} and ΔP_l are pressure drops at the evaporator, vapor line, condenser, and liquid line, respectively. Fig. 22 shows the experimental results for the isothermal characteristic of micro CPLs with total length of 50 mm and 70 mm. The comparison results for the isothermal characteristic between the micro CPL with working fluid and the one without working fluid are shown in Fig. 22.

One of the methods used to evaluate the performance of the micro CPL in the present study is by measuring the thermal resistance, R (°C/W), which is defined in equation (5):

$$R = \frac{\overline{T_e} - \overline{T_c}}{Q} \tag{5}$$

Where $\overline{T_e}$ is the mean wall temperature at the evaporator, $\overline{T_c}$ is the mean wall temperature at the condenser of the micro CPL, and Q is the thermal load (W) imposed on the evaporator.

The test condition in Fig. 22 was the weak heat dissipation at the condenser. That is, the cooling water was not circulated at the condenser in order to investigate only the normal operating characteristic of the micro CPL by phase change of the working fluid. In the case of the heat pipe with the mechanism of vapor-liquid phase change, the heat pipe shows isothermal characteristics which transfer a lot of heat in small temperature difference between the evaporator and the condenser. Therefore, the normal operating state could be confirmed by measuring the temperature difference between the evaporator and the condenser when small power is input to the evaporator. In Fig. 22, the micro CPL with working fluid shows lower thermal resistance than the micro CPL without working fluid in both cases of total length being 50 mm and 70 mm. This means that the fabricated micro CPL in the present study operates normally through the operating mechanism of vapor-liquid phase change. In the case of the total length of 50 mm, the micro CPL with working fluid shows lower thermal resistance about half of that of the micro CPL without working fluid. In the case of the total length of 70 mm, the micro CPL with working fluid shows lower thermal resistance about a third of that of the micro CPL without working fluid. This means that although the total length is increased from 50 mm to 70 mm, the micro CPL with working fluid operates normally by vapor-liquid phase change. However, the thermal resistance of the micro CPL increased when the total length was changed from 50 mm to 70 mm.

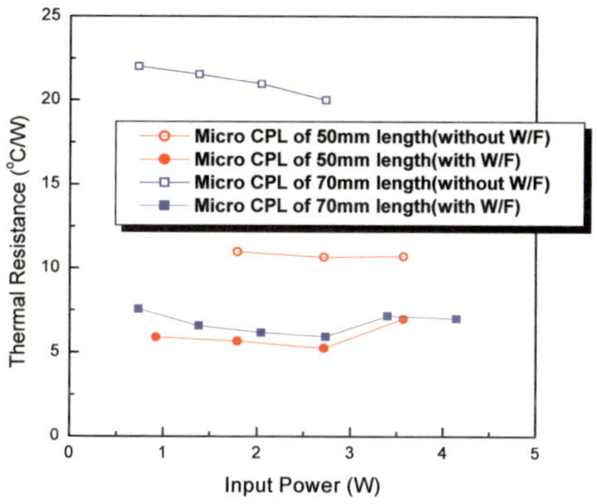

Fig. 22. Comparison of thermal resistance between the flat plate type micro CPL with working fluid and the one without working fluid

Fig. 23 shows the comparison results for the heat transfer rate between micro CPLs with working fluid with total length of 50 mm and 70 mm. In Fig. 23, the input power was not the maximum heat transfer rate; the heat transfer rate supplied to the evaporator was within the wall temperature of 120 °C at the evaporator. In the figure, the heat at the condenser was dissipated to the environment by the circulation of the cooling water. Through this experiment test, it the amount of heat that can be transferred by the fabricated micro CPL within the limited evaporator temperature could be investigated. The heat transfer rate of 7.5 W was obtained within the thermal resistance range of 6.8–19.9 °C/W in the case of the total length of 50 mm. Meanwhile, the heat transfer rate of 6.1 W was obtained within the thermal resistance range of 11.7–19.2 °C/W in the case of the total length of 70 mm. The thermal resistance increased and the heat transfer rate decreased when the total length was increased from 50 mm to 70 mm. The operating mechanism of the flat plate micro CPL developed in the present study was not known in detail. Furthermore, the amount of working fluid and the structural design of the micro CPL were not optimized, therefore further study is needed in the future.

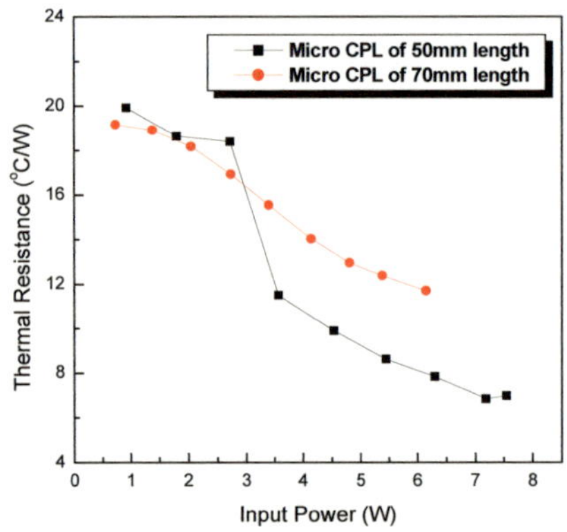

Fig. 23. Heat transfer rate according to increasing input power

4.4 Flow visualization of the micro CPL

Fig. 24 shows some images obtained by the visual inspection. They were captured on arbitrary time while the micro CPL is operating. Figs. 24(b), (c), and (d) show the fluid flow patterns in the path of the condenser. The fluid flow patterns in the micro CPL were very active during the time the results of Fig. 24 are being obtained. Although any change in the evaporator and the vapor line filled with vapor could not be seen with the naked eye, we

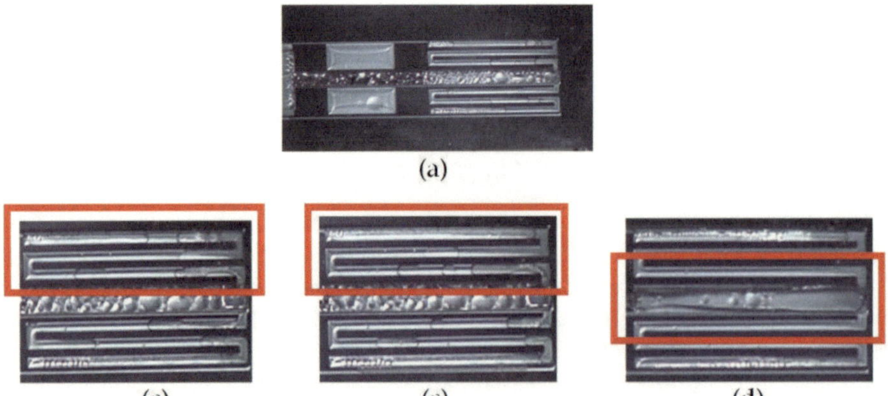

Fig. 24. Flow patterns at the condenser: (a) top view of the condenser; (b) (c) plug flow patterns on low or middle heat flux (1-6 W), respectively; (d) annular flow pattern on high heat flux (over 7 W)

could see the fluid flow phenomenon wherein the liquid and non-condensed vapor flow together. An undesirable phenomenon wherein the vapor transported from the evaporator was condensed on the top and bottom walls in the vapor line was observed with the naked eye. The activity of two-phase flow patterns increases as the input power supplied to the evaporator is increased. The fluid flow pattern was plug flow, wherein the vapor and liquid bridge move in order, in low power (1-3 W) and middle power (4-6 W). The fluid flow pattern changed from being plug flow to annular flow in high power (7-7.5 W). The plug flow in the middle power range has larger velocity than that in the low power range. The micro CPL shows the continuous circulating flow pattern over the entire power range. The liquid drops created on the bottom and top walls at the vapor line should be removed since they may increase the pressure drop in the vapor flow.

5. Commercialization of the MHP and micro CPL

The tubular type MHP, which was considered in chapter 3, can be used in any applications and may also be packaged for high heat flux applications. The FPMHP, which was fabricated by Al extrusion, was designed with consideration of capillary force. However, for the purpose of the commercialization of the FPMHP, not only should the capillary force be considered, but also the securing of the inner space. Furthermore, the fabrication cost and fabrication process limit should also be considered. Fig. 24 shows a commercialized model of the FPMHP, which is designed with consideration of the commercial viewpoint. It may be applied to various fields like display, electronic package, automobile, and optic industry.
Flat plate micro CPL, which was considered in chapter 5, may be applied to slim mobile electronic devices. The fabrication of a structure similar to design can be obtained. However, for wider application, the fabrication of micro CPL using metal, instead of silicon and glass, is needed. The cost and process of fabrication should be considered for commercialization as well. Fig. 25 shows a micro CPL model fabricated by metal for commercialization. It is composed of only two layers, compared to that considered in chapter 5 which has three layers. The most important factor is reserving the inner space for fluid flow in the case of the commercialized model shown in Fig. 25, which has thickness of less than 1 mm and is composed of only two layers.

Fig. 25. FPMHP considering commercialization

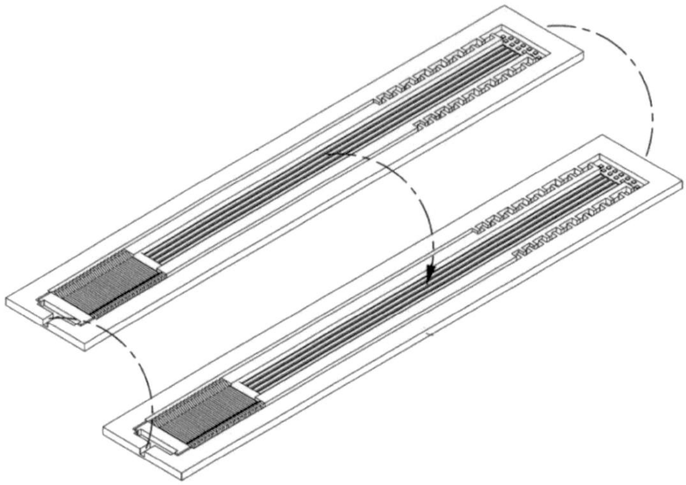

Fig. 26. Micro CPL considering commercialization

6. Conclusions

The characteristics, design, fabrication and thermal performance of MHPs and micro CPLs were investigated.
Firstly, MHPs with polygonal cross section applicable to electronic units with thin structure were manufactured and tested. The high productivity and simple manufacturing process were also considered for future applications. The manufactured MHP showed good isothermal property over the total length, and the temperature difference between the evaporator and the condenser was about 4–6 °C. The inclination angle had a slight effect on the thermal performance, and the thermal characteristic was stable from the top heating mode to the bottom heating mode. The effect of the total pipe length on the thermal performance of the triangular MHP was dominant. In the case of the triangular MHP, the overall heat transfer coefficient was enhanced by about 92% when the total length was decreased from 100 mm to 50 mm for 3 W of thermal load. The heat transfer limit of the triangular MHP was 7 W, which is 1.6 times larger than the 4.5 W heat transfer limit of the rectangular MHP. The heat transfer limit, which was the function of the operating temperature, increased when the operating temperature was increased. The maximum heat transfer limit of the triangular MHP was 10 W for the operating temperature of 90 °C. In the present study, the heat transfer limit was 1.7–2.1 times larger than that of Moon (Moon et al., 1999) for the operating temperature of 60–80 °C. The manufactured MHP in the present study exhibited superior heat dissipation capacity and thus can be widely used in integrated electronic units as a cooling module.
Secondly, the flat plate type micro CPL with thickness of 1.5 mm was designed, and its fabrication technology was developed through the present study. The micro CPL was designed to have an evaporator, a vapor line, two liquid lines, and a condenser in flat plate shape, ensuring a large space for the vapor flow. In particular, the evaporator was designed to have two-step grooves in order to secure the space for the vapor flow and prevent the backflow of bubbles. The individual fabrication processes technologies for each plate of

silicon and glass were developed. Particularly, the bonding technology of the fill tube on the glass top plate was completed by the fragment silicon on which the circular type metal bands were deposited. The filling technology of the working fluid into the micro CPL under vacuum condition was completed by the conventional method of filling after vacuuming. Through the performance tests for the fabricated micro CPLs with total length of 50 mm and 70 mm, it was confirmed that micro CPLs operate normally through the phase-change heat transfer of the vapor and liquid. The thermal resistance of the micro CPL increased and the heat transfer rate decreased within the wall temperature of 120 °C at the evaporator when the total length increased from 50 mm to 70 mm. Through the visual study, it was observed that the fluid flow pattern of the micro CPL was plug flow in the low (1-3 W) and middle (4-6 W) power, and annular flow in the high power (over 7 W). The velocity of the fluid flow increased according to the input power. Further study on determining the operating mechanism of the flat plate type micro CPL and optimizing the structural design is needed in the future.

7. References

A. Faghri, "Heat Pipe Science and Technology," *Talor & Francis*, 1995

A. Hoelke, et al., "Analysis of the Heat Transfer Capacity of a Micromachined Loop Heat Pipe," *ASME 1999*, Vol. 3, 1999, pp.53-60

B. R. Babin, et al., "Steady-State Modeling and Testing of a Micro Heat Pipe," *ASME J. of Heat Transfer*, Vol. 112, No. 3, August, pp. 595~601, 1990

D. Wu, et al., "Investigation of the Transient Characteristics of a Micro Heat Pipe," *AIAA J. Thermophysics Heat Transfer*, 5(2), April, pp. 129~134, 1991

F. M. Gerner, "Flow Limitation in Micro Heat Pipes," *AFSOR Final Report*, No. F49620-88-6-0053, Wright-Patterson, AFB, Dayton, OH, 1989

G. P. Peterson, "An Introduction to Heat Pipes: Modeling, Testing and Applications," *Wiley: New York*, NY, 1994

H. Xie, et al., "The Use of Heat Pipes in the Cooling of Portables with High Power Packages," *Thermacore Co., Technical Note*

J. Kirshberg, et al., "Cooling Effect of a MEMS Based Micro Capillary Pumped Loop for Chip-Level Temperature Control," *ASME 2000*, MEMS Vol.2, 2000, pp.143-150

J. S. Suh, et al., "Friction in Micro-Channel Flows of a Liquid and Vapor in Trapezoidal and Sinusoidal Grooves," *Int. J. of Heat & Mass Transfer*, Vol. 44, 2001, pp.3103-3109

K. S. Kim, S. H. Moon, C. G. Choi, "Cooling Characteristics of Miniature Heat Pipes with Woven-Wired Wick," *11th Int. Heat Pipe Conf.*, Japan, Sep. 1999

L. Meyer, et al., "A Silicon-Carbide Micro-Capillary Pumped Loop for Cooling High Power Devices," *19th IEEE Semi-Therm Symp.*, 2003, pp.364-368

M. C. Zaghdoudi, et al., "Theoretical Investigation of Micro Heat Pipes Performance," *10th Int. Heat Pipe Conf.*, Germany, Sep. 21-25, F-9, 1997

R. Hopkins, et al., "Flat Miniature Heat Pipe with Micro Capillary Grooves," *Transaction of the ASME*, Vol. 121, pp. 102-109, 1999

S. H. Moon, G. Hwang, H. G. Yun, T. G. Choy, "Operation Performance of Miniature Heat Pipe with Composite Wire Wick," *IMAPS 2001*, pp. 207-211, 2001

S. H. Moon, G. Hwang, H. G. Yun, "Improving Thermal Performance of Miniature Heat Pipe for Notebook PC Cooling," *Microelectronic Reliability*, Vol.42, No.1, 2002

S. H. Moon, et al., "An Experimental Study on The Performance Limitation of a Micro Heat Pipe with Triangular cross-section," *11th Int. Heat Pipe Conf.*, Japan, Sep. 1999

S. H. Moon, et al., "Heat Transport Performance of Micro Heat Pipe with Cross Section of Polygon," *IMAPAS 2002, Int. Symposium on Microelectronics*, Session WP4, 2002

S. H. Moon, et al., "Manufacturing and Thermal Performance of the Flat Plate Micro Heat Pipe," *IMAPS ATW on Thermal Management for High Performance Computing Telcom/Wireless*, 2002

T.P. Cotter, "Principles and Prospects for Micro Heat Pipes", *Proceedings of the 5th Int. Heat Pipe Conference*, 1984.

The Investigation of Influence Polyisobutilene Additions to Kerosene at the Efficiency of Combustion

V.D. Gaponov[1], V.K. Chvanov[1], I.Y. Fatuev[1], I.N. Borovik[2], A.G. Vorobiev[2], A.A. Kozlov[2], I.A. Lepeshinsky[2], Istomin E.A.[2] and Reshetnikov V.A.[2]
[1]*OAO "NPO Energomash"*
[2]*Moscow Aviation institute (State Technical University)*
Russia

1. Introduction

Liquid rocket engines reached high efficiency at presence. Next improvement of they energetic, mass and reliabilities characteristics is labor-intensive and high expensive process. It is famous, that addition of polymers to carbonhydogen fuels decrease substantially hydraulic losses at the friction in pipelines and aggregates of engines. Fulfilled in "NPO Energomash" the programme investigation influence of additions polyisobutilene to kerosene at the hydraulic tests exploitated engines was showed, that the decrease of the hydraulic losses may be more 20% [7]. The use of this effect lets or increase pressure in the combustion chamber at constant heat intensivity of the turbine or to increase the resource of the engine at the base decrease heat intensivity of turbine.

The question regarding influence addition at the combustion efficiency stated not clear. This investigation for full-sized engines though doesn't required fabrication new material part, but is completed and expensive process analogically fire tests of the engine.

Most likely mechanism influence of addition at combustion efficiency may be pulverization of liquid fuel. The program investigation of this mechanism was developed at department 202 of MAI [3]. This program included two steps. The first step was directed at obtaining characteristics of pulverization one from mixed head liquid rocket engine of small thrust MAI-202K, working at kerosene and gaseous oxygen. Characteristics of pulverization of the mixing head at clean kerosene and kerosene with additions were diagnosed by dispersal of drops, obtained at automatically measurement system. Method dispersal measurement was based at change intensify projecting at the screen reflected from drops laser ray.

The second step consist fire tests of engine MAI-202 with seven swirl injectors mixing head and oxygen curtain. Tests were fulfilled at the fire stand of department 202 MAI, at the same regime of work, but at different fuels: clean kerosene and kerosene with additions 0.05-0.01% polyisobutilen.

In the article detail materials are introduced about results as hydraulic, so and fire tests, measured equipment, design of mixing head, characteristics of pipeline. Combustion efficiency was obtained as ratio of experimental value mass flow complex β^{exp} to thermodynamic value mass flow complex β^{t}.

2. Composition and structure of test stand

Experimental investigation of influence 0.05% polyisobutilene additions to kerosene was fulfilled at the test-bad № 72-2 department 202 MAI for fire tests liquid rocket engines of small thrust (LRE STh) at ecological clean propellants [2]. Hydraulically pipe line of kerosene is selection pipes from stainless steel of variable diameter (4-16mm) total length 8.12m. Pipe line connects kerosene tank with investigated mixing head and consists control valve, filters (net 7 and 70 micro meters), sensors of mass flow, pressure and temperature (Fig. 1).

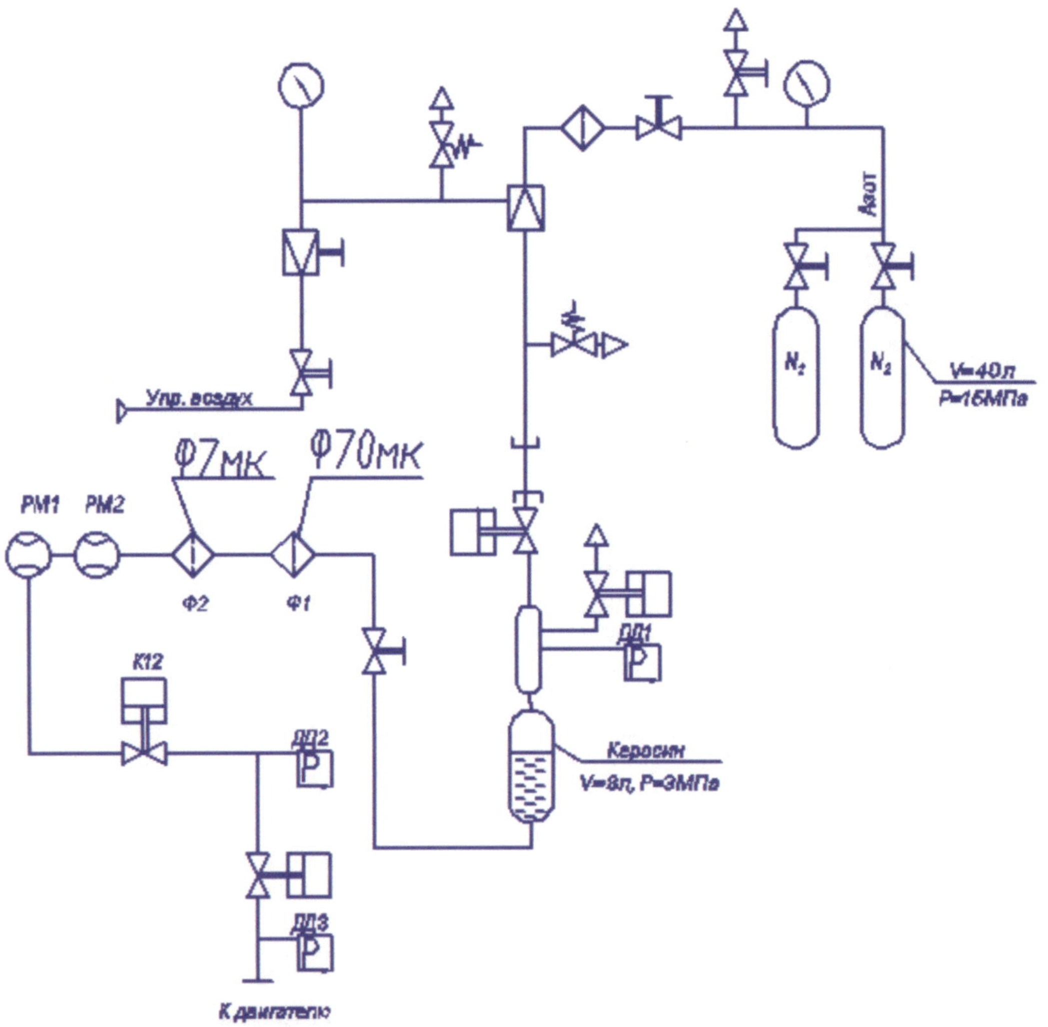

Fig. 1. Kerosene feeding system.

Take into account fire danger of mixture drops of kerosene with oxygen, for the obtaining characteristics of pulverization the special drops-trap was designed and fabricated. Scheme of this drops-trap is showed at Fig. 2.

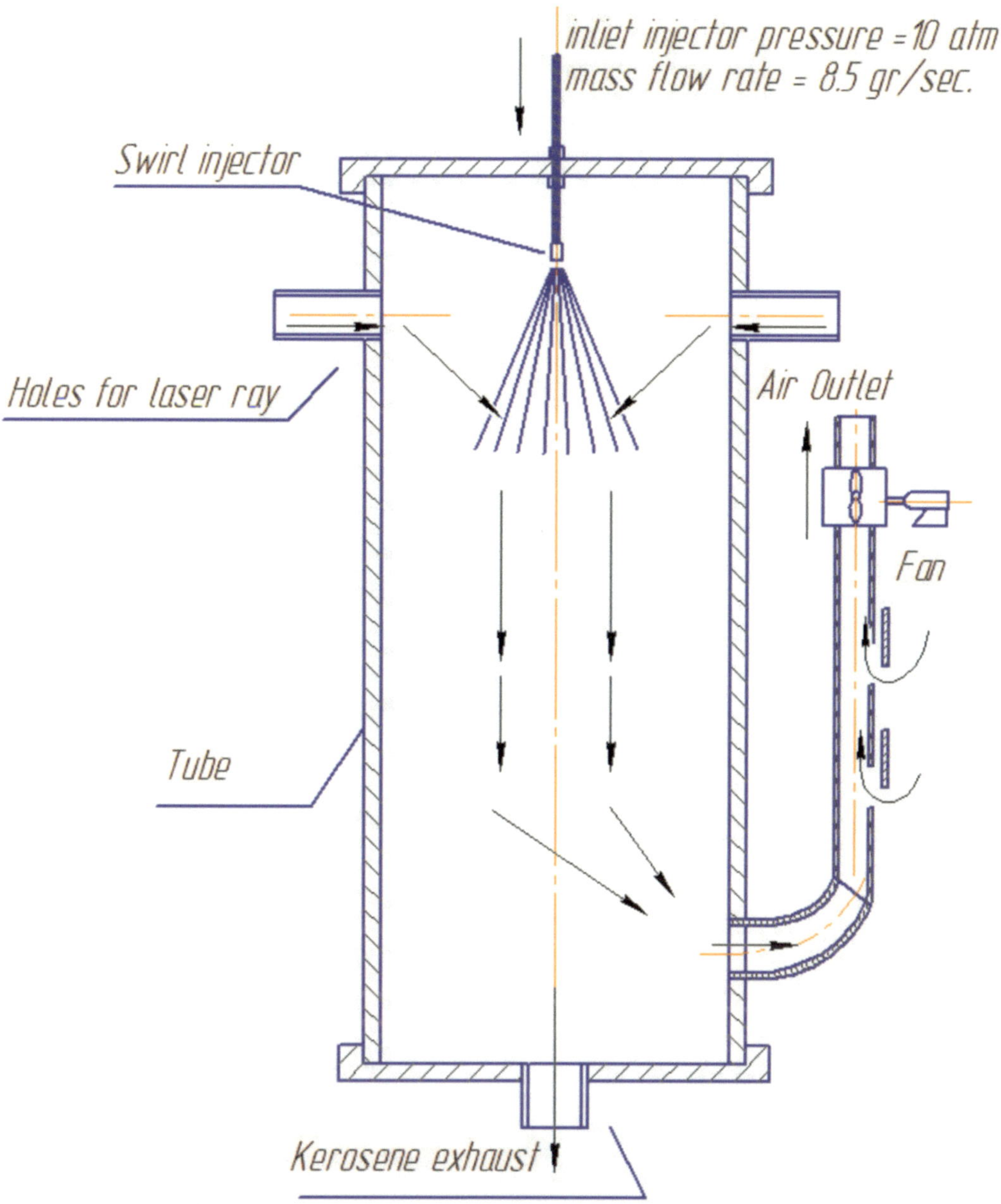

Fig. 2. Scheme of drops-trap.

Drops-trap consists from tube diameter 400mm, upper top with mounted kerosene pipe with injector (or mixing head), two diameterally opposite orifices for registration quality of pulverization, low lid with branch pipe drain of kerosene and system of forced extraction mixture with fan in explosive-protected fulfillment.

Vertical position of drops-trap corresponds vertical position of tested engine and guarantees the same influence of gravitation forces at the torch of pulverization. Photos of drops-trap are presented at Fig. 3. Photos of working laser system during test presented at Fig. 4.

Fig. 3. Photo of drops-trap.

Fig. 4. Laser ray goes through the spray.

System of distanced laser measurer guarantees registration diagram of dispersion of aerosol cloud at the distance till 2 meters from measured volume. In result of mathematical computing sizes and concentration parts are definite (fase-dispersia composition).Structure scheme of measured system is presented at Fig. 5.

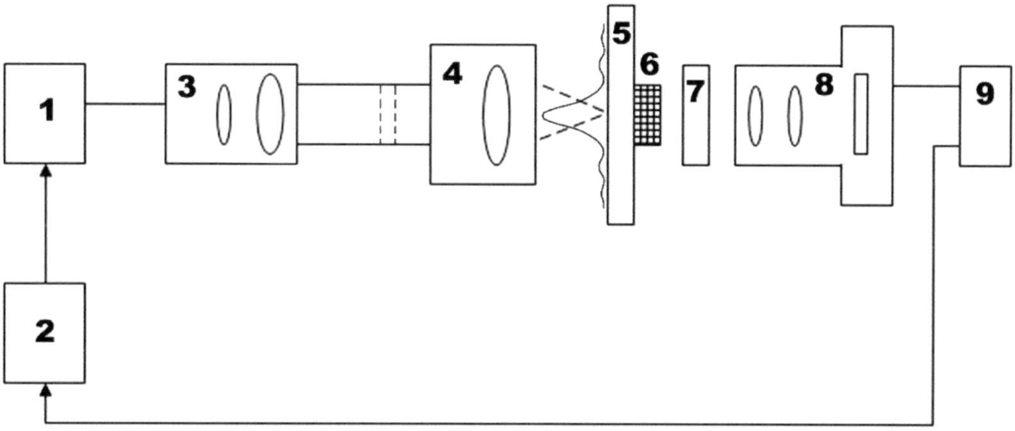

1 - bloc of laser radiators;
2 – control bloc of laser radiators;
3 - bloc of transfer optical system;
4 – bloc of accepted optical system;
5 –light dispersion screen;
6 –mask;
7 – filter;
8- digital camera;
9 – computer.

Fig. 5. Structure scheme of a base distanced laser measurer:

Bloc of transferred lenses 3 contains field diaphragm and, some times, collimator forming probe-rays and sizes of measured volume. Bloc of entranced optics contains Furie-linses, having focus-distance 50-100 cm and light diameter 10-20 cm.

Focus distance of lenses, entranced in collimator, is changed from 10mm till 20cm.Because of small sizes of parts(2-10micron)and big distances(till 2m) diameter most information part of spatial specter, in which about 90% energy dispersed radiation is consisted , is obtained more 20 cm.

Therefore, in order to fix this specter directly at photo-matrix, it is necessary to fabricate it's specially, but it go to big expenditures. In order to fix this spatial specter by series digital camera 8, in frequency plate bloc of entrance optics was mounted light dispersion screen 5, which visualizes the spatial specter. In order don't spoil matrix of camera 8 by direct laser ray, behind screen 5 sometimes expediently to place mask 6, absorbed direct laser radiation. Sometimes, in order to decrease light-dispersion between elements of screen 5, mask is placed before screen. For decrease influence of background light before lenses of camera 8 may to place interference light-filter 7. For two lengths wave of laser radiation light-filter may by changed or special to fabricate. The spatial specter, fixed by digital camera 8, goes in computer 9, where with help of special software the sizes and concentrations parts of aerosol are calculated. Control of laser radiations is realized across computer 9 (for increase of the mobility notebook is used).

Transferred bloc of measurer contains half-conductor laser 1(Fig. 6) (length wave 650 nm, type of laser KLM-650/20) and field diaphragm 2mm,wich decrease diameter of laser ray, tested measured volume with drops of pulwerizated.

Light, distributed in direct direction 4 and under small angles, put at Furie-linse 5, in focal surface of which is placed screen 6.Focus distance of Furie-linse equal 500mm, light diameter 200mm. Screen was fabricated from glass, mated with one side and thickness 3mm, sizes 300x300mm.

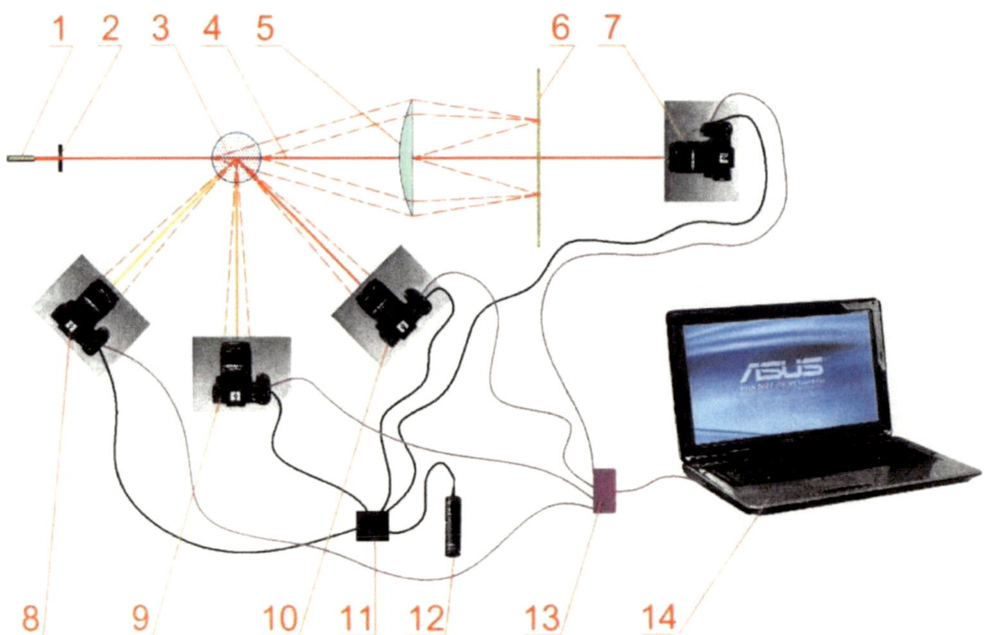

Fig. 6. Principle scheme of system for measurement distribution drops at sizes.

The distribution of light intensivity in focal surface of linse (part of indicatrisse, corresponding small angle of dispersion) is fixed by digital camera 7 (Canon EOS 1000P). Furie linse 5, screen 5 and digital camera 7 entrance in composition reception optoelectronic bloc, intended for measurement characteristics dispersion of kerosene cloud in diapason of measurement 1-10micron. For diapason of measurement 0.2-2micron in reception bloc are introduced additional: digital camera8, fixed opposite dispersion, camera 9, fixed dispersed light under straight angle, and camera10, fixed light , dispersed in direct direction, but at angles more 100°(at angles, going out from limits of small angle dispersion).Montage of cameras 8 and 10 under angles regarding main optical axis α and (180°- α) and lets use for measurement aerosol sizes method asymmetric of indicatrisse of dispersion. This does more simple the treatment of results. Registration of dispersed light by all cameras fulfilled simultaneously. This reaches by use commutater 11 and control bench of photography 12. Control of cameras 7-10 is realized across USB-divider13 and computer 14. It is possible and hand regime control of cameras. Information entrances across USB- divider 13 in computer 14 fixed these cameras and is treated.

If diameter nozzle of model injector, used during test, equal 0.8mm, that quantity drops with diameter less 2 micron will be not significant, therefore the use cameras 8, 9 and 10 in test not certainly.

3. The investigation of influence polyisobutilene additions to kerosene at the dispersion of pulverization

For the treatment obtained drops and calculation parameters of dispersion of aerosol cloud the program is used, developed in MAI at department 201 and realized at the base packet MATLAB (Fig. 7).

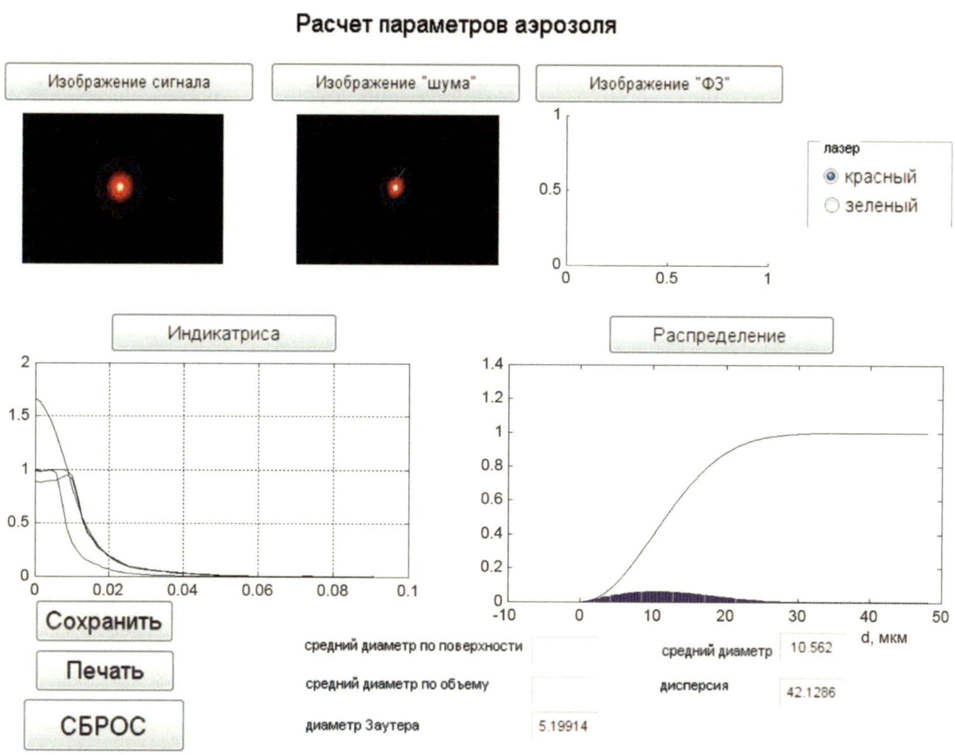

Fig. 7. Interface program calculation of parameters aerosol dispersion.

Laser device is montage at two controlled by altitude tables, placed at different sides from drops-trap (Fig. 8). Axes of scanner ray go in orifices of drops-trap.

In order to separate factor influence of molecules polyisobutilene at the pulverization, the cone of pulverization of single swirl injector with geometrical characteristic $A = Rr_{nozzle}/\left(nr_{in}^2\right) = 3.2$ and nozzle diameter 0.8 mm was analyzed. Distance from nozzle cut off till surface of scanning consists 60 mm.

Estimation of quality of dispersion for common surface area defined by Sauter's diameter parameter. Diameter Sauter is diameter of thermometric drop which has volume\surface area coefficient equal middle volume\surface area coefficient calculated of whole drops in volume.

Fig. 8. Laser device mount in the fire stand.

Results investigation pulverization of centrifugal injector are presented in Table 1.

Presence of addition	*Mass flow g/s*	*Pressure of pressurizat ion,atm*	*Pressure before valve,atm*	*Pressure before injector, atm*	Δp_{inj} *atm*	Δp_{pipe} *line, atm*	*Diameter Sauter's, micron*
Without of addition	4,45	4.5	4.8	3.83	0.97	0.7	5.199
Without of addition	5.0	5.44	5.77	4.72	1.05	0.67	4.56
With addition	4.2	12.0	4.6	3.55	1.05	8.4	3.48
With addition	5.38	14.8	6.44	5.44	1.0	9.36	2.94

Table 1.

Analysis of obtained results lets to approve, that 0.05% addition in polyisobutilene in kerosene to improve quality of pulverization (at equal mass flow of kerosene), about this show decrease of diameter Sauter approximately in 1.5 times. We may white better quality of pulverization during fire tests, because by antypressure in combustion chamber quality of pulverization will be better [5].

It is well known, that median diameter of drops during the pulverization of liquid by swirl injector is obeying the dependence [6]:

$$d_m/d_c = 47.8\Big/\left(A^{0.6}\,\Pi^{0.1}\,\mathrm{Re}^{0.7}\right),$$

$$\Pi = \eta_L^2/\rho_L\sigma_L d_c\,,$$

$$\mathrm{Re} = \rho_L W_e d_c / \eta_L \text{ ,}$$

where ρ_L, η_L, σ_L - density, dynamic viscosity coefficient and surface tension of liquid, A - geometrical characteristic of injector, W_e - equivalent velocity of flow, d_c - nozzle diameter of injector. Then $d_m/d_c \sim \sigma_L^{0.1}\eta_L^{0.5}$, so fineness of pulverization improve when viscosity and surface tension are decrease in case if geometrical characteristic of injector and pressure drop not change.

Therefore, there is interest of directly measurement of viscosity and surface tension of kerosene with additions of polyisobutilene for physical interpretation of pulverization process.

4. Investigation influence of polyisobutilene addition to kerosene at the hydraulic resistance of pipe line

For normal kerosene the mass flow rate is 12 gr/sec for feed pressure 10 atm in filter/no filter cases. For kerosene with polyisobutilene addition the mass flow is 4 gr/sec if filter exist on the pipe line and 16 gr/sec if filter absent on the pipe line.

Result (Fig. 9) show that 7 micron filter increase hydraulic loss even if length of pipe line is not big. If case using this kind of filters in LRE pipe lines the hydraulic loss will be sizeable in the cooling jacket of combustion chamber. Without 7 micron filter the hydraulic loss of pipe line increase. The influence of 70 micron filter on hydraulic loss not defined.

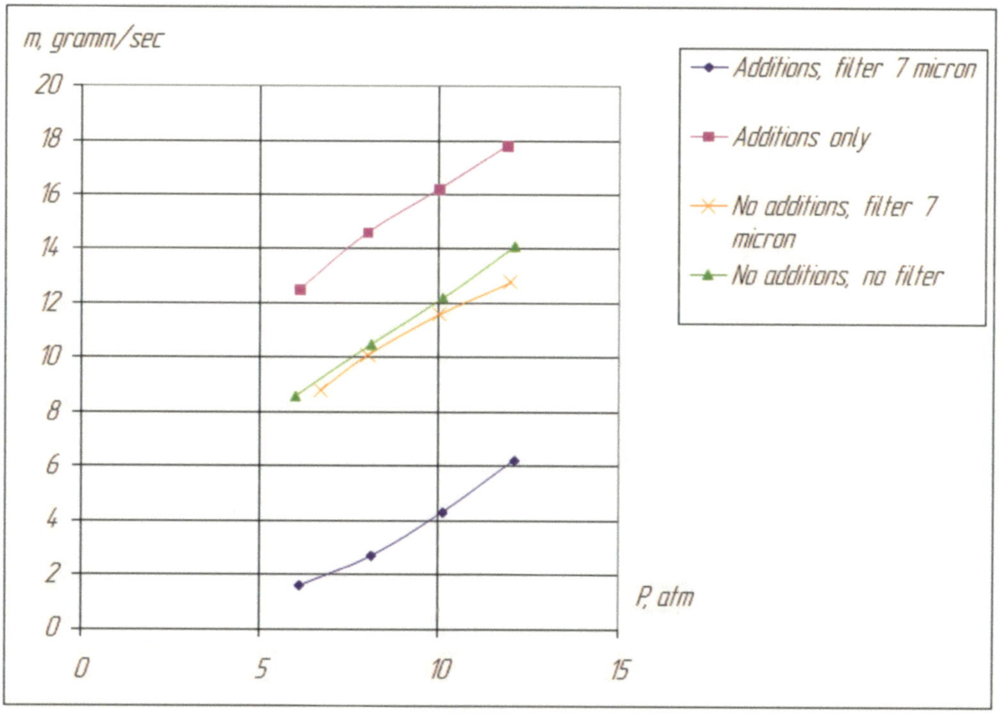

Fig. 9. Feeding pressure/mass flow rate dependency.

5. Design of LRE of small thrust for fire tests

Combustion efficiency of propellant in the combustion chamber depends not only from quality pulverization of injector. It depends and from a lot of additional factors: mass flow ratio, number of injectors and scheme its placement at mixing head, combustion chamber pressure, system of inner cooling and others. It is clear, that for separation influence of addition to kerosene it is necessary to fulfill two fire tests at the same engine and the same regime parameters.

For realization fire tests with the use addition of polyisobutilene to kerosene in MAI was developed engine MAI-200-7OK at propellant gaseous oxygen and kerosene (see Fig. 10 - Fig. 12) [3, 8, 9].

A small number of experiments and the difference in the initial parameters of the experiments cannot define exact dependency of polyisobutilene additions on the quality of the spray component.

To determine the effect of polyisobutilene additions on the quality of the spray component is necessary to continue experimental work for a set of statistics.

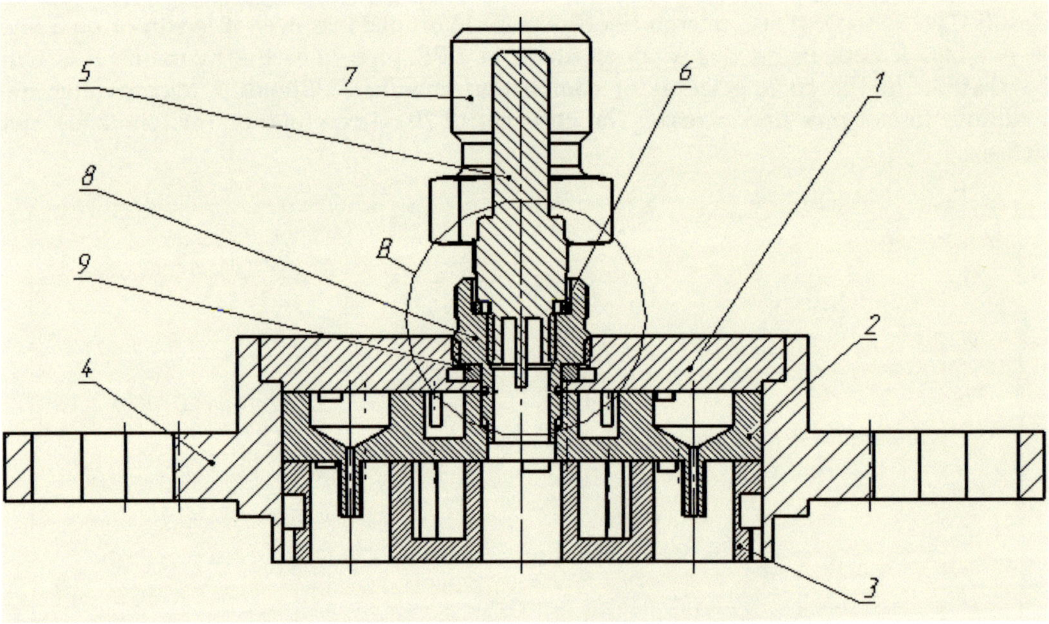

Fig. 10. Drawing of mixing head. 1.top of head, 2.plate of fuel, 3.plate of oxygen,4.flange, 5.candle of ignition, 6. compression of candle, 7.connecting pipes for components, 8.bush of candle, 9.compression of the bush.

Demountable head of engine was fulfilled at technology soldering of plates, bloc from witch place in body-flange. Head has 6 bipropellant swirl injectors with inner stage of fuel and outside open stage of oxidizer. Components go in any injector across two tangencies channels. Ignition of mixture is realize by electrical discharge in the electrical candle, placed in central part of the head. Candle is placed in open volume, in which oxidizer and fuel go during tangential channel like for injectors. Control mass flow of kerosene in volume of

ignition is realized with help of bush with orifices given section. Candle works from electrical high voltage discharger with frequency 200 Hz.
Fuel goes across connecting pipe in central collector and distributed on radial grooves to tangential channels (Fig. 13).
Oxidizer goes in the head across connecting pipe, later goes across plate of fuel, enters in central collector and distributed at radial grooves to tangential channels (Fig. 14).Part of oxidizer from radial grooves at special orifices goes at inner (curtain) cooling.

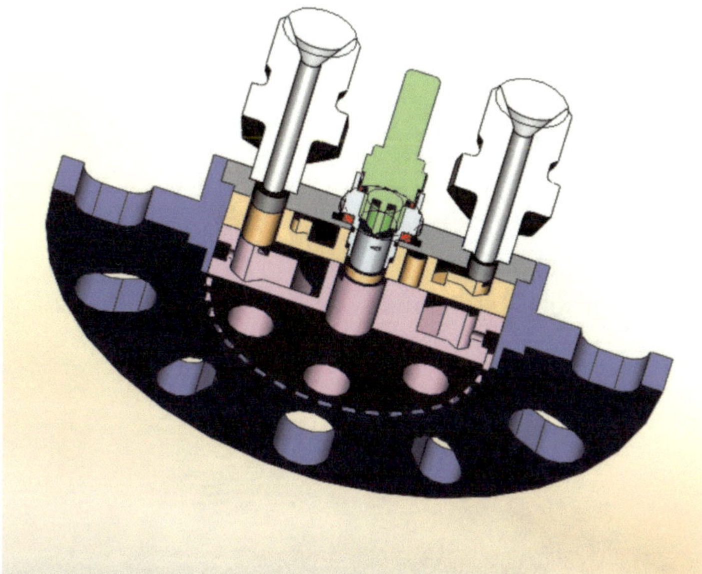

Fig. 11. Model of mixing head.

Fig. 12. Photography of mixing head.

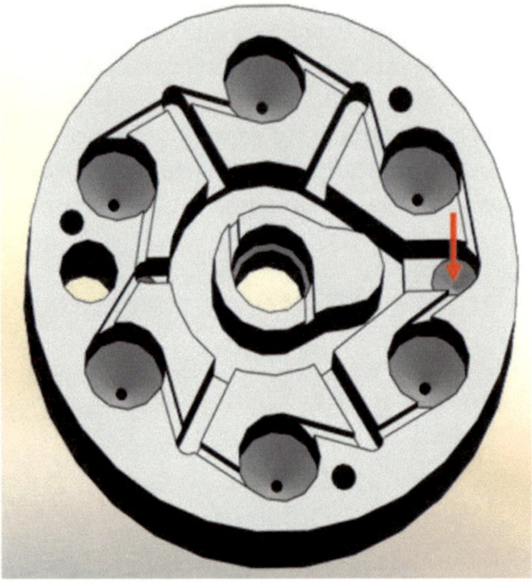

Fig. 13. Plate of fuel. Red arrow - inlet of fluid.

Fig. 14. Plate of oxidizer. Red arrow - inlet of gas.

Mixing head connect with combustion chamber across compressed cupper ring. The fighting of bolts at flange is fulfilled with help of dynamometer key, in order to except unevenness compressed forces. Before fire tests the compressing of engine is fulfilled: combustion chamber in critical area is condensed by flange with central rubber cone. Flanges of head and condensed element are jammed by studs. During compressing the compression of connection chamber with mixing head and placement of candle are examined.

6. Investigation influence of polyisobutilene addition to kerosene at the combustion efficiency kerosene - oxygen propellant

Fire tests were fulfilled at the combustion chamber with short nozzle part (Fig. 15). That methodic of test lets to exam workability of own combustion chamber during long works without the use high expenditure vacuum equipment.

In process of experiment mass flow complex $\beta_S = f_2(\dot{m}_\Sigma, k_m) = p_k F_c / m_\Sigma$ (or C*) of engine MAI-200-7OK is defined for 2 cases:

- standard kerosene;
- kerosene with 0.05% polyisobutilene addition.

The value of $\varphi_{pk} = \beta_{theory} / \beta_{\exp}$ show the efficiency of chamber process.

Main measured values were: stable combustion chamber pressure and mass flow of oxidizer and fuel.

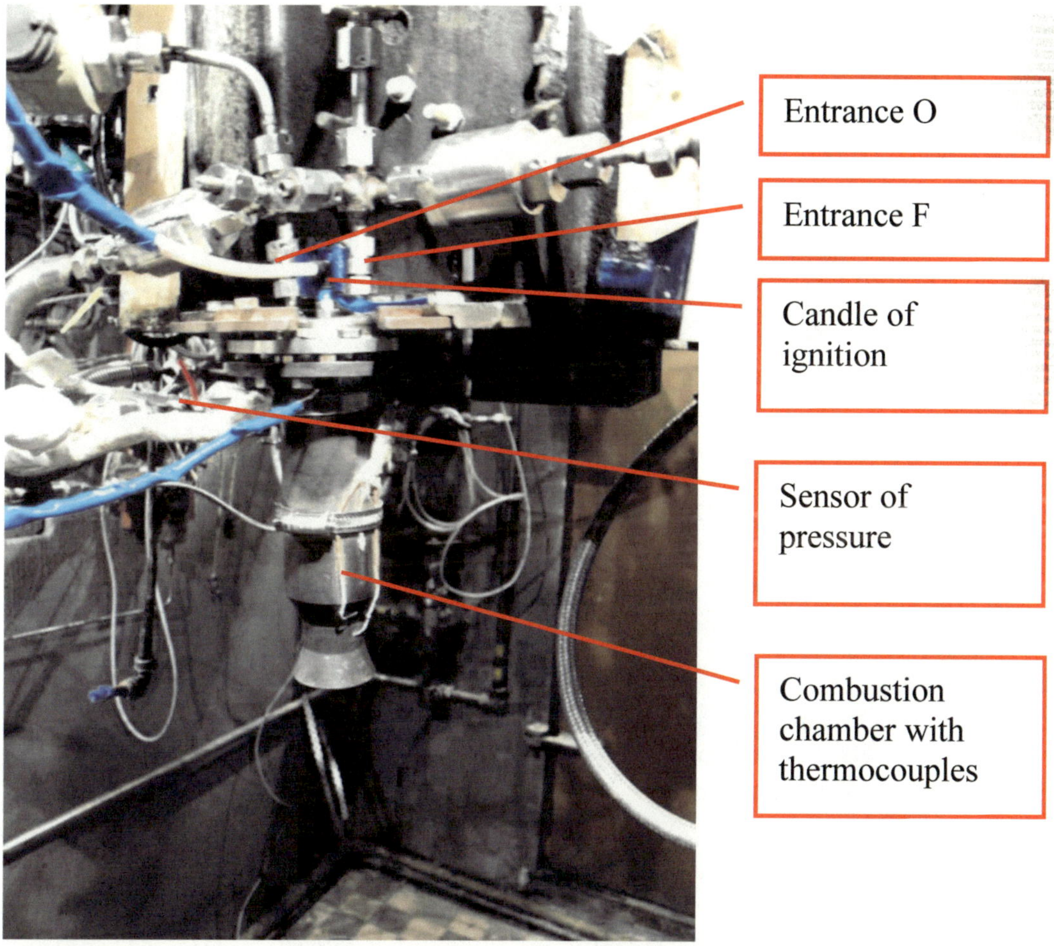

Fig. 15. Engine at the working zone of test bench.

Tests of engine are fulfilled in two stage:

- tuned tests (duration < 0.5 sec);
- pass tests (duration < 5 sec) (Fig. 16);.

Results of pass tests are presented in Table 2.

Kerosene \ Parameter	m_o , gr/s	$m_г$, gr/s	k_m	$p_к$, atm	$\beta^Э$, sec	β^T , sec	φ_{pk}
Without addition	24,3	5,7	4,2	4,09	148	172	0,86
With addition	26	6,8	3,8	4,14	146	169	0,86

Table 2. Base middle parameters of fire tests.

Fig. 16. Photo of working thruster.

In Fig. 17 - Fig. 20 presented the result of tests. The calculation of theoretical value of mass flow complex (C*) is fulfilled with help of program complex "Astra-M". The low value of mass flow complex (C*) because during test the combustion chamber pressure is 4.5 atm and less then nominal value of engine (10 atm). The difference in efficiency of chamber pressure depend on small diff of mass flow rate.

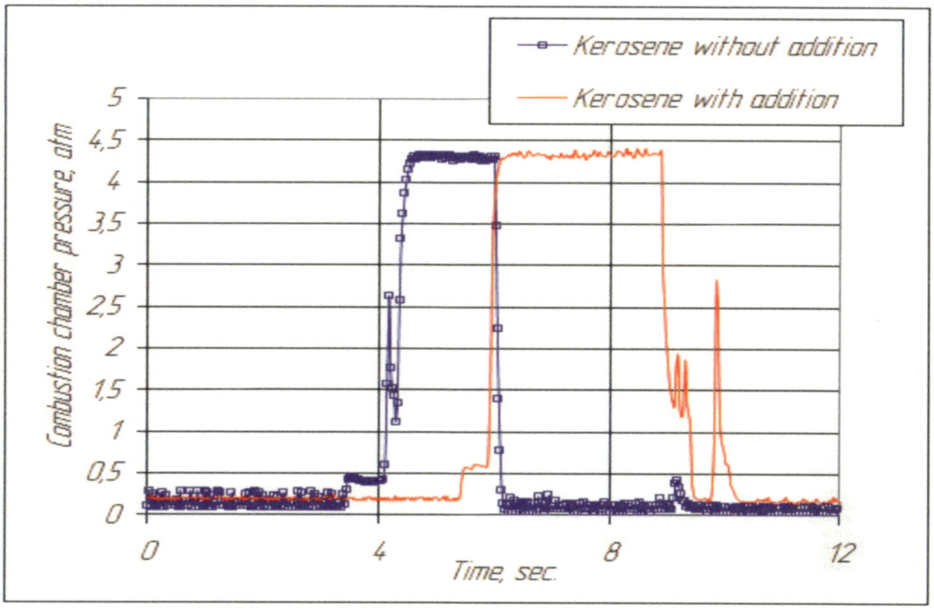

Fig. 17. Combustion chamber pressure.

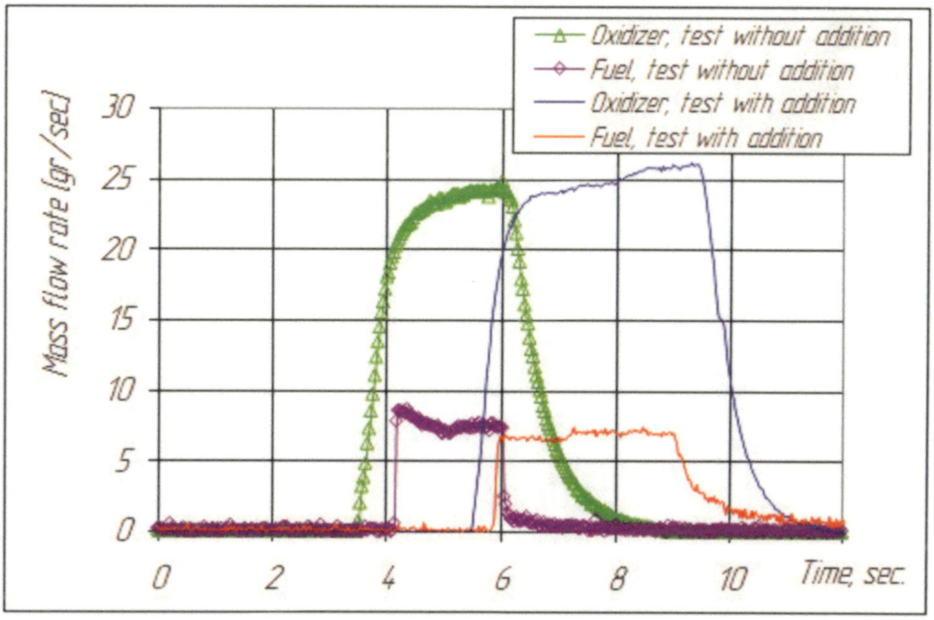

Fig. 18. Mass flow rate.

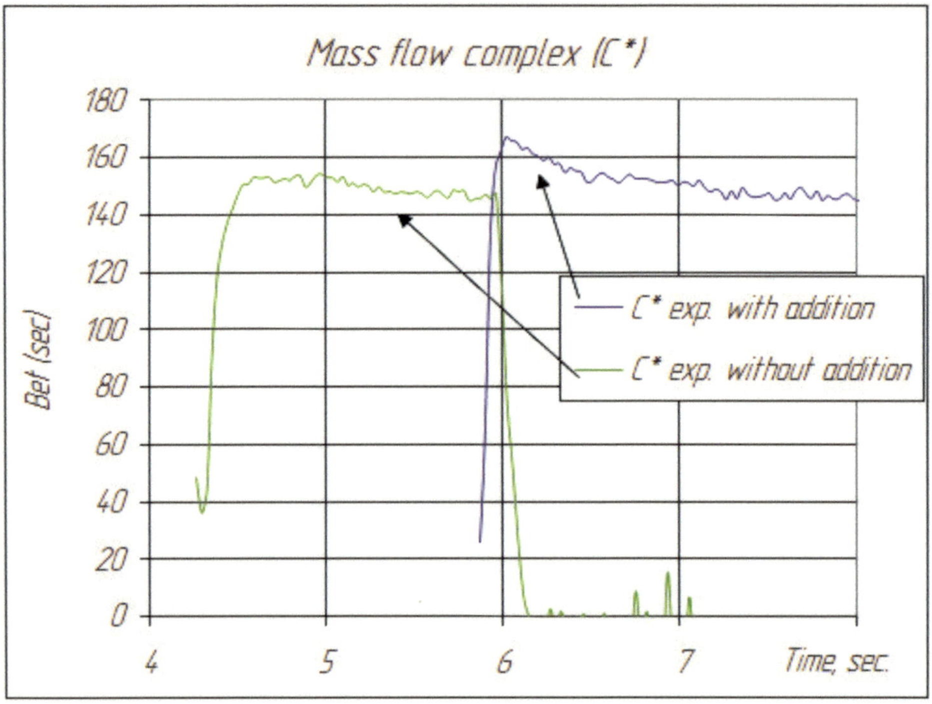

Fig. 19. Mass flow complex (C*).

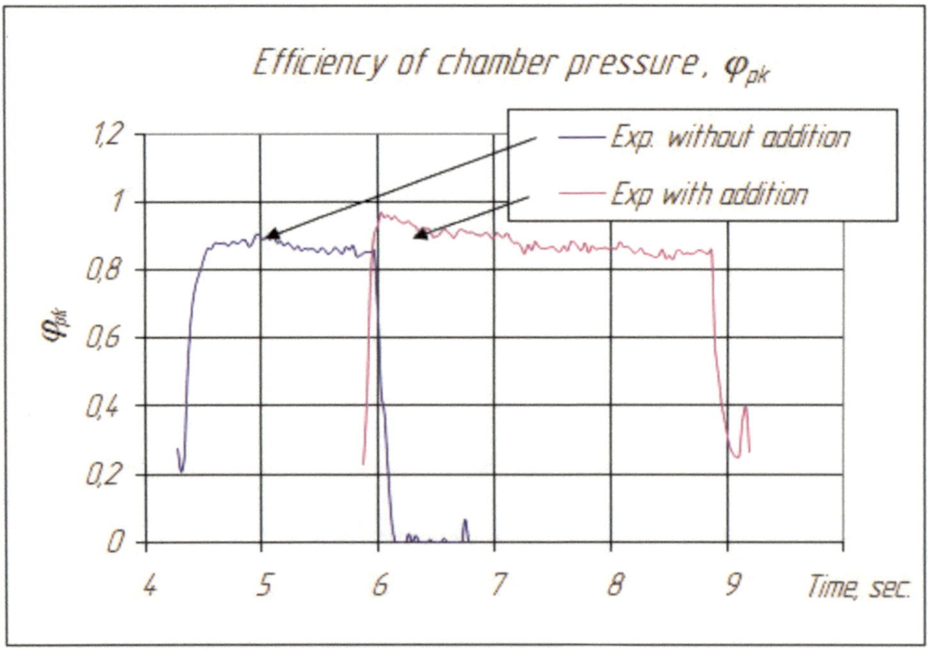

Fig. 20. Efficiency of chamber pressure φ_{pk} .

The result show that addition polyisobutilene to kerosene (0.05%) don't influence at combustion efficiency φ_{pk} of propellant in limit mistake of measurements.

7. Conclusion

Completed the initial investigation of the effect 0.05% addition polyisobutilene to kerosene at characteristics of pulverization and combustion efficiency of propellant kerosene + gaseous oxygen is fulfilled. Fire tests were fulfilled at liquid rocket engine of small thrust MAI-200-7OK.

Estimation quality of pulverization of swirl injector at value Sauter diameter was shown , that addition improved on smallness of pulverization. A small number of experiments and the difference in the initial parameters of the experiments cannot define exact dependency of polyisobutilene additions on the quality of the spray component. To determine the effect of polyisobutilene additions on the quality of the spray component is necessary to continue experimental work for a set of statistics.

A significant effect of additions of polyisobutilene on hydraulic resistance of the fine-mesh filter defined. Fine-meshed filter (mesh size 7mkm) gives a significant increase in hydraulic losses even on a small length of line of the stand. Without the filter pressure drop line decreased.

Effect of additions of polyisobutilene on the combustion effectively of the engine on experimental mode not detected. Firing tests will be continued on different modes to proof and get new results.

8. References

[1] Alemasov V.E., Dregalin A.F., Tishin A.P. Teorija raketnyh dvigatelej. - M.: Mashinostroenie, 1989. - 464 p.

[2] Vorob'ev A. G., Borovik I. N., Hohlov A. N. i dr. Modernizacija ispytatel'nogo ognevogo stenda dlja issledovanija rabochih processov v zhidkostnyh raketnyh dvigateljah malyh tjag na jekologicheski chistyh komponentah topliva. Vestnik MAI, T. 14, №1, 2010.

[3] Kozlov A.A., Abashev V.M. Raschet i proektirovanie zhidkostnogoraketnogo dvigatelja maloj tjagi. - M.: MAI.2003. - 36 p.

[4] Kozlov A.A., Borovik I.N., Vorob'ev A.G. Razrabotka programmy jeksperimental'nyh issledovanij rabochego processa v kamere sgoranija ZhRD pri rabote na kerosine s polimernymi prisadkami poliizobutilena i provedenie pervogo jetapa jeksperimental'nyh issledovanij. Tehnicheskij otchet po Kontraktu № 30610 - 02020/ 980 - 09 - 187 mezhdu MAI i OAO NPO «Jenergomash» im. akademika V.P. Glushko (2 jetap). Moskva, 2009.

[5] Kozlov A.A., Borovik I.N., Vorob'ev A.G. Tehnicheskij otchet po Kontraktu № 40440 - 02020/ 980 - 10 - 190 mezhdu MAI i OAO NPO «Jenergomash» im. akademika V.P. Glushko (3 jetap). Moskva, 2010.

[6] Kudrjavcev V.M. Osnovy teorii i rascheta ZhRD. Moskva, «Vysshaja shkola», 1975.

[7] Chvanov V.K., Fatuev I. Ju., Gaponov V.D., Sternin L. . Uluchshenie harakteristik raket-nositelej pri dobavlenii k toplivu vysokomolekuljarnyh prisadok. Dvigatel', № 6 (42), 2005.

15

Synthesis of Novel Materials by Laser Rapid Solidification

E. J. Liang, J. Zhang and M. J. Chao
Zhengzhou University
China

1. Introduction

High power lasers have been widely used in industry as well as in laboratory for materials surface heat treatment, cladding, welding, cutting, thin film deposition by laser ablation and so on (Bogue, 2010; Chao & Liang, 2004; Wang et al., 2008; Kruusing, 2004), but they are seldom used in the synthesis of pure bulk materials. In recent years, we explored the synthesis of pure bulk materials with a high power CO_2 laser (Liang et al., 2007; 2007; 2008; 2009; Zhang et al., 2010). It is shown that a variety of materials can be successfully synthesized by laser rapid solidification (LRS). The materials synthesized by LRS exhibit unique microstructures, superior properties which may not be realized by traditional synthetic methods. Compared to the commonly used solid state reactions and wet chemical routes which are usually severe time and energy wasting or require expensive precursors, the laser synthetic technique provides a new and rapid method for the production of materials, with which tens of grams of a sample can be produced in a few or tens of seconds. In this paper we address the synthesis and characteristics of negative thermal expansion materials and ionic conductive materials using LRS. Particular attention will be paid to the unique microstructures, special or controlled phase formation and related superior properties of the materials synthesized by LRS which may not be obtained by other methods. The oriented crystalline growth dictated by heat transfer directions and the particular phases formed at high temperatures in the molten pool and pressures induced during the rapid solidification process will be discussed. Besides, many factors such as laser power, scan speed and cooling environments are shown to affect the laser rapid solidification rate and hence the pressures induced. With the help of experimental results, the influence of these factors on the cooling rate, pressures induced and the phases of final products are revealed.

2. Synthesis of negative thermal expansion materials by LRS

It is well known that the vast majority of materials expand on heating and contract on cooling at widely different rates. This can cause a variety of problems in applications such as delamination of layers or cracking of connections, temporary or permanent device failure due to strains induced by expansion and contraction. Materials with opposite thermal properties, namely contract on heating and expand on cooling are particularly desired to facilitate the possibility to engineer materials with controllable overall negative, zero or positive coefficient of thermal expansion by composite them with positive thermal expansion materials. Negative thermal expansion (NTE) materials have therefore received considerable interests (Liang, 2010) since the discovery of negative thermal expansion of ZrW_2O_8 from 0.3 to 1050 K (Mary et al., 1996). However, the synthesis of negative thermal expansion materials such as ZrW_2O_8, HfW_2O_8 and etc. are quite tedious and several days or even weeks are usually needed by traditional solid state reactions or wet chemical routes. Laser rapid solidification has recently been employed for the synthesis of negative thermal expansion materials by us. Here we address the unique microstructures and special phase formation of the materials correlated with the heat transfer and pressures induced by this special technique.

2.1 Microstructures

Figs. 1a and 1b show the SEM images of the surface and cross-section of ZrW_2O_8 synthesized with 500 W laser power and 3 mm/s scan speed, respectively. It is obvious that the sample is composed of nano-threads/rods exhibiting oriented growth and distribution on both the surface and cross section. Enlarged SEM pattern (Fig. 1b) shows that the nano-threads are composed of smaller nano-crystallites. The orientation of the nanostructures reflects the directions of heat transfer since the growth of the nanostructures is against the heat flow direction.

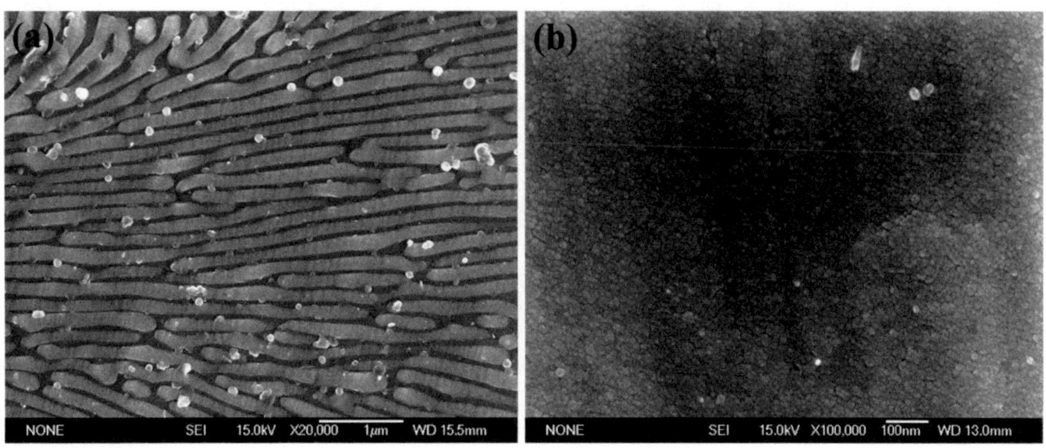

Fig. 1. SEM images of the ZrW_2O_8 block synthesized by LRS with 500 W laser power at 3 mm/s scan speed: (a) surface; (b) cross section.

The laser synthetic route is a rather rapid process, during which the raw materials are heated to melt immediately upon illumination of the laser beam and form a molten pool where the chemical reaction occurs. The product solidifies rapidly as the laser beam moves ahead. The whole process from the laser starting to illuminate to the solidification of the final product completes in only a few or tens of seconds. The heat transfer was mainly directed from the top surface to the bottom and also governed by the moving direction of the laser beam as the laser energy was absorbed by the top layer of the raw materials. The unique microstructures of the samples produced in the laser synthetic route can be attributed to the relatively oriented crystalline growth governed by heat transfer directions in the liquid droplet-like molten pool.

2.2 Pressure and compressive stress induced during LRS

Another distinct difference of the LRS with respect to solid state reactions is the pressure or stress induced in the rapid solidification process. The stress induced in the sample is proportional to solidification rate R, which is proportional to the temperature difference between the molten pool T_m and the ambient T_a, and inversely proportional to the time span Δt for the sample to cool to the final temperature (Liang et al., 2007).

$$P = AR = A\,(T_m - T_a)/\Delta t \qquad (1)$$

where A is a coefficient related to the factors apart from the laser processing parameters such as the surrounding medium. Because the NTE materials expand thermodynamically on cooling, the stress induced in the rapid solidification process is compressive instead of tensile. High pressure neutron diffraction studies showed that an irreversible structural transition from α (cubic) to γ (orthorhombic) phase for ZrW_2O_8 starts at 0.21 GPa and finishes at about 0.5 GPa (Perottoni et al., 1998; Ravindran et al., 2001; Mittal et al., 2003). Eq. 1 may be tested and different phases may be synthesized by tuning the compressive stress.

In order to ensure the formation of a molten pool for sufficient reaction and rapid synthesis, T_m must equal to or higher than the highest melting point of the raw materials. Since the laser solidification is a natural process, T_a is then taken as the room temperature. According to Eq. (1), in order to decrease P, one has to decrease $\Delta T = T_m - T_a$ and to increase Δt. The compressive stress is predominately governed by the cooling rate which is closely related to laser processing parameters such as laser power, scan speed and scan mode. The most convenient and effective way to reduce the cooling rate is to increase Δt. This can be realized by reducing the scan speed since heat influence of the molten pool on the nearest solidifying part persists longer with a lower scan speed.

Figs. 2a and 2b show the Raman spectra of ZrW_2O_8 synthesized by using 800 W laser power at 6mm/s and 1mm/s scan speeds, respectively. It is clear that the γ (orthorhombic) phase dominates with 6 mm/s scan speed as revealed by the splitting of the Raman bands around 800 cm^{-1} while the α (cubic) phase appears with 1 mm/s scan speed as indicated by the presence of the character Raman bands at 904 and 934 cm^{-1} (Liang et al., 2007; 2007).

2.3 The effect of cooling in water

The surrounding medium is also expected to influence the heat transfer rate since different media has different thermal conductivities. Figs. 3a and 3b show the Raman spectra of ZrW_2O_8 synthesized by using a 500 W laser power at 3mm/s scan speed and cooled naturally in air and in water, respectively. It is obvious that the sample cooled naturally in air is dominated by the γ phase while that cooled in water exhibits both characters of α and γ phases. The appearance of the α phase in the water-cooled sample reveals that the cooling in water reduces the compressive stress of the sample.

Fig. 4 shows the SEM images of the surface and cross-section of the sample cooled in water. Compared to the microstructures of the sample cooled naturally in air (Fig. 1), the crystallites in the sample cooled in water are not so densely packed. It seems that the cooling in water produces a tensile stress which cancels the compressive stress, leads to plentiful micro-cracks and loosens the structure. The α phase formation can be understood by the canceling of the compressive stress.

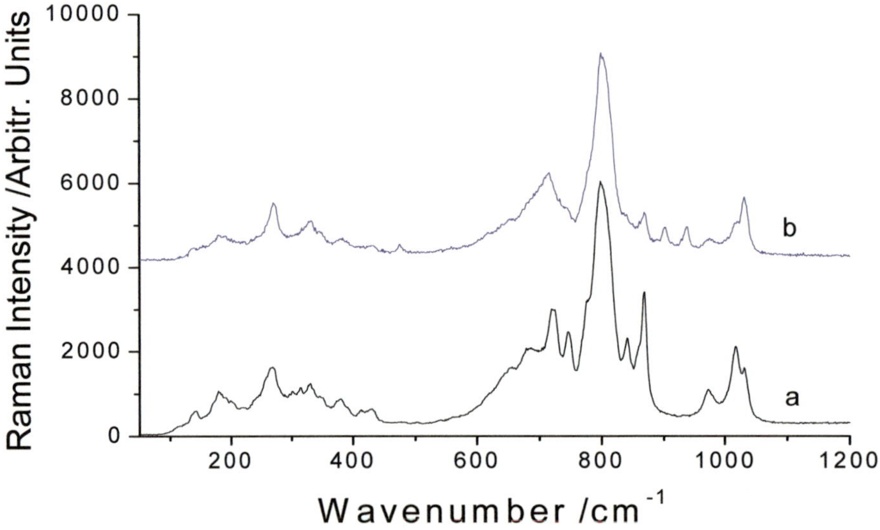

Fig. 2. Raman spectra of ZrW_2O_8 synthesized with 800 W laser power and different scan speed: (a) 6 mm/s and (b) 1 mm/s.

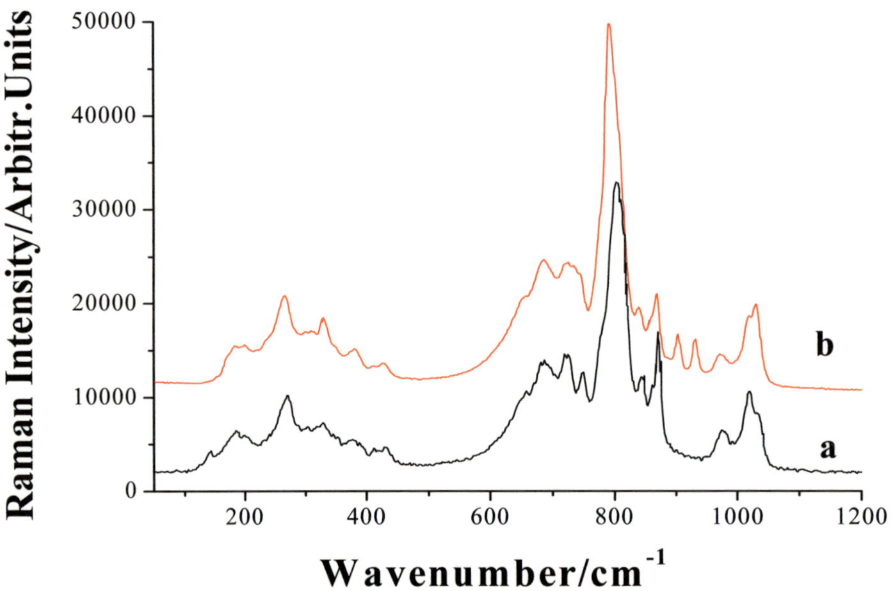

Fig. 3. Raman spectra of ZrW_2O_8 synthesized with 500 W laser power and 3 mm/s scan speed cooled (a) in air and (b) in water.

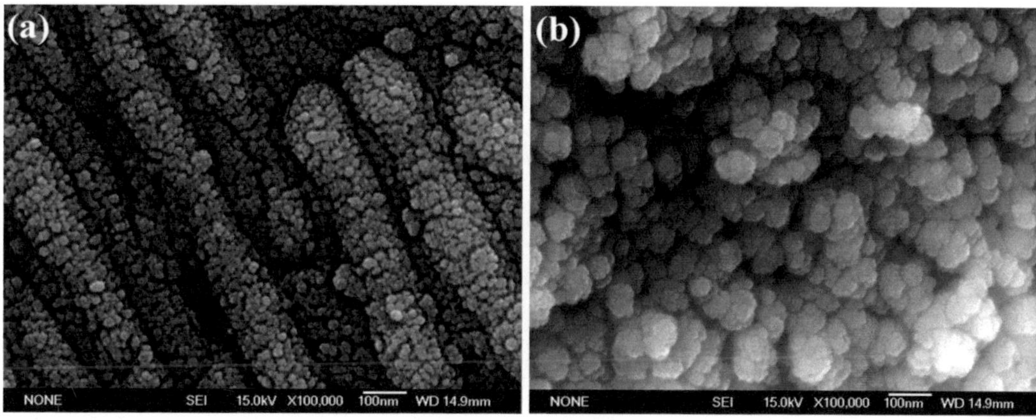

Fig. 4. SEM images of the ZrW_2O_8 block synthesized by LRS and cooled in water.

Fig. 5 shows the mass ratio of the γ to α phase with 500 W and 600 W laser power at different scan speeds, respectively. The content of the γ phase show a general increases while that of the α phase decreases with increasing the scan speed. Since the γ phase can only be produced under pressures ≥ 0.21 GPa, the increase in the γ phase content with scan speed confirms that the compressive stress increases with cooling rate. Besides, at higher scan speed (≥2mm/s), a higher laser power results in a lower content of the γ phase. A higher laser power may result in higher temperatures in the molten pool, but at the same time more heat will be transferred to the part behind which slows the cooling rate of the solidifing part. This leads to a lower compressive stress and consequently lower content of the γ phase. However, the situation is reversed at lower scan speed (1mm/s), i. e. the lower the laser power is, the lower the content of the γ phase generates. The temperature increasing in the molten pool with laser power seems to play a more important role at lower scan speed, resulting in a higher content of the γ phase with higher laser power.

HfW_2O_8 can also be synthesized by LRS. The optimum synthesis conditions for HfW_2O_8 was shown to be around 700 W laser power and 1 mm/s scan speed. These requirements are much stricter than those for the synthesis of ZrW_2O_8 due to the higher melting point of HfO_2 than that of ZrO_2. It is deduced that the pressure induced in the solidification process for the synthesis of HfW_2O_8 is around 0.6 GPa since the sample is dominated by α phase with only minor content of γ phase (Liang et al., 2007). Besides ZrW_2O_8 and HfW_2O_8, NTE materials of $Y_2W_3O_{12}$ and $Y_2Mo_3O_{12}$ can also be synthesized by LRS (Liang et al., 2008; Yuan et al., 2009).

3. Synthesis of ion conductive perovskite oxides by LRS for SOFC applications

Solid oxide fuel cells (SOFCs) can directly convert chemical energy of fuels to electrical energy with advantages of high electrical efficiency, fuel versatility and low pollution emissions. The perovskite oxide, strontium- and magnesium-doped $LaGaO_3$ (LSGM), exhibits a higher ionic conductivity than the conventional Y_2O_3-stabilized ZrO_2 electrolytes over a wide range of oxygen partial pressure. This makes $LaGaO_3$-based oxides the most promising candidates as electrolytes for intermediate temperature SOFCs. However, the synthesis of a pure single phase material of LSGM is a rather difficult task. For example, the solid state reactions suffer from severe time and energy wasting while the wet chemical methods require expensive metal alkoxide precursors, great care in mixing the precursors to

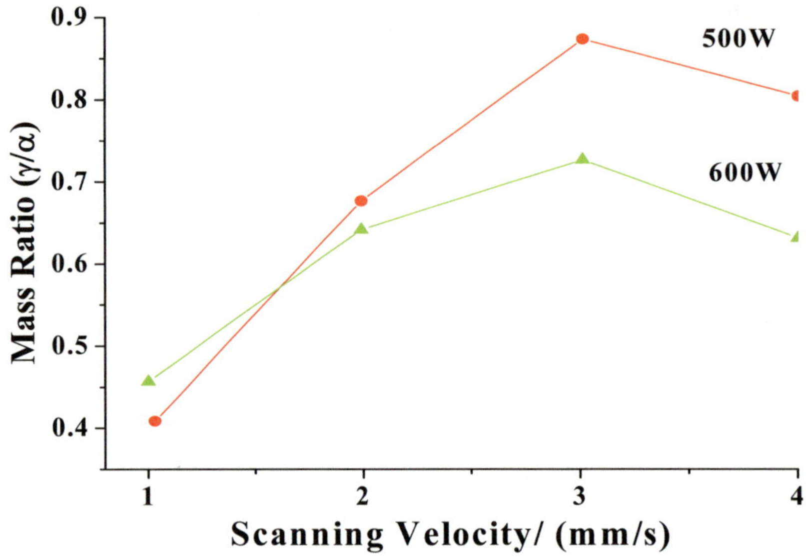

Fig. 5. The mass ratio of the γ to α phase at different scan speeds.

achieve the desired stoichiometry and tedious pretreatment before the final calcinating step. Due to the narrow composition range for the stability of the perovskite phase, small deviations from the ideal composition would result in secondary phases which deteriorate the performance of the electrolyte. We have recently demonstrated that the LRS technique can be used to the synthesis of $La_{0.9}Sr_{0.1}Ga_{0.8}Mg_{0.2}O_{3-\delta}$ (LSGM) (Zhang et al., 2010) and $La_{0.8}Sr_{0.2}Ga_{0.83}Mg_{0.17-x}Co_xO_{2.815}$ with x=0, 0.05, 0.085, 0.10 and 0.15, respectively.

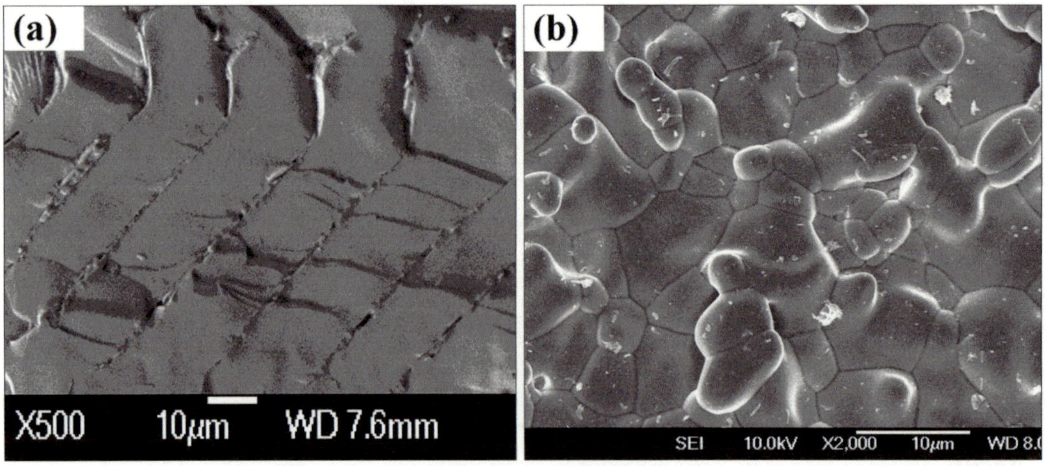

Fig. 6. SEM images of the LSGM samples prepared by (a) LRS and (b) solid state reactions at 1600 °C.

Figs. 6a and 6b show the typical SEM images of the fractured cross-sections of $La_{0.9}Sr_{0.1}Ga_{0.8}Mg_{0.2}O_{3-\delta}$ (LSGM) samples prepared by LRS and solid state reactions, respectively. It is evident that the microstructure of the sample by LRS is characterized by relative orderly arranged and densely packed blocks while that prepared by solid state reactions consists of densely packed irregular shaped globose grains. The unique microstructures of the samples produced in the laser synthetic route are attributed to the relatively oriented crystalline growth governed by heat transfer directions.

Although both samples have similar density (98.5 % by LRS and 96.9% by SSR), the sample prepared by LRS exhibits much superior conductivities (0.027, 0.079 and 0.134 Scm^{-1} obtained at 600, 700 and 800 °C) to the sample prepared by solid state reactions (0.019, 0.034 and 0.041 Scm^{-1}) (Zhang et al., 2010). Both XRD analysis and Raman spectroscopic study suggest that the sample prepared by LRS crystallized in an orthorhombic and that by solid state reactions in a monoclinic phase.

The samples $La_{0.8}Sr_{0.2}Ga_{0.83}Mg_{0.17-x}Co_xO_{2.815}$ with high purity were also prepared by LRS. It is shown that that Co-doped LSGMs exhibit unique spear-like or leaf-like microstructures (not shown here) and superior oxide ion conductivity. The electrical conductivities of $La_{0.8}Sr_{0.2}Ga_{0.83}Mg_{0.085}Co_{0.085}O_{2.815}$ are measured to be 0.067, 0.124 and 0.202 Scm^{-1} at 600, 700 and 800°C, respectively, being much higher than those of the same composition by solid state reactions (0.026, 0.065, 0.105 Scm^{-1}).

The unique microstructures of the samples prepared by LRS should account mainly for their superior electrical properties to those of the samples prepared by solid state reactions. The relatively oriented and densely packed ridge-like (for LSGM) or leave-like (Co-doped LSGM) grains with large and regular sizes in the samples by LRS greatly reduce the scattering probabilities and thus increase the mean free path or the mean free time of charge carriers during the drift motion.

It can be speculated from the appearances and SEM images that the starting materials were sufficiently molten in the molten pool. Since the melting points of the raw materials La_2O_3, $SrCO_3$, Ga_2O_3 and MgO are about 2315, 1497, 1740 and 2827°C, respectively, the temperature of the molten pool is expected to be above 2830°C. The sufficiently high temperature ensured sufficient melting of the raw materials and consequently rapid and uniform reactions.

4. Conclusion

LRS has been used to the synthesis of NTE and oxide ion conductive materials for SOFCs. Special characters of the LRS are the directed heat transfer and rapid solidification. The heat transfer is mainly directed from the top surface to the bottom and also governed by the moving direction of the laser beam as the laser energy is absorbed by the top layer of the raw materials. The samples synthesized by LRS exhibits usually unique microstructures which can be attributed to the relatively oriented crystalline growth governed by heat transfer directions in the liquid droplet-like molten pool. It is also shown that a compressive stress induced in the rapid solidification process can be large enough for the generation of the γ phase ZrW_2O_8. Due to the rapid solidification from the molten pool, highly densely-packed blocks of the samples can be easily achieved, in contrast to traditional solid state reactions where sintering additives are usually required to achieve high density of samples. The densely packed unique microstructures and perhaps also the spectial phases of the electrolyte samples prepared by LRS make them superior in electrical properties to those of the samples prepared by solid state reactions.

5. Acknowledgment

This work was supported by the National Science Foundation of China (No. 10974183)

6. References

Chao, M. J. & Liang E. J. (2004). Effect of TiO_2-doping on the microstructure and the wear properties of laser-clad nickel-based coatings, *Surf. Coat. Techn.* Vol. 179, No. 2-3, (Febrary, 2004), pp. 265-271, ISSN 0257-8972

Bogue, R. (2010). Fifty years of the laser: its role in material processing, *Assembly Automation*, Vol. 30, No. 4, (April, 2010), pp. 317-322, ISSN 0144-5154

Kruusing A, Underwater and water-assisted laser processing:Part 1—general features, steam cleaning and shock processing, *Optics and Lasers in Engineering*, Vol. 41, No. 2, (Febrary, 2004) pp. 307-327, ISSN: 0143-8166

Liang, E.J.; Wu, T. A.; Yuan, B.; Chao, M. J. & Zhang, W. F. Synthesis, microstructure and phase control of zirconium tungstate with a CO2 laser, *J Phys D Appl Phys*.Vol. 40, No. 10, (May, 2007), pp. 3219-3223, ISSN: 0022-3727; Liang, E. J.; Wang, S. H.; Wu, T. A.; Chao, M. J.; Yuan, B. & Zhang, W. F. Raman spectroscopic study on structure, phase transition and restoration of zirconium tungstate blocks synthesized with a CO_2 laser, *J Raman Spectrosc*,Vol. 38, No. 9, (September, 2007) , pp. 1186-1192, ISSN: 0377-0486; Liang, E. J.; Wang, J. P.; Xu, E. M.; Du, Z. Y. & Chao, M .J. Synthesis of hafnium tungstate by a CO_2 laser and its microstructure and Raman spectroscopic study, *J Raman Spectrosc.*, Vol 39, No. 7, (July, 2008), pp. 887-892.; Liang, E. J.; Huo, H. L.; Wang, Z.; Chao, M .J. & Wang, J. P. Rapid synthesis of $A_2(MoO_4)_3$ ($A=Y^{3+}$ and La^{3+}) with a CO_2 laser, *Solid State Sci.*, Vol. 11, No. 1, (January,2009), pp. 139-143, ISSN: 1293-2558 ; Liang, E. J.; Huo, H. L. & Wang, J. P. Effect of water species on the phonon modes in orthorhombic $Y_2(MoO_4)_3$ revealed by Raman spectroscopy, *J Phys Chem C*, Vol. 112, No. 16, (April, 2008), pp. 6577-6581, ISSN:1932-7447; Liang, E. J. Negative Thermal Expansion Materials and Their Applications : A Survey of Recent Patents, *Recent Patents on Mat Sci.*,Vol. 3, No. 2, (May, 2010), pp. 106-128, ISSN:1874-4648

Mary, T. A.; Evans, J. S. O.; Vogt, T. & Sleight, A. W. Negative thermal expansion from 0.3 to 1050 Kelvin in ZrW2O8, *Science*, Vol. 272, No. 5258, (April, 1996), pp. 90-92, ISSN: 0036-8075

Mittal, R.; Chaplot, S. L.; Kolesnikov, A. I.; Loong, C. K. & Mary, T. A. Inelastic neutron scattering and lattice dynamical calculations of negative thermal expansion in ZrW_2O_8, *Phys. Rev. B*, Vol.68 No. 5, (August, 2003), pp. 054302, ISSN: 1098-0121

Perottoni, C. A. & da Jornada J A. H.,Pressure induced amorphization and negative thermal expansion in ZrW_2O_8, *Science*, Vol. 280, No. 5365, (May, 1998), pp. 886-889, ISSN: 0036-8075

Ravindran, T. R.; Arora. A. K. & Mary, T A. High-pressure Raman spectroscopic study of zirconium tungstate, *J. Phys: Cond. Matter*, Vol 13, No. 50, (December, 2001), pp. 11573-11588, ISSN: 0953-8984

Wang, D. S.; Liang, E. J.; Chao, M. J. & Yuan, B. Investigation on the Microstructure and Cracking Susceptibility of Laser-Clad V2O5/NiCrBSiC Coatings, Surf. Coat. Techn. Vol. 202, No. 8. (January, 2008), pp. 1371-1378, ISSN 0257-8972

Yuan, C. ; Liang, Y. ; Wang, J. P. & Liang, E. J. Rapid Synthesis and Raman Spectra of Negative Thermal Expansion Material Yttrium Tungstate, *J Chin Ceram Soc.*, Vol 37, No. 5, (May, 2009), pp. 726-732, ISSN: 0454-5648

Zhang, J.; Liang, E. J. & Zhang, X. H. Rapid synthesis of $La_{0.9}Sr_{0.1}Ga_{0.8}Mg_{0.2}O_{3-\delta}$ electrolyte by a CO2 laser and its electric properties for intermediate temperature solid state oxide fuel cells, *J. Power Sources*, Vol. 195, No. 19, (October, 2010), 195: 6758-6763, ISSN:0378-7753

16

Problem of Materials for Electromagnetic Launchers

Gennady Shvetsov and Sergey Stankevich
Lavrentyev Institute of Hydrodynamics Novosibirsk
Russia

1. Introduction

During the last twenty years, considerable attention of researchers working in the areas of pulsed power, plasma physics, and high-velocity acceleration of solids has been given to electromagnetic methods of accelerating solids. These issues were the subject of more than twenty international conferences in the U.S. and European countries. Papers on this topic occupy an important place in the programs of international conferences on pulsed power, plasma physics, megagauss magnetic field generation, etc. The increased interest of the world scientific community in problems of electromagnetic acceleration of solids to high velocities is due to the high scientific and practical importance of high-velocity impact research. Accelerators of solids are used to study the equations of state for solids under extreme conditions, simulate the effects of meteorite impact on spacecraft, investigate problems related to missile defense, test various artillery systems and weapons, etc.

Information on the development and current status of research on electromagnetic methods for high-velocity acceleration of solids in the United States, Russia, France, Germany, Greate Britain, China and other country can be found in reviews (Fair, 2005, 2007; Shvetsov et al., 2001, 2003, 2007; Lehmann, 2003; Haugh & Gilbert, 2003; Wang, 2003).

For high-velocity accelerators of solids, the most important are two characteristics and answers to the following two questions: 1) what absolute velocities can be achieved in a particular type of launcher for a body of a given mass? and 2) what is the service life of the launcher? An analysis of existing theoretical concepts and available experimental data has shown that the most severe limitations in attaining high velocities and providing acceptable service life of electromagnetic launchers are thermal limitations due to the circuit current. A number of crisis (critical) phenomena and processes have been found that disrupt the normal mode of accelerator operation and lead to the destruction of the accelerated body or accelerator or to the termination of the acceleration.

In electromagnetic plasma armature railguns, one of the main factors limiting the projectile velocity is the erosion of rails and insulators, leading to an increase in the mass accelerated in the launchers, an increase in the density of the gas moving in the channel, an increase in viscous friction, and a decrease in the dielectric strength of the railgun channel, which can cause a secondary breakdown in the channel with the formation of a new arc and setting an additional mass of gas in motion, etc. The main factor responsible for the intense erosion of materials is their heating by the radiation from the plasma armature to temperatures above the melting and vaporization temperatures of the materials.

In coil guns, Joule heating by the current results in a reduction in the mechanical strength of projectiles up to its complete loss during melting. Magnetic forces can lead to deformation and fracture of the inductor and accelerated body and other phenomena.

The main problem limiting the attainment of high velocities in metal armature railguns is the problem of preserving the sliding metallic contact at high velocities. An increase in the current density near the rear surface of the armature, due mainly to the velocity skin effect, leads to rapid heating, melting, and vaporization of the armature near the contact boundary. The development of these processes result in a rapid transition to an arc contact mode, enhancement of erosion processes, reduction or termination of the acceleration, and destruction of the barrel and accelerated body (Barber et al., 2003).

One of the necessary conditions for the implementation of crisis-free acceleration is the requirement that the elements of the launcher and accelerated body be heated below the melting point throughout the acceleration. The heating limitation condition implies restrictions on the maximum value of the magnetic field strength and the maximum linear current density in electromagnetic launchers and to a limitation on the velocity.

In the chapter, the velocity to which a solid of a given mass can be accelerated at a certain distance provided that, during acceleration, the temperature of the rails and accelerated body does not exceed certain values critical for the type of launcher and material used is considered the ultimate velocity in terms of the heating conditions or simply the ultimate velocity.

An analysis has shown that the ultimate velocities can be substantially increased by using composite conductors with controllable thermal properties and by optimizing the shape of the current pulse. Thus, the problem of materials and thermal limitations for electromagnetic launchers of solids is central to the study of their potential.

This chapter presents the results of studies of thermal limitations in attaining high velocities in electromagnetic launchers; analyzes the possibility of increasing the ultimate (in terms of heating conditions) velocities of accelerated solids in subcritical modes of operation of electromagnetic launchers of various types (plasma armature railguns, induction and rail accelerators of conducting solids) taking into account the limitations imposed on the heating of the launcher and accelerated body during acceleration; and investigates various ways to increase the ultimate kinematic characteristics of launchers through the use of composite conductors of various structures and with various electrothermal properties as current-carrying elements.

2. Problems of materials in plasma-armature railguns

In analyzing various physical factors that limit the performance of plasma-armature railguns, it is convenient to use the concept of the critical current density I^* / b (I^* is the current in the circuit, b is the width of the electrodes) above which these factors begin to manifest themselves. This was apparently first noted in (Barber, 1972). Estimates show (Barber, 1972; Shvetsov et al., 1987) that the smallest value of I^* / b is obtained from the condition that the current flowing in the circuit must not lead to melting of the electrodes and, consequently, to high erosion.

Investigation of the ultimate capabilities of the erosion-free operation of plasma-armature railgun requires, first of all, knowledge of the plasma armature, railgun, and power supply characteristics necessary for this operation regime. As shown by experimental studies (Hawke & Scudder, 1980; Shvetsov et al., 1987), plasma-armature properties (length l_p, average density ρ_p, impedance r_p) differ only slightly from some typical values $r_p \sim 1$ mohm, l_p~(5÷10)·b, and $\rho_p \sim 10 \div 30$ kg/m³ in both different experimental setups and in the acceleration process. Thus, for a given accelerator channel cross section and a given projectile mass, the only parameters that can be varied to control the development or slowing of erosion processes are the linear current density in the accelerator I/b and the thermophysical properties of the electrode material.

In analysis of the possibilities of increasing I^*/b, the question naturally arises as to whether composite materials can be used for this purpose. Prerequisites for increasing I^*/b are the well-known fact of high erosion resistance of composite materials in high-current switches (in 1.5-3 times higher the resistance of tungsten) and the assumption that in a railgun, the plasma armature interacts with the electrodes in the same way as in high-current switches (Jackson et al., 1986). A number of papers have reported experiments with electrodes coated with high-melting materials such as W-Cu, W/Re-Cu, Mo-Cu, etc. (Harding et al., 1986; Shrader et al., 1986; Vrabel et al., 1991; Shvetsov, Anisimov et al., 1992). It has been noted that the coated copper electrodes (W-Cu, W/Re-Cu) offer advantages over uncoated ones for use in rail launchers under the same conditions.

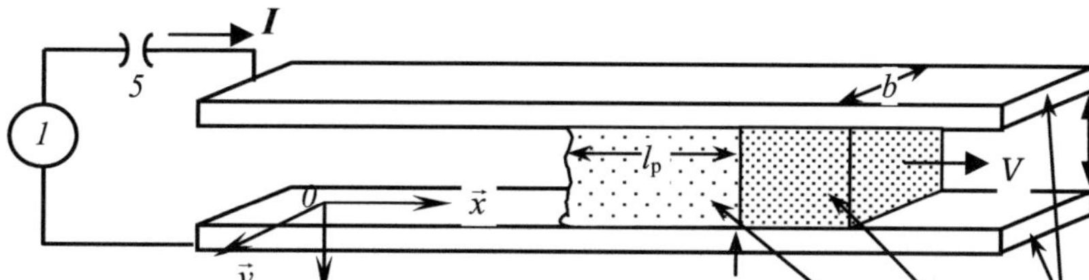

Fig. 1. Schematic diagram of the plasma-armature launcher of solids. *1* – power supply, *2* – electrodes, *3* – plasma armature, *4* –projectile, *5* – switch.

We will analyze the possibilities of increasing the critical current density by using composite electrodes in conventional plasma-armature railguns. A schematic diagram of a plasma armature launcher of dielectric solids is shown in Figure 1, where *1* is the current source, *2* are electrodes, *3* is the plasma armature, *4* is the projectile, *5* is the switch, l_p is the length of the plasma armature, b is the electrode width, and h is the distance between the electrodes. When the switch *5* is closed, a current starts to flow in the circuit, producing an electromagnetic force which accelerates the plasma armature and the projectile.

We will assume that changes in the electrode temperature are only due to the effect of the heat flux from the plasma. As shown in (Shvetsov et al. 1987), if the temperature change due to Joule heating is neglected, the error in determining the surface temperature is usually not more than a few percents. The problem of determining the temperature in some local neighborhood of a point x_0 on the electrode surface can be regarded as the problem of heating of a half-space $z > 0$ which generally has inhomogeneous thermophysical properties by a heat flux q acting for a time $\Delta t(x_0)$ equal to the time during which the plasma armature passes over the point x_0. This problem reduces to solving the heat-conduction equation with a given initial temperature distribution and boundary conditions:

$$\rho c \frac{\partial T}{\partial t} = \mathrm{div}(k\,\mathrm{grad}T)$$

$$T(x,y,z,t \le t_0(x_0)) = T_0 \tag{1}$$

$$-k\frac{\partial T}{\partial z}\bigg|_{z=0} = q(t),\ T\big|_{z\to\infty} = T_0,\ t_0(x_0) \le t \le t_0(x_0) + \Delta t_0(x_0)$$

where in the general case density ρ, heat capacity c, and thermal conductivity k may be functions of x, y, and z depending on temperature T. T_0. is the initial temperature, $t_0(x_0)$ is the time of arrival of plasma armature to the point x_0.

Following (Powell, 1984; Shvetsov et al., 1987) let us consider that the armature moves as a solid body with constant mass, length l, and electric resistance r. Neglect the variation in the internal thermal energy of plasma armature and assume that all energy dissipating in it uniformly releases through the surface limiting the volume occupied by plasma. In this case, all released energy is absorbed in the channel of the railgun as if the release had happened in a vacuum.

These assumptions make it possible to establish a simple connection between the total current I through plasma armature and the intensity of heat flux q from its surface:

$$q = \frac{r_p I^2}{S}, \tag{2}$$

where S is the area of plasma armature surface.

The dynamics of plasma armature and the projectile is determined by integrating the equations of motion:

$$\frac{dV}{dt} = \frac{\lambda}{2m} I^2,\ \frac{dL}{dt} = V \tag{3}$$

where λ is the inductance per unit-length of railgun channel, m is the sum of mass of plasma and projectile, and V is projectile velocity.

The critical current density is determined under the condition that the temperature at any point x_0, y on the electrode surface ($z = 0$) during the acceleration does not exceed the critical temperature T^* of the electrode material (the melting temperature for homogeneous materials or the melting or evaporation temperatures for one of the components of composite material).

Dependences of the current density on time or the distance L traveled by the plasma armature can be obtained by simultaneously solving equations (1)–(3).

A similarity analysis for the thermal problem (1) shows that the maximum temperature $T_{\max}(K)$ of a homogeneous electrode and a composite electrode consisting of a mixture of fairly small particles depends only on the magnitude of the thermal action $K = q\sqrt{\Delta t}$. The magnitude of the thermal action K^* at which the electrode surface reaches the critical temperature $T_{\max}(K^*) = T^*$ depends only on the thermophysical properties of the electrode material and can be regarded as a characteristic of the heat resistance of the material heated by a pulsed heat flux (Shvetsov & Stankevich, 1995).

The maximum projectile velocity in plasma-armature railguns subject to the electrode heating constraint is achieved when the shape of the current pulse provides a constant thermal action at each point of the electrode surface and the magnitude of this action is equal to the heat resistance of the electrode material. It is established that the dependences of the critical current density and the ultimate projectile velocity on the traveled distance $I^*(L)/b,\ V(L)$ for a railgun accelerator with electrodes made of an arbitrary material X are linked to the corresponding dependences for the same accelerator with copper electrodes by the relations

$$\left(\frac{I^*(L)}{b}\right)_X = \alpha\left(\frac{I^*(L)}{b}\right)_{Cu}, \quad V_X(L) = \alpha V_{Cu}(L) \tag{4}$$

where the coefficient $\alpha = \alpha_X / \alpha_{Cu} = \left(K_X^* / K_{Cu}^*\right)^{2/3}$ characterizes the relative heat resistance of the material X with respect to the heat resistance of copper.

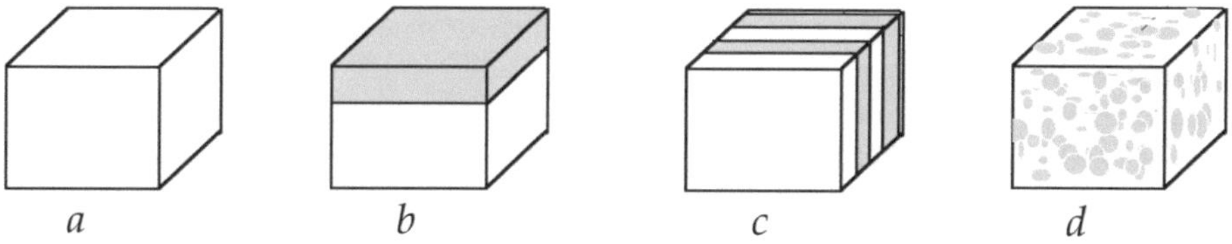

Fig. 2. Electrode structures. *a*) homogeneous electrode, *b*) coated electrode, *c*) multilayer electrode with vertical layers, *d*) composite electrode consisting of a mixture of powders.

Cu	Mo	W	Al	Ta	Re	Cr	Fe	Ni
1.0	1.17	1.38	0.55	0.99	0.99	0.87	0.69	0.78

Table 1. Homogeneous metals

An analysis was made of the heat resistance and critical current density for electrodes of various structures: a homogeneous electrode (Fig. 2, *a*), an electrode with a high-melting coating (Fig. 2, *b*), an electrode with vertical layers of different metals (Fig. 2, *c*), and a composite electrode consisting of a mixture of particles (Fig. 2, *d*).

Calculations of the coefficient of relative heat resistance of homogeneous electrodes of metals such as W, Mo, Re, Ta, Cr, Ni, Fe, and Al showed that only tungsten and molybdenum electrodes can compete with copper ($\alpha_W = 1.38,\ \alpha_{Mo} = 1.17$), and for other metals $\alpha < 1$ (Table 1).

An increase in the heat resistance of electrodes coated with a high-melting material (Fig. 2, b) is possible if the thermal conductivity of the base material is higher than the coating thermal conductivity and the heating rate of the base at a given heat flux is lower than that of the coating. The maximum increase in heat resistance is achieved at an optimal coating thickness at which the temperatures of the surface and the interface between the materials simultaneously reach the values critical to the coating and base materials. The optimum coating thickness depends on the heat flux and the duration of heat pulse; therefore, to maintain the highest possible linear current density for a given pair of materials, the coating thickness along the electrode should decrease according to a definite law. The results of calculations of the relative heat resistance of copper electrodes coated with various metals are presented in Table 2.

	W-Cu	Ta-Cu	Mo-Cu	Re-Cu	Cr-Cu	Os-Cu
α_1	1.443	1.188	1.299	1.197	1.124	1.312
α_2	1.628	1.358	1.445	1.344	1.202	1.487

Table 2. Coated electrodes

The calculations were performed for two cases: in the first, it was assumed that during the time of passage of the plasma armature, both materials remain in the solid phase (α_1), and in the second case, melting of the base to a depth equal to the coating thickness (α_2) was allowed. One can see that with the use of copper electrodes with an optimized thickness of the tungsten coating, the heat resistance coefficient (and the maximum velocity) increases to a value $\alpha_{W/Cu} = 1.45$ under the maximum heating to the melting temperature, and to a value $\alpha_{W/Cu} = 1.68$ in the case where during the travel time of the thermal pulse, the copper base is melted to a depth equal to the coating thickness and the surface tungsten layer remains in the solid phase.

Analysis of the problem of heating of electrodes with vertical layers (Fig. 2, *c*) and composite electrodes consisting of a mixture of particles (Fig. 2, *d*) by a heat flux pulse shows that for electrodes of this type, the heat resistance cannot be increased if the maximum temperature of the components does not exceed the melting temperature. However, if we assume that during melting of one of the materials, the matrix consisting of the higher-melting material remaining solid prevents the immediate removal of the melt from the electrode surface, then, for such structures, the critical temperature will be the melting temperature of the material forming the matrix or the evaporation temperature of the lower-melting material.

The heat resistance and the relative heat resistance coefficient was calculated for a number of combinations of metals with various volume contents ε and 1- ε by numerically solving the thermal problem (1) of heating of two-component composite materials with infinitely small sizes of the components. Temperature dependences of the volumetric heat capacity and thermal conductivity of composites and the latent heat of melting for the lower-melting material were taken into account. The results of some calculations are given in Table 3. The upper and lower values correspond to the maximum and minimum estimates of the thermal conductivity of the composite.

ε	Re-Cu	Mo-Cu	W-Cu	Ta-Cu	W-Mo	W-Re
0.25	1.792 1.417	1.825 1.675	1.829 1.720	1.781 1.504	1.428 1.426	1.183 1.126
0.5	1.543 1.160	1.622 1.456	1.632 1.509	1.528 1.237	1.416 1.413	1.264 1.182
0.75	1.253 0.992	1.399 1.288	1.420 1.339	1.241 1.047	1.402 1.400	1.330 1.263

Table 3. Composite electrodes consisting of a mixture of powders

Figure 3 shows curves of $V(L)$ (Fig. 3,*a*) and $I^*(L)/b$ (Fig. 3, *b*) for copper electrodes obtained for a inductance per unit-length of railgun channel $\lambda = 0.3\ \mu H/m$, plasma-armature resistance $r = 10^{-3}$ ohm, a total mass of the projectile and plasma of 1 g, and a channel cross-section of 1×1 cm. Curves *1-3* correspond to plasma armatures 5, 10, and 15 cm long.

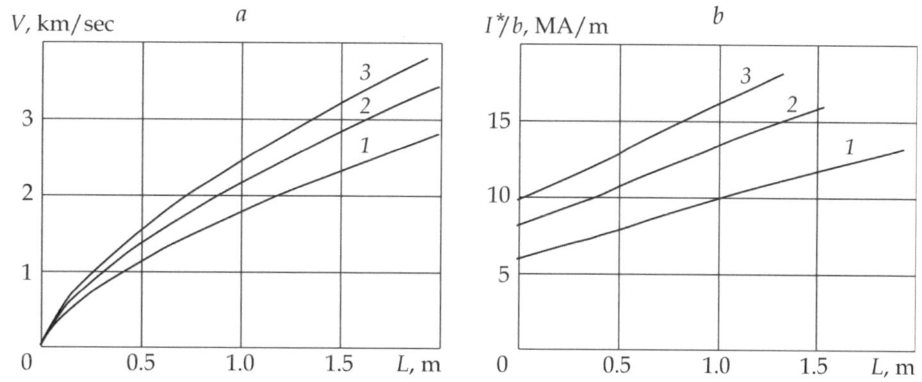

Fig. 3. Velocity (*a*) and critical current density (*b*) vs. plasma piston position in the railgun channel for copper electrodes. The numbers *1*, *2*, and *3* correspond to plasma length equal 5, 10, 15 cm.

Using electrodes with heat resistance twice the heat resistance of copper can lead to a factor of two increase in the critical current density and velocity, which (as seen from the figure and scale rations (4)) provides projectile velocities of 3-4 km/sec over an acceleration distance of 1 m and velocities of 5-7 km/sec over an acceleration distance of 2 m in the regime without significant erosion of the electrodes. It can be concluded that the use of composite materials is promising for achieving high velocities in plasma-armature railgun accelerators of solids.

3. Ultimate kinematic characteristics of conducting solids accelerated by magnetic field

A factor limiting the attainment of high velocities during acceleration of conducting projectiles by a magnetic field is the Joule heating of conductors to temperatures above the melting point of the material. This can lead to loss of the mechanical strength of the conductors, change in their shape, and, ultimately, failure. The requirement that the conductors should not melt during acceleration imposes restrictions on the maximum permissible amplitudes of the accelerating magnetic fields, thus limiting the maximum velocity to which a conductor of given mass can be accelerated over a specified acceleration distance (Shvetsov & Stankevich, 1992).

3.1 Formulation

To estimate the limits of the induction acceleration method, it is sufficient to consider the problem of the ultimate (in terms of the heating conditions) kinematic characteristics of infinite conducting flat sheets (Fig. 4) accelerated by magnetic pressure in the absence of resistance. In this section, we consider the acceleration of homogeneous sheets (Fig. 4, *a*), multilayer sheets (Fig. 4, *b*), and sheets containing a layer of composite material with electrothermal properties varying across the layer thickness (Fig. 4, *c*).
At the initial time ($t = 0$), the velocity of the sheet $V = 0$, its temperature is T_0, and a magnetic field is absent in the sheet. In general, the electrothermal properties of the sheet (electrical conductivity σ, density ρ, specific heat c, and thermal conductivity k) can depend on the x coordinate of the temperature T. For magnetic fields typical of induction accelerators, the magnetic permeability μ of materials will be equal to the magnetic permeability of vacuum

μ_0. Heat transfer between the sheet and the surrounding medium and the compressibility of the sheet are neglected. We assume that the change in the internal thermal energy of the sheet is determined by Joule heating and heat transfer.

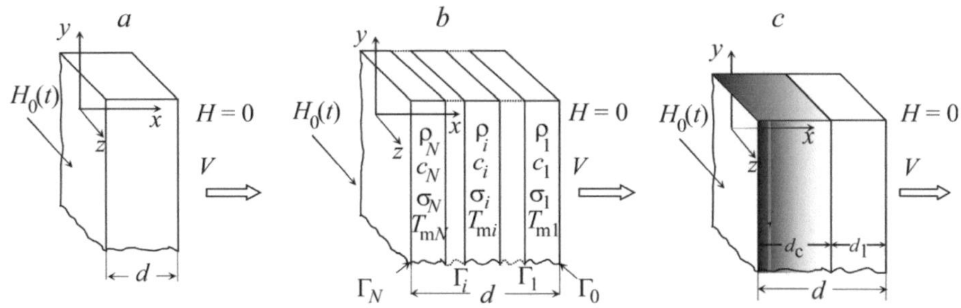

Fig. 4. Structure of accelerated sheets.

In Cartesian coordinates attached to the sheet (the boundaries of the sheet correspond to the planes $x = 0$ and $x = d$), the distributions of the magnetic field $H(x,t)$ and temperature $T(x,t)$ in the sheet depend only on the *x* coordinate and time *t*, and are described by the equations of magnetic field diffusion and heat conduction with the initial and boundary conditions:

$$\frac{\partial H}{\partial t} = \frac{1}{\mu}\frac{\partial}{\partial x}\left(\frac{1}{\sigma}\frac{\partial H}{\partial x}\right) \tag{5}$$

$$\rho c\frac{\partial T}{\partial t} = \frac{\partial}{\partial x}\left(k\frac{\partial T}{\partial x}\right) + \frac{1}{\sigma}\left(\frac{\partial H}{\partial x}\right)^2 \tag{6}$$

$$H\big|_{t=0} = 0,\ T\big|_{t=0} = T_0,\ H\big|_{x=0} = H_0(t),\ H\big|_{x=d} = 0,\ \left.\frac{\partial T}{\partial x}\right|_{x=0} = \left.\frac{\partial T}{\partial x}\right|_{x=d} = 0\,.$$

The time dependence of the magnetic field is assumed to be known and given by the relation $H_0(t) = H_a h_0(\tau)$, where $\tau = t / t_0$ (t_0 is a characteristic time).
For sheets consisting of several layers of materials with different electrothermal properties, it is assumed that at the internal boundaries between the layers, where the properties of the medium undergo a discontinuity, the continuity of the magnetic, electrical, and thermal fields is preserved.
The velocity of the sheet *V* and the distance *L* traveled by it are determined by integrating the equations of motion:

$$M\frac{dV}{dt} = \frac{\mu_0 H_0^2(t)}{2},\ \frac{dL}{dt} = V \tag{7}$$

where $M = \int_0^d \rho\, dx$ is the mass of the sheet per unit area of its surface (*d* is the sheet thickness).

The heating constraint is given by the requirement that during acceleration of the sheet at a given distance L, the heating of any component of the sheet materials be not higher than its melting temperature. Under this constraint and for a given function $h_0(\tau)$, the maximum velocity of the sheet of given structure in the general case is determined by solving the optimization problem consisting of choosing the maximum allowable (in terms of the heating conditions) values of the magnetic field amplitude H_a and the acceleration time t_0 that ensure the achievement of the maximum velocity over a given acceleration distance.

Similarity analysis for system (5) - (7) shows that for a sheet of arbitrary structure, it is sufficient to determine the maximum velocity as a function of the sheet mass $V(M,L)$ or thickness $V(d,L)$ for any one acceleration distance L. For any other distance L', these functions can be found using the relation:

$$V'(M',L') = aV(M,L), M' = aM, V'(d',L') = aV(d,L), d' = ad, \quad a = \left(\frac{L'}{L}\right)^{1/3} \tag{8}$$

3.2 Homogeneous sheets

For a homogeneous "thin" sheet (in this case, the acceleration time is much longer than the time of magnetic field penetration into the sheet), direct integration of equations (5)-(7) gives (Knoepfel, 1970)

$$V_0 = \frac{\mu_0 J}{2\rho^2} M, \tag{9}$$

where $J = \int_0^t j^2 dt = \int_{T_0}^{T_m} \rho c \sigma dT$ is the current integral. In this case, the ultimate velocity of the sheet does not depend on the magnetic field pulse shape $H_0(t)$ and the acceleration distance and are determined only by the electrothermal properties of the sheet material and the sheet mass per unit area or the sheet thickness.

For "thick" sheets (in this case, the time of magnetic field penetration into the sheet is much greater than the acceleration time), the ultimate velocity in terms of the heating conditions can be determined from the asymptotic relation (Shvetsov & Stankevich, 1994)

$$V_\infty = \psi \sqrt{\Delta Q_m \frac{L}{M}}$$

where ΔQ_m is the change in the thermal energy density of the sheet material under heating to the melting temperature and ψ is a coefficient which depends on the form of the function $h_0(\tau)$. If the magnetic field increases monotonically with time during the acceleration, $\psi = 1.1\text{-}1.2$.

Figure 5 shows the results of numerical calculations of the dependence $V(M,L)$ (curve *1*) and the asymptotic dependences $V_0(M)$ and $V_\infty(M,L)$ (curves *2* and *3* for the approximations of "thin" and "thick" sheets, respectively) for a copper sheet, $L = 1$ m, and a linearly increasing magnetic field. The dependence $V(M,L)$ is characterized by a velocity maximum which is reached for a certain mass or thickness of the sheet. The maximum velocity and the optimum mass depend weakly on the magnetic field pulse shape and are determined mainly by the electrothermal properties of the material and the acceleration distance.

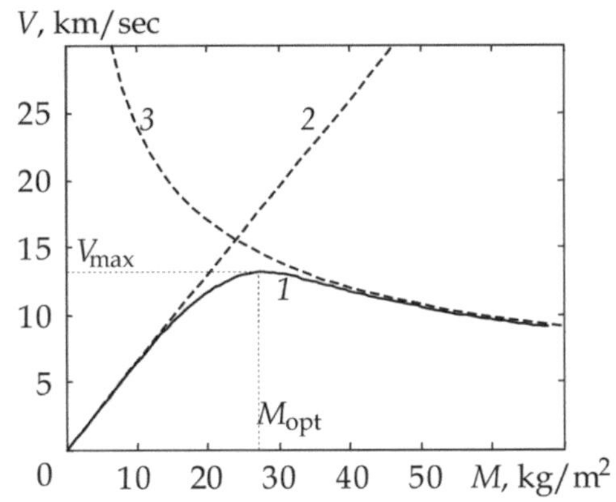

Fig. 5. Dependence $V(M)$ (curve *1*) for L = 1 m. Curve *2* and curve *3* are asymptotic dependences obtained in the approximations of "thin" and "thick" sheets, respectively.

Figure 6, *a* shows curves of the ultimate velocity versus sheet mass for Cu, W, Ti, Be, Fe, Mo, Ag, Au, and Fig. 6, *b* shows curves of the ultimate velocity versus sheet thickness for the same materials. It is seen that from the point of view of providing the maximum velocity, different materials can be optimal, depending on the given mass or required thickness of the sheet.

A characteristic feature of the dependences $V(M,L)$, $V(d,L)$ (Fig. 6 *a*, *b*) is the presence of a velocity maximum which is reached for a certain sheet thickness $d_{opt}(L)$ or linear mass $M_{otp}(L)$ that are optimal for each material. A decrease in the maximum velocity of the sheets for $M > M_{opt}$ is related to the localization of the region of maximum heating near the sheet surface, where most of the current flows because of the time of field diffusion is much greater than the time of acceleration of the sheet.

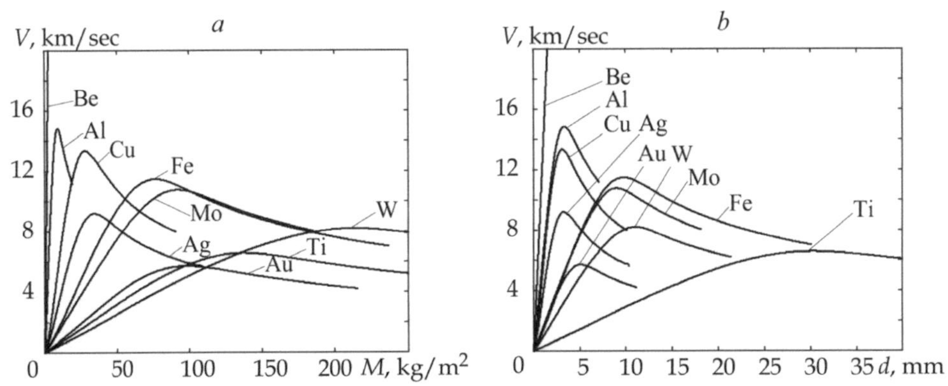

Fig. 6. Ultimate velocities vs. sheet mass (*a*) and vs. sheet thickness (*b*), for $L = 1$ m

3.3 Multilayer sheets

In some papers (Karpova et al.,1990; Shvetsov & Stankevich, 1992, Zaidel' , 1999), it was noted that the use of heterogeneous conductors with electrical conductivity discreet or continuously increasing with distance from the surface of the sheet can decrease their local heating considerably.

In this subsection, we analyze the possibility of increasing the ultimate velocity of solids accelerated by a magnetic field by using multilayer conducting sheets consisting of several layers of materials with different electrothermal properties (Fig. 4, *b*).

A simple analytical method for optimizing the sheet structure can be obtained if we assume that the electrical properties of materials does not depend on temperature, neglect the thermal conductivity of materials, and consider steady-state solutions of equations (5) - (7) that correspond to the acceleration of sheets in an exponentially growing magnetic field $h_0(\tau)=e^{\tau}$.

An analysis has shown that for a given set of layer materials, the optimal structure (the sequence of layer materials and thicknesses) is the one in which the melting temperature in each layer is reached simultaneously by the time the sheet has traveled a given distance.

Solving equations (5) and (6) under the above assumptions, we obtain:

$$h_i=\sqrt{h_{i-1}^2-q_i \mathrm{cth}^2\lambda\left(1-\frac{q_{i-1}\sigma_i}{\sigma_{i-1}q_i}\right)}, \tag{10}$$

$$\Delta\xi_i=\frac{\sqrt{\sigma_1}}{\lambda\sqrt{\sigma_i}}\ln\left(\frac{h_i+\mathrm{cth}(\lambda)\sqrt{q_i}}{h_{i-1}+\mathrm{cth}(\lambda)\sqrt{q_{i-1}\sigma_i/\sigma_{i-1}}}\right), \tag{11}$$

where $q_i=\rho_i c_i(T_{mi}-T_0)/\rho_1 c_1(T_{m1}-T_0)$, $\Delta\xi_i=d_i/d_1$, h_i is a dimensionless magnetic field, and $\lambda=\sqrt{t_0/\mu_0\sigma_1 x_1^2}$ is an invariant parameter. In expressions (10) and (11), we set $h_1=1$, $i=2,\dots,N$, where N is the specified number of layers. In this case, as can be seen from (10), the sequence of layer materials should be chosen to reduce the values $\Delta Q_{\mathrm{m}}/\sigma$ in the direction of magnetic field diffusion.

Using the heating constraint and the equation of motion (7) and taking into account the similarity relations (8), we have:

$$V(\lambda)=\left(\frac{\mu_0\sigma_1 L}{2}\left(\frac{\Delta Q_{\mathrm{m1}}}{\bar{\rho}\lambda\Delta\xi\theta(\lambda)}\right)^2\right)^{1/3} \tag{12}$$

$$d(\lambda)=\left(\frac{4(\lambda\Delta\xi)^2 L\bar{\rho}\theta(\lambda)}{(\mu_0\sigma_1)^2\Delta Q_{\mathrm{m1}}}\right)^{1/3} \tag{13}$$

where $\theta(\lambda)=(\mathrm{cth}(\lambda)/h_N)^2$, h_N is the dimensionless magnetic field on the inner surface (x = 0) of the multilayer sheet calculated by expression (10), $\Delta\xi=1+\sum_{i=2}^{N}\Delta\xi_i$, and $\bar{\rho}$ is the average density of the multilayer sheet.

Figure 7 shows curves of the ultimate velocity versus linear mass of multilayer sheets calculated using analytical relations for an acceleration distance of 1 m (the sequence of materials is indicated in the figure).

It can be seen that the use of multilayer sheets allows a considerable increase in the ultimate velocity in terms of the heating conditions, compared to homogeneous sheets.

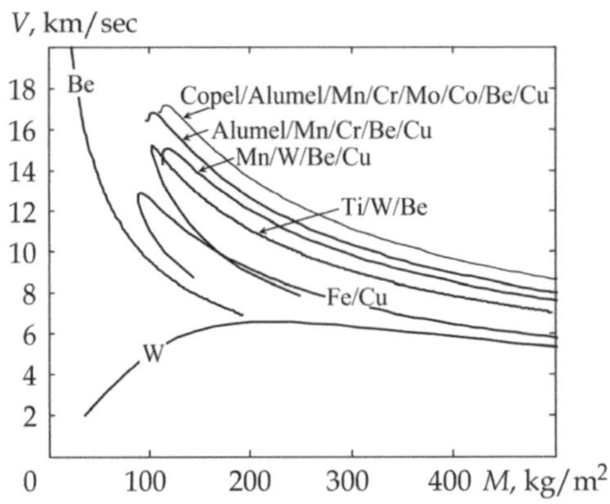

Fig. 7. Ultimate velocity versus mass of multilayer sheets for some sequences of layer material (indicated in the figure).

3.4 Sheets with a composite layer

Let us consider the possibility of increasing the ultimate kinematic characteristics of sheets which contain a composite material layer with electrical conductivity continuously increasing in the direction of magnetic field diffusion (Fig. 4, c).

Generally, we assume that the accelerated sheet of thickness d comprises two layers in contact: a composite layer of thickness d_c consisting of a mixture of two materials (first and second) with different electrothermal properties and a homogeneous layer of thickness d_1 made of the first material (Fig. 4, *c*). Below, the subscripts 1 and 2 are used to denote the parameters of the first and second materials, respectively. Let the electrical conductivity of the first material be higher than the conductivity of the second material $\sigma_1 > \sigma_2$, and let the electrical conductivity at different points of the composite layer be changed as a result of change in the volume concentration $\varepsilon(x)$ of the first material (the x coordinate is reckoned from the sheet surface in contact with the field). Furthermore, the characteristic sizes of the particles comprising the composite are so small that it is possible to ignore the variations in the magnetic and thermal fields due to the discrete dependence of the electrothermal properties of the composite material on the coordinates. Thus, the averaged properties of the composite material are assumed to depend continuously on the x coordinate according to the distribution of the volume concentration $\varepsilon(x)$ at $0 \le x \le d_c$ and $\varepsilon(x) = 1$ at $d_c < x \le d$.

The density ρ and the heat capacity per unit volume C for an arbitrary composite material can be obtained from the relations

$$\begin{aligned} \rho(x) &= \rho_1 \varepsilon(x) + \rho_2 (1 - \varepsilon(x)), \\ C(x) &= \rho_1 c_1 \varepsilon(x) + \rho_2 c_2 (1 - \varepsilon(x)). \end{aligned} \tag{14}$$

At the same time, the dependence of the averaged electric conductivity σ on the volume concentration ε can be determined only for a composite material of known structure or experimentally. Below, we assume that the composite layer has a layered or fibrous structure (the direction of the fibers coincides with the direction of the current), then, we have

$$\sigma(x) = \sigma_1 \varepsilon(x) + (1 - \varepsilon(x)) \sigma_2 \,. \tag{15}$$

The optimum distribution of the volume concentration $\varepsilon(x)$ that ensures uniform heating can be obtained in analytical form using the steady-state solutions of system (5)-(7) admissible for $h_0(\tau)=e^{\tau}$. The optimum law of variation of electrical conductivity in this layer can be found from the condition that the temperatures at each point of this layer reach a certain critical temperature at the end of acceleration. Using the dimensionless variables $\tilde{C}=C/C_1$, $\tilde{\sigma}=\sigma/\sigma_1$, and $\xi=x/d_1$ and the function $y(\xi)=\sqrt{\tilde{C}(\xi)/\tilde{\sigma}(\xi)}$, from (14) and (15) we obtain:

$$\tilde{\sigma}(y^2)=\frac{\tilde{C}_2-\tilde{\sigma}_2}{y^2(1-\tilde{\sigma}_2)+\tilde{C}_2-1}$$

$$\varepsilon(y^2)=\frac{\tilde{C}_2-y^2\tilde{\sigma}_2}{\tilde{C}_2-1+y^2(1-\tilde{\sigma}_2)} \tag{16}$$

From the solution (5) and (6), for the $y(\xi)$ we obtain (Shvetsov & Stankevich, 2003):

$$y'=\frac{\mathrm{d}y}{\mathrm{d}\xi}=\lambda\sqrt{\int_1^{y^2}\tilde{\sigma}(y^2)dy^2+\theta_0^{-1}(\lambda)}$$

$$\xi(y,y(0),\lambda)=\int_y^{y(0)}\frac{dy}{y'(y,\lambda)} \tag{17}$$

where $\theta_0(\lambda)=\left(\mathrm{ch}\,\lambda\,/\,\mathrm{sh}\,\lambda\right)^2$. The thickness of the composite layer ξ_c can be determined by using $y=1$ as the lower limit of integration in expression (17).
The average density of the sheet is determined from (14), (16) and (17):

$$\bar{\rho}(y(0),\lambda)=\frac{1}{\Delta\xi}\int_0^{\Delta\xi}\rho\,\mathrm{d}\xi=\frac{1}{1+\Delta\xi_c}\left(\int_1^{y_0}\frac{(\rho_1\varepsilon(y)+\rho_2(1-\varepsilon(y))\mathrm{d}y}{y'(y,\lambda)}+\rho_1\right) \tag{18}$$

Relations between the ultimate velocities of sheets with composite layers and the sheet mass or thickness can be derived using equations (12) and (13) in which

$$\theta=\left(\frac{\tilde{C}_2-\tilde{\sigma}_2}{1-\tilde{\sigma}_2}\ln\left(\frac{y^2(0)(1-\tilde{\sigma}_2)+\tilde{C}_2-1}{\tilde{C}_2-\tilde{\sigma}_2}\right)+\theta_0^{-1}(\lambda)\right)^{-1},$$

average density is defined by (18) and $\Delta\xi=1+\Delta\xi_c$ is defined by (17).
Figure 8 shows curves of ultimate velocity versus sheet thickness calculated using the above analytical relations for a sheet consisting of a Cu/Fe composite layer and a homogeneous copper layer (curves 3, 4, and 5) and curves of ultimate velocity versus thickness for homogeneous sheets of iron and copper (curves 1 and 2). For curves 4 and 5, the electrical conductivity of iron was decreased by a factor of 10 and 100, respectively.
The calculations were performed for the electrothermal properties of the materials averaged over the temperature range from room temperature to the melting point of Cu.
It should be noted that for each value of the mass per sheet unit area *M* or sheet thickness *d*, there is an optimal distribution $\varepsilon(x)$ that provides the attainment of the maximum velocity over a given acceleration distance.

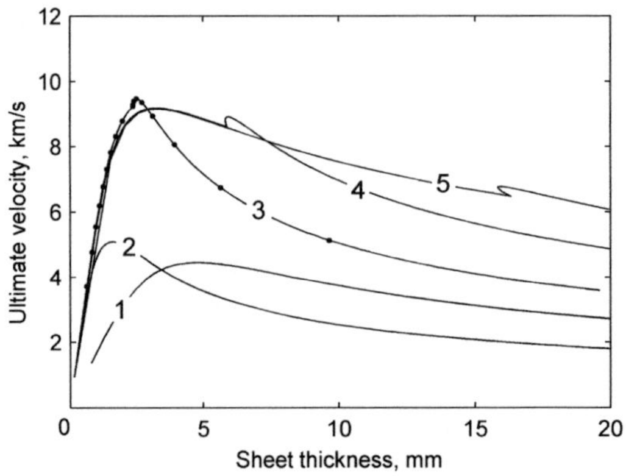

Fig. 8. Ultimate velocity versus sheet thickness for $L = 0.1$ m.

As can be seen (Fig. 8), for a certain sheet thickness $d_{\text{inflection}}$, each of curves 3, 4, and 5 has a point of inflection. The segments of the curves before the points of inflection correspond to the sheet consisting only of a composite layer. The segments of the curves behind the points of inflection were obtained for the sheet consisting of a composite layer and a homogeneous layer. On curve 3, dark circles show the points at which the volume concentration of Cu changes by 0.1 if $d \le d_{\text{inflection}}$ or the relative thickness of the composite layer $\nu = \xi_c / \Delta\xi$ changes by 0.1, if. For $d > d_{\text{inflection}}$ a certain thickness of the sheet in the neighborhood of the point of inflection, up to three different structures of the sheet that ensure its uniform heating can exist. One can see that the ultimate velocity of the sheet containing an optimized composite layer increases by a factor of about two when the total thickness of the sheet exceeds the thickness of the homogeneous copper layer for which its maximum ultimate velocity is attained.

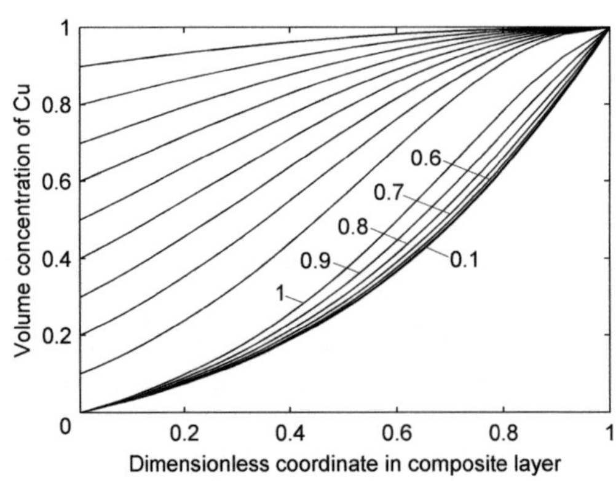

Fig. 9. Distribution of the volume concentration of Cu in a Fe/Cu composite layers

Figure 9 shows the optimum distributions of copper concentration in the composite layer consisting of copper and iron. These distributions correspond to the points on the Cu/Fe curve in Fig. 8. The curves with zero surface concentration correspond to the acceleration of the sheet consisting of a composite layer and a homogeneous copper layer. The numbers at these curves show the relative thickness of the composite layer. The curves with nonzero surface concentration of copper were obtained for a sheet consisting only of a composite layer. One can see that when the relative thicknesses of the composite layer are smaller than 0.7, the optimum profiles of copper distribution in the composite layer practically do not differ from each other.

It can be seen from Fig. 9 that as the thickness of the sheet decreases, the composite layer becomes similar in properties to a homogeneous sheet of the material with high electrical conductivity (first material). For a small values of d, the curve of $V(d)$ (Fig. 8) approaches the asymptote to a "thin sheet" (9) determined for the sheet of the first material. But the ultimate velocity of the composite sheet enters this asymptote for larger values of d than those for the homogeneous sheet. Accordingly, its maximum ultimate velocity can be much higher than the maximum ultimate velocity of the homogeneous sheet. On the other hand, with increase in the thickness of sheets containing a composite layer of any pair of materials, the ultimate velocity always decreases, even in the case of artificial decrease in the electrical conductivity of the first material, as is the case for curves 4 and 5 in Fig. 8. However, for larger d, the ultimate velocity of sheets containing a composite layer is larger than the ultimate velocity of homogeneous sheets made of the materials constituting the compact. Furthermore, it is evident that with decrease in σ_2, the increase in ultimate velocity becomes more considerable.

The analysis performed showed that sheets containing a composite layer with electrical conductivity increasing in the direction of magnetic-field diffusion can be used to advantage to improve the ultimate (under hearting conditions) kinematic characteristics of accelerators. Thus, the ultimate velocity of a sheet containing a composite layer of Fe and Cu is about twice that of homogeneous iron and copper sheets. As the electrical conductivity of iron decreases by a factor of 100, the ultimate velocity can increase by a factor 3.

The analysis showed that increasing the ratio of electrical conductivities of the compact constituents, one can achieve a considerable increase in ultimate velocity compared to a homogeneous sheet. One would expect that use of conducting and nonconducting materials in combination may open up fresh opportunities. However, in this case, to ensure microuniform heating of the composite material, one would need to decrease the characteristic particle size in the compact and/or to use a material with high thermal conductivity as an insulator.

The analysis performed does not cover all aspects of the use of composite materials as current-carrying projectiles accelerated by a magnetic field. In particular, the thermomechanical and strength properties of the compact constituent materials should apparently be chosen in a special manner to ensure the integrity (nonfailure) of the projectile during acceleration.

4. The ultimate kinematic characteristics of railguns with a metal armature

The velocity skin effect (VSE) is a principal factor that limits the use of a metallic armature in electromagnetic railguns in the regime with sliding metallic contact (Young & Hughes, 1982; Thornhill et al., 1989). A sharp increase in the current density due to the VSE at the contact boundary leads to fast heating of the armature in this region in excess of its melting temperature. Metallic contact is lost, and transition to the acceleration regime with plasma

contact occurs. Some undesirable consequences may be failure of the armature, a change in its ballistic characteristics, and enhanced erosion of the rails, which reduces the life time of the EM accelerator. Furthermore, ejection of the eroded material into the interelectrode space behind the accelerated body can result in shunting of the current and deterioration in acceleration. As the ultimate velocity for the regime with sliding metallic contact, Long and Weldon (Long & Weldon, 1989) proposed to consider as an ultimate velocity for sliding metallic contact such a value of the velocity V so that the metallic armature can be accelerated providing that its maximum temperature does not exceed its melt point. For traditional homogeneous materials, the ultimate velocity is usually lower than 1 km/sec (Long & Weldon, 1989; Shvetsov & Stankevich, 1992), and this currently limits the use of conducting solids in railguns.
A considerable number of papers deal with the search for methods of increasing the critical (ultimate) velocity, or, in other words, decreasing the current concentration due to the VSE.
This subsection is concerned with analyzing the ultimate velocity versus projectile mass at fixed acceleration distance for various methods of decreasing the current density at the rail-armature interface. The analysis is performed by numerical solution of the system of equations of unsteady magnetic-field diffusion and unsteady heat transfer in a two dimensional formulation. Homogeneous and multilayer projectiles and homogeneous rails and rails with a resistive coating are considered.

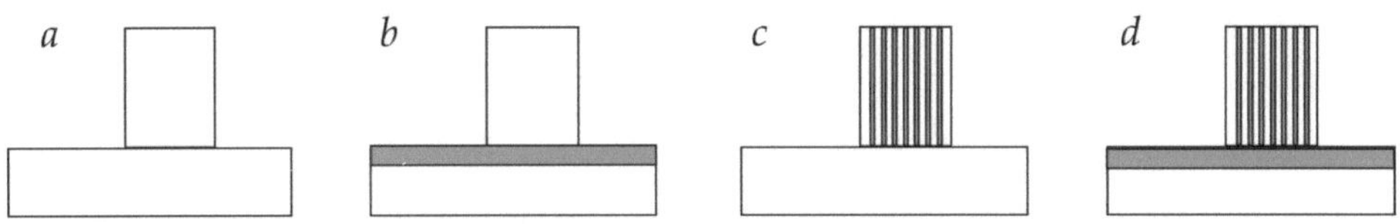

Fig. 10. Configuration of calculation regions. Homogeneous armature and rails (*a*), homogeneous armature and rails with a resistive layer (*b*), multilayer armature and homogeneous rails (*c*), and multilayer armature and rails with a resistive layer (*d*).

4.1 Formulation of the problem

We consider the acceleration of homogeneous and multilayer conducting solids in electromagnetic launchers with homogeneous rails and rails with a high resistive layer (Fig.10). Accelerated bodies will be called armature or projectile, as is common in the literature. The time-dependent distributions of the magnetic field and the temperature of the armature and rails were determined by numerical solution of the system of unsteady equations of magnetic-field diffusion and heat transfer in a two-dimensional formulation. Neglecting the effects associated with the system finiteness in the direction z and displacement currents, these equations in a moving frame of reference connected with the armature can be written in the form:

$$\mu_0\left(\frac{\partial H}{\partial t}-V\frac{\partial H}{\partial x}\right)=\frac{\partial}{\partial x}\frac{1}{\sigma}\frac{\partial H}{\partial x}+\frac{\partial}{\partial y}\frac{1}{\sigma}\frac{\partial H}{\partial y} \tag{19}$$

$$\rho c\left(\frac{\partial T}{\partial t}-V\frac{\partial T}{\partial x}\right)=\frac{\partial}{\partial x}k\frac{\partial T}{\partial x}+\frac{\partial}{\partial y}k\frac{\partial T}{\partial y}+\frac{1}{\sigma}\left(\frac{\partial H}{\partial x}\right)^2+\frac{1}{\sigma}\left(\frac{\partial H}{\partial y}\right)^2. \tag{20}$$

It was assumed that the electrothermal properties of materials do not depend on temperature and there is ideal electric and thermal contact on the boundaries between the armature and the rails and between the resistive coating and the support, the continuity conditions for the magnetic field, temperature, the tangential component of the electric field, and the normal components of the current density and heat flux are satisfied. The magnetic field in the railgun channel was assumed to be known and equal to the linear current density through the armature.

The velocity of the armature V and the distance L traveled by it are determined by integrating the equation of motion:

$$M\frac{dV}{dt}=\frac{\lambda}{2}\left(\frac{I}{b}\right)^2, \quad \frac{dL}{dt}=V \tag{21}$$

4.2 Homogeneous armature and resistive layer

The effect of a resistive layer on the ultimate kinematic characteristics of a homogeneous armature was studied in a series of calculations for three materials of the layer with considerably different electrical conductivities: titanium ($\sigma = 1.8\cdot10^6$ (Ohm·m)$^{-1}$), Copel alloy (Ni: 42.5 - 44%, Mn: 0.1 - 0.5%, Cu: the rest, $\sigma = 0.21\cdot10^6$ (Ohm·m)$^{-1}$), and graphite ($\sigma = 0.04\cdot10^6$ (Ohm·m)$^{-1}$). The thickness of resistive layer d was 0÷1.2 mm. Armatures made of aluminum, copper, and tungsten were examined. The acceleration distance was 1 m.

Use of a resistive layer has an ambiguous effect on the rate of change in the maximum armature temperature, and hence, and the ultimate velocity. The ultimate velocity can both increase and decrease, depending on the thickness and conductivity of the layer, the electrothermal properties and dimensions of the armature, and the specified acceleration distance.

Two regimes are typical of heating in the armature. In the first of this, the change in the maximum temperature in the armature is primarily determined by Joule heating of the armature, and the second regime occurs when the armature is heated as a result of increase in the temperature of the contact boundaries due to Joule heating of the resistive layer.

The increase in the maximum temperature in the armature due to Joule heating of the resistive layer proceeds mainly in the initial stage of acceleration. As the armature is accelerated, the heating of the resistive layer decreases, and so does the maximum temperature of the armature. With a further increase in the velocity, the Joule heating of the armature due to current concentration caused by the velocity skin effect becomes more intense, and the maximum temperature begins to increase again.

Figure 11 *a, b* shows the dependences of the ultimate velocity on the thickness of the resistive layer for a aluminum armature (Fig. 11, *a*), copper armature (Fig. 11, *b*) with various lengths in the direction of motion and for various materials of coatings calculated for an acceleration distance of 1 m. The figures on the curves denote the lengths of armatures in mm. The continuous curves refer to the resistive coating of graphite, the dotted curves to titanium and the dashed curves to Copel.

For all the armature materials studied, use of resistive coatings of titanium and Copel leads to an increase in the ultimate velocity compared to the case where rails without coating are used. A titanium layer ensures an only 15%÷20% increase in the velocity for acceleration of an Al armature (Fig.11, *a*). The dependences for W and Cu armatures accelerated on rails with a titanium coating show the same relative increment of the velocity as for an Al armature. Unlike a titanium coating, a Copel layer increases the ultimate velocity by a factor of 2 or 3. As can be seen from Fig. 11, the ultimate velocity decreases with increase in the length of the armature. This dependence of the ultimate velocity of the length of the armature is typical of cases where Joule heating of the armature, whose intensity is determined by the current concentration due to the VSE, plays a predominant role.

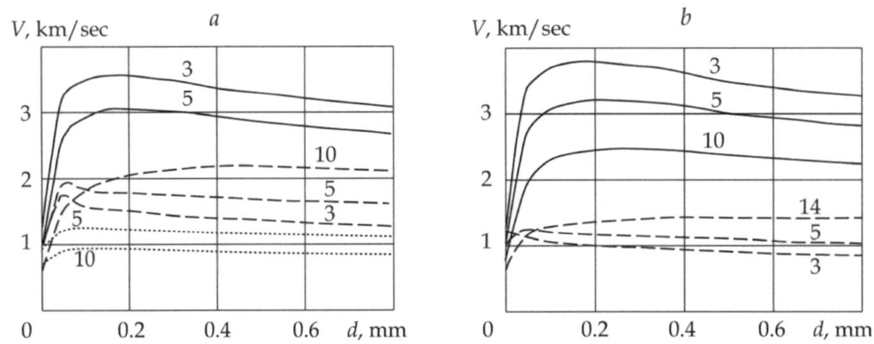

Fig. 11. The dependences of the ultimate velocity on the thickness of the resistive layer for a plane aluminum armature (*a*) and copper armature(*b*).

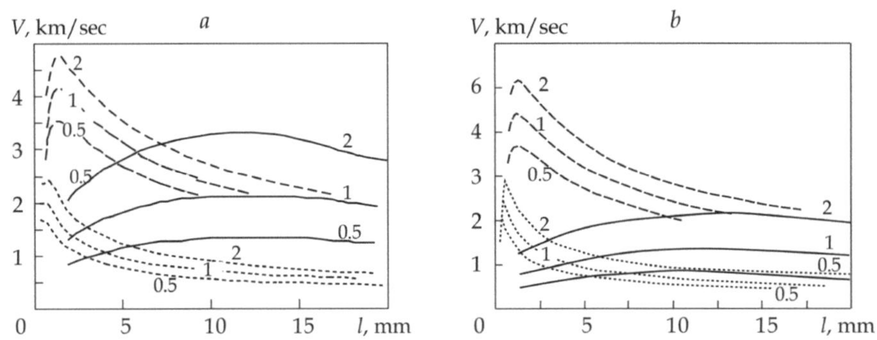

Fig. 12. The dependences of the ultimate velocity on the length of an armature.

Figures 12 *a*, *b* give dependences of the ultimate velocity on the length of an armature calculated at various acceleration distances L = 0.5 m, 1 m, and 2 m for homogeneous armatures of Al (*a*) and Cu (*b*). The dotted curves in these figures show the dependences obtained for the copper rails without a resistive layer. The dashed curves show the dependences obtained using a Copel resistive layer, and the solid curves correspond to a graphite resistive layer. The figures on the curves denote the acceleration distance (m).
It is seen that a reasonable choice of the coating material can lead to a significant (severalfold) increase in the ultimate velocity.

4.3 Multilayer armature and homogeneous rails

Let us consider the ultimate kinematic characteristics of multilayer armatures with orthotropic conductivity (4.3.1) and multilayer armatures with alternating layers of materials with high and low conductivity (4.3.2) during acceleration in railgun.

4.3.1 Multilayer armature with insulating layers

An increase in the ultimate velocity in terms of the heating conditions can be achieved through the use of a multilayer armature consisting of a set of alternating conducting and insulating layers (Fig. 10, *c*) (Shvetsov & Stankevich, 1997). The insulating layers provide rapid penetration of the field from the middle part of the armature to the interface, due to the infinite rate of field diffusion in these layers. This increases the size of the region of current flow through the interface, thus reducing the current density and armature heating rate. This effect is the greater the smaller the thickness of the layers and the larger the distance between the rails *h*.

To illustrate the potential of this type of armatures, we consider an armature with orthotropic conductivity, assuming infinitely small sizes of the insulating and conducting layers. Using Faraday's law and taking into account that at the center of symmetry of the armature (for $y = h/2$), the x component of the electric field $E_x = 0$ and that the tangential component of the electric field is continuous on the contact boundary, we obtain the equation describing the distribution magnetic field in the armature (Shneerson et al., 1996):

$$\frac{\partial H}{\partial t} = \frac{1}{\mu \sigma_{\text{armature}}} \frac{\partial^2 H}{\partial x^2} + \frac{2}{\mu \sigma_{\text{rail}} h} \left. \frac{\partial H}{\partial y} \right|_{y=0}$$

The effect of the singularities on the contact boundary due to both the relative motion of the conductors and the passage of the current around the corner point is described by the last term on the right side of this equation, which can be made arbitrarily small by choosing a sufficiently large value of *h*.

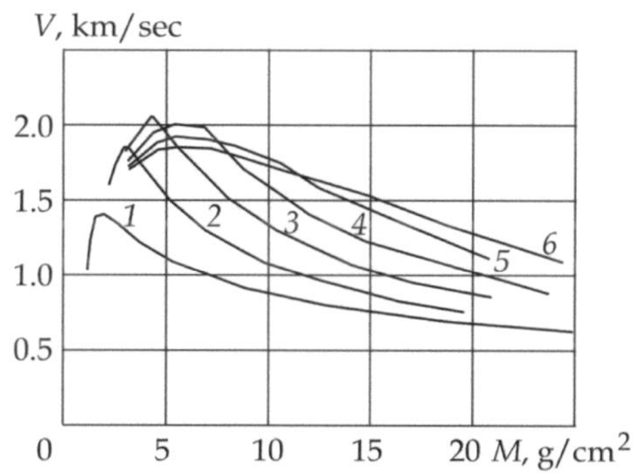

Fig. 13. Multilayer tungsten armature. Curves 1 to 6 correspond to numbers of layers 1, 4, 6, 10, 16, and 22, respectively.

Figure 13 shows the ultimate velocity of a tungsten armature versus its mass for various numbers of tungsten layers with insulating layers between them. The bore cross-section is $b \times h = 2\ \text{cm} \times 2\ \text{cm}$, and the acceleration distance is 1 m. The curves labeled 1 to 6 correspond to numbers of layers of 1, 4, 6, 10, 16, and 22.

One can see a shift of the maximum toward large masses and a total increase in the velocity for large masses.

Figures 14 give dependences *V*(*d*) obtained for armatures composed of conducting layers of aluminum (Fig. 14, *a*) and tungsten (Fig. 14, *b*). Curves 1, 2, and 3 correspond to a multilayer armature (*N*=22) with $h = b = 1$, 2, and 4 cm, respectively; curve 4 corresponds to a homogeneous armature; and curve 5 to induction acceleration of the same multilayer armature for which the equations of motion (21) are valid. Indexes a, b, c correspond to acceleration distances 0.5 m, 1 m, and 2 m respectively. The ultimate velocity of the multilayer armature in the railgun can be seen to be much higher than that of the homogeneous armature; however, for the given range of h values, it remains much lower (about two times) than the velocity achieved during induction acceleration.

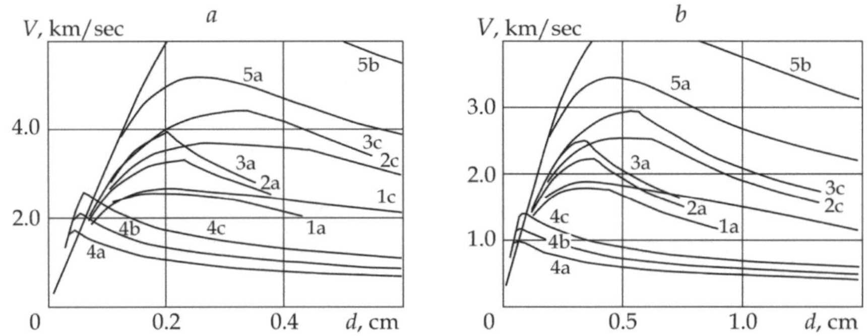

Fig. 14. The dpendances of the ultimate velocity for aluminum (*a*) armatures and tungsten (*b*) multilayer armatures.

A comparison was made of the ultimate kinematic characteristics for two projectile configurations: a multilayer armature 2 with orthotropic electrical conductivity and a nonconducting armature of stabilized structure 3 (Fig. 15, *a, b*).

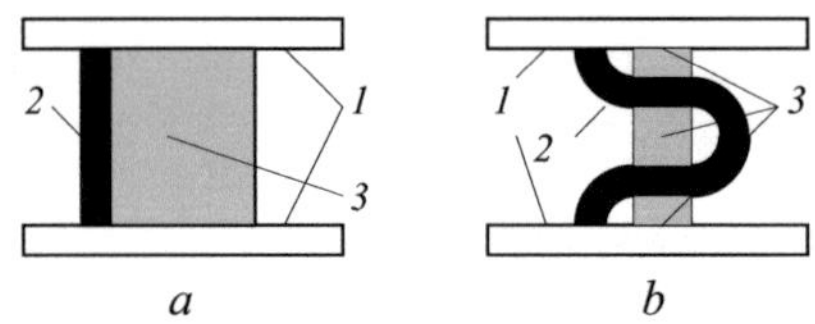

Fig. 15. (*a*) plane multilayer armature, (*b*) shaped multilayer armature. *1* rails; *2* armature; *3* supporting structure.

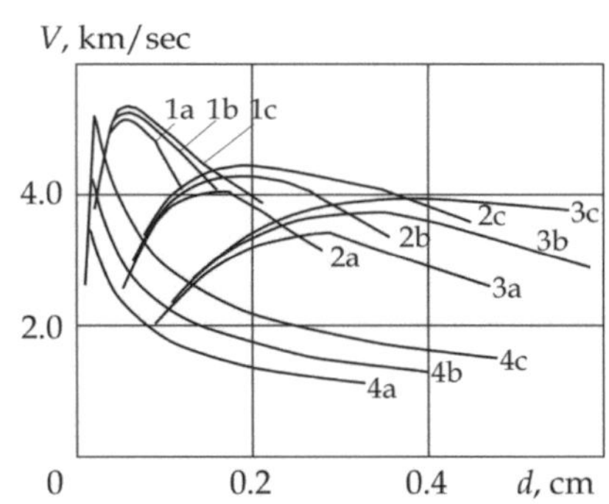

Fig. 16. Shaped aluminum armature. Curves 1 to 3 correspond to f=1, 0.5, 0.3. Curves 4 are for homogeneous armature.

Let us consider some ultimate velocities of a multilayer armature of mass M_a which has a nonconducting supporting structure of mass M_s to provide stability of its motion. Suppose that the ratio $f = M_a / M$ remains constant with a change in the total mass $M = M_a + M_s$. Figure 16 shows dependences *V(M)* for an armature with conducting tungsten layers (*N*=22). Curves 1, 2, and 3 correspond to values *f* = 1, 0.5, and 0.3, respectively; curves 4 correspond to a homogeneous flat armature; letters a, b, and c denote acceleration distances *L* = 0.5, 1, and 2m, respectively. One can see that in the range of large masses, the ultimate velocity of the shaped armature far exceeds the ultimate velocity of the flat armature.

4.3.2 Multilayer armature with high and low electroconductivity layers

Figure 17 presents dependences *V(M)* for an armature composed of copper and titanium alloy layers ($\sigma = 1.8 \cdot 10^6 (\Omega \cdot \mathrm{m})^{-1}$). The curves labeled 1, 2, 3, and 4, correspond to values of *N* = 1, 4, 6, 10. One can see that the maximum value of the ultimate velocity of this armature exceeds the ultimate velocity of the homogeneous copper armature, but the increase in the velocity achieved for this armature is smaller than that for the multilayer armature with insulating layers.

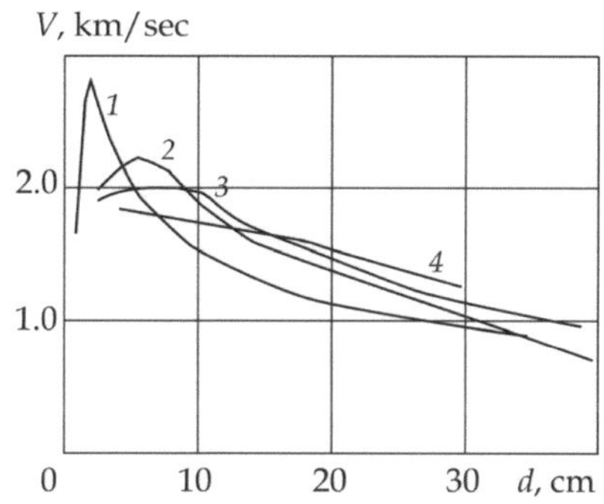

Fig. 17. Copper and titanium alloy composed armature. The curves 2 to 4 correspond to *N* = 4, 6, 10.

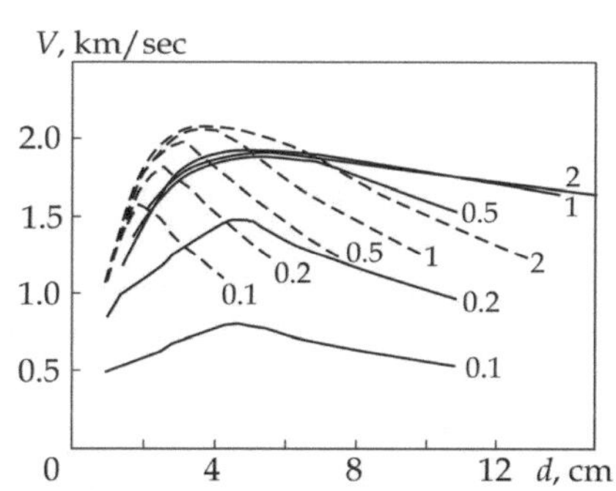

Fig. 18. Ultimate velocity as a function of the multilayer tungsten armature length for acceleration distances of 0.1, 0.2, 0.5, 1, and 2 m.

4.4 Multilayer armature and rails with resistive layer

In railguns with homogeneous rails, the ultimate velocities of armatures consisting of alternating conducting and nonconducting layers, far exceed the ultimate velocities of homogeneous armatures (see Fig. 13 and 14). In addition, for armatures with orthotropic conductivity, the maximum ultimate velocity is reached for greater armature lengths compared to homogeneous armatures.

The curves of the ultimate velocities of a multilayer tungsten armature versus its length l calculated for acceleration distances of 0.1, 0.2, 0.5, 1, and 2 m are given in Fig. 18. The dotted curves show the calculation results for a multilayer armature and rails without a coating (see Fig. 10, *c*), and the solid curves show the results for the same armature and rails with a resistive Copel alloy coating (see Fig. 10, *d*). The figures on the curves indicate the acceleration distances in meters. It is evident that, indeed, there is a shift of the maximum ultimate velocity toward larger values of l, and this maxima is rather flat.

It is important that in this case, the ultimate velocity does not depend on the acceleration distance for $L > 0.5$ m. The above is true for Al and Cu armatures with orthotropic conductivity, and for $L \geq 0.5$ m. This indicates that the dimensions of the region of the contact boundary through which the current passes are determined mainly by the properties and dimensions of the resistive coating (in this case, Copel alloy).

The results obtained indicate that a resistive coating can be used to advantage to decrease the current concentration in the armature due to the velocity skin effect. This considerably decelerates armature heating near the contact boundaries. As a result, the ultimate velocity to which the armature can be accelerated in a channel of a specified length with retention of solid metal contact with the rails can be increased by a factor of $2 \div 4$ and the kinetic energy of the armature can be increased by a factor of $4 \div 16$ compared to the case of rails without a coating. With a further decrease in the conductivity of the resistive layer, the current concentration in the armature decreases, but, in this case, overheating and failure of the resistive layer can take place.

The ultimate maximum velocity of a multilayer armature increases with an increase in the number of layers and bore height h. For $h = 2 \div 6$ cm and an acceleration distance of 2 m, the ultimate velocity can reach $2 \div 4$ km/sec for multilayer copper and aluminum armatures and $1.8 \div 2.8$ km/sec for multilayer tungsten armature. These ultimate velocities are $2 \div 3.5$ times higher those of a homogeneous armature having the same mass.

Thus, the analysis shows that the ultimate velocities attained for homogeneous materials can be considerably increased by changing the structure and thermophysical properties of projectile and rail materials.

It should be noted that the use of resistive coatings in a number of features that can considerably lower the attainable velocities. First, the velocity of magnetic-field diffusion along a resistive coating is higher than the diffusion velocity in the armature. As a result, the armature current flows along the contact boundary in the opposite direction to the rail current. Interaction of these currents gives rise to a magnetic-pressure force that repels the contact surfaces. If special precautions are not taken, this can lead to a loss of metal contact between the projectile and the rails. Since the repelling force decreases as the magnetic field penetrates into the armature and the armature velocity increases, one method for overcoming this problem is to use a resistive coating with conductivity decreasing in a predetermined manner in the direction of motion. Second, the resistive layer fails under the considerable thermal stresses caused by sharp temperature variations on the boundaries of the resistive layer. These stresses can be reduced using a resistive material with high thermal conductivity, melting point, and mechanical strength.

5. Comparison between 2D and 3D electromagnetic modeling of railgun

The results presented in fourth subsection have shown that the ultimate (with respect to heating conditions) kinematic characteristics of launchers depend greatly on the electrothermal properties and the structure of the materials used, the projectile weight, the acceleration dynamics determined by the shape of the current pulse, and the acceleration distances. They can be considerably increased (severalfold) by using multilayer and composite conductors as current-carrying elements of launchers and by optimizing the current pulse shape.

However, these results were obtained by numerical modeling of launchers in a two-dimensional spatial formulation. Thus, it is unclear how the maximum current density on the contact boundary, its dependent armature heating rate, and the ultimate kinematic characteristics in a real launcher differ from the values obtained by two-dimensional modeling. The purpose of this section is to compare results of 2-D and 3-D calculations of armature heating for various armature shapes, matched acceleration dynamics, and total-current distribution curves.

5.1 Formulation of the problem

Comprehensive three-dimensional modeling of a rail launcher requires considerable computational resources; therefore, in the present work, we confine ourselves to a consideration of the launcher region which includes the armature and part of the rails in immediate proximity to the armature (Fig. 19). The symmetry of the problem allows the model to be simplifies to one fourth of the armature and rails. This area is of greatest interest because it is in the neighborhood of the armature that the distribution of the magnetic field and current has a substantially three-dimensional form.

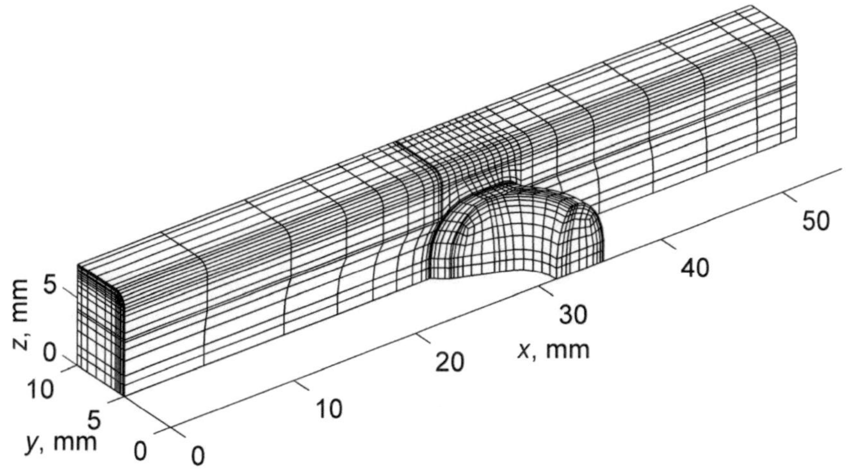

Fig. 19. Model of an electromagnetic railgun.

Further simplification of the problem can be achieved by taking into account that, even at relatively low velocities of armature motion (~50 m/sec), the currents flow primarily in a thin surface layer on the armature and their distribution is determined mainly by the shape of the armature and rails and by the velocity of the armature. In this area, the current density distribution can found using steady-state solutions of the Maxwell equations. In addition, we assume that, during acceleration, ideal electrical and thermal contact between the armature and rails is preserved and the electrothermal properties of the materials remain constant (do not depend on temperature).

From Maxwell equations we obtain a system of equations for the vector and scalar potentials ($\boldsymbol{A}, \varphi$), which, for the region of conductors (armature and rails), in a moving frame of reference connected with the armature, can be written as follows:

$$\boldsymbol{j} = -\frac{1}{\mu_0}\nabla^2 \boldsymbol{A} = -\sigma(\nabla\varphi - \boldsymbol{V}\times\nabla\times\boldsymbol{A}) \tag{22}$$

$$\nabla\cdot\boldsymbol{j} = \nabla\cdot(\sigma(\nabla\varphi - \boldsymbol{V}\times\nabla\times\boldsymbol{A})) = 0 \tag{23}$$

In the surrounding nonconducting space, where the current density, $\mathbf{j} = 0$ instead of (22) we have:

$$\nabla^2 \boldsymbol{A} = 0 \tag{24}$$

The temperature distribution in the armature and rails is given by the solution of the unsteady heat-conduction equation:

$$\rho c\left(\frac{\partial T}{\partial t} + \boldsymbol{V}\cdot\nabla T\right) - \nabla\cdot k\nabla T = \frac{\boldsymbol{j}^2}{\sigma} \tag{25}$$

The electromagnetic part of the problem, in fact, contains one boundary condition, which requires that the components of the vector potential vanish at an infinite boundary. At all inner boundaries between the conductors and the surrounding nonconducting space, the continuity conditions for the vector-potential components and their derivatives are satisfied.

The boundary conditions for the heat-conduction equation (25) specify the absence of heat transfer with the surrounding medium during acceleration and reduce to Neumann conditions at all boundaries of the conductors.

5.2 Numerical solution procedure

Equations (22), (23), (25) were approximated by the Galerkin weighted residual finite-element method in the conductors, and the boundary element method was used outside the conductors. The numerical technique is described in more detail in (Shvetsov & Stankevich, 2009).

Owing to the stationary character of the electromagnetic part of the posed problem, the solutions for the electromagnetic and thermal problems can be obtained separately. At the first stage, for a certain set of armature velocities V_i ($i = 1,...N$, $V_i = |\mathbf{V}|$) and the total current in the launcher I, we solved problem (22)-(24) and found the distributions of the current density in the armature and rails $\mathbf{j}(\mathbf{r}, V_i)$ and the accelerating magnetic pressure forces $F(V_i)$. The times t_i (that correspond to the velocities V_i) were calculated by integrating the equation of motion $M\mathrm{d}V / \mathrm{d}t = F(V)$, where M is the total mass of the accelerated projectile and $F(V)$ the interpolated dependence of the accelerating magnetic force. After that, we solved the armature heating problem using the time dependence of the current density interpolated over the known values of $\mathbf{j}(\mathbf{r}, t_i)$ in the right side of Eq. (25) (Shvetsov & Stankevich, 2011).

To compare 3-D and 2-D modeling results, in addition to 3D modeling, we solved the problem of armature and rail launcher heating in a 2-D formulation (19), (20). The computation domain had the geometry of the middle section of a 3-D launcher (plane $z = 0$ in Fig. 19). The value of the magnetic field in the launcher channel was set equal to the linear current density I / b (b is the rail width). In the two-dimensional formulation, the time dependence of the armature velocity obtained in the 3-D calculations was used. For a comparison with the 3-D modeling results, we solved the two-dimensional problem in two steps: first calculated the current density distributions as a function of the armature velocity and then solved the unsteady thermal problem.

5.3 Results

In the first step, we calculated the current density distributions and magnetic field, the accelerating magnetic force, and the inductance per unit length for an electromagnetic rail launcher with a 1×1 cm channel cross-section, copper rails 1.5 cm wide and 0.5 cm thick (Fig. 19). To study the effect of the armature shape and its electrothermal properties on the current density distribution, heating dynamics, and inductance gradient, we considered various armature shapes: rectangular, standard C-shaped, cylindrical, modified C-shaped, saddle-shaped (Figs. 20 *a, b, c, d, e*, respectively) made of copper and an Al alloy.

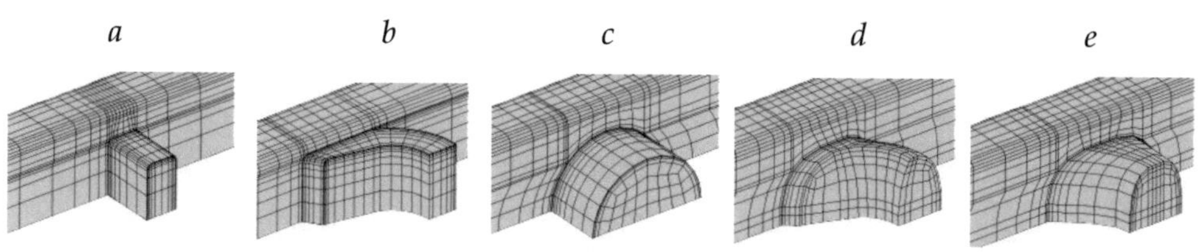

Fig. 20. Armature shapes. *a*) rectangular, *b*) standard C shaped, *c*) cylindrical, *d*) curved C shaped, *e*) saddle-shaped.

The calculations show that the maximum current density is reached on the perimeter of the rail-armature interface. This results, first, from the presence of the geometrical singularity at the armature-rail interface and, second, from the velocity skin-effect, which pushes the current toward the rear boundary of the perimeter of the contact interface. Here it should be noted that the presence of the edges formed by the intersection of the flat lateral surfaces of the armature favors an additional increase in the current density at the points of intersection of the edges with the rail surface (Satapathy et al., 2007). The position of the point at which the maximum absolute value of the current density is reached depends on the shape, electrical conductivity, and velocity of the armature. This point can be located on the frontal, rear or any other part of the perimeter. We note that, regardless of the material for rectangular and cylindrical armatures, the current density maximum is always located on the rear part of the interface. As the armature velocity increases, the velocity skin effect begins to play a dominant role and the current density maximum, which may initially be located on the frontal part of the curve, is shifted along the perimeter of the interface to its rear part. This process was observed for all the remaining armature shapes studied. The armature velocity, at which this shift begins, depends on the electrical conductivity of the armature. For armatures of identical shape, the lower the electrical conductivity of the armature, the higher the velocity at which the shift of the current density maximum begins.

In two-dimensional modeling of the armature current density distribution under the adopted boundary conditions, the maximum current density is always reached at the point of contact of the rear boundary of the armature to the rail boundary. Thus, the 2D model does not allow the current density distribution to be calculated correctly for armatures for which the maximum current density is reached (in a certain velocity range) on the frontal part of the contact boundary, in particular, for C-shaped armatures. However, by modifying the shape of such an armature by rounding its leg and the frontal side (Fig. 20, *c*), it is possible to achieve a more uniform current density distribution along the perimeter of the contact interface with transition of the current density maximum to rear boundary at velocities of 100 m/sec for the copper armature and 200 m/sec for the aluminum armature.

In the second step, the technique described above was used to calculate the nonstationary temperature distribution in the armature. It was assumed that the armature is accelerated in a rail accelerator at a constant total current equal to 100 kA.

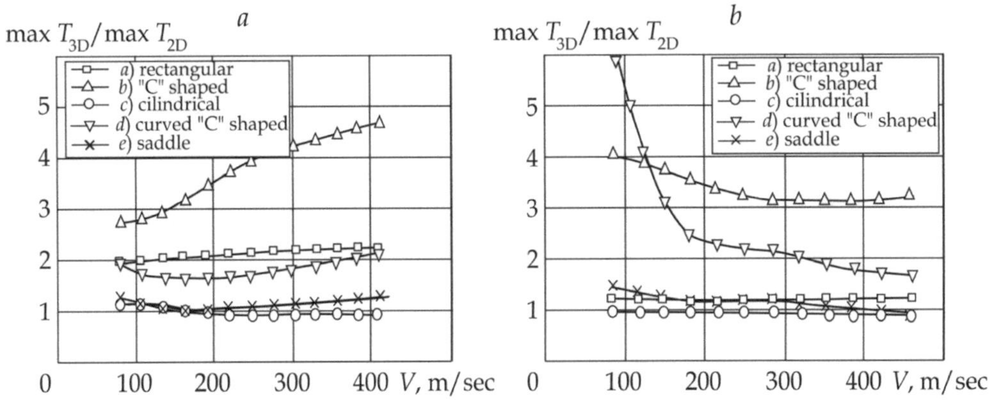

Fig. 21. Ratio of the maximum armature temperatures obtained by 3D and 2D calculations versus Cu (*a*) and Al (*b*) armature velocity

The ratio of the maximum armature temperatures obtained by the 3D and 2D calculations ($\vartheta(V) = \max T_{3D} / \max T_{2D}$) for armatures of various shapes versus velocity are given for copper armatures in Fig. 21, *a* and for armatures made of aluminum alloy in Fig. 21, *b*. The mass of copper armatures of various shapes was *a*) 2.66g, *b*) 3.54 g, *c*) 6.95 g, *d*) 4.13 g, *e*) 5.43 g. The mass for armatures of aluminum alloy *a*) 0.82g, *b*) 1.08 g, *c*) 2.13 g, *d*) 1.27 g, and *e*) 1.67 g.

It is evident that for almost all armature shapes (except for the copper cylindrical armature) $\vartheta(V)$ is larger than unity. The ratio of the maximum armature temperatures obtained in the 3-D and 2-D calculation depends weakly on the velocity, especially at high armature velocities. The value of this ratio is determined primarily by the armature shape. A significant difference between the 3D and 2D calculation results for the maximum temperature is observed for the C shaped (Fig. 21). This is due to the fact that, in these cases, the region of the maximum heating at low velocities is located on the frontal contact interface, which is not described by two-dimensional modeling. For the cylindrical and saddle-shapes armatures, the temperatures ratio $\vartheta(V)$ is close to unity.
From the results it follows that, for armatures of some geometrical shapes (in our case, these are cylindrical and saddle shapes), a two-dimensional description of the armature heating is in reasonable agreement with that in a three-dimensional description. Hence, one can hope that the severalfold increase in the ultimate (under heating conditions) kinematic characteristics predicted by the two-dimensional modeling may be achieved for real rail accelerators of solid bodies.

6. Conclusion

A study was made of the effect of the structure and electrothermal properties of the materials of the electrodes and projectiles on the ultimate kinematic characteristics of three types of electromagnetic launchers: plasma armature railguns, coilguns, and railguns with conducting bodies. Homogeneous, layered, and composite materials were considered. The study shows that by controlling the structure and electrothermal properties of electrode materials and projectiles, it is possible to significantly reduce the thermal constraints and increase the ultimate projectile velocities by a factor of two to four.

7. References

Barber J. P. (1972). The acceleration of macroparticles and a hypervelocity electromagnetic accelerator, *Ph.D. Thesis,* The Australian National University.

Barber J. P., Bauer D. P, Jamison K., Parker J. V., Stefani F. & Zielinski A. A. (2003). Survey of armature transition mechanisms, *IEEE Transaction on Magnetics,* Vol. 39, No. 1, part 1. (January,2003), pp. 47–51, ISSN 0018-9464.

Fair H. D. (2005). Electromagnetic Launch Science and Technology in the United States Enters a New Era, *IEEE Transaction on Magnetics,* Vol. 41, No. 1, (January, 2005), pp. 158-164, ISSN 0018-9464.

Fair H. D. (2007). Progress in Electromagnetic Launch Science and Technology, *IEEE Transaction on Magnetics,* Vol. 43, No. 1, (January, 2007), pp. 93-98, ISSN 0018-9464.

Jackson G. L., Farria L. K. & Tower M. M. (1986). Electromagnetic railgun extended-life bore material test results, *IEEE Transaction on Magnetics,* Vol. 22, No. 6, (December 1986), pp. 1542–1545, ISSN 0018-9464.

Harding J. T., Kaplan R. B., Pierson H. O., Tuffias R. H. & Upshaw J. L. (1986). Chemically vapor deposited materials for railguns, *IEEE Transaction on Magnetics,* Vol. MAG-22, No. 6, (December 1986), pp. 1506–1509, ISSN 0018-9464.

Haugh D. C. & Gilbert S. (2003). U. K. Electric Gun National Overview, *IEEE Transaction on Magnetics,* Vol. 39, No.1, (January, 2003), pp. 18-21, ISSN 0018-9464.

Hawke R. S. & Scudder J. K. (1979). Magnetic propulsion railguns: their design and capabilities. *Proceedings of 2nd International Conference on Megagauss Magnetic Fields Generation and Related Topics,* Plenum Press, New York and London, ISBN 0-306-40461-3, Washington, D.C. May 30-June 1, 1979, pp. 297–311.

Karpova I. M., Semakhin A. N., Titkov V. V. & Shneerson G. A. (1990). Analysis of Methods of Lowering Heating of and Thermal Stresses in the Coils in High pulsed Magnetic Fields, *Proceedings of 5th International Conference on Megagauss Magnetic Fields Generation and Related Topics,* Nova science Publisher, New York, ISBN 0-941743-86-1, Novosibirsk, USSR, July 3 - 7, 1989.

Knoepfel H. (1970) *Pulsed High Magnetic Fields,* North Holland, ISBN 0471885320, Amsterdam—London.

Lehmann P. (2003). Overview Of the Electric Launch Activities at the French-German Institute of Saint-Louis (ISL), *IEEE Transaction on Magnetics,* Vol. 39, No.1, (January, 2003), pp. 24-28, ISSN 0018-9464.

Long G.C., Weldon W.F. (1989). Limits to the velocity of solid armature in railgun, *IEEE Transaction on Magnetics,* Vol. 25, No. 1, (January, 1989), pp. 347-352, ISSN 0018-9464.

Powell J. D. (1984). Thermal-energy transfer from arc to rails in arc-driven railgun, *IEEE Transaction on Magnetics,* Vol. MAG-20, No. 2, (February, 1984), pp. 395–398, ISSN 0018-9464.

Satapathy S., Watt T & Persad Ch. (2007) Effect of Geometry Change on Armature Behavior, *IEEE Transaction on Magnetics,* 2007. Vol. 43, No. 1. (January, 2007), pp. 408-412, ISSN 0018-9464.

Shneerson G. A., Karpova I. M. & Titkov V. V. (1996). The Inverse Problem of Superfast Contact Theory for an Accelerated Body with Orthotropic Conduction, *Journal of Applied Mechanics and Technical Physics,* Vol. 37, No. 2, pp. 200-205, ISSN 0021-8944.

Shrader J. E., Bohn A. J. & Thompson J. G. (1986). Railgun experimental results due to varying bore and arc materials, and varying the number of barrel turns, *IEEE Transaction on Magnetics,* Vol. MAG-22, No. 6, (November 1986), pp. 1739–1741, ISSN 0018-9464.

Shvetsov G. A., Anisimov A. G., Bashkatov Yu. L. & Chistyakov V. P. (1987) Railgun accelerators of macroparticles, Part 1: General characteristics. *Proceedings of 4th International Conference on Megagauss Magnetic Fields Generation and Related Topics,* Plenum Press, New York, 1987, ISBN 0-306-42574-2, Santa Fe, New Mexico, USA, July 14–17, 1986, pp. 775–794, ISBN 0-306-42574-2.

Shvetsov G. A., Anisimov G. A., Stankevich S. V. et al. (1992). Interaction between plasma piston and railgun electrodes, *Proceedings of 8th IEEE International Pulsed Power Conference,* IEEE Publishing Service, 1992, ISBN 0-7803-0176-5, San Diego, CA., June 16-19, 1991, pp. 771–777.

Shvetsov G. A. & Stankevich S. V. (1992). Ultimate Velocities of Magnetically Driven Plates, *Proceedings of 6th International Conference on Megagauss Magnetic Fields Generation and*

Related Topics, Nova Science Publisher Inc., 1992, ISBN 1-56072-160-X, Albuquerque, New Mexico, Nov. 8-11, 1992. pp. 852-857.

Shvetsov G. A. & Stankevich S. V. (1994). Ultimate Velocities of Plates Accelerated by Magnetic Field, *Journal of Applied Mechanics and Technical Physics*, Vol. 35, No. 3, (May, 1994), pp. 336-344, ISSN 0021-8944.

Shvetsov G. A. & Stankevich S. V. (1995). Critical current density in railguns with composite electrodes, *IEEE Transaction on Magnetics*, Vol. 31, No. 1, (January 1995), pp. 237-242, ISSN 0018-9464.

Shvetsov G. A. & Stankevich S. V. (1997). Ultimate kinematic characteristics of armatures with ortotropic and anisotropic electroconductivity, *IEEE Transaction on Magnetics*, Vol. 33, No. 1, (January 1997), pp. 266-271, ISSN 0018-9464.

Shvetsov G. A., Rutberg P. G. & Kolikov V. A. (2001). Problems, Results and Prospects of Electric Launch in Russia, *IEEE Transaction on Magnetics*, Vol. 37, No.1, part I of two part, (January, 2001), pp. 42-45, ISSN 0018-9464.

Shvetsov G. A., Rutberg P. G. & Savvateev A. F. (2003). Results of Resent Research on Electromagnetic Launch Technology in Russia, *IEEE Transaction on Magnetics*, Vol. 39, No.1, (January, 2003), pp. 29-34, ISSN 0018-9464.

Shvetsov G. A. & Stankevich S. V. (2003). Ultimate Kinematic Characteristics of Composite Solids, *IEEE Transaction on Magnetics*, Vol. 39, No.1, (January, 2003), pp. 327-331, ISSN 0018-9464.

Shvetsov G. A., Rutberg P. G. & Budin A. V. (2007). Overview of Some Resent Research in Russia, *IEEE Transaction on Magnetics*, Vol. 43, No.1, (January, 2007), pp. 99-106, ISSN 0018-9464.

Shvetsov G. A. & Stankevich S. V. (2009). Effect of Shaper of Metal Solids on their Joule Hearting in Electromagnetic Rail Launchers, *Journal of Applied Mechanics and Technical Physics*, Vol. 50, No. 2, (Mach, 2009), pp. 342-351, ISSN 0021-8944.

Shvetsov G. A. & Stankevich S. V. (2011). Three-dimensional numerical modeling of the joule hearting of various shapes of armatures in railgun, *IEEE Transaction on Plasma Science*, Vol. 39, No. 1, Part I of two parts, (January, 2011), pp. 456–460, ISSN 0093-3813.

Thornhill L. D., Batteh J. D. & Brown J. L. (1989). Armature Options for Hypervelocity Railguns, *IEEE Transaction on Magnetics*, Vol. 25, No. 1, (January, 1989), pp. 552-557, ISSN 0018-9464.

Wang Y., Cheng S. & Zheng P. (2003). Widely Developing Electric Launch Technology in China, *IEEE Transaction on Magnetics*, Vol. 39, No.1, (January, 2003), pp. 39-41, ISSN 0018-9464.

Vrable D. L, Rosenwasser S. N. & Korican J. A. (1991). Design and fabrication of an advanced, lightweight, high stiffness, railgun barrel concept, *IEEE Trans. on Magnetics*, Vol. 27, No. 1, (January, 1991), pp. 470-475, ISSN 0018-9464.

Young F. J. & Hughes W.F., (1982). Rail and armature current distribution in electromagnetic launchers, *IEEE Transaction on Magnetics*, Vol. 18, No 1, (January,1982), pp. 33-41, ISSN 0018-9464.

Zaidel' R. M. (1999). Composite electrodynamic liner, *Journal of Applied Mechanics and Technical Physics*, Vol. 40, No. 5, (September, 1999), pp. 777-783, ISSN 0021-8944.

Selective Catalytic Reduction NO by Ammonia Over Ceramic and Active Carbon Based Catalysts

Marek Kułażyński
Wrocław University of Technology
Poland

1. Introduction

The need for environmental protection is an indisputable objective. This is particularly important wherever environmental burden has become so high that the environment is no longer capable of self-purification. Such situation exists in our country. A major problem is the protection of the atmosphere.
The main pollutants emitted into the atmosphere include carbon monoxide (CO), sulphur dioxide (SO_2), nitrogen dioxides (NO_2), hydrocarbons (CH), and particulates.
Share of individual sectors of the industry in the total emissions is not identical. It is demonstrated by Fig. 1.

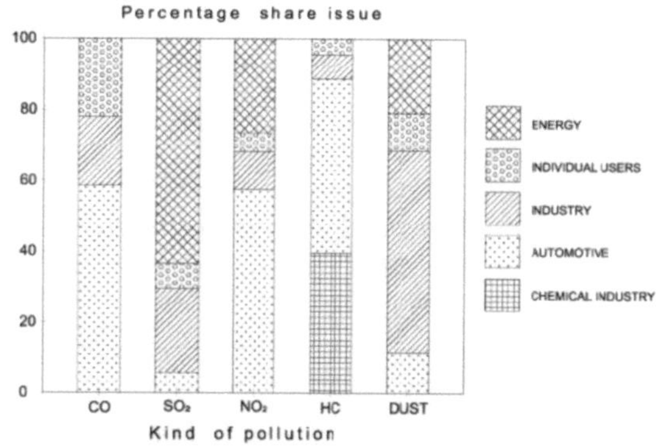

Fig. 1. Share of primary industries in emissions of toxins and particulates.

Although it is difficult to compare the harmfulness of each of the toxins to one another, it is assumed that the relative impact of NO_x : CO : HC on the human body is like 100 : 1 : 0.1. It follows that nitrogen oxides are the most harmful for the human body. According to the data presented in figure 1, nitrogen dioxides are emitted mostly by transport, followed by the power industry and heavy and light industries. On the other hand, sulphur compounds are particularly dangerous for the environment. Here, the ratio is different because these compounds are emitted mainly by the power industry, followed by heavy and light industries, and then households.

The first method of combat is to reduce emissions by lowering energy consumption and fuel consumption per unit of energy produced. However, it is also obvious that although the above processes are essential, they are slow and demand constant disproportionate increase of expenses. In such case it becomes necessary to act in other directions, i.e. active and passive control of environmental pollutants.

Active methods include changes in the combustion process, but especially changes in the fuel, including its desulphurisation. However, fuel desulphurisation is an extremely expensive process and can only be used in the situations where fuel consumption is relatively small and there are practically no other methods of solving the problem.

Fuel desulphurisation does not solve the second problem, which is emission of nitrogen oxides. Here, the most adverse effects are produced by coal-burning devices. This is due to high combustion temperatures occurring in the process. In this case design changes (active methods) do not provide major results.

Much better results are obtained by the introduction of design changes in the processes of combustion of hard and brown coal in the so-called dry processes. The obtained results are not as good as in the case of newly built systems, but they are still significant (particularly with respect to hard coal combustion).

Changes with active methods do not result in achievement of target values - present and future emission standards. Therefore, passive methods must be used, particularly catalytic methods.

Composition of exhaust gases, including their concentrations of toxic components, varies widely. It depends on the type of fuel and the combustion process.

While emissions of sulphur oxides depend on its content in the fuel, nitrogen oxides produced in the combustion process depend, among other, on the following factors: combustion temperature, concentration of reagents (oxygen and nitrogen) during the combustion, contact time of reagents, especially in the high temperature zone, type of furnace equipment and fuel type and the quality of its mixture with air.

At present nitrogen oxide emissions can be limited by means of:

- processing and refining of fuel,
- limiting the amount of nitrogen oxides produced in the combustion process,
- removing nitrogen oxides from exhaust gases.

The first direction is feasible when it comes to crude petroleum, but in the case of coal it is unlikely to be used in the near future, because it is ineffective and requires building of a fuel refining industry.

The next two directions are currently being used and developed on a large scale in many highly industrialised countries. Nitrogen oxides are reduced by 10 to 80% depending on the type of fuel, type of boiler, and the applied method. The third direction is very effective since it reduces the nitrogen oxide content in exhaust gases by 70 to 95%.

At present the methods of catalytic selective reduction with the use of ammonia as a reducing factor are the most widely used. The process is described as a selective one because ammonia has greater chemical affinity to nitrogen oxides than to oxygen.

In this method nitrogen oxides are converted to nitrogen and water, i.e. neutral components of the atmosphere. Yield of reaction depends on: the temperature, type of catalyst, ratio of ammonia to nitrogen oxides and gas flow rate through the catalyst layer. The effectiveness of the process is primarily determined by the catalyst activity.

Nitrogen oxides are reduced by ammonia selectively on catalysts prepared with the use of noble metals (Pt, Rh, Pd) and metal oxides (V_2O_5, TiO_2, MoO_3). Effective catalysts used in

SCR reactors are catalysts deposited on honeycomb ceramic monoliths, containing longitudinal ducts with square or round cross-section [1-4].

The main advantages of such solution are:

- low resistance of gas flow through the catalyst bed,
- small catalyst volume,
- storage of ammonia in catalysts, which ensures high flexibility of operation under variable load conditions,
- small losses of ammonia,
- resistance to poisoning,
- possibility of using spent catalysts as a raw material in the ceramic industry.

2. Nitric oxides

Depending on the combustion process, waste gases differ in chemical composition, concentration of toxins, dispersion of particulate matter, and temperature. The composition of exhaust gases may differ, just as there may also exist differences in the techniques of removal of their toxic components.

The primary toxic components of exhaust gases that must be removed are nitric oxides and sulphur dioxide.

Removal of nitric oxides is facing two major difficulties arising from the very nature of the process.

Nitric oxides created in the processes of industrial combustion consist almost entirely of nitrogen oxide NO (90%). Nitrogen oxide is very poorly soluble in water. Consequently, the methods of waste gas scrubbing face the problem of conversion of nitrogen oxide to oxides (by oxidation), which, on the other hand, dissolve better.

The second problem is the presence of oxygen in exhaust gases. Oxygen is present in the combustion process in excess (3-12%), ensuring optimum fuel combustion and preventing formation of carbon monoxide, soot, and boiler corrosion. However, excess oxygen hinders reduction of nitrogen oxides obtained with the use of chemical reducing agents because they react more readily with free oxygen than with oxygen from nitrogen oxides. Still, that problem can be resolved by means of catalysis.

Selective Catalytic Reduction (SCR) - enables reduction of nitrogen oxides using ammonia in the presence of a catalyst to form nitrogen and water. At the entrance to the reactor the exhaust gases must be mixed to the maximum possible extent with ammonia.

Nitrogen oxide (NO) is formed from water and nitrogen, present in fuel and atmospheric air. During the combustion of pulverized coal, over 80 % of nitrogen oxides are formed from nitrogen present in fuel. Natural gas contains approx. 0.5% nitrogen, fuel oils - approx. 0.1-0.2% nitrogen, and carbon - up to 2 % nitrogen.

Nitrogen oxide (NO) turns into nitrogen dioxide (NO_2) in the presence of oxygen in the air, with the speed of reaction depending on the concentration of nitrogen oxide.

Combustion processes produce nitrogen oxide (NO) whereas nitrogen dioxide (NO_2) is formed by oxidation of nitrogen oxide in atmospheric air. In addition to nitrogen oxide (NO) and nitrogen dioxide (NO_2), boiler flue gases also contain nitrous oxide (N_2O). The greatest amount of nitrous oxide is formed during combustion of coal, and the least amount - during combustion of natural gas. Nitrous oxide participates in reactions destroying the ozone layer of the Earth, thus contributing to the formation of the greenhouse effect. Specifically, it absorbs infrared radiation, preventing cooling of the Earth during the night.

Some of nitrogen oxides formed during combustion are decomposed into oxygen and nitrogen by coke formed at the same time in the process of pyrolysis. This process occurs with high intensity during fluidal combustion and, in addition to low combustion temperature, contributes to the generation of minimum amounts of nitrogen oxides in this type of combustion. Boiler flue gases containing NO_x consist of approx. 95% nitrogen oxide (NO) and approx. 5% nitrogen dioxide (NO_2). Concentration of nitrogen oxides in boiler flue gases depends on the type of furnace, the temperature inside it, the method of fuel combustion, the type of fuel, the excess air ratio,
and the boiler load.

Nitrogen oxides formed in the boiler combustion chamber can be divided into:

- thermal,
- fuel,
- fast.

Thermal nitrogen oxides are formed from nitrogen contained in atmospheric air during the combustion of each fuel at very high temperatures. Fuel nitrogen oxides are formed from nitrogen contained in fuel and their formation depends on the type of fuel and the method of its combustion. Fast nitrogen oxides are formed from nitrogen contained in atmospheric air, primarily during combustion of gaseous fuels, and their formation depends mainly on the excess air ratio.

Fluidal combustion at a temperature of 800-1000°C is accompanied by formation of fuel nitrogen oxides. Spatial combustion (in pulverized-fuel boilers) at a temperature of 1300°C is also accompanied by formation of mainly fuel nitrogen oxides, but with an increase in temperature their amount diminishes whereas thermal nitrogen oxides appear, which above the temperature of 2100°C constitute the only oxides. In the temperature range of 1300-2100°C fast nitrogen oxides are also produced in the amount of 7-10% of the total amount of formed nitrogen oxides. At temperatures above 2300°C (low-temperature plasma) thermal nitrogen oxides are formed.

In order to reduce formation of nitrogen oxides, temperature of the flame cone must be lowered, oxide content in the combustion zone must be reduced, and the duration of fuel staying in the high-temperature zone must be shortened.

With the above methods, the amount of formed nitrogen oxides can be reduced by no more than 40-50% which, however, is insufficient to meet the requirements of European standards. To comply with the standard, two methods are used: selective catalytic reduction (SCR) and selective non-catalytic reduction (SNCR).

3. Methods of denitrification of exhaust gases

Catalytic reduction of nitrogen oxides by ammonia in the presence of a catalyst
The reduction results in the formation of nitrogen and water:

$$4NO + 4NH_3 + O_2 \Rightarrow 4N_2 + 6H_2O$$

$$2NO_2 + 4NH_3 + O_2 \Rightarrow 3N_2 + 6H_2O$$

$$6NO_2 + 8NH_3 \Rightarrow 7N_2 + 12H_2O$$

The catalyst load is measured according to the exhaust gas flow rate, i.e. the amount in Nm^3 passing through 1 m^3 of catalyst over 1 hour. Obviously, the lower the load, the higher the

effectiveness of the process of exhaust gas denitrification. Catalysts can be plate type or honeycomb type.
A plate catalyst is made of high-grade stainless steel with active mass, consisting of titanium oxides (TiO_2), vanadium (V_2O_5), tungsten (WO_3) or molybdenum (MoO_3).
It is highly resistant to erosion, has high mechanical and thermal strength, causes small pressure losses, and has a low propensity for clogging. It can operate in areas with high particulate concentrations, i.e. in front of an installation for particulate removal and desulphurisation of exhaust gases.
Ceramic honeycomb catalyst has an identical active layer, but it works well in areas of low particulate emissions. Consequently, it must be placed behind the installation for particulate removal and desulphurisation of exhaust gases. However, in order to ensure proper operating conditions for the catalyst, exhaust fumes must be additionally heated up because they are cooled down in the desulphurisation installation. The optimum operating temperature of the catalysts is 300-450°C if they are connected in front of an air heater, and 280-380°C if they are connected in front of the flue. A catalyst operates between 2 to 3 years in an area with high particulate concentration, and between 4 to 5 years in a clean area. 1 MW of power plant capacity requires approx. 1 m^3 of catalyst. With up to 95% effectiveness, it is the most effective of all the methods in use. However, this is the most expensive method in terms of investment and operation. Sizes of commercial catalysts with honeycomb structure and square meshes (grid cross-section) are shown in Table 1. Additionally, various manufacturers offer catalysts in the form of corrugated plates.

Determination	**Sizes (mm)**	
	mesh	**wall thickness**
Gas-fired boiler	3 to 6	0.5 to 1.6
Oil-fired boiler	6 to 8	1 to 1.5
Coal-fired boiler	6 to 10	1 to 2

Table 1. Dimensions of industrial catalysts with the honeycomb cross-section.

After passing through the electrostatic precipitator, the particulate content in exhaust gases does not exceed 50 mg/m^3 Although catalyst holes practically never become clogged, fine particulate matter deposits on the surfaces of its walls, deactivating the device. The problem is solved by selection of a catalyst with proper resistance to abrasion, mesh sizes, and wall thickness.
Selective non-catalytic reduction (SNCR) of nitrogen oxides by ammonia.
It is a variation of the first method but without the use of a catalyst.
It has 50% effectiveness but it is cheaper in terms of investment and operation than the previous one. Ammonia reacts with nitrogen oxides at a temperature of 800-1000°C without a catalyst, producing nitrogen and water. At other temperature ranges the reaction occurs very slowly and ammonia enters the flue. When the boiler load changes, it is accompanied by changes in the temperature of the exhaust gases and its distribution in the boiler.
If ammonia is injected at a certain point where the existing temperature is suitable for the occurrence of the reaction, then with a change in the boiler load - and thus a change of the temperature at that point - the reaction will not occur.
Irradiation of hot exhaust gases (at a temperature of 900 °C) by electron beam.

Free radicals formed during irradiation of exhaust gases by electron beam react with NO_x and SO_2 molecules, creating ammonium nitrate and ammonium sulphate.
The DESONOX method of combined desulphurisation and denitrification of exhaust gases.
The essence of the method is catalytic oxidation of sulphur dioxide to sulphur trioxide, of which sulphuric acid is produced, while nitrogen oxides are also catalytically reduced to nitrogen (with the SCR method). This method offers 95% desulphurisation and 90% denitrification of exhaust gases. It is free of sewage and waste while the produced sulphuric acid is of commercial grade.
The Bergbau Forschung-Uhde method.
In this method sulphur dioxide is absorbed from exhaust fumes by special active coke, obtained from hard coal. Ammonia is fed to the absorber and reacts with nitrogen oxides without a catalyst. Active coke is regenerated at a temperature of 400°C in the desorber, from which gas rich in sulphur dioxide outflows and is used in sulphuric acid production.
Exhaust gases that passed through the desulphurisation installation and electrostatic precipitators for the capture of particulate matter have a temperature below 100°C. This temperature is too low for effective operation of the catalyst. It follows that exhaust gases must be heated up to appropriate temperature. However, in the case of old system designs there is often not enough place to incorporate the appropriate heating devices (not to mention the energy costs of such heating).
Therefore, there is no choice but to use catalysts that could operate efficiently at waste gas temperatures, particularly considering the fact that the amounts of gases that must be heated up pose a serious energy problem that puts into question the efficiency of the power acquisition system.
Low-temperature catalysts could also be used in the removal of nitrogen oxides from various technological processes [1-11].

4. DeNOx carriers and catalysts

4.1 The process of selective catalyst reduction (SCR) of nitric oxides with ammonia

Catalysts of denitrification of exhaust gases from power boilers must meet several requirements relevant to users. They should be characterised by:

- **Stability**

a. thermal resistance:

The catalyst should maintain its activity at a temperature up to 500°C for a long period of time under the operating conditions of an industrial boiler.

b. resistance to poisoning:

Acid centres are poisoned mostly by alkali metal ions while centres in oxidation reactions are poisoned mainly by arsenic oxide. Therefore, catalysts should be selected that are resistant to the above poisons. Active components, e.g. CuO, Fe_2O_3 or carriers react with gas components (SO_3 etc.). That problem was resolved through the use of catalysts based on vanadium pentoxide deposited on titanium dioxide. The results of some studies have shown that vanadium-titanium catalysts can be promoted with some alkali metal salts, e.g. sodium sulphates and lithium sulphates, whereas potassium sulphate content had a negative impact on their activity. On the other hand, it was determined that the negative impact of some poisons on catalytic activity occurred only in the absence of SO_2 and disappeared in its presence. Also of note is the observation that a catalyst can be completely regenerated by washing it with water.

c. resistance to abrasion:

In the case of gases containing large amounts of particulates, a catalyst is subject to abrasion. In general, abrasion resistance is inversely proportional to catalytic activity. Therefore, it is important for industrial catalysts to be resistant to abrasion, and when a catalyst is poisoned especially in its surface layer, catalytic activity is maintained with gradual abrasion of the surface (poisoned) layers.

- **High activity over a wide range of temperatures of the process**

The temperature of exhaust gases depends on changes in the boiler load but, despite this, the effectiveness of denitrification must be maintained at the same level. Vanadium catalysts deposited on TiO_2 show highest activity at lower temperatures, in the range of 300 - 400 °C, whereas WO_3 on titanium dioxide or V_2O_5 WO_3 on titanium dioxide show highest activity at somewhat higher temperatures.

Low conversion of SO_2 to SO_3

Composition of the gases depends on the type of burnt fuel. Gases from the burning of coal and heavy heating oils contain SO_2, SO_3, and particulates. The denitrification catalyst should cause minimum oxidation of SO_2 to SO_3. In the course of this reaction there is increased corrosion of the apparatuses and deposition of acid ammonium sulphate, as a result of reaction of SO_3 with ammonia below the crystallisation temperature at the subsequent apparatuses of the system. For this reason, vanadium pentoxide is being partially replaced in the catalyst by other metals, e.g. tungsten trioxide. Thanks to this, catalysts are obtained that enable acquisition of large conversions of nitrogen oxides at minimum oxidation of sulphur dioxide.

Small pressure drop and low particulate retention on the catalyst bed

Despite the use of different types of electrostatic precipitators to remove particulates from exhaust gases, they contain from a few tenths of a milligram to several grams of particulates per cubic meter of exhaust gases. This causes clogging of catalyst bed in the form of various types of granulates, extrudates, or spheres [12].

Selection of DeNOx catalyst carrier

Over the course of more than a dozen years, many types of catalysts have been tested in a number of laboratories and in some cases the method of their manufacture was patented. For example, according to Japanese researchers [7] the examined denitrification catalysts can be classified by the type of carrier used, as shown in Table 2.

Determination	Type of carrier		
	TiO_2	Fe_2O_3	Al_2O_3
Activity	high	average	average
Resistance to SO_2	high	low *	low **
Selectivity	high	high (low surface temp.450°C)	high (low surface temp.450°C)
Oxidation of SO_2	low	high	low
Regeneration	possible	impossible	impossible

*formation of $Fe_2(SO_4)_3$ ** formation of $Al_2(SO_4)_3$ * * * removal of deposited NH_4HSO_4

Table 2. Comparison of DeNOx catalyst carriers.

The presented data suggest that the best DeNOx catalyst carrier is titanium dioxide. Titanium carrier can be prepared with the use of several methods. A commonly used method is precipitation of TiO_2 by $TiCl_4$ hydrolysis with water [13].

Inomata and associates prepared both crystallographic forms of titanium dioxide: anatase and rutile by hydrolysis of, respectively, titanium sulphate or titanium chloride. Mixed anatase and rutile compositions are obtained by calcination of commercial titanium dioxide. In general, titanium dioxide has a small specific surface area. As a result of the so-called flame hydrolysis of $TiCl_4$, a high-purity (over 99.5%) carrier is obtained, with crystallite size of the order of 10-30 nm., specific area of approx. 55 m^2/g, and approx. 75% anatase content (the rest consists of rutile). This is a commercial product by Degussa [14]. Rhone-Poulencs, on the other hand, produces TiO_2 by precipitation from titanium sulphate solutions. The product obtained this way, with the surface area of approx. 100 m^2/g and the crystallite size of the order of 300 nm., consisted exclusively of contaminated anatase with approx. 2% sulphate ions. Table 3 shows physicochemical properties of carriers formed from the two types of titanium dioxide discussed above. As we can see, compared to the carrier obtained by the flame method, the carrier obtained from precipitated titanium dioxide is characterised by almost twice as big specific surface area, somewhat greater porosity, and bimodal character of the porous structure.

TiO_2	flame	precipitated
Crystalline phase	75% anatase, 25 % rutile	100% anatase
Specific surface area [m^2/g]	48	92
Pore volume [m^2/g]	0.34	0.40

Table 3. Comparison of the properties of carriers formed by extrusion from different types of titanium dioxide (Shape: cylinders; Diameter: 4 mm; Length: 4 mm).

Carriers from titanium dioxide obtained by the flame method maintain their properties up to the temperature of approx. 400°C, after which there is a gradual reduction of the specific surface area and porosity as well as recrystallisation of anatase to rutile and an increase in the size of pores. At a temperature of approx. 700°C the carrier contains only rutile, the specific surface area shrinks to under 20 m^2/g, and porosity does not exceed 0.1 ml/g.

By choosing the calcination temperature of the carrier, the ratio of anatase to rutile content can be regulated. Also, use of calcination temperatures higher than 400-500°C may lead to significant changes in its properties and the porous structure. The duration of calcination also exerts some influence on the properties of the carrier, but it is less significant. Carriers from precipitated TiO_2 are more stable, they maintain anatase structure up to approx. 900°C, but starting from approx. 400°C there is also a gradual reduction in porosity and the specific surface area, although this process is much slower than previously. Above the temperature of 800°C there is a clear sintering of pores, the bimodal structure disappears - sintering occurs in smaller pores (8 nm.) while bigger pores shrink in diameter (300 nm.).

Haber and associates [15] developed a method for obtaining very fine crystalline anatase with the specific surface area of the order of 120 m^2/g by hydrolysis of titanium butoxide (IV). Aluminium and silicon carriers initially used to produce catalysts of nitrogen oxide reduction came mainly from typical industrial production and then techniques were developed for homogenous precipitation, i.e. carrier precipitation from solutions, when the process takes place simultaneously in the whole mass. For example, Shikada et al. [16] used that method to produce a silicon-titanium carrier. Urea dissolved in acidified solution of sodium metasilicate and titanium tetrachloride decomposes during heating and the released ammonia increases the pH of the solution in a controlled manner and causes precipitation.

Those so-called mixed carriers are characterised by higher mechanical strength and thermal stability as well as exhibit interesting properties due to their diversified surface acidity.

The activity of the DeNOx catalysts used in the installations can be improved by a reduction of diffusion resistance in the catalyst pores [17]. This new type of catalyst is based on a titanium-silicon carrier. Although other researchers also used a titanium-silicon carrier [18], it emerged that catalyst activity can be increased thanks to the acquisition of a bimodal structure and provision of adequate mechanical strength of the monolith. According to Solar et al. [19] a titanium-silicon carrier combines the benefits of both types of oxides: introduction of silica ensures acquisition of the appropriate porous structure, while titanium oxide layered on pores makes the carrier exhibit its surface properties. After deposition of vanadium the obtained $V_2O_5/TiO_2/SiO_2$ catalyst maintains its properties at a temperature much higher than its normal operating temperature, i.e. in the range of 350 to 380°C.

There are also reports [20] of high activity of DeNOx catalysts whose carrier is silica, on which a few percent of TiO_2 were deposited by impregnation in order to stabilise vanadium oxide on the surface of carrier (prevention of agglomeration). A catalyst of this type shows high activity in the reaction of reduction of nitrogen oxides by ammonia. At a temperature below 200°C they are excellent catalysts of the DeNOx process [21, 22], may form compositions with a titanium-vanadium catalyst, are active in a wider range of process temperatures, and are more resistant to deactivation [23]. The method of production described above is very similar in the case of catalysts without zeolites. An important difference is the deposition of active metal on zeolite by means of exchange. The applied metals are mostly copper and iron, but also other transition metals, including noble metals. The zeolites most commonly used for this purpose are mordenite and ZSM-5, but other zeolites are also appropriate and cited in the literature.

Ion exchange of zeolite should be made before zeolite is mixed with other components in the first stage.

According to Boer et al. [24] the main components of the DeNOx catalyst carrier are titanium dioxide and zeolites, which constitute a homogenous structure. Attempts were also made to deposit the active layer on the carrier surface, the so-called "washcoat", but this solution did not find wider practical application [25]. Apart from TiO_2, which is the primary carrier component, preferably in the form of anatase, and the previously-mentioned silica [19, 26], transition metal oxides are also added to the formed carrier. [27]. An important role is fulfilled by various types of inorganic additives introduced together with TiO_2, e.g. fibreglass and glass powder, diatomaceous earth, silica gel, aluminium oxide, and titanium dioxide in the form of sol or gel. Those additives reduce the propensity of extruded monoliths to crack during the subsequent thermal operations and ensure its adequate mechanical strength. Organic additives may contain polyvinyl alcohol, starch, polymers, and waxes as binding and surface-active agents. Some of TiO_2 may be thermally pre-treated (calcinated), which also prevents monolith cracking. During the mixing of those carrier precursors, vanadium compound may also be introduced. Only after thorough dry homogenisation water is added and the mixture is kneaded until a uniform mass is obtained [25]. The next stages of the carrier production are slow drying, thermal decomposition of organic binders, and final calcination at a temperature in the range of 400-650°C if it already contains vanadium pentoxide to prevent its deactivation by sintering, or to more than 700°C for maximum mechanical strength.

Deposition of the active phase

Impregnation

Active metals can be deposited on the carrier during the process of kneading of the carrier precursor mass by the introduction of appropriate metals to their salt solutions, followed by formation of the mass prepared in that way. This ensures uniform distribution of the active phase in the whole catalyst mass. The simplest way of depositing the active phase on the finished carrier is impregnation. Impregnated carriers are most often solutions of nitrates or metal acetates.

Much attention was devoted to the preparation of vanadium-titanium catalysts. Such catalysts can be prepared e.g. by wet impregnation of titanium carrier with titanium meta vanadium in oxalic acid solution, followed by calcination at a temperature in the order of 500°C [28, 29]. Vanadium pentoxide was deposited in the same way on aluminium oxide carrier [30]. Saleh et al. prepared V_2O_5/TiO_2 (anatase) catalyst by dissolving vanadium pentoxide in aqueous solution of oxalic acid and saturating titanium carrier [31].

A comprehensive review of the methods of deposition of various active metals on carriers and preparation of DeNOx catalysts was presented by H. Bosch and F. Janssen in their work on the catalytic reduction of nitrogen oxides [32]. In that publication the authors mention a number of methods of applying vanadium on a monolithic carrier by means of vanadium oxalate [11] and other vanadium salts, e.g. ammonium metavanadate from aqueous solutions [33, 34]. On the other hand, catalysts containing tungsten, WO_3/TiO_2, are prepared by impregnation of the carrier with aqueous solution of ammonium paratungstate, followed by drying and calcination.

Vanadium catalysts on silica were prepared by its impregnation with solution of ammonium metavanadate. In the case of commercial silica gels, titanium dioxide was first deposited on their surface in such way that in the first stage the carrier was saturated with titanium sulphate solution and then immersed in ammonia solution, thereby precipitating titanium hydroxide on the surface of pores. After washing and thermal treatment vanadium pentoxide was deposited by impregnation [35].

In some studies attempts were made to prepare vanadium-titanium catalysts using non-aqueous solutions of $VOCl_3$. In this method vanadium oxychloride reacts with surface OH groups. Bond and Konig deposited $VOCl_3$ dissolved in petrol on anatase with small surface [36]. Vanadium catalysts on titanium oxides, silicon oxides, and aluminium oxides were also prepared by impregnating the appropriate carrier with $VOCl_3$ solutions in CCl_4 [10] or by passing gaseous $VOCl_3$ over the carrier, TiO_2 [28].

Single-stage preparation

Catalysts can also be prepared by simultaneous precipitation of the carrier and the active phase. Catalysts of the WO_3/TiO_2 type or the $WO_3/$ Fe_2O_3 type were prepared by mixing hydrogel of titanium hydroxide or ferric hydroxide with aqueous solution of ammonium paratungstate, followed by thermal treatment [37]. Vanadium catalyst on titanium dioxide was prepared with the sol-gel method using hydrolysis of their organic derivatives of tetra-1-amylenes of titanium and vanadium [38]. This group of methods can also include the previously discussed ways of preparation of vanadium-titanium and other catalysts involving the introduction of salts of active metals to the mixture of carrier precursors before their kneading and formation.

Types of catalysts used

It has been established that some catalysts deposited on carriers made of aluminium oxides or iron oxides e.g. Fe_2O_3 - SnO_2, Fe_2O_3, WO_3 or Fe_2O_3 deposited on Al_2O_3 or V_2O_5 deposited on Al_2O_3 were characterised by high activity in reaction of denitrification of exhaust gases. However, those catalysts were losing their activity due to formation of sulphates during research on pilot systems for the purification of exhaust gases containing sulphur oxides. On the other hand, catalysts on titanium oxide as the carrier demonstrated not only high activity and selectivity, but also resistance to sulphur poisoning [39].

Indeed, TiO_2 does not react with either SO_3 or SO_2 at a temperature above 200°C and because of this it maintains its structure for a long time in an environment of gases containing those oxides. On carriers made of titanium oxide, the active components are mainly V_2O_5, MoO_3, and WO_3 and in some cases also Fe_2O_3, CoO, NiO, MnO_2, Cr_2O_3, and CuO [40]. Catalysts of this type are active in DeNOx reactions at a gas temperature of between 200 and 500°C. For example, V_2O_5/TiO_2, a typical DeNOx catalyst, ensures under specific process conditions almost 100% reduction of nitrogen oxides with ammonia in the temperature range of 220-425°C. After the temperature on the catalyst bed exceeds 430°C, reduction of nitrogen oxides rapidly decreases. Under the same conditions of the reduction process, the use of another monolithic catalyst, but with a completely different composition, containing zeolite - TiO_2 + SiO_2/Fe_2O_3 + Fe – mordenite, a 95% reduction of oxides can be obtained at catalyst temperature range of between 375 - 600°C. Significant differences can also be observed in the activity of zeolite catalysts, which differ from each other only by the type of replaced metal [25].

Copper catalyst [9.2% Cu-mordenite/6.92% CuO/+ 8% silicon binder] enabled obtainment of over 95% conversion of nitrogen oxides in the temperature range of 225-440°C, whereas a catalyst of the composition, but containing 4.70 % Fe_2O_3 instead of copper, showed a similar degree of conversion at a temperature range of 310-560°C.

In industrial installations of DeNOx there are certain operational problems. At a process temperature of under 200°C there is a noticeable deposition of acid ammonium sulphate in the catalyst pores. Therefore, in the case of exhaust gases containing sulphur oxides, the process temperature must be maintained at over 230°C. On the other hand, at a temperature over 400°C there is a noticeable increase in oxidation of SO_2 to SO_3. Since V_2O_5 is the main promoter of the reaction of SO_2 oxidation, at such time mainly the TiO_2 - MoO_3 or TiO_2 - WO_3 catalysts are used with minimum content or even elimination of V_2O_5 from the catalyst. In such arrangement, a catalyst operating mostly in the gas temperature range of the order of 300 - 400°C can be operated for a long time without disturbances caused by deposition of acid ammonium sulphate on its surface and pores.

It should be noted that catalysts containing only vanadium show the highest activity, approx. 95 % conversion of nitrogen oxides at a temperature of 300-350°C. The maximum activity of DeNOx catalysts containing small amounts of V_2O_5, of the order of 1%, and approx. 10% WO_3 occurs in the temperature range of 380 - 450°C [7]. Catalysts containing only 3% of vanadium pentoxide on titanium dioxide offer 95% conversion of nitrogen oxides at a temperature of approx. 380°C. Further increasing of the active phase content no longer increases conversion of nitrogen oxides, but causes a few percent increase in SO_2 oxidation to SO_3. In the case of catalysts containing tungsten (e.g. 10% WO_3) a 95% conversion at a temperature of 380°C can already be achieved at less than 1% content of vanadium pentoxide (in such case conversion of SO_2 to SO_3 does not exceed 1%). Tungsten catalyst deposited on titanium dioxide oxidises SO_2 only to a small degree. With continued operation the scope of oxidation increases. According to Morikawa, this is caused by deposition of vanadium on the catalyst by exhaust gases together with ash [41].

The relationships presented above concern only catalysts on a titanium carrier. According to Shikad et al. [35], over 95% reduction of nitrogen oxides on the V_2O_5/TiO_2-SiO_2 or V_2O_5/SiO_2 catalysts requires more than 10% content of the active phase (at a temperature of 200°C). The use of the first of those catalysts at a 20% content of V_2O_5 enables almost complete reduction of nitrogen oxides at a temperature below 200°C. Much smaller activity was exhibited by vanadium catalysts on aluminium oxides or silicon oxides [32]. Similarly to titanium carriers, the optimum calcination temperature for mixed titanium and silicon carriers falls in the range of 350-400°C. A higher processing temperature gradually reduces catalyst activity, which is presumably due to a reduction of its specific surface area [16].
Table 4 shows the dynamics of development of the SCR systems, specifying the installations at power stations for hard coal and only for boilers with dry slag, situated in Germany.

No	Name of the power station	Power unit capacity	System	Provider
1	Reinhafen,	550 MW	High dust	Steinmüller (STM)
2	Reuter - West, Units D + E,	2 x 300 MW	High dust	Balcke - Diirr (B-D)
3	Reuter, Units 1 + 2 ;	2 x 50 MW	Tail End	Lentjes
4	Hannover - Stocken, Units l +2;	2 x 375 MW	Tail End	Uhde - Lentjes
5	GKM Mennheim -Neekarau, Unit 7,	475 MW	High dust	EVT
6	Heyden,	800 MW	High dust	Uhde - Lentjes
7	Farge,	325 MW	High dust	Uhde - Lentjes
8	Mehrurn / Hannover,	642 MW	High dust	Uhde - Lentjes
9	Weiher,	707 MW	High dust	Steinmüller
10	Volklingen,	210 MW	High dust	KWU

Table 4. SCR installations at selected hard coal-fired power stations (Germany)

In some cases non-selective catalytic methods are also used. Here, the reducer can be hydrogen or methane. Those methods, referred to briefly as NSCR, are associated with considerable consumption of the reducer, because it also reacts with oxygen present in exhaust gases. This leads to disproportionately large consumption of the reducer, which is not economically viable.
In general, SCR are optional equipment - an addition to the primary methods. Such solution allows for a significant reduction of the amount of ammonia fed to exhaust gases, it reduces contamination of the catalyst, air heater, etc.; it also reduces the speed of catalyst poisoning.
In the SCR method the evaporated ammonia at a temp. of approx. 200°C is blown into boiler exhaust gases by air. Reduction of NO_x in catalysts proceeds according to the following major reactions:

$$4\,NO + 4\,NH_3 + O2 \Rightarrow 4\,N_2 + 6\,H_2O$$

$$6\,NO_2 + 8\,NH_3 \Rightarrow 7\,N_2 + 12\,H_2O$$

In the case of large boilers, problems may arise in connection with introduction of sprayed ammonia to the exhaust stream in order to obtain uniform concentration and direct the exhaust stream so that the catalyst is uniformly loaded. Apart from the main reactions, there are also adverse associated reactions:

$$4\,NH_3 + 3\,O_2 \Rightarrow \rightarrow 2\,N_2 + 6\,H_2O$$

$$4\,NH_3 + 5\,O_2 \Rightarrow 4\,NO + 6\,H_2O$$

$$NH_3 + SO_3 + H_2O \Rightarrow NH_4HSO_4$$

The first two reactions occur after the temperature of the exhaust gases goes above 400°C and result in increased demand for ammonia. At temperatures below 330°C and in the presence of SO_3, a third reaction takes place in the exhaust gases, where acid ammonium sulphate is formed that deposits in pores of the catalyst surfaces, causing a reduction of the catalyst activity. Acid ammonium sulphate has the dew point at a temperature of 150°C and deposits in liquid state on the rotating elements of air heaters at a temperature range of 150°C to 250°C, which may primarily lead to the clogging of LUVO, but also to its corrosion. To mitigate the negative effects, special solutions are used in revolving heaters (specially shaped plates) as well as effective cleaning devices.

4.2 Preparation of ceramic carriers and catalysts

Preparation of the carrier

Fig. 2 shows schematic diagram of production of a monolithic catalyst carrier

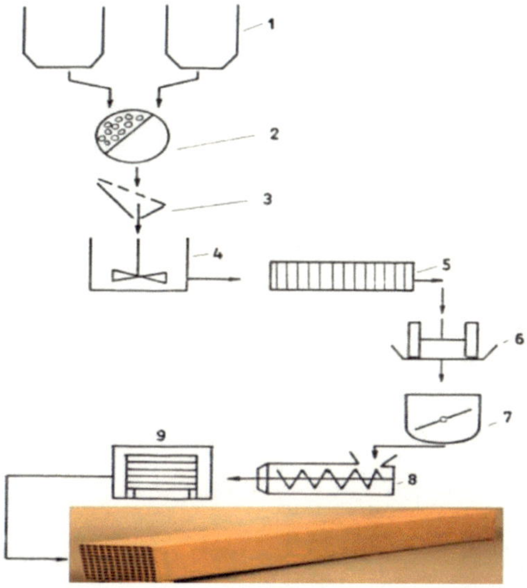

Fig. 2. Schematic diagram of production of monolithic carriers. 1-Raw material dispensers, 2-Grinder, 3-Sieve, 4-Tank with agitator, 5-Press, 6-Crusher, 7-Agitator, 8-Belt press, 9-Dryer.

Manufacturing of a carrier involves preparation of aluminosilicate mass, fragmentation, and selection of appropriate sieve fraction (aluminosilicate desludged and fragmented under 0.05 mm.). Degree of fragmentation of raw material affects the forming properties of the mass. It will also affect the quality of the final product - monolithic carrier. The next stage is mixing of aluminosilicate with additives such as lubricants and plasticizers, followed by forming of the obtained mass.

Forming of the carrier after mixing of the mass in a z-shaped mixer. Such method of preparation of the mass ensured uniform saturation with plasticizers of grain agglomerates and de-aeration of the mass. Kneaded mass was directed to the forming operation. Fig 2 shows a diagram of the extruder die for forming a monolithic carrier.

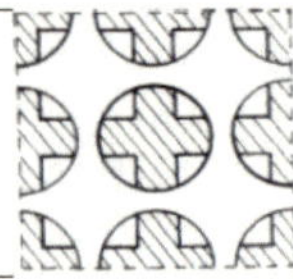	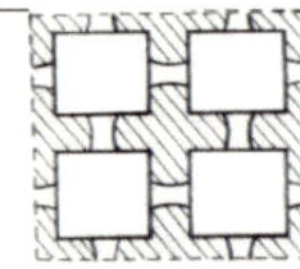
View from the side of mass entry into the extruder die	View from the side of monolith exit from the extruder die

Fig. 2. A diagram of the extruder die for forming a monolithic carrier

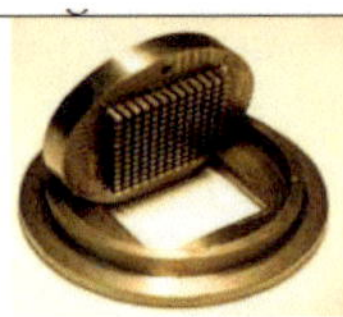

Fig. 3. A view of the extruder die for forming a monolithic carrier [42]

Fig. 4. A view of the exit of a monolithic carrier profile from the press.

The extrudate coming out of the extruder die was cut off after it achieved the appropriate length, giving the formed material its final shape. After pre-conditioning at room temperature, but no later than after 1 hour, honeycomb-structured profiles, the so-called "green monoliths", were subjected to the appropriate process of drying in a microwave dryer. Microwave action enables taking of water molecules to the carrier surface while blowing hot air carries the emitted water away from the monolith surfaces.

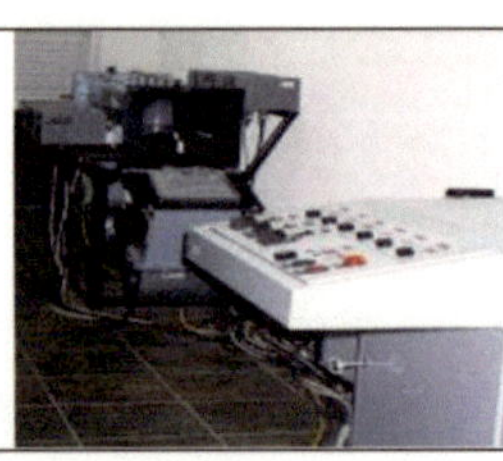

Fig. 5. Drying of creaming carrier in a microwave dryer.

The obtained dried profiles are then subjected to the process of calcination. In the conducted experiments monolithic carrier was put into a furnace and subjected to calcination at a temperature of 800°C with the temperature increase of 50 deg/h. After the final temperature was achieved, they were kept in it for 4 hours. The obtained monolithic carrier was characterised by good mechanical properties.

Preparation of the catalysts

The method o preparing catalysts:

- Determination of carrier absorptiveness.
- Preparation of the appropriate amount of solution of salts of a given metal for saturation.
- Deposition of the salt solution of a given metal on the carrier.
- Drying at room temperature in the open air for 24h.
- Drying at a temperature of 110°C for 12h.
- Calcination of the catalyst to a temperature of 500°C and maintaining it at that temperature for 4h.

4.3 Testing of catalysts

The basic characteristics examined by manufacturers and users of catalysts are the activity and/or the so-called flashpoint, pressure drop, resistance to abrasion and crushing, lifetime, chemical composition, resistance to poisoning, grain shape and size, bulk density, porosity, specific surface area, and thermal stability. Some of those properties, e.g. grain shape and size or pressure drop on the bed are of secondary importance in the case of use of catalysts on a honeycomb-shaped carrier, while other concern all types of heterogonous catalysts, and description of the methods of their determination is available in the standardisation literature.

From the user's perspective, of greatest importance is the activity of the catalyst. Determination of that property is relatively challenging because this is a speedy and strongly exothermic reaction, which causes huge temperature changes in the catalyst sample. Therefore, it is difficult to determine the reaction speed (or activity) as a function of temperature. Additionally, the catalyst exhibits changes in activity characteristic of hysteresis, i.e. differences in reaction speed depend on the direction from which we arrive at selected parameters of the process. This phenomenon is particularly evident at a temperature of the order of 450°C and sometimes several days of tests are required to determine the actual balance. The activity is generally determined as conversion under certain conditions and with a fixed catalyst volume or as an activity relative to a standard catalyst.

A measurement connected with the activity is the flashpoint or the threshold temperature of the catalyst reaction. This is a very important property of the catalyst because it shows the minimum gas temperature at the inlet to the reaction, below which the reaction slows down or stops. At working installations this property is also a function of gas speed, reactor geometry, characteristics of heat exchange, catalyst operation stage, and accuracy of temperature measurement.

Under laboratory conditions the activity of catalysts is usually determined by using model mixtures. Composition of gas, i.e. SO_2, NO, O_2, and N_2 content is determined at the inlet and outlet from the reactor, e.g. with the chromatographic method etc.

Based on this method (Preparation of the catalysts) 3 manufactured units of a monolithic catalyst have been produced in industrial conditions (approx. $1m^3$ each), based on an aluminosilicate carrier: Cupric catalyst - Cu/natural aluminosilicate, Manganic catalyst – Mn/natural aluminosilicate, Mixed cupric manganic catalyst-CuMn/natural aluminosilicate.

Active metals were placed on the carrier by impregnation of water solutions of cupric nitrate and manganic acetate. Chemical composition of natural clay used for carrier prepared is presented in Table 5.

Chemical composition	Parameters
SiO_2	54 – 56 wt.%
Al_2O_3	37 – 39 wt.%
TiO_2	max. 1,0 wt.%
Fe_2O_3	2,2 – 2,7 wt.%
CaO	max. 0,4 wt.%
MgO	max. 0,6 wt.%
Na_2O + K_2O	1,5 – 2,1 wt.%

Table 5. Chemical composition of natural clay from deposit of Lower Silesia Poland

Mineralogical composition: Kaolinite-min. 72 wt.%; Ilite- max. 23 wt.%: Quartz- max. 3,0 wt.%; Colour -Light beige

The laboratory apparatus for testing the activity prepared catalysts is presented on Fig 6. The research was conducted on the laboratory flow equipment using the model gas of determined chemical composition corresponding to that of a waste gas from power plant. Tests of activity were performed with the apparatus shown in figure 6 (apparatus for denitriding tests).

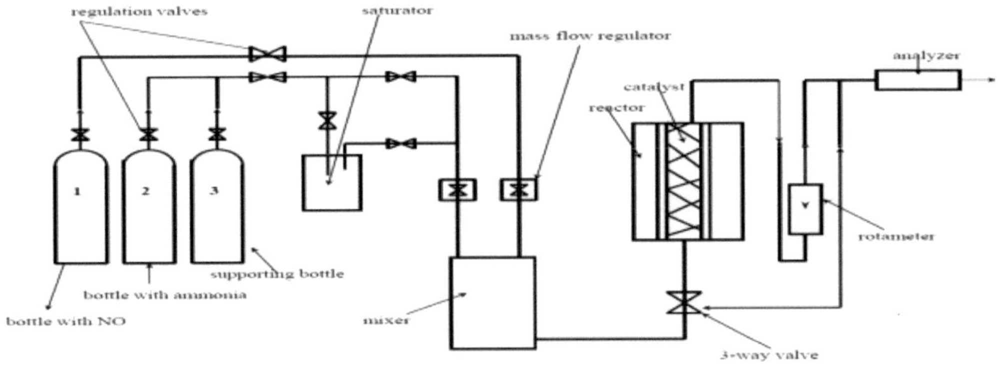

Fig. 6. Schematic diagram of laboratory apparatus for testing catalyst

The apparatus consisted of model gas feeding system and dosage system, catalytic reactor and analyzer for determining the content of nitric oxide in the gas. The reactor was supplied with the mixture of nitrogen and air enriched with water steam. The mixture contained, depending on the type of test, a specific amount of nitric oxide. The concentration of NO was measured at the entrance and exit of the reactor.

10 cm^3 of the tested catalyst (grain fraction 0,6 – 1 mm) was put in the electrically heated reactor (Fig. 6). The mixture of gases from the mixer was directed into the SCR reactor. The reactor was equipped with an electrical heating jacket powered by auto transformers. The temperature of the reactor at the entrance and at the exit of the catalyst bed was measured by means of thermocouples connected with the temperature regulator. Additionally, the reactor was equipped with an isolation mantle in order to provide isothermal conditions inside it. Gases coming out of the reactor went into an analyzer through the filters. Nitric oxide contents at the entrance and at the exit of the catalyst bed was determined with using of MSI 2500 analyzer.

Activity of catalyst - reaction of denitriding the combustion gases - is determined according to the following formula:

$$\alpha = 100(c_o - c_k)/c_o \quad [\%]$$

c_o concentration of NO_x before reactor
c_k concentration of NO_x after reactor
The conditions of the process were: temperature range from 150 °C to 500 °C, volume speed of the gas flow was GHSV=3000 $m^3/m^3 \bullet h^{-1}$, oxygen content in the model gas = 6 %, NO content within the range ~ 500 ppm, ratio NH_3/NO equal to 1, water vapor content in the model gas ~ 1 %, Model gas and gas leaving the reactor was analyzed using the MSI 2500 analyzer.
The results of investigation of catalyst activity of 8,78 wt% CuO; 3,63 wt% MnO; and 8,78 wt% CuO with 3,63 wt% MnO catalysts are presented in Fig.7.

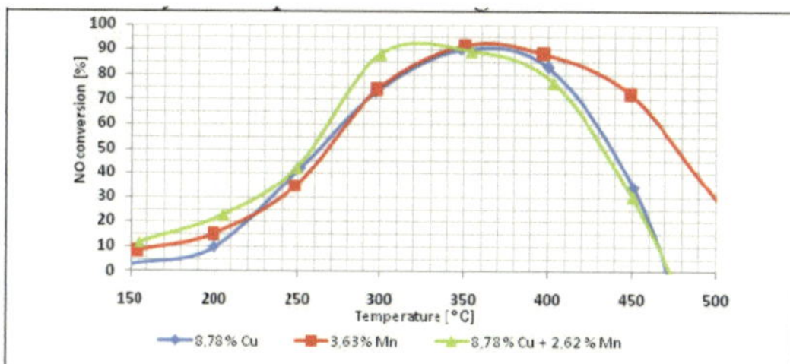

Catalyst activity of 8,78 wt% CuO; 3,63 wt% MnO; and 8,78 wt% CuO with 3,63 wt% MnO catalysts

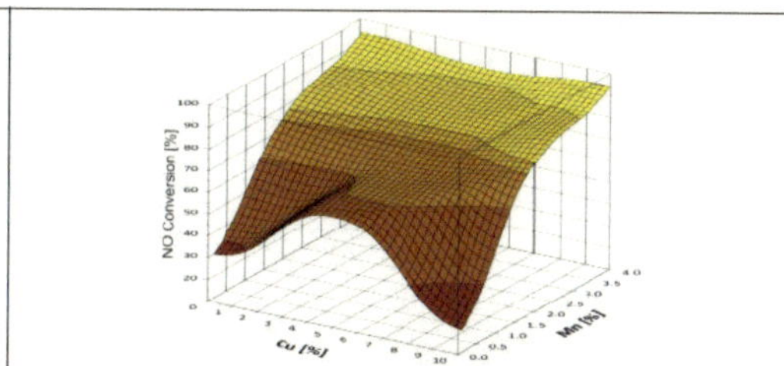

Isolines of nitrogen oxide conversions depending on the content of Cu an Mn at a temperature of 350°C

Fig. 7. Catalyst activity

As we see (Fig. 7- left) in range of temperatures above 400°C take places ammonia combustion reactions and take place decreasing of NO conversion.

4.4 Catalyst deactivation

The aging of the catalyst is caused by several mechanisms acting simultaneously, which can be divided into three groups:

- thermal deactivation,
- mechanical deactivation,
- chemical deactivation.

Thermal deactivation is caused by raising the catalyst temperature to about 600-650°C, which causes irreversible degradation of the carrier. Presumably this is connected with a change of the porous structure of the catalyst as well as blocking of some of the pores by molten components of the active phase. There is no doubt that the porous structure of the catalyst changes, and its specific surface area diminishes while the average pore radius increases. There are also crystallographic changes: amorphous silica crystallises to form a-cristobalite.

Mechanical deactivation is caused by blockade of the gaps between catalyst granulates (or channels in the monolith) by particulates carried by gas. This type of deactivation depends mainly on the purity of the gas fed to the reactor. Chemical deactivation or catalyst poisoning are usually regarded as rapid loss of activity caused by reaction of trace impurities with the catalyst. However, there are many substances that react with its components resulting in a reduction of the activity or deterioration of its mechanical strength. Most known catalyst poisons, such as arsenic oxides, nitrogen oxides, carbon monoxide, lead, and mercury are harmless in small quantities. Hydrogen chloride and chloride acting for a longer time can cause a loss of catalyst activity [23].

4.5 Methods of installing catalysts

The level of the required catalyst temperatures determines where catalysts are incorporated. Generally speaking, catalytic systems can be installed on boiler exhaust gas lines irrespective of other installations, e.g. desulphurisation of exhaust gases (DESOX). However, in practically all exhaust gas purification solutions, the DENOX and DESOX systems are designed in a comprehensive manner. This creates various possibilities for locating SCR installation in the exhaust gas line downstream the boiler.

Fig. 7-12 shows installation options for the SCR systems. Fig. 7 shows the most common version of SCR installation, the so-called high dust (with high ash content in exhaust gases).

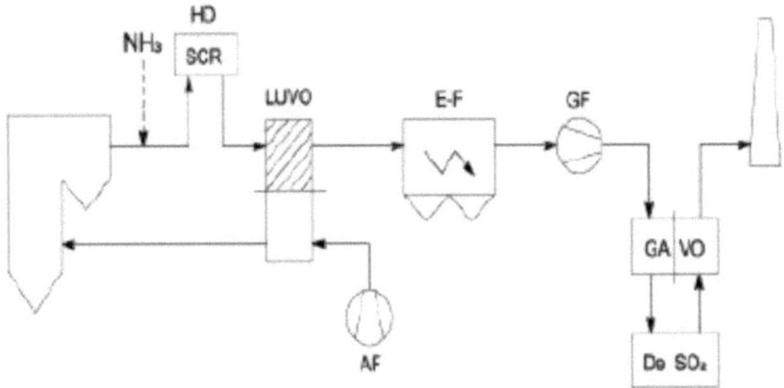

Fig. 7. "High dust" system with desulphurisation and denitrification.

Fig. 8 presents a "Low dust" only for organisation purposes. It is used exclusively in the USA, in boilers with hot electrostatic precipitator.

The systems presented in Fig. 9 and 10 also constitute "low dust" solutions but due to their incorporation at the end of the exhaust gas line they are commonly referred to as "Tail End".

Fig. 9 shows a very interesting "High dust" concept with the so-called DENOX - LUVO, in which heating elements in special execution have catalytic properties.

The use of RAH (LUVO) as a two-function device by extending its heat-exchanging function to include a catalyst function, would result in a considerable reduction in the cost of implementing the SCR method.

The proposed [43-45] way of using the regenerative air heater (RAH-SCR) as a catalyst would eliminate the necessity of building a SCR reactor, as the existing RAH could then be used. The trials will enable to compare the activity levels of industrial-scale manufactured catalysts in reduction of NO_x using ammonia. A series of trials is also anticipated, during which fly-ash from boilers will be added to flue gas. It will enable researchers to assess durability and time-based changes of reduction efficiency of the catalysts at variable ash loads (fly-ash which may 'pollute' the catalyst).

Figures 12 show detailed diagrams of the most interesting SCR systems. Ammonia injection is placed at least 3 m upstream of the catalyst.

Fig. 13 shows balance of ammonia at the individual devices of the HD system. It is very important that sulphur oxides are removed from exhaust gases prior to the process of selective catalytic reduction, because it prevents the formation of ammonium sulphates that can form in the following reactions:

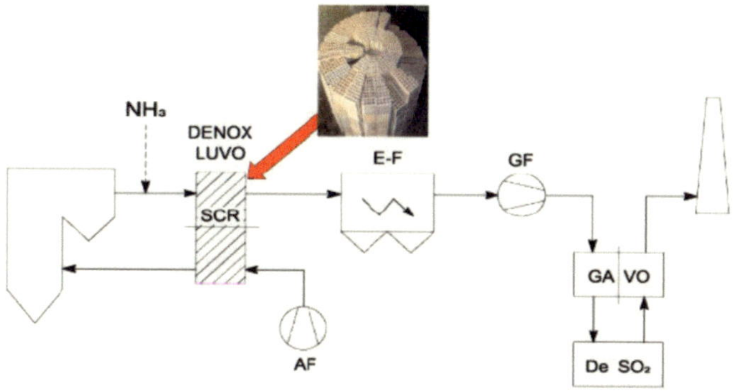

Fig. 8. "High dust" system with DENOX LUVO.

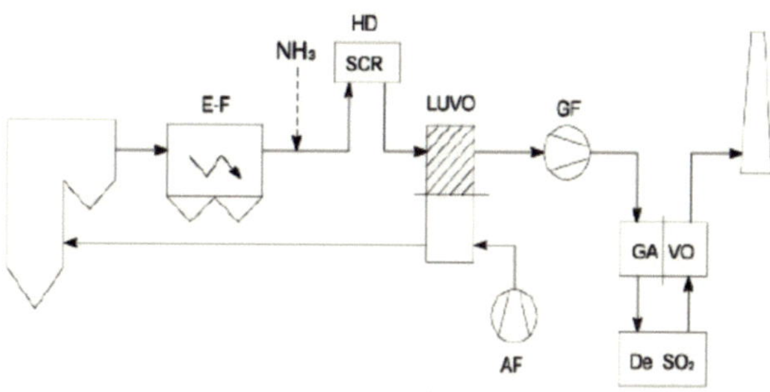

Fig. 9. Classic "Low dust" system of SCR with desulphurisation.

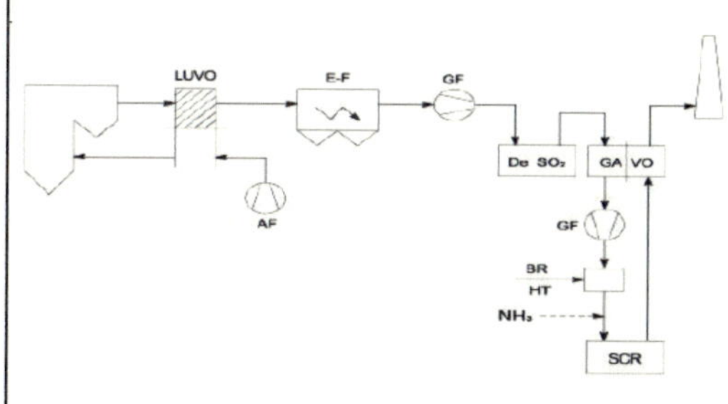

System downstream of desulphurisation

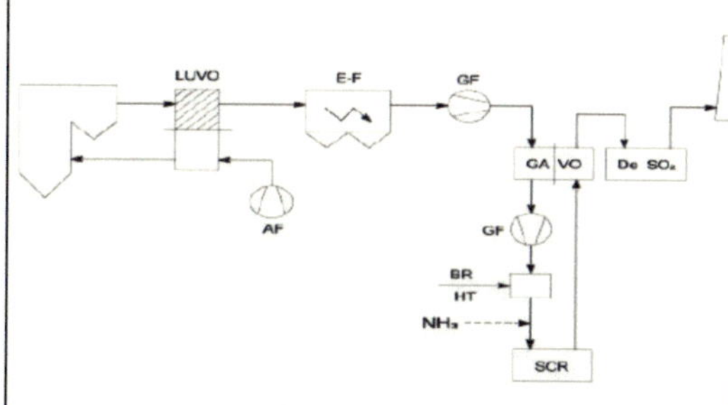

System upstream of desulphurisation

Fig. 10. "Tail end" system

$$2\,SO_2 + O_2 \Rightarrow 2\,SO_3$$

$$2\,NH_3 + SO_3 + H_2O \Rightarrow (NH_4)_2SO_4$$

$$NH_3 + SO_3 + H_2O \Rightarrow NH_4HSO_4$$

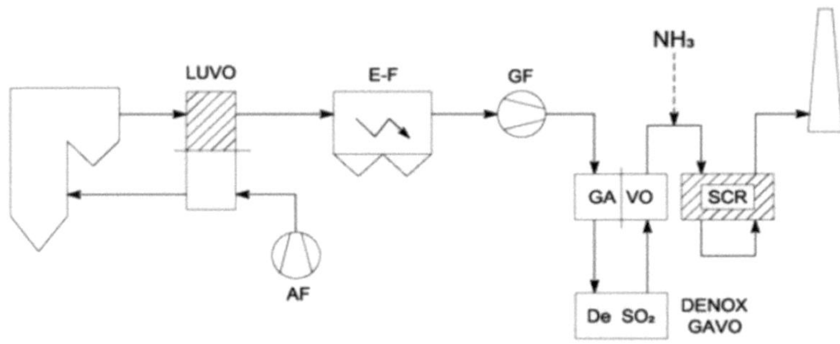

Fig. 11. "Tail end" system downstream of desulphurisation with DENOX GAVO.

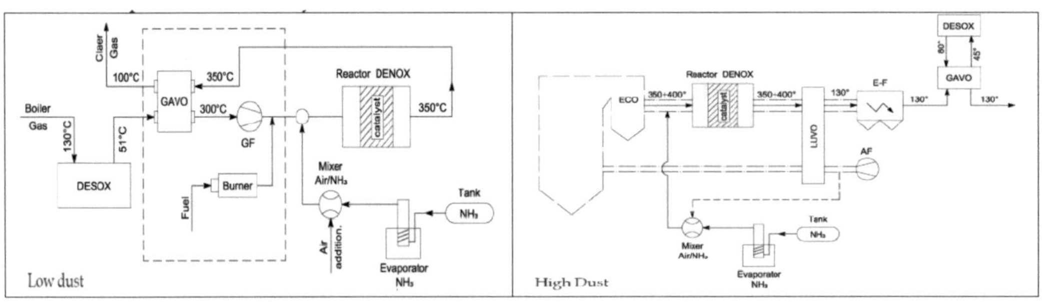

Fig. 12. DENOX installation

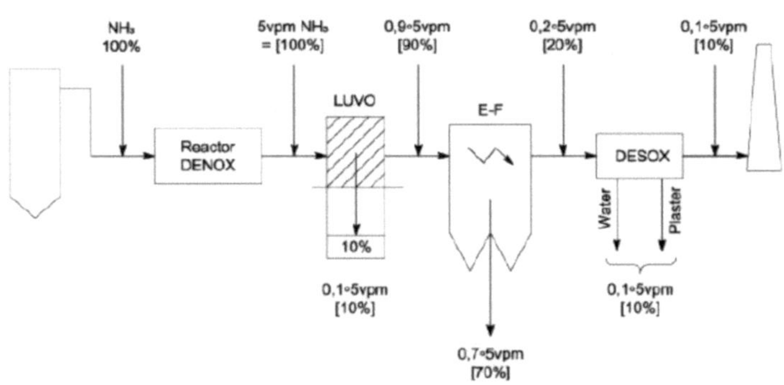

Fig. 13. NH_3 balance at SCR – "HD"

The forming ammonium sulphate causes clogging of catalyst beds and corrosion of the SCR installation. This is an especially serious problem at power stations fuelled by bastard coal. The effects of particulate matter, which may cover the catalyst surface, is very clearly visible on the following photographs (Fig. 14)

Initially, catalytically-active components of catalysts were pure noble metals, such as: platinum, rhodium, and palladium. The benefits of application of those metals were:

- high activity,
- resistance to deactivation.

However, the high price of noble metals lead to the development of cheaper and equally effective catalysts, such as:

- catalysts based on transition metal oxides (V_2O_5, NiO) deposited on carriers (TiO_2, Al_2O_3, SiO_2, ZrO_2),
- zeolites containing Cu,
- metals such as Fe, Mn, or Cu deposited on carbon and mineral-carbon carriers.

In monolithic shape | Catalyst destroyed

Fig. 14. View of a catalyst SCR.

Currently, more and more attention is devoted to carbon materials and their use in catalytic processes. Especially strong interest is generated by active carbon as a carrier of the active phase of catalysts. It has several important benefits thanks to which it has been widely used is different fields of chemistry. Active carbon is characterised by a very well-developed surface (from 1000 to 1500 m^2/g), diverse pore diameters, and high adsorption capacity. For these reasons, it is used as a component of water purification filters or as the main component of canisters in gas masks, capturing such hazardous substances as: some organic compounds, sulphur oxides, hydrogen chloride, ammonia, hydrogen cyanide, or nitrogen oxides. Active carbon is also used in medicine, administered by the oral route in some cases of poisoning. As a carrier of the active phase of catalysts, it is used e.g. in organic chemistry. In industrial processes, such as SCR, active carbon can be connected with mineral compounds, which significantly increases its mechanical strength as an active phase carrier.

The catalytic method with the use of ammonia was developed by Englehard Corporation in the United Stated in 1957. The first SCR installations used platinum catalyst for nitrogen oxide reduction. However, its use was abandoned due the fact that the reduction reactions took place at temperatures similar to the flashpoint of the ammonia and air mixture. Currently, most catalysts in use are made of a carrier - titanium oxide (TiO_2) and the active phase - tungsten oxide (WO_3) and/or vanadium oxide (V_2O_5).

In the process of selective catalytic reduction of nitrogen oxides, in addition to the active phase the temperature of the process is also important. It is recommended to use temperatures in the range from 300°C to 400°C, so optimum conversion of nitrogen oxides can be obtained. At temperatures above 450 °C ammonia is burnt to NO.

$$4NH_3 + 5O_2 \Rightarrow 4NO + 6H_2O$$

On the other hand, too low process temperature (under 200 °C) may lead to the formation of ammonium nitrites and nitrates. When exhaust gases contain a lot of particulates (approx. 20 g particulate per 1 m^3) a reactor is used, in which the fuel jet is directed vertically downwards to the catalyst bed. Typically, three or four catalyst layers are used, placed over each other (Fig.16). Each layer consists of certain number of buckets containing catalysts, which facilitates ongoing replacement of spent catalysts starting from the top of the reactor. This way each bed is periodically subjected to the process of cleaning of deposits by means of overheated water vapour.

The technique of manufacture of catalytic monoliths was first perfected in Japan. For dusty exhaust gases two types of flow profiles were developed: plate and honeycomb.

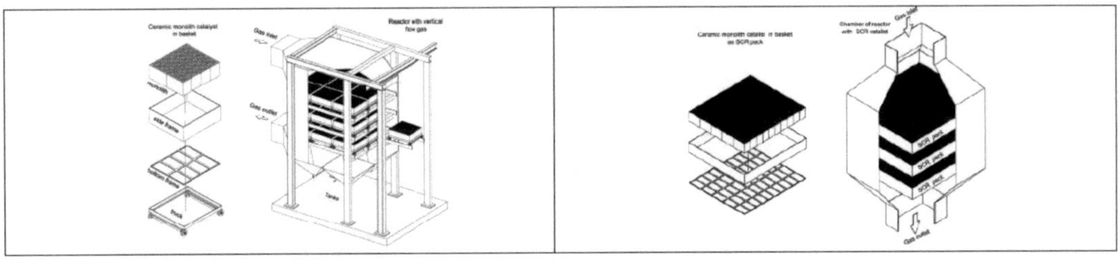

Fig. 16. Diagram of the SCR reactor (right - the element and package of ceramic catalysts)

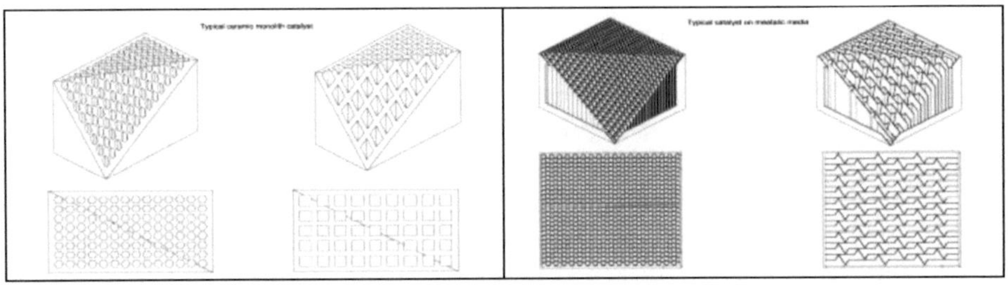

Fig. 17. Types of industrial catalysts in use.

In Japanese metal catalysts the primary material is titanium oxide TiO_2 mixed with glass fibre, covered on the outside with tungsten and vanadium pentoxide.
Besides high investment cots, catalysts containing those heavy metals also create serious problems connected with their storage after they are spent. Catalysts have a lifetime of 2 to 7 years. In their search for cheaper solutions, German companies developed their own iron-chrome (Didier) and ceramic (Mannesman) catalysts. Spent ceramic catalysts are powdered and used as a raw material for the production of new catalysts.
Plate catalysts are made in the form of blocks similar to the so-called baskets of heating elements of rotating air heaters. Plate catalyst blocks are assembled as packages (modules), as shown in Fig. 17.
Honeycomb catalysts are made in the form of small-scale elements, which are then assembled into modules.
Catalysts are installed in the exhaust duct in 3 or 4 layers, with a gap between them to incorporate cleaning devices (blowers).
Table 5 shows SCR device suppliers to illustrate the dynamics of development of the catalyst manufacturing industry in Germany supplying the power industry. Table 6 shows types of catalysts used in the Japanese power industry. In table 5 of note are unmentioned catalysts based on carbon chemistry (active carbon, active coke). This technology will be presented later in the text.
Catalytic methods, included among non-waste methods, due to lack of waste are an alternative to the waste methods. They are characterised by a high degree of exhaust gas purification (simultaneous removal of NO_x and SO_2) and achievement of a commercial product in the form of concentrated sulphuric acid, sulphur, or other products. They involve catalytic oxidation of SO_2 to SO_3 and three-stage condensation of exhaust gases: at the first

stage - with condensation of concentrated sulphuric acid, at the second stage – after moistening of sulphuric acid of a lower concentration, at the third stage – after moistening of hydrogen chloride and hydrogen fluoride.

Supplier	Licence	Type of catalyst
Deutsche Babcooc	Kawasaki (Jap.)	Honeycomb-type based on titanium oxide (TiO_2)
Deutsche Rauchgas	Babcoc- Hitachi (Jap.)	Plate-type based on titanium oxide
Dider	own (Germany)	Iron-chrome
EVT	Mitsubishi (Jap.)	Honeycomb-type based on titanium oxide (TiO_2)
Flakt	Hitachi - Zosen (Jap.)	Plate-type based on titanium oxide
GEA	Engelhard	Honeycomb-type based on titanium oxide (TiO_2)
Linde	Norton (USA)	Powder catalyst
Knauf Research Cottrel	USA	Plate-type or honeycomb-type based on titanium oxide (TiO_2)
Steinmuller	Ishikawajima	Honeycomb-type based on titanium oxide (TiO_2)
Mannesmann	own	Molecular sieve, ceramic
Holter- Lurgi	Hitachi - Zosen (Jap.)	Plate-type based on titanium oxide
Thyssen	MHI	Honeycomb-type based on titanium oxide (TiO_2)
Uhde	Bergbau-Forschung	Active coke
Uhde- Lentjes	Babcoc - Hitachi (Jap.)	Plate-type based on titanium oxide

Table 5. Companies offering SCR devices

Name of the power station	Power unit capacity Mw	Type of catalyst	Concentration of NO,[mg/Nm^3]		SCR effectiveness, %
			input	output	
Takehara	2x250; 700; 2x250;	plate	700-500	134-96	81
Shimono Seki	175	honeycomb	840	360	57
Shin Ube	156	honeycomb	800	280	65
Mizushima	156	plate	700	240	65
Saijo	156; 250;	honeycomb	760-660	260-180	65-70

Table 6. SCR systems in larger carbon units in Japan

It should be emphasized that those are the methods of the future and they already have their applications in the world, e.g. Münster in Germany and Vendsyssel power station in Denmark. However, they are characterised by an extensive centre of catalytic oxidation and reduction of gas contaminants as well as centres of condensation of separated contaminants with complex devices. They do not require the use of sorbents and they provide end-products with specific commercial properties.

5. DENOSOX method of simultaneous removal of SO_x and NO_x from exhaust gases by catalysts

The DESONOX method was devised and developed as a relatively simple method of treatment of exhaust gases, in which the end-product is sulphuric acid. The method was used for desulphurisation and denitrification of exhaust gases at power unit No 3 of the heat and power generating plant in Münster (Germany) (Heizkraftwerk der Stadtwerke Münster GmbH).

The DESONOX method was devised and developed with the following objectives:

- removal of sulphur dioxide and nitrogen oxides,
- generation of sulphuric acid from the removed sulphur dioxide, which can be used in industry.

Figure 18 shows a simplified diagram of the DESONOX installation. The DESONOX technology includes utilisation of two toxic components of waste gases in installations:

- nitrogen oxides (using the DENOX technology),
- sulphur dioxide (using the DESOX technology).

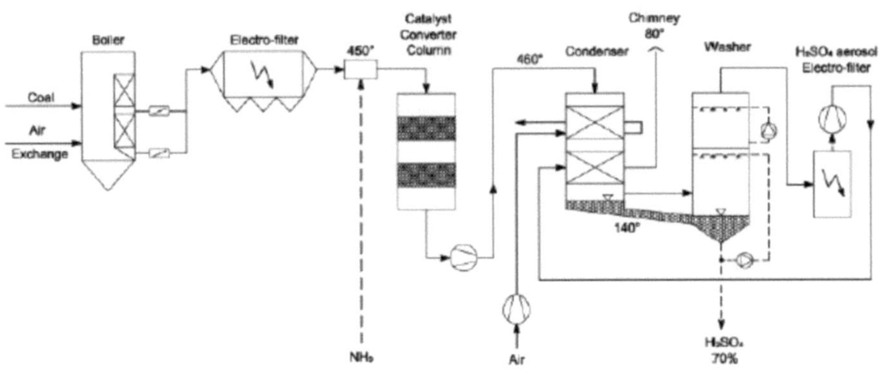

Fig. 18. DESONOX installation diagram.

The DENOX technology

The DENOX technology utilises the phenomenon of selective catalytic reduction (SCR). It uses catalysts in whose presences - by means of ammonia, forming a reducing atmosphere - nitrogen oxides are reduced to pure nitrogen and water, i.e. components neutral to the atmosphere. The temperature of waste gases, after passing through the heat exchanger and the electrostatic precipitator, amounts to approx. 160°C. This temperature is too low for the work of the catalyst. Therefore, waste gases pass through a recuperator where they are heated up to 450°C. In some cases, in order to obtain that temperature, it is necessary to use an additional flame heater, powered e.g. by natural gas. Heated gases are enriched with ammonia. The mixture of warm exhaust gases and ammonia is directed to the catalytic converter. The catalyst is deposited on ceramic monoliths arranged in 1-2 layers.

Above the monolith layers there is an element for uniform distribution of gases throughout the cross-section of the converter. The converter has a modular structure. A single module often consists of up to 100 monoliths. The number of monoliths depends on the size of the system. For example, a boiler with the capacity of 200 MW requires approx. 200 m³ of catalytic bed.

The degree of conversion of nitrogen oxides varies between 85 and 98% while the degree of conversion of sulphur dioxide remains at the previous level of approx. 90%. The effectiveness of the DESONOX process depends primarily on the temperature of exhaust gases. If we assume at least 80% degree of conversion, then the reduction of nitrogen oxides with such efficiency is already present at temperatures above 390°C, whereas the same degree of SO_2 conversion is obtained at a temperature above 440°C.

Hence, the working temperature of the catalyst in the DESONOX system is assumed to be equal to 450°C. Raising of that temperature has a positive effect on the efficiency of the process. The only adverse effect of lowering the temperature below the adopted value is a reduction of the process effectiveness, starting with a reduction of the SO_2 conversion. A positive effect, which makes it possible to maintain the process temperature at a set level, is the fact that the oxidation of SO_2 to SO_3 is an exothermic reaction. It also goes to show that the higher the concentration of sulphur in carbon, the greater the amount of heat generated during oxidation of sulphur dioxide and the easier it is to maintain the adopted parameters of the process. One drawback of the technology is the need to feed ammonia to the system.

The DESOX technology

Utilisation of sulphur dioxide occurs at the second stage of the catalytic converter in the DESOX process. After waste gasses pass through the first bed, they still contain approx. 8 % of excess oxygen. Therefore, in the second bed, sulphur dioxide can be oxidised in the presence of a catalyst - at temperatures of the order of 450°C - to sulphur trioxide, according to the following reaction:

$$5O_2 + 1/2\, O_2 \Rightarrow SO_3$$

Unlike sulphur dioxide, sulphur trioxide is readily soluble in water, which is present in exhaust gases in the form of vapour arising during reduction of nitrogen oxides with ammonia, and binds to it to form sulphuric acid.

$$SO_3 + H_2O \Rightarrow H_2SO_4$$

Sulphuric acid it taken from the gaseous form to liquid form in the heat exchanger through which exhaust gasses pass, and is stored in a settler. Exhaust gases then pass to the washer, where the remaining sulphuric acid is removed by sprinkling with a shower of sulphuric acid, previously collected in the settler of the heat exchanger.

Ay residue of the sulphuric acid deposits on this electrostatic precipitator. The waste gases upstream the electrostatic precipitator have a temperature of approx. 50°C. This temperature may be too low, therefore waste gases are additionally heated to a temperature above 80°C. Sulphuric acid may be derived either directly in concentrated form (concentration over 70 %) or taken to gypsum after reaction with lime.

6. Methods of simultaneous removal of sulphur and nitric oxides on adsorbents and carbon catalysts

At the current state of technology, the methods whose process include carbon sorbents seem to be the easiest, the most technologically mature, and the most economically promising solution.

A precursor of these technologies is believed to be the UNITIKA method developed in Japan in the 1970s, which uses sorption of sulphur dioxide by active carbon with simultaneous reduction of nitrogen oxides to free nitrogen. The reducer was ammonia and the process was taking place in the temperature range of 200-250°C. The catalyst used in the form of fixed bed was regenerated by gas at a temperature of 300-350°C. Sulphur dioxide was recovered from regenerated gases in a separate device by means of absorption. The applicability of that method was limited by the required low particulate content in purified gases (under 0.2 g/m^3).

A more technologically mature method, brought to the stage of an industrial solution, was also developed in Japan by SUMITOMO Heavy Ind., and involved simultaneous disposal of SO_2 and nitrogen oxides on a fixed bed of appropriately prepared coke and in the presence of ammonia supplied to the gases. The process takes place at significantly lower temperatures than in the UNITIKA method, of the order of 120-150°C, where sulphur dioxide is oxidised to trioxide, and then is retained in sorbent pores in the form of sulphuric acid and its ammonium salts. Simultaneously, nitrogen oxide is reduced with ammonia to nitrogen and water vapour.

Because adsorbed sulphuric acid or its ammonium salts reduce the catalyzing ability, sorbent is regenerated in a stream of inert gas, at temperatures of 350-600°C, with emission of sulphur dioxide and carbon, water vapour, and nitrogen. Gases containing 10-30% SO_2 are converted to sulphuric acid or elemental sulphur.

The process was implemented on a large industrial scale a the Matsushima installation, where removal of 95% of sulphur dioxide and 40-50% of nitrogen oxide was achieved. That method became the starting point for the development of a similar process under the name Bergbau-Forschung (BF). The primary sorption factor is active coke while gaseous ammonia is the reducer of nitrogen oxides. In accordance with this method an installation was built (by UHDE) for two brown coal-fired power units at the Arzberg power station.

The main benefits of catalysts based on carbon sorbents are their sorption of sulphur dioxide to a technically usefully degree as well as the ability to realise the process of reduction of nitrogen oxide with ammonia under relatively low temperatures 80-130°C. In this scope, typical of energy waste gases, these catalysts are characterised by high efficiency. Their adsorption and catalytic properties were used in the process of removal of nitrogen oxides and sulphur from gases in practically one operation, at one reactor, and at similar parametric conditions.

The raw materials used to obtain active carbons contain carbon and include hard coal, brown coal, peat coal, charcoal, coconut shells [46], industrial waste [47] and organic compounds containing nitrogen, polyacrylonitriles, and melamine and urea resins [48-51]. The initial mechanical processing of the raw material involves: grinding of the output product; addition of binding agents, for example aluminosilicate [52], tar, and refinery waste [47, 53-55]; thickening; drying; initial thermal treatment; and giving the appropriate form, usually that of granulate. The obtained material is subjected to the process of activation, usually thermally at a properly selected high temperature by contact with inert gases or gases containing ammonia [50, 56-58] or water vapour [49].

The product obtained this way is a mixture of amorphous and microcrystalline structures, with the cavities formed inside it constituting a system of channels of different dimensions – micropores with a diameter of up to 20 A, mesopores 20-500A, and macropores of greater diameters. Macropores primarily determine the total volume of pores. The inner specific surface area formed this way usually amounts to between 500 and 1500m^2/g.

Carbon sorbents should exhibit the following properties:

- inner volume as large as possible to ensure maximum absorptiveness of sulphur dioxide or sulphuric acid,
- high proportion of micropores, responsible for a large expansion of the specific surface area, of decisive importance in the process of catalytic reduction of nitrogen oxide with ammonia,
- low resistance to gas flow,
- high mechanical resistance to crushing and abrasion;
- defined quantitative and qualitative share of ash,
- high flashpoint together with low susceptibility to spontaneous combustion,
- selectivity with respect to catalysed reaction.

The patent literature usually provides very sketchy descriptions of the processes of production of appropriate carbon sorbents, mostly from hard or brown coal as starting raw materials [47, 52-55,59,60]. One of the patents describes [54] preparation of active formed coke by subjecting ground coal with the granulation of under 1 mm and containing 10-45% of volatiles to the process of fluidisation at a temperature of 180-270°C in the presence of 2-10% of oxygen. The product with the addition of binding substances is used to form granulate with the diameters of 4-9 mm, which is subjected to a temperature of 650-950°C.

Another method of obtaining sorbent [47] involves use of soot and waste from kerosene processing installations, which are mixed together with molasses as a binding agent, extruded, dried, carbonised, and thermally activated. Hard product, with the pore volume of the order of 60 $cm^3/100g$, exhibiting high absorptiveness of nitrogen oxides (over 6% NO_x by weight), was used to for their removal from waste gases. There are also reports [86-89] of preparation of a new type of sorbent by charring some nitrogenous organic compounds (polyacrylonitriles, urea and melamine resins) in a stream of air or inert gas, at temperatures of 300-1200 °C, and their subsequent activation by water vapour. The obtained sorbent, with the specific surface area of up to 1000 m^2/g, additionally modified with vanadium compounds (up to 2% V), exhibited high purification efficiency of up to 90% for SO_2 and up to 100% for NO.

Many patents point to the advisability of modification of carbon sorbents by addition of ammonium salts of hydrohalic acids [61-66] or by metal oxides of Cu, Cr, Ni, Mn, Fe, Co, V, Mo and others. This increases the ability to catalyse NO reactions [51] and, as a result, most sorbents modified this way are used in the processes of exclusive removal of NO. Patent applications [67-71] suggest they may also be used in the processes of joint removal of sulphur and nitrogen oxides. Activation of such sorbents is carried out by prior contact with inert gasses containing ammonia at an elevated temperature of 500-950 °C, over a period of approx. 10 minutes [56-58].

Some patents demonstrate more research to increase efficiency of carbon sorbents, among others toward further lowering of the working temperature or improvement of their mechanical characteristics. An example of such efforts is the patent information [72] on preparation of network catalysts covered with carbon material by the introduction of coke into the cellular structure of the Al_2O_3, TiO_2 type or isomorphic aluminosilicates. The currently obtained carbon sorbents have well-developed specific surface area up to 1500 m^2/g [49, 73], bulk density in the range of 500-900 g/dm^3 [54, 73], specific volume of pores of 40 to 60 $cm^3/100$ g [47, 54], and exhibit absorption of sulphur dioxide at a temperature of 120 °C from 10 to 15% in relation to the sorbent weight [53, 57, 59]. However, as previously mentioned, they are characterised by relatively low permissible gas load values, usually amounting to 500-1000 $m^3/m^3 \cdot h$ [58, 63, 74]. Their mechanical strength, especially if a movable bed is used, is also considered to be insufficient [75].

6.1 The Bergbau - Forschung/Uhde (BF) method

The process is carried out in a two-stage reactor with a movable bed of active coke by carrying out reduction of nitrogen oxide by ammonia with simultaneous adsorption of sulphur dioxide at temperatures of 120-150 °C. The installation is situated in the power production line after the air heater and electrostatic precipitators. The process consists of a three-stage procedure -adsorption, desorption, and the section of sorption gas processing.
The adsorber design is subdivided into two stages, which fulfil the functions of, respectively, SO_2 adsorption and NO reduction; the regeneration stage takes place in a separate reactor. Both reactors and the adsorber are connected by movable bed of active coke, forming a closed circuit. Adsorption: During the first stage of the adsorber, sulphur dioxide is stopped in the amount of approx. 90%, which in the presence of water vapour and oxide forms adsorptionally bound sulphuric acid. At this stage the sorbent also stops HCl, HF, heavy metals, and fly ash. At the second stage of adsorption, after ammonia is added to the gases, nitrogen oxide is reduced.

Desorption:

Coke from the adsorber is fed pneumatically or transported or mechanically transported to the desorption system. Heating through membrane by hot exhaust gases to 400°C causes desorption of sulphur dioxide; the resulting gases contain approx. 30% of SO_2. After desorption coke is cooled by air, also through membrane, and returned to the second stage of the adsorber.

Gas processing for desorption:

Gases leaving the desorber and containing in addition to sulphur dioxide also CO_2, water vapour, and nitrogen are subjected to separate processing into sulphuric acid, elemental sulphur, or liquid SO_2.
Figure 19 shows a simplified diagram of the Bergbau-Forschung (BF) method with the omission of the section of gas preparation for adsorption and the section of sorption gas processing, containing up to 20% SO_2.

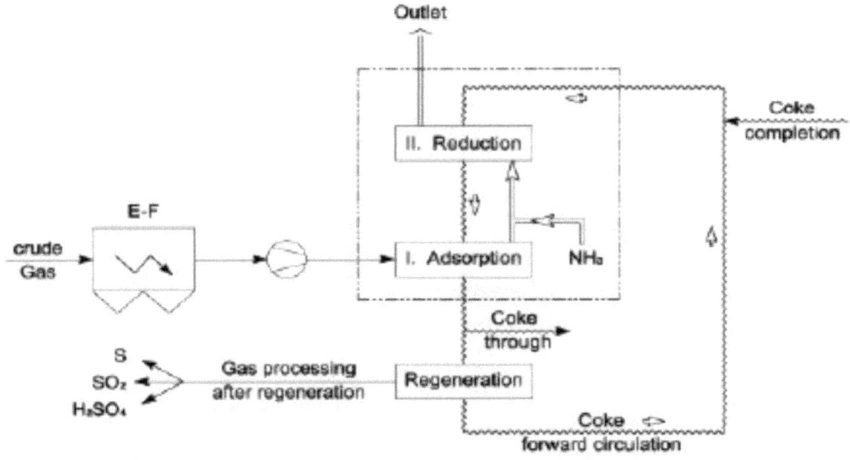

Fig. 19. Diagram of the Bergbau-Forschung process.

Purified gases leaving the electrostatic precipitator section of the power production line are sucked in by a blower and compressed to the pressure corresponding to the conditions in the adsorber. The optimum process temperature (120 °C) is set by water injection using compressed air or steam in the column not shown on the drawing, upstream the adsorber and in the additional steam exchanger. Gases introduced into a two-stage adsorber flow horizontally through movable bed of active coke, and then to the second stage of the process, selective reduction of nitrogen oxide with ammonia. On the other hand, regenerated coke from the desorber passes through the container situated at the top of the tower first to the second stage of the process of reduction, and from there it lowers gravitationally and passes to the first stage. Coke with SO_2 adsorbed on it is collected from the first stage at the bottom of the adsorber and is directed to desorber. In this way, the movable bed of active coke forms a closed circuit between the adsorber and the regenerating unit.

Purified gases, leaving the adsorber at a temperature of 120°C are discharged through the flue into the atmosphere. Heat losses due to emission through the adsorber walls and in smoke flues are offset by the heat of reaction. The dew point of the sulphuric acid is not exceeded anywhere along the exhaust gas line to the flue and reheating of the gases is unnecessary.

Gases leaving the regeneration system contain approx. 20% of SO_2, water vapour, carbon dioxide, nitrogen, HCl and HF, and heavy metals. After purification of the gases by means of sorption with the so-called "Halex" mass, the gases are converted to sulphuric acid, elemental sulphur, or liquid SO_2, depending on the variant of the procedure.

Chemical mechanism of the process

In the Bergbau-Forschung (BF) process, active carbon acts both as an adsorbent, and as a catalyst. In the absence of ammonia, sulphur dioxides as well as oxygen and water vapour contained in gases are adsorbed on the active surface of coke. Later in the process they undergo catalysed transformation to sulphuric acid, which remains adsorbed in pores of the sorbent:

$$SO_2 + 1/2 O_2 + H_2O \Rightarrow H_2SO_4$$

Simultaneously to this reaction, nitrogen dioxide which is present in gases in 5-10% of the total amount of NO_x, is rapidly reduced:

After the addition of ammonia, favourable conditions are created for the reduction of nitrogen oxides to free nitrogen and water vapour:

$$6\,NO + 4\,NH_3 \Rightarrow 5\,N_2 + 6\,H_2O \quad \text{and} \quad 6\,NO_2 + 8\,NH_3 \Rightarrow 7\,N_2 + 12\,H_2O.$$

Sulphur dioxide in the presence of active coke reacts with ammonia to form ammonium sulphate:

$$SO_2 + 2\,NH_3 + 1/2\,O_2 + H_2O \Rightarrow (NH_4)2SO_4$$

The individual salts are similarly formed, reacting with sulphuric acid adsorbed in pores:

$$NH_3 + H_2SO_4 \Rightarrow NH_4HSO_4 \quad \text{and} \quad NH_3 + NH_4HSO_4 \Rightarrow (NH_4)2SO_4$$

In the process of thermal regeneration, at a temperature above 300°C, adsorbed sulphuric acid reacts with carbon to form carbon dioxide and sulphur dioxide. The reaction goes through surface-formed CO oxides:

$$2\,H_2SO_4 + 2\,C - 2SO_3 + 2\,C + H_2O \Rightarrow 2\,SO_2 + 2\,H_2O + 2\,CO \quad \text{and} \quad 2\,CO \Rightarrow C + CO_2$$

Decomposition of ammonium salts goes in the opposite direction. On the other hand, ammonia reduces sulphur trioxide formed by decomposition of sulphuric acid and surface oxides of CO according to the following reaction:

$$2\,NH_3 + 3\,CO \Rightarrow N_2 + 3\,H_2O + 3\,C$$

thus reducing carbon losses.

Process of adsorption on carbon sorbents

Depending on the sulphur content in fuel, SO_2 concentration in exhaust gases varies between 500 and 2000 ppm; depending on the type of boiler and the manner of conducting the process, the amount of nitrogen oxides in gases stays in the range of 500-1500 ppm.

The amount of chlorine and fluorine compounds is much lower; the amount of volatile particulates is of the order of 150 mg/m^3. These values and temperature in gases upstream the reactor affect the physical and chemical conditions of the execution of the purification process, with active carbon performing both adsorptive and catalytic functions.

The mechanism of reduction of nitrogen oxide with ammonia in the presence of sulphur dioxide on active carbon, adopted by Richter [76], and the conclusions from laboratory-scale experiments in this area [77,78] clearly indicated the advisability of initial lowering of the SO_2 concentration in purified gases, using excess ammonia with respect to the total of SO_2 and NO, ensuring adequate contact time by increasing the height of the bed layer, and maximally lowering the temperature of the process.

In experiments referred to by Knoblauch [78], conducted on a fixed bed of active coal, attention was drawn to the distribution of sulphuric acid and its ammonium salts in the bed, as well as distribution inside it of areas of individual reactions, including the reaction of reduction of nitrogen oxide.

The mechanism of the process presented by Richter [76] and the results of experiments cited, inter alia, by Knoblauch, were the basis for the decision to use a two-stage model of the process of SO_2 adsorption and NO reduction, carried out in a suitably designed reactor.

In the first stage of adsorption, the primary processes of sulphur dioxide sorption take place inside pores of active coke. At this stage over 90% of the total amount of SO_2 is stopped, as well as HCl, HF, heavy metals, volatile particulates, and the total amount of NO_2. The middle part of the adsorber, designed in the form of a mixing chamber, ensures uniform concentration of SO_2 in gases upstream the second stage. At the same time, nozzles supply ammonia, which, in order to prevent formation of streams, is pre-mixed at a ratio of 1:25 with purified gas.

At the second stage nitrogen oxide is catalytically reduced at temperatures of 90 to 150 ^{o}C. Ammonia is adsorbed and then reacts on the coke surface according to the total reaction:

$$6NO + 4NH_3 \Rightarrow 5N_2 + 6H_2O$$

Additionally, there is binding of residual sulphur dioxide; neutral and acid ammonium salts are formed and deposit on the surface layer of sorbent. Purified gases as discharged through the flue to the atmosphere.

Some solutions [79] provide for a three-stage adsorption system, where the first and second stages of SO_2 adsorption and NO reduction have been supplemented with a third stage, adjoining the second one and powered with part (50%) of the sorbent leaving the first stage, which contains mainly adsorbed sulphuric acid. Such solution is designed to limit ammonia losses in gases leaving the installation.

Parameters of the adsorption process

On the basis of numerous data, contained in patent information, findings of studies conducted on an increased scale on existing pilot installations, as well as on the basis of bidding information of Bergbau-Forschung [75], we can attempt to identify the parameters characterising the process using carbon sorbents. For example, patent information [80] provides some data about the process executed on a Japanese pilot installation for the amount of gas V=1400 m^3/h, with the SO_2 and NO_x content of, accordingly, 2900 and 500 ppm. Ammonia was supplied to gas prior to adsorbers. Inertness of active carbon bed in the adsorber was 4.6 and 1.8 m^3; the process was carried out in two stages.

Based on these data and assuming an average bulk density of the sorbent d=0.700 kg/m^3, the amount of sorbent can be estimated at the respective stages, sorbent load with the GHSV gas, and the duration of stay of active carbon [h] in the adsorber:

Duration of stay of coke in the adsorber approx. 200 hours, including on the second stage for 150 hours [75, 81, 82]. The flow rate of coke in the adsorber approx. 0.1 m/h.

The installation provides for the use of a third adsorption stage, whose task is to remove residual ammonia from gases leaving the adsorber. This stage, which is connected directly with the second one, is supplied with sorbent from the first stage in the amount of 50% of coke supplied to the adsorber.

Regeneration of the carbon sorbent

The process of regeneration of active coke, saturated with sulphuric acid and its salts, takes place mostly be means of thermal distribution at temperatures above 300 °C.

Knoblauch presented [78] the results of experiments on thermal regeneration of active carbon, heating it in a stream of helium in a differential reaction at a rate of 10 deg/min. Initially, secretion of physically adsorbed water vapour is observed. As the temperature increases, desorption of sulphur dioxide and a two-stage decomposition of ammonium sulphate takes place according to the following reaction:

$$(NH_3)2SO_4 \Rightarrow NH_4HSO_4 \text{ and } NH_4HSO_4 \Rightarrow NH_3 + SO_3 + H_2O$$

with some of the ammonia released at the first stage being oxidised with surface oxides. At a temperature of approx. 500°C acid ammonium sulphate decomposes with release of sulphur dioxide, ammonia, and water vapour to the gas phase. At temperatures above 500°C increasing amounts of nitrogen, carbon dioxide, and carbon monoxide start to appear in the exhaust gases.

There are several variants of the process depending on the form of contact of the solid phase with gas, mobile bed of sorbent or the fluidal system, and direct method of supplying heat energy. In the first solutions by Bergbau-Forschung [78], hot sand was used as the heating medium, heated separately to the temperature of 600-650°C, which was mixed with coke leaving the adsorption system. Under these conditions sulphuric acid and sulphates were reduced. Loss of carbon in the regeneration process, causing a change in the configuration of sorbent pores, simultaneously lead to an increase in its absorbing and catalytic capacity by an increase of the effective catalytic area. In another variant [83], thermal decomposition of sulphuric acid and sulphates was achieved by hot sorption gases, additionally heated in a separate exchanger to the temperatures of 300-600° C. The process was carried out in a fluidal system.

In a recently proposed solution, a three-section tube desorber was used on a large industrial scale [75, 81]. Active coke from the first adsorption stage, totally free of particulates, passes through an intermediate tank to the upper part of the desorber, which consists of three parts. In the actual upper desorption part, coke moves gravitationally through the tubes, heated through membrane to the temperatures of 400-450°C. The source of heat are hot exhaust gases, produced in a separate combustion chamber.
In the middle part of the apparatus, sulphur dioxide desorbs from the bed, passing to the gases discharged outside as a so-called "rich" (containing up to 30% of SO_2) desorption gas. In the lower part coke is air-cooled through membrane to approx. 100°C. After subgrain is separated on the sieve and the missing content is filled in, coke is directed to the upper part of the second stage of the adsorber. The operation of thermal regeneration of sorbent constitutes a significant power load for the process. The literature signals [84-86] attempts at regeneration of the sorbent at a lower temperature by washing with water; however, this process results in very diluted solutions of sulphuric acid and sulphates.
Other attempts at regeneration of carbon sorbents by means of inert gases containing in their composition ammonia and at an elevated temperature of the order of 250-450°C usually concerned a process that realised only sorption of nitrogen oxides [50, 70, 87].

Variants of the process

In a classical system of simultaneous removal of sulphur and nitrogen oxides according to the Bergbau-Forschung method, the purification installation is located in the power production line directly downstream of the electrostatic precipitators and such system does not require additional heating of the gases.
Because active coke, in addition to catalytic properties, may provide sorption functions, nitrogen oxide will be removed together with "residual" sulphur dioxide, which leaves desulphurisation installation in the amount of approx. 400 mg/m^3. The resulting ammonium sulphate has an adverse affect on the catalytic activity of coke. This necessitates periodic regeneration of the sorbent, but in very small amounts, therefore desorber dimensions may be only slightly decreased [88].
Gases containing sulphur dioxide emitted in the regeneration process are returned to the desulphurisation unit, thus increasing the total effect of SO_2 removal. With this solution and when desulphurising with lime milk, the system for processing of post-regenerated gases is not used, and the only product of the process is gypsum.
A similar solution is proposed by H. Petersen, which uses the Bergbau-Forschung licence. The purpose of the procedure is to obtain liquid SO_2 with the omission of gypsum production. Sulphur dioxide is absorbed by means of the NaOH solution, whose pH stabilises with the addition of appropriate organic compound. Blowing with air desorbs SO_2 from the post-absorption solution, resulting in gases where it is present in high amount. After drying and cooling the gases are subjected to separate processing. On the other hand, NO reduction is carried out on the bed of active coke in the same way as in the BF process, with periodic coke regeneration and returning of the gases to the desulphurisation stage after regeneration.
Despite obvious benefits, the presented variants have not as yet been implemented on a large industrial scale. The patent literature indicates a number of proposed changes to some fragments of this process. These changes concern supplementation of the sorbent composition to give it different qualities or properties, the method of conducting basic operations - adsorptions or the number of apparatuses on the technological diagram, and the search for reducers other than ammonia.

Some of the patents [73] suggest the possibility of obtaining much higher gas loads of sorbent than e.g. in the case of active coke without modifying additives, which is often associated with the need to use higher temperatures [70, 87, 89, 90]. It is also proposed that the composition of carbon sorbents is supplemented with substances havening alkaline functions, for example hydroxides and alkaline earth carbonates [91-96], with the patent by Ishikawajima Harima Heavy Ind. [94] suggesting that a process of NO reduction can be conducted without ammonia.

One of the publications considers the advisability of NO reduction using active carbon saturated with urea [97]. Besides the most common form of operation with the use of a fixed or movable bed, there is perceived possibility of conducting adsorption in the fluidal phase [98, 99]. Similarly, the previously described methods of thermal regeneration - by mixing sorbent with hot sand [78], heating by inert hot gases in a fluidal system [70, 77, 83, 87, 100, 101] and by way of membrane heating [75, 88, 81], as well as the two-stage method described in one of the patents [102] and attempts at regeneration by washing with water or appropriate solutions [84-86] - may determine different shaping of the whole technological process. Different adsorber designs represent two patents [103, 104]. Several patents propose replacement of ammonia as an NO reducer by carbon monoxide or hydrocarbons [61, 89, 105, 106] as well as hydrogen sulphide [104].

7. The manufacturing of CARBODENOX catalysts on the basis of monolithic carbon carrier

Active carbon based catalysts elaborated by EKOMOTOR Ltd. (Poland) are sufficiently active to realise SCR reaction at low temperature, from 100 to 200°C. They are especially useful for application in these processes at which flue gases temperature is lower than 200°C. Above 200-220°C and in the presence of oxygen (in air) active carbon catalyst is oxygenated and therefore higher process temperature is limited. This type of carbon catalyst after exploitation can be easily utilised e.g. by combustion. In comparison to titania based ceramic SCR catalysts active carbon based catalysts are relatively cheaper. Active carbon based catalysts are capable to adsorb SO_2 and other chemical compounds from the flue-gases. It is necessary to said that they show appreciably higher specific surface area, from 200 to 800 m^2/g and pore volume, from 0.2 to 0.8 dm^3/kg. For instance titania based catalysts are characterised by specific surface area lower than 100 m^2/g and pore volume 0.15 - 0.30 dm^3/kg. **Active carbon based SCR catalysts should be operated after ESP or between preheaters and ESP but always after desulfurization process**. DeSONOx combined process is also possible with using the same active carbon based catalytic material but with using different active phases and different temperatures and **deSOx have to be the first step of the process.**

High efficiency of denitrification of flue gases can be accomplished as a result of utilisation of carbon catalysts within the temperature range 100-200°C. The possibility of a high efficiency of gas purification at relatively low temperature range, close to temperatures of flue gases exiting from the electrostatic precipitator, makes the process very attractive particularly for domestic power stations equipped predominantly with "cold" electrostatic precipitator. Therefore, the new carbon-based catalysts will result in elimination of preheating stage of flue gases prior to their classic SCR processes [107]. The economic advantages of application of these catalysts are very obvious.

The application of active carbons additionally enables an effective removal of halide species, which are particularly harmful for the environment. In comparison to the grain shaped catalysts the honeycomb monolithic catalysts exhibit appreciably lower pressure drop, the cleaning operations are easier and more seldom, in the end the plugging risk is lower than in the case of the grained catalysts.

Active carbon based catalysts and adsorbents which are commonly applied all over the world in the form of spherical tablets or granules create high pressure drop along the catalyst bed and require the dust separation and application of small gas flow rates.

Active carbon monoliths can be effectively utilised in all operations where active carbon is being applied as a granulate (adsorption in gases and liquids, catalysts, catalyst supports). In comparison to grain catalysts the "honeycomb" structure guarantees developing of high geometric specific surface of catalysts per volume unit while pressure drop (low flow resistance) is low. This structure assures also an uniform gas flow, appropriate temperature distribution and gives the possibility to apply high linear flow rate of flue gases without excessive pressure drop.

Monolithic form of catalyst ensures its resistance against deactivation by dust fines contained in the cleaned gas. Due to the fact that such catalysts can be easy regenerated, extending their period of exploitation (life time), assures the operation at relatively high dust concentration, and reduces the operation costs by limitation the number of demanded ventilation and gas conditioning equipment. Active carbon monoliths can be manufactured with using of the special types of coal (e.g. 34 type) or carbonaceous material which are susceptible for forming and retaining the monolithic form after thermal treatment. The additional specific property of the monolithic material is low thermal expansion coefficient.

On the basis of own technologies EKOMOTOR Ltd. (Poland) has manufactured carbon monoliths of "honeycomb" structure. It was found that the active carbon having such a structure exhibits unique properties both as a sorbent and as a support for catalysts. Its sorption properties can be fully utilized for gas and liquid purification. An active carbon can also be applied as a support in manufacturing of catalysts for low temperature selective catalytic reduction (SCR) of nitrogen oxides with ammonia and of catalysts for desulfurization as well.

In relation to other technologies of flue gases cleaning, the catalytic methods are recognized as wasteless and costs of their operation are low. Preliminary studies of catalytic cleaning of flue gases shown that the application of catalysts manufactured from active carbon leads to the apparent lowering of temperature of cleaning process. It was found that efficiency of flue gases desulfurization was within the range of 60 - 80% whereas efficiency of denitrification reached above 75% when active carbon catalysts were applied even within the range of temperature of 100 - 190°C. Such a high purification extent of flue gases at relatively low temperatures makes the process very attractive from the point of view of energy consumption. In the case of carbon-based catalysts it is not necessary to pre-heat flue gases prior to the desulfurization and denitrification as it has to be performed in the case of standard ceramic catalysts. In the later, required temperature of the process is in the range of 300 - 450°C. The remarkable reduction of economic costs is therefore obvious when carbon catalysts are used.

The manufacturing of novel catalysts of "honeycomb" structure from active carbon in the laboratory scale was the result of previously performed investigations. These catalysts appeared to be an unique achievement even in the world scale. It is mainly due to the fact, that the elaborated and developed catalysts for low temperature gas purification are resistant to deactivation by dust fines contained in the cleaned gas. Such a form of a modified active carbon exhibiting thin wall structure with a longitudinal channels creates very low flow resistance. Due to the fact that such catalysts can be easy regenerated, extending their period of exploitation (life time), assures the operation at high dust concentration, and reduces the operation costs by limitation the number of demanded ventilation and gas conditioning equipment.

Catalysts and adsorbents based on active carbon are commonly applied all over the world in the form of spherical tablets or granules create high pressure drop along the catalyst bed and require the dust separation and application of small gas flow rates. Active carbon monoliths can be effectively utilized in all operations where active carbon is being applied as a granulate (adsorption in gases and liquids, catalysts, catalyst supports).

Geometry of fabricated catalyst of the "honeycomb" structure guarantees its highest developing of specific surface per a unit of volume. This structure assures also an uniform gas flow, appropriate temperature distribution and suitable residence time in the catalyst layer. Moreover, monolithic carbon catalysts except of being remarkably active have an essential virtue of being cheap. According to the preliminary cost analysis, these catalysts are expected to be considerably cheaper in relation to standard ceramic catalysts employed for high temperature catalytic desulfurization and denitrogenation of flue gases. High efficiency of desulfurization (60-80%) and denitrification (above 75%) of flue gases can be accomplished as a result of utilization of carbon catalysts within the temperature range as low as 120-190°C. The possibility of such high efficiency of gas purification within a relatively low temperature range, close to temperatures of flue gases exiting the electrofilter, makes the process very attractive particularly for power stations equipped predominantly with "cold" electrofilters. Therefore, the new carbon-based catalysts will result in elimination of preheating stage of flue gases prior to their desulfurization and denitrification processes. The economic advantages of application of these catalysts are very obvious.

The CARBODENOX catalysts are supported on the carrier of the same type – "honeycomb" structure monoliths of active carbon. As carbon plays a very important role in changes occurring on the catalyst when it is functioning, the division into the carbon carrier and the catalyst placed on the carrier must be regarded conventionally. Based on literature analysis, it was decided that the research should use hard gas-coke coal type 34 coming from the polish coal mine "NOWY - WIREK".

Tables 7 and 8 show the results of the technical and elemental analysis, of the petrographic composition, and of the carbon structure parameters determined from the X-ray diffraction method.

W^a	A^a	V^{daf}	C^{daf}	H^{daf}
1.9	6.1	33.4	85.9	5.0

Table 7.Technical and elemental analysis of the gas-coke coal from the coal mine "Nowy Wirek" [%]. The symbols show as follows: W^a - analytic moisture, A^a - ash content, V^{daf} - volatile matter content counted as dry and ash-free matter, C^{daf} - carbon content counted as dry and ash-free matter, H^{daf} - ash content counted as dry matter

Vitrinite [%]	Exinite [%]	Micrinite [%]	Fuzynite [%]	Mineral matter [%]	R_o mean	d_{002} [nm]	L_c [nm]	L_a [nm]
66.1	6.3	3.8	20.6	3.2	0.92	0.36	0.87	1.36

Table 8. Petrographic composition and structure parameters of coal from the "Nowy Wirek" coal mine. The symbols show as follows: $R_{o\ mean}$- average light reflecting power, d_{002}- distance between crystal planes,L_c- crystallites height, L_a- crystallites diameter

Table 9 shows coke properties of the gas-coke coal from the "Nowy Wirek" coal mine, which was used in the research.

RI	SI	Dilatometric properties					Plastic properties			
		t_I	t_{II}	t_{III}	a	b	t_1	t_{max}	t_3	F_{max}
		°C			%		°C			deg angle/min
63	4.5	373	417	435	28	15	370	338	454	178

RI - Roga agglomeration number (agglomeration capability), SI - free-swelling index, Dilatometric properties in the Arnu-Audibert method (t_I - softening point, t_{II} - contraction temperature, t_{III} - dilatation temperature, a – contraction, b – dilatation), Plastic properties of the Griesler method
t_1 - softening point
t_{max} - temperature of maximal plasticity
t_3 - temperature of the end of plasticity
F_{max} - maximal plasticity

Table 9. Coke properties of the gas-coke coal from the "Nowy Wirek" coal mine. The symbols are as follows:

Chemical composition of natural clay used for carrier prepared is presented in Table 5.
The technology of production of CARBODENOX catalysts covers two basic stages:

- manufacturing of the carrier,
- manufacturing of the catalyst on the produced carrier.

Active carbon based catalysts can be manufactured from type 34 hard coal and carbonaceous like additives which are susceptible for carbonisation. The carbon catalysts produced out of the basic types of materials: gas– coke hard coal type 34, natural aluminosilicate, active metals salts (for example: ferric, cupric and manganese nitrate).
Coal is a basic material used for obtaining monoliths out of active carbon shaped into block of "honeycomb" structure. The following substances are put on the surface area of monoliths depending on their use cupric oxide, ferric nitrate, manganese nitrate.
The block diagram of manufacturing of catalysts shaped into block of "honeycomb" structure used for low-temperature cleaning of combustion gases are presented below (Fig.20).
The three types of catalysts can be used in the process of low–temperature cleaning of combustion gases: ferric oxide (3,5 wt%) based catalyst, copper oxide (3, 5 wt%) based catalyst, copper (3,5 wt%) and manganese (3, 5 wt %) oxides based catalyst. The carrier is the same for all catalysts. Geometry of catalysts based on monoliths of honeycomb structure is presented in table 10.

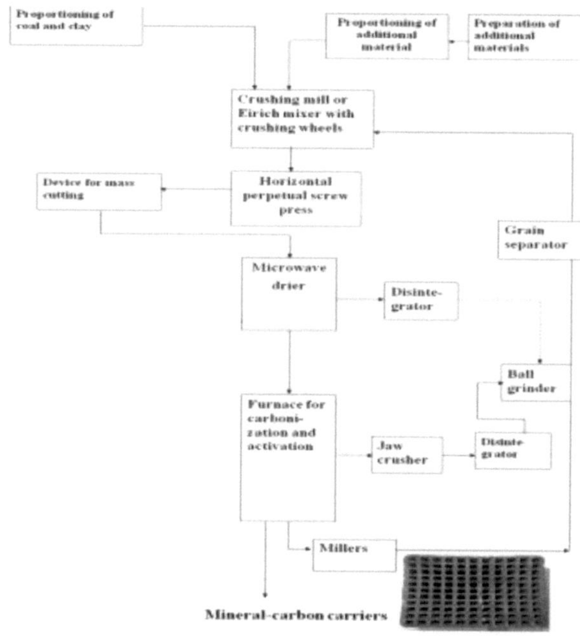

Fig. 20. Block diagram of manufacturing of the carrier

Parameter	Dimension	Typical monoliths
Determined draw hole	-	5.0
Dimensions of the cross-section (length of side)	Mm	98
The number of draw holes	-	11 x 11
External wall thickness	Mm	3,0 - 4.0
Internal wall thickness	Mm	2,2 - 2,8
Draw hole size	Mm	4.5
Open space	%	31,5
The development of the surface after carbonization and activation	m^2/g	600 -800

Table 10. Geometry of catalysts based on monoliths of honeycomb structure.

Carbon monoliths of "honeycomb" structure were obtained with the following structural parameters: specific surface of micropores (for pore radius below 1.5 nm): 40-200 m^2/g; specific surface of mezopores (for pore radius within 1.5-50 nm): 20 - 160 m^2/g; specific surface of macropores (for pore radius above 50 nm): 20 - 80 m^2/g; total porosity: 0.3 - 0.6 cm^3/g.

The above mentioned catalysts were prepared by wet impregnation method. It means that carbon monoliths were dipped in the suitable concentration solution of active metal salts. Fe $(NO_3)_2$; Cu $(NO_3)_2$; Mn $(NO_3)_2$.

After each impregnation the monoliths were dried at ambient temperature and 110°C. Removal of water occurs at 100°C – 115°C. After the monoliths impregnated with nitrates are dried, they are calcined at 400°C in oxygen-free conditions. There is a possibility of using the furnaces (used for carbonisation and activation of the carrier) for calcination process of the catalyst. It must be remembered, however, that aggressive gassing waste containing huge amount of nitrogen oxide (NO_x) are emitted during the calcination process of the CARBODENOX catalyst and it must be reduced. Calcination step was carried out at 400°C for 4 hours in nitrogen stream. In the case of Cu-Mn/C catalyst this operations was repeated twice. New, freshly-produced catalysts of the selective reactivity of catalytic reduction of nitric oxide with ammonia, require conditioning before the test starts. It is advisable to condition the catalyst for 72 hours in testing conditions. The quality of produced catalysts must be estimated by estimation of the geometric shape as well as regards activity of the catalysts. In order to estimate activity of the catalysts the monoliths selected from produced mass must be loaded into the testing flow micro-reactor reactor and undergo a test of activity. The activity of prepared catalysts was determined with testing method of a selective catalytic reduction of nitric oxide by ammonia. operating in the way shown in Fig. 6 was used to carry out the research. The conditions of the test (in temperature range: 100 - 200°C):

Oxygen content in the model gas:	8%
Nitric oxide contents:	1000ppm
GHSV:	3 000 $m^3/m^3 \cdot h^{-1}$
Mole ratio NO : NH_3	1:1

The estimation of catalyst activity was carried out by determination of the conversion of nitric oxide on the surface of the tested catalysts in dependence on catalyst bed temperature. As catalyst activity indicator can be used NO_x conversion at temperature 180 °C, (temperature of flue gases in the case of applying of cold electro-precipitator) [108- 113].The results of activity some prepared catalyst were presented in Fig. 21

Scheme of SCR reactions on active carbon catalyst:

1. Small quantity of NO is reduced by carbon support:

$$2\,NO + 2C \Rightarrow N_2 + 2\,CO \qquad 2\,NO + O_2 \Rightarrow 2\,(NO_2)_{ads.}$$

$$2\,(NO_2)_{ads.} + 2C \Rightarrow N_2 + 2\,CO_2$$

2. More of NO from exhaust gases is reduced by ammonia:

$$4\,NO + 4\,NH_3 + O_2 \Rightarrow 4\,N_2 + 6\,H_2O \qquad 6\,NO_2 + 8\,NH_3 + O_2 \Rightarrow 7\,N_2 + 12\,H_2O$$

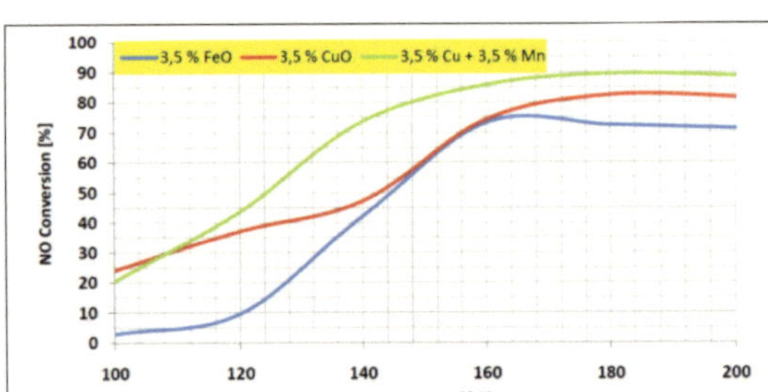

Catalyst carbon activity of: 3, 5 wt % FeO, 3, 5 wt % CuO, 3, 5 wt % CuO and 3, 5 wt % MnO catalysts

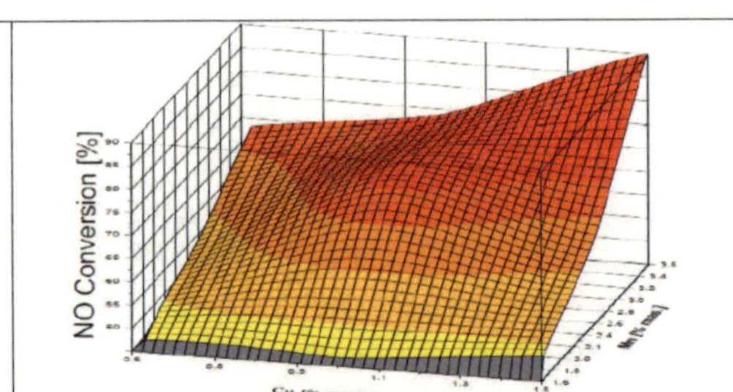

Isolines of nitrogen oxide conversions depending on the content of Cu an Mn at a temperature of 180°C

Fig. 21. Carbon based catalyst activity

8. Conclusion

The rapid development of industry results in an increase in the emission of sulphur and nitrogen oxides into the atmosphere. The issue becomes even more complex due to the gas temperature and dustiness. From among the currently known technologies used for simultaneous elimination of both sulphur and nitrogen oxides the dominant role seems to be played by the processes employing carbon sorbents. The methods for which it is not necessary to preheat combustion gases will always be more cost-effective. Their main advantage is the possibility of carrying out SO_2 adsorption and NO reduction at low process temperatures, approximately 90-130°C, namely within the range of gas temperatures behind electrostatic precipitators in the majority of boilers. It enables smooth incorporation of the purifying installation in the existing energy system, without the necessity of additional gas preheating. Concurrently, in connection with the positive thermal and catalytic effect, the thermal balance of the process is also positive and the temperature of purified gases is adequate for releasing them into the atmosphere. Therefore, the installation does not upset energy relationships in the existing system.

Another considerable advantage of the technologies based on carbon sorbents is the high effectiveness of their operation, in both sulphur (95-100%) and nitrogen removal processes (75-80%), where the sorbent activity increases with time. The Bergbau-Forschung method, representative of this group of technologies, is characterised by its flexibility in the case of a variable motion of the power unit (insensitivity to switching the unit on and off) and capacity for adaptation to variable concentrations of SO_2 in the purified gases, i.e. to a varying sulphur content in the fuel. In the case of high concentrations of SO_2 in gas, a two-stage adsorber system is used, which concurrently meets the requirement for the minimum ammonia consumption and reduces the negative impact exerted by SO_2 on the course of nitrogen oxide reduction. In the case of a low content of sulphur dioxide (gas fuel) a one-stage system is sufficient, and hence the dimensions of the installation and energy input can be significantly reduced.

Unlike the selective catalytic NO reduction methods, the technologies employing carbon sorbents are not sensitive to gas contamination with chlorine, arsenic and mercury compounds and alkalis, which in practice are removed completely.

Since no data are available from large-scale industrial installations it is difficult to present any reliable characteristics of these methods in terms of their cost-efficiency. In the SCR method a half of capital expenditure is related to the purchase of a catalyst. BF method is commensurate with the expenditure incurred for the SCR method. Disadvantages of the BF method include a high demand for active carbon, resulting from, among others, admissible gas loading on sorbents, which is significantly lower than in the case of the SCR methods, considerable consumption of active coke (up to 50% by weight) caused by subgrain formed when a movable bed is used, increased resistance of gas flow through moving beds and large dimensions of the adsorber and regenerator in the case of a two-stage system. The last one of the listed disadvantages, combined with difficult location problems, restricts the use of this method to single power units, with the power output up to 200 MW. The presented information, gathered from the literature review, concerning catalytic methods of combustion gas purification shows clearly that research on this problem goes in many directions and it is aimed at working out technological solutions tailored to the local raw material conditions as well as universal ones.

9. Acknowledgments

Financial support by MNiSzW (Project 344083/Z0306-W3) is gratefully acknowledged.

10. References

[1] Konieczyński, J. ,*Cleaning of tail gases,* Silesian Technical University Press no. 1468, Gliwice, 1990 (in polish).

[2] Kucowski, D.; Laudyn, D.; Przekwas. M. *Power industry and environment al protection,* WNT Warszawa, 1993 (in polish).

[3] Warych, J.,*Cleaning of industrial tail gases,* WNT Warszawa, 1988 (in polish).

[4] Kułażyński, M.; Trawczyński, J.; Walendziewski, J. *Catalytic decomposition of nitric oxide.* Pol. J. Environ. Stud. vol. 6, Supl., 89-82, 1997

[5] Kułażyński, M.; Bratek, K.; Walendziewski, J. *Optimization of an active phase composition in the low-temperature nitric oxide reduction catalyst,* In Polish Journal of Chemical Technology. vol. 9, no 3, 33-37, http://dx.doi.org/10.2478/v10026-007-0049-0 ,2007

[6] Kułażyński, M.; Bratek, K.; Bratek, W., *Reduction of NO_x by ammonia over active carbons obtained from waste ion exchange resin.* Global Symposium on Recycling, Waste Treatment and Clean Technology. REWAS '04, Madrid, September 26-29, 2004. Vol. 3/Ed. by I. Gaballah , Warrendale, Pa : The Minerals, Metals and Materials Society; San Sebastian : Inasmet, [2004]. pp.2879-2880, 2004

[7] Kułażyński, M.; Trawczyński, J., *Low temperature selective catalytic reduction of nitric oxide with ammonia.* Catalysis and adsorption in environmental protection. International conference, Szklarska Poręba, Poland, October 13-15, 1994 Wrocław: Oficyna Wydaw. PWroc., 1994. 43-47, Prace Naukowe Instytutu Chemii i Technologii Nafty i Węgla Politechniki Wrocławskiej. Konferencje ; nr 7, 1994

[8] Kułażyński, M., *Studies on catalysts for Denox process.* Catalysis and adsorption in environmental protection. International conference, Szklarska Poręba, Poland, October 13-15, 1994 Wrocław: Oficyna Wydaw. PWroc., 1994. pp. 69-75. Prace Naukowe Instytutu Chemii i Technologii Nafty i Węgla Politechniki Wrocławskiej. Konferencje ; no 7, 1994

[9] Kułażyński, M., *Optimization on the composition of catalyst for simultaneous rejection of carbon monoxide and nitrogen oxide from engine exhaust gases.* Catalysis and adsorption in environmental protection. International conference, Szklarska Poręba, Poland, October 13-15, 1994 Wrocław: Oficyna Wydaw. PWroc., 1994. pp. 225-233.Prace Naukowe Instytutu Chemii i Technologii Nafty i Węgla Politechniki Wrocławskiej. Konferencje ; no 7, 1994

[10] Kułażyński, M.; Trawczyński, J.; Walendziewski, J. *Selective catalytic of nitrogen oxides by LPG.* Catalysis and adsorption in fuel processing and environmental protection. II International conference, Szklarska Poręba, Poland, September 18-21, 1996 Wrocław: Oficyna Wydaw. PWroc., 1996, 127-133, Prace Naukowe Instytutu Chemii i Technologii Nafty i Węgla Politechniki Wrocławskiej. Konferencje, 1996

[11] Kułażyński, M.; Radomyski, B.; Trawczyński, J.; Walendziewski, J. *Catalytic decontamination of NO from engine exhaust gases.* Internal combustion engines, Warsaw-Poznań, 5-8 September 1995 Warsaw: Institute of Aeronautics, 1995. 291-296., Journal of KONES ; vol. 2 no 1, 1995

[12] Nakatsuji, T.; Miyamoto, A. Catal. Today, 10, 21, 1991

[13] Wong, W. C. Ind. Eng. Chem. Prod. Res. Dev., 25, 179, 1986
[14] Bankman, M. at all Catal. Today, 14, 225, 1992
[15] Haber, J. ;Kozłowska, A.; Kozłowski, R., J. Catal., 102, 52, 1986
[16] Shikadai, T. ;at all, Ind. Eng. Chem. Proc. Res. Dev., 20, 91, 1991
[17] Beeckman, J.; Hegedus, L. L., Ind. Eng. Chem. Res., 30. 969, 1991
[18] Odenbrand, C. U. I. ;at all, Appl. Catal., 18, 335, 1985
[19] Solar, J.P. ;at all, Catal. Today, 14, 211, 1992
[20] Bjorklund, R. B. ;at all, J. Catal., 128, 574, 1991
[21] Kiovsky, M. J. ;at all, Ind. Eng. Chem. Prod. Res. Dev., 19, 218 (1980),
[22] US Pat. 4 778 665, 1988
[23] US Pat. 4 663 300, 1987
[24] Boer, F. P. ;at all, Chemtech, 312, 1990
[25] Łachman, I. M.; Wiliams, J. L., Catal. Today, 14, 317, 1992
[26] Groeneveld, M. J. ;at all, Proc. of 9th International Congress of Catalysis, Vol. 4, 1743, Calgary, Canada, 1988
[27] US. Pat. 4 085 193, 1978
[28] Miyamato, A. ;at all, J. Phys. Chem., 85, 2366, 1981
[29] Miyamato, A. ;at all, J. Phys. Chem., 86, 2945, 1982
[30] Inamata, M. ;at all, J. Phys. Chem., 87, 754, (1983),
[31] Saleh, R. T. ;at all, J. Catal., 98, 102, 1986
[32] Bosch, H.; Janssen, F., Catal. Today, 2, 4, 404, 1987
[33] Morikawa, S. ;at all, Chem. Lett, 251, 1981
[34] Morikawa, S.;at all, Proc. of 1 Conf. on Catal., Berlin, 1984, III - 661, 1984
[35] Shikada, T. ;at all, J. Chem. Tech. Biotech., 33A, 446, 1983
[36] Bond, G. C. ;Konig., P. J. Catal., 77, 309,1982
[37] Imanari, M.;at all, Proc. of the 7* International Congress on Catalysis, Elsevier, Amsterdam, 1981, 841, 1981
[38] Pearson, I. M. ;at all,., Ind. Eng. Chem. Pro. Res. Dev, 22, 381, 1983
[39] Natsuda, S. ;at all, J. Air Pollution Control Assoc., 28, 360, 1978
[40] Nakijama, F., Catal. Today, 10, 1, 1991
[41 Morikawa, S. ;at all, Buli. Chem. Soc. Jpn., 55, 2254, 1982
[42] Pl. Pat. 172062 B1
[43] Wejkowski, R.; Wojnar, W.; Kułażyński, M.; Walendziewski, J.; Pronobis, M.; Walewski, A.; Ostrowski, P.; Litka, R.; Ciukaj, Sz. *Selective catalytic reduction scr in rotary air heater, New Technologies combustion and clearing flue gases.* Wydawnictwo Politechniki Śląskiej Gliwice 2010, 368-388, 2010 (in polish)
[44] Kułażyński, M.; Pronobis, M.; Walewski, A.; Wejkowski, R.; Wojnar, W., *Selective catalytic reduction scr in rotary air heater*) In-Rynek Energ. 2008, 6, 82-87, 2008 (in polish)
[45] Pronobis, M.; Wejkowski, R.; Kułażyński, M., *NOx control for pulverised coal fired boilers.* Pol. J. Environ. 2009 vol. 18 no. 1B, 183-187, 2009
[46] Jap. Pat. 58-133 820, publ. 09.08.83.
[47] Srivastava, A. C., Singh, B. N., *An activeded carbon for removal NO_x from gases,* In Fert. echnl. 1979, 16, 3-4, 250, 1979
[48] Jap. Pat. 54-139 880, publ. 30.10.79.

[49] Nashiyama, A.; at all, *New type of active carbon catalyst for simultaneous removal of SO_x and NO_x*, In Buli. Chem. Soc. Jap., 1980, 53, 11, 3356-60, 1980
[50] Jap. Pat. 80-51438, publ. 15.04.80.
[51] Hagimara, H; at all, *Kagaku gyutsu kenkyucho hokoko,* In J. Nat. Chem. Lab. Ind. 1983, 78, 9, 427-32, 1983
[52] Jap. Pat. 74- 23 190, publ. 01.03.74.
[53] Hoffmann, V., *Activated coke will reduce emissions in Arzberg,* In Mod. Power Syst. ,1986, 6, 71-75, 1986
[54] Ger. Pat. 3036 531, publ. 27.05.82.
[55] Jap. Pat. 82-100 910, publ. 23.06.82.
[56] Ger. Pat. 790, publ. 25.04.74.
[57] US. Pat. A 3926 590, publ. 16.12.75.
[58] Jap. Pat. 52-19 557, publ. 28.05.77.
[59] Richter, E.; at all, *Regenerative processes for SO_2 and NO_x removal from off gases of chemical and power plants* In Vie Congr. Mond. Qual. Air, Paris 16-20 mai 1983, Textes Conf. Vol.3_ Paris. 511-8, 1983
[60] Ger. Pat. 3443 686, publ. 05.06.86.
[61] Jap. Pat. nr 74-43 870, publ. 25.04.74.
[62] Jap. Pat. 50-23 091, publ. 25.01.75.
[63] N.N., *Low temperature catalyst for dry-type flue-gas denitrification* ,In CEER, chem. Econ. and Eng. Rev. 1975, 7, 10, 48, 1975
[64] US. Pat. 3961 020, publ. 01.06.76.
[65] N.N., Chem. Age, 1976, 112, 2496/7, 21, 1976
[66] Jap. Pat. 53-141, 190, publ. 08.12.78.
[67] Ger. Pat. 2911 712, publ. 25.09.80.
[68] Jap. Pat. 50-81, 728, publ. 20.06.80.
[69] Jap. Pat. 56-02 828, publ. 13.01.81.
[70] Ger. Pat. 3039 477, publ. 08.05.82.
[71] Ger. Pat. 3342 500, publ. 11.06.85.
[72] Jap. Pat. 74-74 696, publ. 18.07.74.
[73] US. Pat. 3887 683, publ. 03.06.75.
[74] Jap. Pat. 58-43 224, publ. 12.03.83.
[75] Weber, E., Hübner, K., Energie, 11, 1986, 38, 4, 10-15,1986
[76] Richter, E.: *Mechanismen der NO-Adsorption und Reduktion an Aktivkohle,* In Chem. Ing Techn, 1983, 55, 12, 98886-7, 1983
[77] Richter, E.; at all, *Simultante Entfernung von SO_2 und NO_x unter der Bendingungen der Rauchgasreiningung von Kraftwerken* In Chem. Ing. Techn. 1980, 52, nr 6, s. 456-7, 1980
[78] Knoblauch, K.; at all, *Application of active coke in processes of SO_2 und NO_x removal from flue gases.* In Fuel 1981, 60, Sept. s. 832-7, 1981
[79] US, Pat. 9469 662, publ. 04.09.84.
[80] Jap. Pat. 58-166 921, publ. 03.10.83.
[81] R. Erath: Das *Bargbau-Forschung/Uhde Yerfahren.* In Staub, Reinhalt. d. Luft, , 45, nr 9, s.56-9, 1985
[82] Wojciechowska, M. *On the catalytic removal of nitrogen oxides,* Catalysis and adsorption in environmental protection. III International conference, Szklarska Poręba, Poland,

September 15-18, 1999 Wrocław: Oficyna Wydaw. PWroc., 1999, ISSN 0324-9867, Prace Naukowe Instytutu Chemii i Technologii Nafty i Węgla Politechniki Wrocławskiej no 56, Konferencje no 9, 55-66, 1999

[83] Knoblauch, K.; at all, Simultane SO_2 und NO_x - *Entfernung aus Rauchgasen durch Adsorptionskatalyse an Aktivkoksen*, In Chem. Ing. Techn. 1985, 57, 3, 239-41, 1985

[84] Ger. Pat. 3429 999: publ. 27.02.86.

[85] Ger. Pat. 3433 093: publ. 20.03.86.

[86] Ger. Pat. 3423 744: publ. 09.01.86.

[87] Jap. Pat. 55-1 617, publ. 13.05.80.

[88] Jungten, H., Richter, E., *Rauchgasreinigung in Grossfeuerungsanlagen*. In Staub, Reinhalt, d. Luft, 1985, 9,8-20, 1985

[89] Jap. Pat. 74-44 969, publ. 27.04.74.

[90] Jap. Pat. 52-3C 144, publ. 05.08.77.

[91] US Pat. 3864 450, publ. 04.02.75.

[92] US. Pat. 4113 839, publ. 12.09.78.

[93] Jap. Pat. 60-105 732, publ. 22.08.81.

[94] Jap. Pat. 83-79 522, publ. 13.05.83.

[95] Jap. Pat. 83-79 523, publ. 13.05.83.

[96] Ger.Pat. 3512 169, publ. 09.10.86.

[97] Jungten, H.; Hoang-Phu, T.; Richter, E., *Simultaneous NO_x-SO_2 Removal by Active Coke.* CCRE - 9 th Coli. of the Working Party Chemical Reaction Engineering of Flue Gas Desulphurization and Denitration, 6/7 March 1985, Bad Soden, 1985

[98] Jap. Pat. 58-43 223, publ. 12.03.83.

[99] Jap. Pat. 58-43 222, publ. 10.03.83.

[100] Jap. Pat 55-17616, publ. 13.05.80.

[101] Takenouchi, S.; at all, Sumitomo Jukikai Giho, 30, 89, 54-9, 1982

[102] Ger. Pat. 3426 913, publ. 30.01.86.

[103] Ger. Pat. 3346 176, publ. 04,07.85.

[104] Jap. Pat. 85-102 922, publ. 07.06.85.

[105] Jap. Pat. 74-15 676, publ. 12.02.74.

[106] US. Pat. 3795 730,publ. 05.03.74.

[107] Pl Pat. PL167912 (B1) Publ. 1995-12-30

[108] Kułażyński, M.; Walendziewski, J.; Kaczmarczyk, J., *Porous structure and reactivity of mixed ceramic-carbon honeycomb catalysts.* In Pol. J. Environ. Stud. vol. 15, 6A, 117-119, 2006 .

[109] Kułażyński, M.; Walendziewski, J., *DENOX activity of mineral carbon catalysts supported on monolithic materials.* In Pol. J. Environ. Stud. vol. 15, 6A, 95-99, 2006.

[110] Kułażyński, M.; Walendziewski, J.; Bratek, K., *Synthesis of mineral-carbon monolithic catalyst supports.* Conference on Porous Ceramic Materials. PCM 2005. Proceedings, Brugge, Belgium, 20-21 October 2005, Brugge : Flemish Institute for Technological Research (Vito), 2005, 6, 2005

[111] Olivares, J.; Kułażyński. M.; Salvador, L.; Walendziewski, J.; Trawczyński, J., *Laboratory and pilot plant performance of novel carbon monolithic catalysts development for selective flue gas d-noxing at low temperature.* Catalysis and adsorption in fuel processing and environmental protection. IV International conference, Kudowa Zdrój, September 18-21, 2002 Wrocław: Oficyna Wydaw. PWroc., 2002, 169-176., Prace Naukowe

Instytutu Chemii i Technologii Nafty i Węgla Politechniki Wrocławskiej. Konferencje, 2002

[112] Trawczyński, J.; Kułażyński, M., *Nitric oxide reduction using active carbon based monolithic catalysts.* Materiaux carbones fonctionnalises a porosite controlee. Seminaire Franco-Polonais GDRE. Centre National de la Recherche Scientique - DRI [i in.], Zakopane, 27-29 septembre 1998 [B.m.: b.w., 1998]. 33-41, 1998

[113] Trawczyński, J.; Kułażyński, M., *Active carbon monoliths as catalyst supports for SCR (Selective Catalytic Reduction) of NOx with ammonia.* Coal science. Proceedings of the Eighth International Conference on Coal Science, [Oviedo, Spain, September 10-15, 1995]. Vol. 2 / Ed. by J. A. Pajares, J. M. D. Tascon Amsterdam : Elsevier, 1995. 1803-1806., Coal Science and Technology; [vol.] 24, 1995